COMPUTATIONAL HYDROLOGY IN FLOOD CONTROL DESIGN AND PLANNING

COMPUTATIONAL HYDROLOGY IN FLOOD CONTROL DESIGN AND PLANNING

THEODORE V. HROMADKA II, PH.D., PH.D., R.C.E.

Director, Water Resources Engineering
Williamson and Schmid
Irvine, California

and

Director, Computational Hydrology Institute
Irvine, California

RICHARD H. MCCUEN, PH.D.

Professor, Department of Civil Engineering
University of Maryland
College Park, Maryland

CHUNG CHENG YEN, PH.D.

Research Hydrologist
Williamson and Schmid
Irvine, California

Mission Viejo, California

Library of Congress catalog number 87-080497

First Lighthouse Publications edition

ISBN 0-914055-05-4

table of contents

CHAPTER 1 HYDROLOGY MODELS IN MUNICIPAL ENGINEERING............................ 1

1.1. Objectives.................................... 1

1.2. Stormflow Determination Methods................ 2

1.3. Method for Development of Synthetic Flood Frequency Estimates............................ 4

1.4. Watershed Modeling Uncertainty.................. 12

1.5. Watershed Runoff................................ 19

1.6. Selection of a Flood Flow Modeling Approach..... 20

1.6.1. Choice of Modeling Category.............. 20

1.6.2. Runoff Hydrograph Generation Techniques.. 36

1.6.3. The Choice of Design Storm............... 38

1.6.4. The Design Storm/Unit Hydrograph Approach................................. 39

1.7. Hypothetical Floods, Balanced Floods, and Design Storm Methods............................ 40

CHAPTER 2 PRECIPITATION................................ 46

2.1. Introduction.................................. 46

2.2. Classification of Precipitation................ 46

2.3. Mechanisms for Cooling.......................... 47

2.4. Measurement and Synthesis of Precipitation Data. 47

2.5. Comparison of Point Rainfall Information....... 48

2.6. Precipitation Depth-Duration-Frequency......... 48

2.7. Event Depth-Duration............................ 54

2.8. Point Precipitation Data and The Design Storm.. 55

2.9. Intensity-Duration-Frequency Curves............ 55

2.10. Depth-Duration-Frequency Relationships......... 57

2.11. Area-Averaged Point Rainfall................... 57

2.12. Depth-Area Relationships....................... 59

2.13. Synthetic Critical Storm Patterns.............. 59

2.14. Modified Composite Storm Pattern............... 66

2.15. Program 2.1: Subroutine RAIN.................. 71

2.15.1. The 24-Hour Nested Design Storm Patterns 73

2.15.2. Depth-Area Adjustment................... 73

2.15.3. SCS Storm Patterns...................... 73

2.15.4. Constant Rational Method................ 81

2.15.5. Linear Interpolation Method............. 81

2.15.6. Program Data Entry Sequences............ 81

2.15.6.1. 24-Hour Nested Design Storm Pattern........................... 83

2.15.6.2. SCS Storm Pattern................. 83

2.15.6.3. Constant Rainfall Intensity for a Given Duration.................. 83

2.15.6.4. Linear Interpolation Method....... 84

2.15.6.5. User Pre-Set Storm Pattern........ 84

2.15.6.6. Design Storm Pattern with Continuous Depth-Area Adjustment.. 84

2.16. Program 2.1................................ 85

CHAPTER 3 ESTIMATION OF EFFECTIVE RAINFALL................ 99

3.1. Introduction.. 99

3.2. Physical Processes: An Overview................ 99

3.3. Phi-Index Method................................. 103

3.4. Constant Proportion Loss Rate................... 104

3.5. Coupled Phi-Index and Constant Proportion Loss Rate Function............................... 105

3.6. Horton Loss Rate................................. 107

3.7. Exponential Loss Rate Function.................. 107

3.8. Initial Abstraction.............................. 107

3.9. SCS Loss Separation.............................. 109

3.10. SCS Hydrologic Soil Groups...................... 111

3.11. Soil Cover....................................... 113

3.12. Watershed Development Conditions................ 119

3.13. Impervious Areas................................. 119

3.14. Program 3.1: Subroutine SLOSS.................. 121

3.14.1. Initial Abstraction Models.............. 121

3.14.1.1. SCS Model.......................... 121

3.14.1.2. Total Depth of Abstraction........ 121

3.14.1.3. Percentage of Precipitation to a Maximum Total Depth of Abstraction 121

3.14.2. Loss Models............................. 123

3.14.2.1. SCS Model.......................... 123

3.14.2.2. Phi-Index.......................... 123

3.14.2.3. Percentage of Rainfall............ 123

3.14.2.4. Percentage of Rainfall Limited by a Phi-Index.................... 123

3.14.2.5. Horton Equation.................. 123

3.14.2.6. Exponential of Rainfall Intensity. 126

3.14.3. Applications of Loss Function Models..... 126

3.14.4. Data Entry Sequences..................... 133

3.14.4.1. SCS Model.......................... 133

3.14.4.2. Total Depth of Initial Abstraction 133

3.14.4.3. Percent of Precipitation to a Maximum Total Depth of Initial Abstraction...................... 133

3.14.4.4. Phi-Index.......................... 134

3.14.4.5. Percentage of Rainfall............ 134

3.14.4.6. Percentage of Rainfall Limited by a Phi-Index.................... 134

3.14.4.7. Horton Equation................... 134

3.14.4.8. Exponential Loss Function......... 134

3.15. Program 3.1.................................. 135

CHAPTER 4 PEAK DISCHARGE ESTIMATION...................... 142

4.1. Introduction.................................. 142

4.2. Time of Concentration Estimation.............. 142

4.3. Rational Method............................... 148

4.4. Runoff Coefficient............................ 149

4.5. Limitations of the Rational Method............ 156

4.6. Modeling with the Rational Method............. 157

4.7. The Link-Node Method.......................... 157

4.8. The Nodal Point Method........................ 158

4.9. The Subarea Summation Method.................... 159

4.10. Confluence of Streams (Junction Analysis)....... 159

4.11. Presentation of Product.......................... 160

4.12. A Regional Master Plan of Drainage (MPD)........ 162

4.13. Application: A City Master Plan of Drainage.... 163

4.13.1. MPD Data Collection...................... 163

4.13.2. Existing Conditions and Ultimate Development Conditions.................. 165

4.13.3. Development of Major Facility Watershed Boundaries.................... 165

4.13.4. Development of MPD Link-Node Schematics. 165

4.13.5. Development of the Link-Node Models..... 169

4.13.6. Calibration of the Rational Method...... 169

4.13.7. Preparation of MPD System Information... 170

4.14. A Rational Method Planning/Design Computer Program.. 170

4.15. Computer-Aided Design Interaction............... 174

4.16. Example 4.2. Rational Method Program Application...................................... 177

4.17. Program 4.1...................................... 187

4.18. Program 4.2...................................... 192

4.19. Program 4.3...................................... 195

4.20. Program 4.4...................................... 195

4.21. Program 4.5...................................... 197

4.22. Program 4.6...................................... 201

4.23. Program 4.7...................................... 205

4.24. Program 4.8...................................... 213

4.25. Program 4.9 220

4.26. Program 4.10 223

4.27. Program 4.11 225

CHAPTER 5 SYNTHETIC UNIT HYDROGRAPH METHODS 227

5.1. Introduction 227

5.2. Unit Hydrograph Terminology 228

5.3. Conceptual Evaluation of Unit Hydrographs 230

5.4. Convolution 232

5.5. Unit Hydrograph Analysis 233

5.6. Forming Synthetic Unit Hydrographs 235

5.7. The SCS Unit Hydrograph 243

5.8. Variation on the SCS Unit Hydrograph 244

5.9. The Espey-Altman Unit Hydrograph 247

5.10. Design Storm Effective Rainfall 248

5.11. Synthetic Runoff Hydrograph Development 248

5.12. The Rational Method as a Unit Hydrograph Method. 250

5.13. Presentation of Product 252

5.14. A Design Storm Runoff Hydrograph Computer Program 252

5.15. Program 5.1: Subroutine SGRAPH 273

5.16. Data Entry Sequences 276

5.16.1. Watercourse LAG Time Model 276

5.16.2. S-Graph Development 277

5.16.3. Review of Computational Results 277

5.16.4. Accepting/Rejecting Computational Results and Output 277

5.17. Program 5.1 278

5.18. Program 5.2 284

CHAPTER 6 RUNOFF HYDROGRAPH ROUTING TECHNIQUES 288

6.1. Introduction 288

6.2. Benefits and Concerns of Detention Storage 288

6.3. Detention Basin Routing Procedure (Modified Puls Method) 290

6.4. Flow-by Channel Model (Runoff Hydrograph Separation) 295

6.5. The Modified Convex Channel Routing Method 295

6.6. The Modified Att-Kin Method 301

6.7. Muskingum Channel Routing 303

6.8. A Pipeflow Routing Model 305

6.9. Hydrograph Translation 306

6.10. Kinematic Wave Routing Approximation 306

6.11. Subroutine FLOWBY 306

6.11.1. Application: Flow-by Basin 309

6.11.2. Data Entry Sequence 311

6.12. Subroutine FLOWTR 312

6.12.1. Application: Flow-through Basin 312

6.12.2. Data Entry Sequence 313

6.13. Subroutine RCHAN 315

6.13.1. Convex Routing Model 316

6.13.2. Pipeflow Routing Model 316

6.13.3. Hydrograph Translation Model 316

6.13.5. Muskingum Routing Model................ 317

6.13.6. Application: Channel Routing Modeling Techniques.................... 317

6.13.7. Data Entry Sequence: Translation....... 319

6.13.8. Data Entry Sequence: Convex Routing.... 321

6.13.9. Data Entry Sequence: Pipe Routing...... 322

6.13.10. Data Entry Sequence: Kinematic Wave.... 322

6.13.11. Data Entry Sequence: Muskingum Routing 323

6.14. Program 6.1.............................. 325

6.15. Program 6.2.............................. 341

CHAPTER 7 CONSIDERATIONS IN CALIBRATING ELEMENTS OF DESIGN STORM MODELS.............................. 370

7.1. Introduction.............................. 370

7.2. Goals and Objectives of Model Calibration....... 371

7.3. Criteria for Model Assessment.................. 373

7.4. Calibrating Unit Hydrographs and Loss Functions. 379

7.5. Least Squares Calibration of Unit Hydrographs... 384

7.6. Fitting Synthetic Unit Hydrographs............. 387

7.7. Calibration of S-Hydrographs................... 388

7.8. Calibration to Flood Frequency Discharges....... 391

7.9. Calibration of Time of Concentration Models..... 393

7.10. Calibration of Channel Routing Parameters....... 397

7.11. Calibration of the Convex Method............... 397

7.12. Calibration of the Modified Att-Kin Method...... 400

7.13. Calibration of the Muskingum Method............ 400

CHAPTER 8 COMPLEX WATERSHED MODELING.......................... 402

8.1. Link-Node Model.................................. 402

8.2. A Watershed Computer Program..................... 403

8.3. Link-Node Model Functions........................ 404

8.4. Data Bank Development............................ 407

8.4.1. Depth-Area Adjustment Curve Sets......... 407

8.4.2. User Specified 24-Hour Storm Pattern..... 407

8.4.3. User Specified Storm Pattern with Continuous Depth-Area Adjustment......... 407

8.4.4. S-Graphs................................. 408

8.4.5. Interactive Procedure.................... 408

8.4.5.1. Depth-Area Curve Sets........... 411

8.4.5.2. User Specified 24-Hour Storm Patterns........................ 413

8.4.5.3. User specified 24-Hour Storm Pattern with Continuous Depth-Area Adjustment........... 414

8.4.5.4. S-Graphs........................ 416

8.4.5.5. Data Bank Summary............... 417

8.4.5.6. Program 8.1..................... 419

8.5. Link-Node Model Development...................... 435

8.5.1. Interactive Procedure.................... 435

8.5.2. Utility Commands......................... 438

8.5.2.1. ADD Command..................... 438

8.5.2.2. SPLIT Command................... 439

8.5.2.3. CLEAR Command................... 439

8.5.2.4. VIEW and PRINT Commands......... 440

8.5.2.5. EXECUTION and EXIT RUNOFF Program........................ 440

8.6.2.6. Program 8.2.................... 441

8.6.2.7. Program 8.3.................... 448

8.6. INPUT/OUTPUT Units................................ 454

8.7. Application of Link-Node Model.................. 454

Bibliography.. 496

Author Index.. 504

Subject Index... 507

preface

Computational Hydrology in Flood Control Design and Planning is a book prepared for use by practitioners. The main emphasis is towards application of the most frequently used computational algorithms in rainfall-runoff modeling techniques. The Book's presentation carries the reader from an extensive literature review of the reported progress in surface runoff modeling, to the development of a computer program which accommodates an entire spectrum of modeling techniques, including various design storm approaches, loss rate submodels, basin and channel routing methods, and subarea runoff generation.

Chapter 1 presents a review of the literature concerning the development, calibration, and application, of surface runoff models. It is an important chapter in that the reader becomes acquainted with some of the reported successes and failures in the use of various types of hydrologic models.

Chapter 2 includes discussions on precipitation, statistical analyses of precipitation, and the more commonly used algorithms in the application of rainfall data to the development of runoff data. Computer codes are developed which provide a variety of design rainfall conditions and approaches.

Chapter 3 focuses upon loss rates and other rainfall extractions which must be considered in correlating rainfall data to runoff data. Because of the variety of loss functions utilized in the several surface runoff modeling approaches, many of the more popular loss rate techniques are included in this chapter. The several loss rate functions are then summarized into computer program form.

Chapter 4 concentrates on the development, calibration, and application of the most commonly used rainfall-runoff predictor technique--the rational method. Computer programs are developed which provide a rational method modeling capability. As an application, a City Master Plan of Drainage is considered which was developed using the Chapter 4 computer software and a personal computer.

The well-known unit hydrograph technique is developed in detail in Chapter 5. Culminating in a design storm/unit hydrograph computer program, a detailed example problem carries the reader through the development of a runoff hydrograph based on the unit hydrograph method.

Channel and basin routing models are considered in Chapter 6. Several channel routing techniques are presented, and corresponding computer codes are developed. Flow-through and flow-by detention basin modeling techniques are also examined. The routing methods of this chapter, combined with the design storm, loss rate, and unit hydrograph techniques of Chapters 2, 3, and 5, provide a complete link-node computer modeling capability for a personal computer environment.

Chapter 7 focuses upon the important unit hydrograph, loss rate, and routing model calibration process. Use of calibration in hydrologic models enables for a better correlation between rainfall and runoff data.

The final chapter presents the total link-node model computer program. The Chapter 8 computer codes utilize the previous computer codes of Chapters 2, 3, 5, and 6 to formulate a comprehensive link-node computer program for use on personal computers.

Should the reader wish to key-in the several computer programs, an estimated time investment of about 200 hours is suggested for program entry, debugging, and testing. Another option is to purchase an available personal computer disc with the Chapters 2, 3, 5, 6, and 8 computer codes already compiled and ready-to-use. Further information regarding the available computer disc is included after the book index. The Chapter 4 computer codes are available in form-field fillout mode by the firm Advanced Engineering Software, 17782 Sky Park Boulevard, Irvine, California, 92714.

Not only will this book be of key interest to practitioners involved with the use of surface runoff models in flood control facility design and planning, but also to water resources engineers and students as a reference to the state-of-the-art in design storm hydrology methods.

hydrology models in municipal engineering

1

1.1. Objectives

A main objective of this work is to present widely used hydrologic modeling methods for estimating synthetic runoff quantities corresponding to severe design storm events in urban watersheds. Such model design storms are usually associated with the 50-year, 100-year, or even more severe return frequency storm events and conditions. Additionally, attention is devoted to workable design storm parameter estimation procedures for the several hydrologic processes. Additional information is supplied by the development of a flood flow frequency curve which represents the peak flowrate as a function of the return frequency. Necessary to any hydrologic method is a calibration procedure to correlate not only the rainfall and runoff output of the model, but also the design storm properties needed to produce a level of protection selected as a policy statement of the governing flood control agency. The modeling and calibration procedures presented herein include many of the most popular methods such as, for example, described in the publication TD-15 (U.S. Army Corps of Engineers, Hydrologic Engineering Center Training Document No. 15, 1982). Finally, FORTRAN computer programs are presented for use in hydrologic design and planning purposes. The computer programs are readily adaptable for use on many currently available microcomputers.

The included computer programs have been used to master plan entire city flood control drainage systems (Hromadka, et al. 1983a). The program presentations were formulated in the development of a hydrology design/planning manual for both the largest county in the mainland United States, San Bernardino, California, (Hromadka and McCuen, 1986a) and the coastal county of Orange, California, (Hromadka and McCuen, 1986b). The runoff hydrograph computer software is based upon the procedures discussed in TD-15, using a synthetic design storm pattern of 24 hours duration, which is constructed from a set of nested unit rainfalls of an identical return frequency, continuously adjusted for depth-area effects due to the watershed size.

For smaller studies (less than about one square mile) or studies involving city master plan of drainage (MPD) storm drain networks (such as where drainage fee statements are needed), link-node computer programs based upon a simpler rational method are provided which afford a computerized MPD data base development.

The hydrologic cycle (see Fig. 1.1.) is thoroughly covered in many texts such as Eagleson (1970). In this book, the hydrologic processes will be treated only as they pertain to urban flood control design and planning. Consequently, the main emphasis will be on design storm precipitation, design storm patterns, infiltration losses and depression storage, and the overall watershed runoff process. The hydrologic concepts of transpiration, groundwater flow, snowmelt, and other processes will not be dealt with thoroughly because these factors are usually not important in urban hydrologic design.

1.2. Stormflow Determination Methods

When studying a watershed for severe storm runoff characteristics, the usual procedure is to collect data on precipitation, soil types, stream discharge, and other hydrologic and geologic characteristics. This data then may be evaluated in accordance with theory presented in standard texts. Although precipitation and streamflow data are available at selected locations throughout the country (for example the U.S. Weather Service and the U.S. Geological Survey), sufficient data are usually unavailable for local watersheds to develop precise hydrologic calculations. More importantly, the long-term effects on flood hydrology due to urbanization of the watershed are usually not precisely represented by the available data. For these reasons, synthetic flood hydrology methods are usually required.

Generally speaking, two basic flood hydrology methods are used to estimate design discharges. The rational method is usually used for small watersheds of less than an aggregate of approximately one square mile. The unit hydrograph method (such as presented in TD-15) is used for larger watersheds or for projects where detailed knowledge of flood hydrographs are required. In this book, both approaches are reviewed and FORTRAN computer programs presented for use in design storm hydrology analysis. The basis of the rational method design/planning computer program (Chapter 4) is given in Hromadka and McCuen (1986a,b), and the unit hydrograph computer program (Chapter 5) is developed in Hromadka et al. (1983a). Other simulation modeling approaches can be incorporated into the provided computer program listings by using a menu-driven main program which branches to the various desired modeling approaches (Chapter 6).

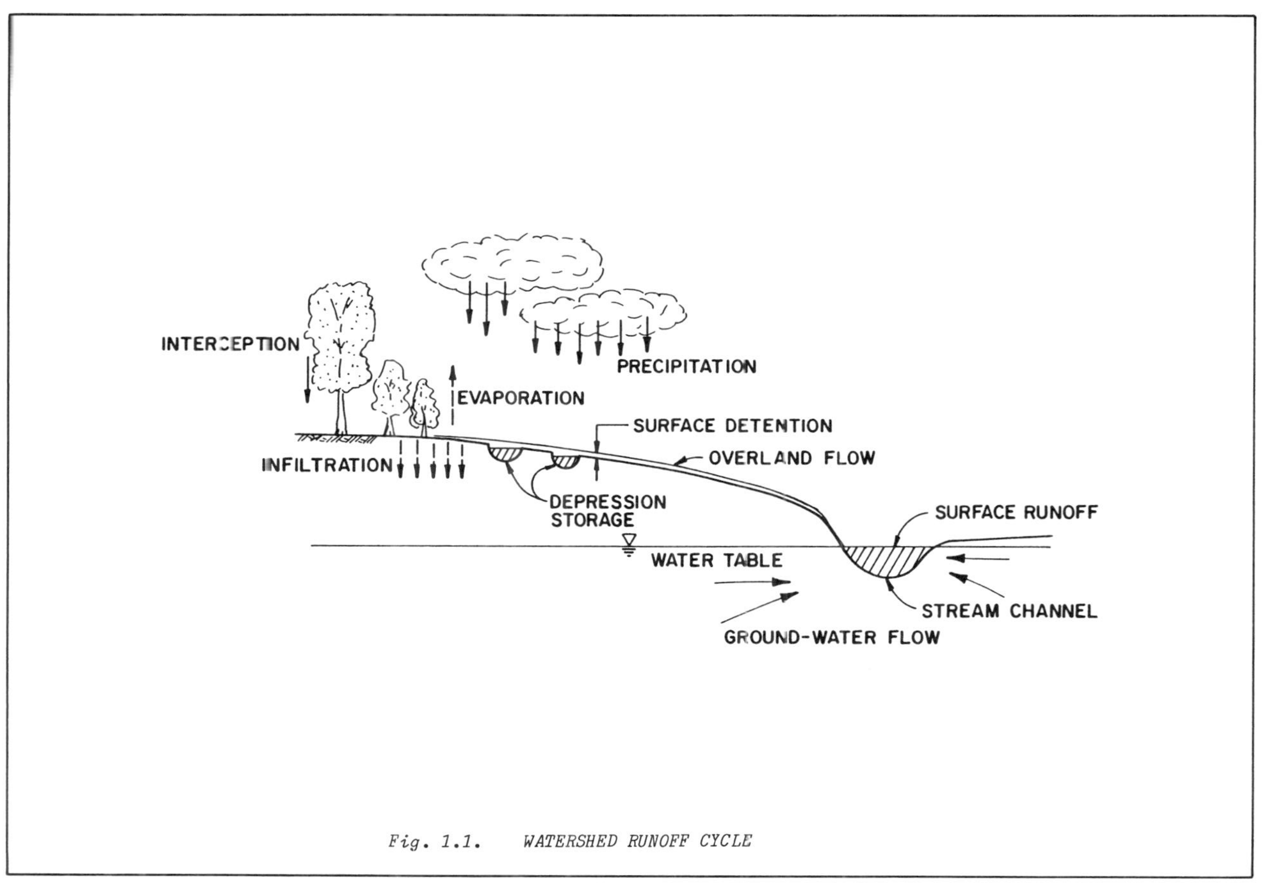

Fig. 1.1. WATERSHED RUNOFF CYCLE

1.3. Methods for Development of Synthetic Flood Frequency Estimates

The uses of flood flow frequency data range from the specification of flood insurance risk relationships to the commonly occurring problem of designing flood control facilities. Typically, however, stream gage data are usually unavailable at the study site; consequently, some type of method is needed to synthesize a flood frequency curve for ungaged streams.

The various types of procedures used to develop flood flow frequency estimates at ungaged locations can be grouped as follows: (1) Data transfer methods, (2) Statistical methods, (3) Empirical equations, and (4) Simulation models.

Because flood flow frequency information is used for various purposes, the hydrologist must be aware of the limitations and factors involved which are associated with each of the groupings of methods. For example, flood flow frequency estimates used for design of flood control facilities often are conservative in that the design discharges are high for the corresponding return frequency. In this fashion, the designer compensates for the unknown reliability of the design flow rate and provides for a factor of safety. For flood insurance studies, however, use of the computed flood flow frequency estimate may be desirable in order to avoid excessively high costs for the corresponding benefit (see U.S. Army Corps of Engineers Training Document No. 11, TD-11, 1980).

Detailed discussions of the several categories of flood flow frequency analysis procedures are contained in TD-11. In that publication, the four groupings of methods are further defined into eight categories as follows:

(I) statistical estimation of peak flowrates

(II) statistical estimation of moments

(III) index flood estimation methods

(IV) transfer methods

(V) empirical equations

(VI) single event methods

(VII) multiple discrete event methods

(VIII) continuous simulation methods.

Advantages and disadvantages of methods in each of these eight categories are discussed in the following paragraphs.

Category I: Statistical estimation of peak flowrate (Q_p) methods use regression equations for determining a specific return frequency of flowrate by correlating stream gage data to watershed characteristics. Ungaged stream flowrate estimates can then be obtained from the regression equations. Table 1.1 (TD-11, 1980) compares the advantages and disadvantages associated with this category of methods.

Category II: The statistical estimation of moments procedure extends the procedures of Category I by correlating the statistical moments of the frequency function developed from stream gage data to watershed characteristics. Table 1.2 (TD-11, 1980) lists the advantages and disadvantages of this category of methods.

Category III: Index flood estimation methods (see Table 1.3) is analogous to the above two categories except that a selected index flood, such as the mean annual event, is used for the development of the necessary statistical relationships for events other than the index event.

Category IV: Transfer methods (Table 1.4) usually refer to the relationships used to estimate flowrates immediately upstream or downstream of a stream gage location. However TD-11 broadens this category to include procedures for the direct transfer of peak flood flow frequency values or frequency functions from similar gage locations to the subject study point.

Category V: Empirical equations are often used for the estimation of peak flowrates. The well-known rational method is an important example of this category. Table 1.5 (TD-11) compares the advantages and disadvantages of this group of methods.

Category VI: Single event methods are the most widely used approach for developing runoff hydrographs which are subsequently used to develop a flood flow frequency curve. Incorporated in this category are the design storm methods which attempt to relate runoff and rainfall frequency curves. Table 1.6 from TD-11 examines several features of this category of methods.

TABLE 1.1
STATISTICAL ESTIMATION OF Q_p
(CATEGORY I)
Reference: TD-11 (1980)

Applicability/Advantages	Limitations/Disadvantages
. Procedures are based on accepted statistical methods.	. Requires knowledge of both statistics and hydrology in derivation and utilization.
. Procedures are available for most of the country.	. Procedures require numerous regression analyses and are time consuming to develop.
. Reliability of the prediction equations is known for gaged areas used in derivation.	. Only provides estimates of specific peak flood flow frequency relationships.
. Estimates are reliable for hydrologically similar basins as those used in the derivation.	. Cannot evaluate effects resulting from modifications in the system (physical works and alternative land use patterns).
. Once developed, the procedure is quick and easy to use.	. Procedures are often misused by application for areas with different stream patterns and other hydrologic characteristics from the gaged locations used in the derivation.
. Permits direct calculation of specific peak flood flow frequency estimates that are individually and statistically derived.	. Cannot adequately evaluate hydrologically unique areas in the region.
. Procedures may be used in conjunction with other procedures such as to provide calibration relationships for simulation models.	. Easy to use and may be used where other methods would be more appropriate.
. Provides a quick check for reasonableness for situations requiring use of other procedures.	. Derivation requires several hydrologically similar gaged basins in the region.
	. Does not assume a distribution; hence reliability confidence limits cannot be calculated.

TABLE 1.2
STATISTICAL ESTIMATION OF Q_p
(CATEGORY II)
Reference: TD-11 (1980)

Applicability/Advantages	Limitations/Disadvantages
. Procedures are based on accepted statistical methods.	. Requires knowledge of both statistics and hydrology in derivation and utilization.
. The entire frequency function is developed from the three moments; means, standard derivation and skew.	. Procedure requires regression analysis for the two or three moments of the frequency.
. Reliability of the prediction equations is known for gaged areas used in derivation.	. May be time consuming to develop.
. Estimates are as reliabile for hydrologically similar basins as those used in derivation.	. Does not calculate specific flood flow frequency events.
. Once developed, the procedure is quick and easy to use.	. Only provides estimates of peak flood flow frequency relationships.
. Procedures may be used in conjunction with other procedures, such as, to provide calibration results for simulation models.	. Cannot evaluate effects resulting from modifications in the system (physical works and alternate land use patterns).
. Provides a quick check for reasonableness for situations requiring use of other procedures.	. Cannot adequately evaluate many complex river systems.
	. Cannot evaluate hydrologically unique areas in the region.
	. Ease of use may result in improper application.
	. Derivation requires several hydrologically similar gaged basins in the region.

TABLE 1.3
INDEX FLOOD ESTIMATE
(CATEGORY III)
Reference: TD-11 (1980)

Applicability/Advantages	Limitations/Disadvantages
. Procedure is easier to develop than other statistical methods, and has only one regression analysis.	. Procedure yields same variance (slope of frequency curve) for all applications.
. Procedures are commonly used and based on accepted statistical methods.	. Probably least accurate of the statistical procedures.
. Reliability of prediction equation for index flood is known for derivation.	. Requires knowledge of both statistics and hydrology in derivation and utilization.
. Estimates are reliable for hydrologically similar basins as those used in derivation.	. May be time consuming to develop.
. Once developed, the procedure is quick and easy to use.	. Only provides estimates of peak flood flow frequency relationships.
. Procedures may be used in conjunction with other procedures, such as to provide calibration results for simulation models.	. Cannot evaluate effects resulting from modifications in the system (physical works and alternative land use patterns).
. Provides a quick check for reasonableness for situations requiring use of other procedures.	. Cannot adequately evaluate many complex river systems.
	. Cannot evaluate hydrologically unique areas in the region.
	. Ease of use may result in improper application.
	. Derivation requires several hydrologically similar gaged basins in the region.

TABLE 1.4
TRANSFER METHODS
(CATEGORY IV)
Reference: TD-11 (1980)

Applicability/Advantages	Limitations/Disadvantages
(WRC Transfer of Q_p)	(WRC Transfer of Q_p)
. Procedure is easy and quick to use.	. Ease of use may result in improper application.
. Provides reliable estimates immediately upstream and downstream of gage location if hydrologic characteristics are consistent.	. Can only be utilized immediately upstream and downstream of gaged area where hydrologic characteristics are consistent.
. Procedure is commonly used and generally acceptable.	
(Direct Transfer)	(Direct Transfer)
. Provides quick estimate where time constraints are binding and other procedures are not applicable.	. Estimates are not accurate enough for most analysis requirements.
. Can readily be used as a check for reasonableness of results from other procedures.	. Cannot be used for modified basin conditions.
. Provides valuable insight as to the regional slope characteristics of the flood flow frequency relationships.	. Can only be used as check in areas where hydrologic characteristics are nearly similar and with drainage areas within the same order of magnitude.

TABLE 1.5
EMPIRICAL EQUATIONS
(CATEGORY V)
Reference: TD-11 (1980)

Applicability/Advantages	Limitations/Disadvantages
• Provides quick means of estimating peak discharge frequency for small areas.	• Generally are not applicable for areas greater than one square mile.
• Concepts can be understood by nonhydrologists.	• Estimate only the peak discharge frequency relationships.
• Suitable for many types of municipal engineering analyses (storm sewers, culverts, small organization impacts, etc.).	• Cannot be used to design storage facilities.
• Familiarity of procedures and use has led to politically acceptable solutions for small areas.	• Cannot adequately evaluate complex systems where timing and combining of flood hydrographs are important.
• Can be used as a check for reasonableness of more applicable procedures in small areas.	

TABLE 1.6
SINGLE EVENT SIMULATION
(CATEGORY VI)
Reference: TD-11 (1980)

Applicability/Advantages

- Generates other hydrologic information rather than peak discharges (volumes, time to peak, rate of rise, etc.).
- Generates balanced floods as opposed to historically generated events which may be biased.
- Enables evaluation of complex systems and modifications to the watersheds.
- Provides good documentation for quick future use.
- Uses fewer parameters than most continuous simulation models.
- Approximates the hydrologic runoff process as opposed to statistical methods
- Procedures are most economical than continuous simulation procedures.
- Calibration procedures are easier than continuous simulation models.
- Models may be calibrated to either simple or complex systems.

Limitations/Disadvantages

- Balanced flood concept is difficult to understand.
- Modeling requires more time, data, and resources (costs) than statistical procedures.
- Hydrologists must understand the concepts utilized by the model.
- Requires calibration to assure rainfall frequency approximates runoff frequency.
- Unit hydrograph assumes a linear relationship with runoff.
- Requires data processing capabilities.
- Procedures greatly simplify the hydrologic process.
- Procedures are generally limited to basins greater than one square mile.
- Parameters are difficult to obtain for existing and modified conditions.
- Difficult to obtain antecedent moisture conditions.
- Depth-area of rainfall varies with drainage area size.

Category VII: By considering a series of important record storm events with a single event method, an approximate flood frequency curve can be developed. The multiple discrete event category (see Table 1.7) of models serves as a blend of the single event category of models and the concept of continuous simulation.

Category VIII: Continuous simulation (or continuous record) models attempt to develop a continuous streamflow record based on a continuous rainfall record. Although in concept this category of models appears to be plausible, the success of these methods has not been clearly established due to the lack of evidence that this approach out performs the much simpler and more often used unit hydrograph procedures of Category VI.

1.4. Watershed Modeling Uncertainty

Watershed runoff is a function of rainfall intensity, the storm duration, the infiltration capacity of the soil, the cover of the soil, type of vegetation, area of the watershed and related shape factors, distribution of the storm with respect to space and time, watershed stream system topology, connectivity and branching, watershed geometry, stream system hydraulics, overland flow characteristics, and several other factors. Because of the dozens of variables which are included in a completely deterministic model of watershed runoff and due to the uncertainty which is associated to the spatial and temporal values of each of the variable mathematical definitions, urban hydrologists often use statistical methods in predicting surface runoff quantities.

With the widespread use of minicomputers and inexpensive microcomputers, the use of deterministic models are commonplace. These models attempt to simulate several of the most important hydrologic variables that strongly influence the watershed runoff quantities produced from severe design storm events. Generally speaking, the design storm and simulation models include approximations for runoff hydrograph generation (coupled with models for estimating interception, evapotranspiration, interflow, and infiltration), channel routing, and detention basin routing. The program user then combines these processes into a link-node schematic of the watershed. Because each of the hydrologic processes involve several parameters, the resulting output of the model (the runoff hydrograph) may be a function of several dozen parameters. In a procedure called calibration, many or all of the parameters are estimated by attempts to duplicate significant historical runoff hydrographs. However, Wood (1976) notes that the watershed model parameter interaction can result in considerable difficulty in optimizing the parameter set. In a

TABLE 1.7
MULTIPLE DISCRETE EVENTS
(CATEGORY VII)
Reference: TD-11 (1980)

Applicability/Advantages	Limitations/Disadvantages
. Concepts are easier to understand than those associated with hypothetical frequency events.	. Requires numerous storm analyses and subsequent event analyses.
. Antecedent moisture conditions are determined.	. Important events may be overlooked.
. Depth-area precipitation problems are eliminated.	. Results may be biased by historic records.
. Evaluates fewer events than continuous simulation models.	. Procedures use simplified hydrologic process.
. Enables evaluations of complex systems and physical modifications in the watershed.	. Requires data processing capabilities.
. Uses fewer parameters than continuous simulation models.	. Parameters are difficult to obtain.
. Approximates hydrologic process as opposed to statistical methods.	. Unit hydrograph assumes linear relationship with runoff.
. Provides good documentation for future use.	. Requires calibration which is more time consuming than single event due to the large number of events that are processed.
	. Procedure is significantly more expensive than single event modeling.
	. Procedures generally not feasible for small study areas, short time constraints, etc.

TABLE 1.8
CONTINUOUS SIMULATION
(CATEGORY VIII)
Reference: TD-11 (1980)

Applicability/Advantages	Limitations/Disadvantages
• Concepts are easily understood. • Concepts are more physically based than other procedures. • Antecedent moisture conditions are automatically accounted for. • Can be used in unique basins where other procedures such as statistical procedure are not applicable. • Process analyses in single computer runs as numerous discrete events. • Can automatically determine annual peak floods at various locations even if their frequencies are different. • Can model the effects of complex systems and physical works.	• The calibration process is extensive and generally must be performed by qualified experienced hydrologists. • Procedures are expensive and time consuming to use, impractical for moderate or small resources allocated projects. • The results may be biased by the use of historic rainfall data. • The procedures require large analytical processing capabilities. • The models typically require a large amount of data to properly define the parameters.

similar deterministic modeling approach for soil systems and soil water movement, Guymon et al. (1981) found that just the normal range of uncertainty associated with laboratory measurement of groundwater flow hydraulic parameters can produce considerable variation in the model output. A detailed analysis of the sensitivity corresponding to a watershed model is given by Mein and Brown (1978).

The complexity of the uncertainty problem can be envisioned by formulating an abstract representation of the watershed model. Figure 1.2 shows a link-node model of a simple watershed configuration. Also shown in the figure is a rain gage and stream gage (both located off-site of the catchment). Consequently, a hydrologic model, when calibrated to the available rainfall-runoff data, is but a statistical correlation between the rain gage and stream gage data and typically represents the correlation of the available data rather than the relationship between the actual rainfall distributions over the catchment and resulting runoffs. The output of the model is the runoff hydrograph defined by

$$R(t,x_i) = R(t,x(m_i, s_i)) \tag{1.1}$$

where t = model time, and x_i = model process function with statistical mean m_i and standard deviation s_i. Many of the process variables utilized in equation (1.1) are described in the following paragraphs.

(1). Storm pattern arrangement- The design storm pattern can be thought of as a finite series of unit rainfalls (e.g., 6-, 10-, 60-minutes unit durations) where the rainfall intensity is assumed constant during each unit rainfall duration. The set of possible permutations of the set is (n!) where n is the number of unit periods, and each member of the set may have an assumed probability of occurrence. By examining a long history of local storm patterns, a general shape of the storm pattern may be statistically described by some type of shape factor (for example, the center of mass location as a percentage of total storm duration). Studying historical storm patterns with respect to a defined shape factor results in a statistical distribution with a corresponding mean and standard deviation.

(2). Rainfall intensity- The statistical representation of peak unit rainfall intensities result in several mathematical interpretations. The rain gage data is usually insufficient to extrapolate a severe storm event (e.g., a 100-year return frequency) to within a precise confidence limit. Consequently, each unit rainfall has a statistical function associated to its average rainfall intensity, and its probability of occurrence with respect to the other rainfalls.

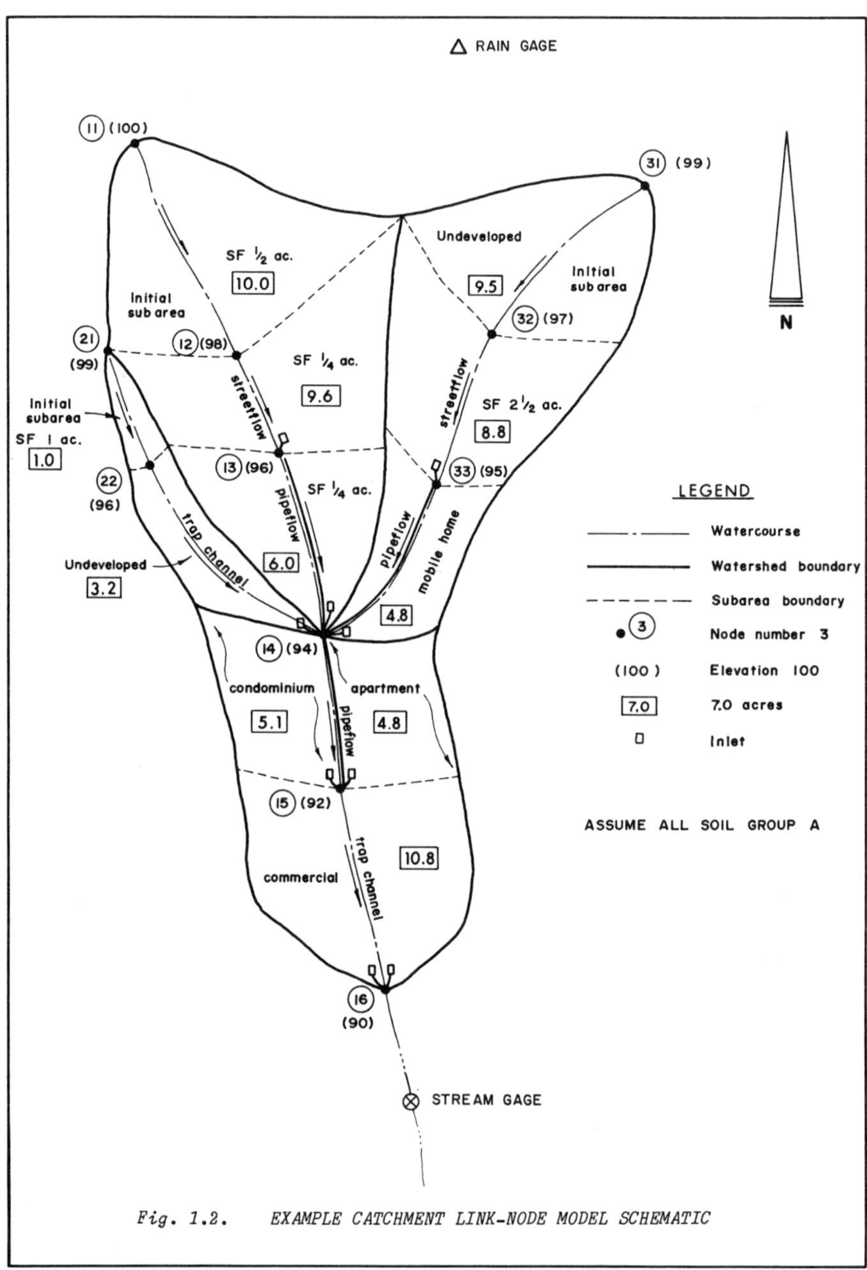

Fig. 1.2. EXAMPLE CATCHMENT LINK-NODE MODEL SCHEMATIC

(3). Point rainfall measurement- Besides the uncertainty in estimating severe storm rainfall magnitudes there exists the uncertainty associated with the measuring and analysis of rain gage data. The rain gage network is usually sparse and many rainfall events either miss a rain gage or the peak storm intensities occur at some location other than at a gage. Additionally, measurement error is introduced due to winds, gage malfunctions, and human measurement error. By studying the rain gage network point rainfall estimates and comparing to actual storm histories, a probabilistic function may be estimated.

(4). Point rainfall depth-area adjustment- Factors to adjust point rainfall values for watershed size involves another probabilistic function which relates the depth-area factors to the storm return frequency, storm duration, location of watershed, and other factors.

(5). Storm movement- In large watershed studies, the movement of a severe storm may be a factor in producing peak runoff flowrate quantities. Associated to this concept is the error in the assumed storm movement and the corresponding probability distribution. Small watersheds, however, may usually exclude this variable.

(6). Antecedent moisture condition- The soil saturation condition at the initiation of the design storm event has a probabilistic distribution relating the soil moisture content of the soil to the design storm's return frequency. This function is further complicated by the different recovery rates of various soil types and soil cover conditions.

(7). Infiltration capacity- This variable alone leads to considerable uncertainty in watershed runoff determination. Errors arise due to the imprecise knowledge of soil cover, condition, landscape practices, compaction, and cultural practices, among other factors. Additional uncertainty occurs from laboratory measurement errors, soil type nonhomogeneity, plant root densities, position of the water table, and soil layering characteristics. Guymon et al. (1981) noted that the doubts associated to deterministic models of soil water flow alone introduce considerable pessimism about deterministic models of more complex processes which involve soil water flow models as a submodel. Additional uncertainty is generated by the infiltration capacity of a soil complex, the relationship of infiltration capacity to the rainfall intensity and antecedent moisture condition, and the proper definition of the capacity function itself.

(8). Groundcover condition- Associated to the condition of the soil is a probability with respect to severe storm occurrence. For example, a burned watershed has a different runoff condition than the unburned meadow in good condition.

(9). Depression storage capacity- Each watershed contains local depressions which essentially store runoff and subtract from the effective rainfall- runoff budget. This depression storage has a probabilistic function which relates the remaining depression storage capacity to the occurrence of the severe storm event.

(10). Runoff hydrograph model uncertainty- This error function attempts to relate several uncertainty factors including overland flow, the unit hydrograph shape, the linearity assumption used with the unit hydrograph, lag time (or time to peak) estimation, and related watershed parameters (e.g., basin factors, slope, etc.).

(11). Channel routing models- An exact channel routing deterministic model is currently unavailable. However by accepting a level of error, reasonable routing methods such as the often-used convex method can be used. The method is inexact and has an associated probabilistic error function. Additional uncertainty is introduced in the routing method parameter estimations and the actual channel definition.

(12). Baseflow and interflow- The interaction between groundwater flow effects and the stream channels introduce several uncertainty variables, each with an unknown error function.

(13). Basin routing models- Analogous to channel routing models, detention basin and reservoir routing models involve a mathematical interpretation of a physical process which has an associated error in precision. Also involved are errors in defining outflow versus depth, and basin storage versus depth relations. Considerable error can be associated with backwater effects in the basin outlet system due to unknown downstream hydraulics.

(14). Submodel combination- Besides the individual process error functions, an error probability exists in the interaction between the several variables. For example, the errors in channel routing may result in the channel peak flowrate being improperly defined and offset (with respect to time) with the arrival of another channel peak flowrate, causing an attenuation of the overall storm peak runoff rate.

(15). Discretization errors- The philosophy on the selection of nodal points and corresponding linkage processes varies between hydrologists and, consequently, different link-node models of a watershed can be produced. Because each of the submodels are imprecise, the total error function is also a function of the number of processes involved and the number and spacing of nodal points. This error includes the error in the hydrologist's interpretation of the watershed runoff characteristics.

Several other error functions can be included in the above list such as hydraulic modeling errors (e.g., backwater effects), stream runoff measurement errors, stream gage data collection and synthesis errors, and so forth. Due to the tremendous uncertainty associated with the watershed simulation approach, a truly precise deterministic model is unavailable. For severe design storm conditions, however, many of the above error functions are eliminated due to the overtaxing of the corresponding physical system. For example, the local design storm condition may include enough rainfall such that the depression storage capacity is exhausted prior to the initiation of the high intensity rainfall. Additionally, due to the difficulty in defining the objective function (a severe design storm runoff hydrograph) the uncertainty involved in watershed modeling is usually acceptable in urban hydrology design studies as long as the results are conservative. Obviously, over-estimation should be reduced whenever reasonable; however, the unknown objective function usually necessitates that reasonably conservative methods and parameters be utilized in order to obtain a higher probability of achieving the accepted level of public protection.

1.5. Watershed Runoff

The general hydrologic processes involved in producing watershed runoff are described by the following scenario: During the early portions of the storm, a quantity of rainfall is trapped by the watershed vegetative cover as interception; if the storm is of light intensity or of short duration, the entire storm rainfall budget may be consumed by interception alone. Should the rainfall quantities exceed the interception capacity of the watershed, rainfall reaches the soil surface to begin filling natural and man-made puddles and surface depressions. Runoff will generally not appear downstream of such depressions until the total volume is occupied by runoff (or a combination of runoff, debris, and sediments) from the upstream tributary areas. These two processes are usually grouped together and called the initial abstraction of the watershed and may be either a significant or negligible factor in the subsequent rainfall-runoff budget. For example, a light storm occurring over a watershed which contains severe obstacles to direct runoff flow (such as gravel pits, dams, or natural sumps) may result in a zero runoff to the downstream watershed. In contrast, a fully developed watershed with total storm drain and connected street systems may have an associated initial abstraction which is negligible; consequently, a severe storm event satisfies the interception and depression storage capacity in the early portions of the storm. The infiltration capacity of the soil depends on several variables including the condition of the soil and the soil moisture content. If the soil infiltration capacity is exceeded by the storm rainfall intensity, overland flow begins and the entire soil surface is covered by a film of water (termed surface detention or detention storage). Typically, surface runoff is associated to the runoff quantities which appear in the watershed's stream system. Should the storm be of long duration (usually days), portions of the infiltrated water may reappear as runoff by lateral groundwater movement into a lower elevation stream channel.

In urban hydrologic studies, flood control systems are typically designed to accommodate a severe storm event which is considered to provide a reasonable level of public protection for the corresponding cost to the community. For such severe storm conditions, the urban watershed initial abstraction is generally assumed to have a negliglible effect on the peak storm runoff quantities. This severe assumption is in anticipation of a prior storm event which depletes the available depression storage capacity or assumes a fully developed condition with total positive drainage. Consequently, in urban design storm studies, several of the watershed model variables may be assumed to have a small impact on the subsequent runoff quantities, and the corresponding watershed modeling approach is significantly simplified.

1.6. Selection of a Flood Flow Modeling Approach

1.6.1. Choice of Modeling Category

Due to the need for developing runoff hydrographs for design purposes, statistical methods such as contained in model categories I-V are usually precluded in watershed hydrologic studies. Consequently, the categories of models available are essentially restricted to categories VI, VII, and VIII. The "single event" models directly transform a design storm (hypothetical causative input) into a flood hydrograph. The "multiple discrete event" models transform an annual series of selected discrete rainfall events (usually one storm for each year) into an annual series of runoff hydrographs whose peak flowrates are used for subsequent statistical analysis. The "continuous record" or "continuous simulation" model results in a continuous record of synthetic runoff hydrographs for statistical synthesis. Each of the above three categories of deterministic models contain various versions and modifications which range widely in complexity, data requirements, and computational effort.

In general, the well-known unit hydrograph design storm approach has continued widespread support among practitioners and governmental agencies involved in flood control design. Such general purpose models include the U.S. Department of Agriculture, Soil Conservation Service or SCS model (1984) and the U.S. Army Corps of Engineers (HEC) hydrology computer program package (see TD-15). In a recent survey of hydrologic model usage by Federal and State governmental agencies and private engineering firms (U.S. Department of Transportation, Federal Highway Administration Hydraulic Engineering Circular No. 19, October, 1984), it was found that "practically no use is made of watershed models for discrete event and continuous hydrograph simulation." In comparison, however, design storm methods were used from 24 to 34 times more frequently than the discrete event or continuous simulation models by Federal agencies and the private sector, respectively. The frequent use of design storm methods appear to be due to several reasons: (1) design storm methods are considerably simpler to use than discrete event and continuous simulation models; (2) it has not been established in general that the more complex models provide an improvement in computational accuracy over design storm models; and (3) the level of complexity typically embodied in the continuous simulation class of models does not appear to be appropriate for the catchment rainfall-runoff data which is typically available. Consequently, the design storm approach continues to be the most often selected for flood control and drainage design studies.

A criterion for classifying a model as being simple or complex is given by Beard and Chang (1979) as the "difficulty or reliability of model calibration.... Perhaps the simplest type of model that produces a flood hydrograph is the unit hydrograph model"...and..."can be derived to some extent from physical drainage features but fairly easily and fairly reliably calibrated through successive approximations by relating the time distribution of average basin rainfall excess to the time distribution of runoff." In comparison, the "most complicated type of model is one that represents each significant element of the hydrologic process by a mathematical algorithm. This is represented by the Stanford Watershed Model and requires extensive data and effort to calibrate."

The literature contains several reports of problems in using complex models, especially in parameter optimization. Additionally, it has not been clearly established whether complex models, such as in the continuous simulation or discrete event classes of models, provide an increase in accuracy over a standard design storm unit hydrograph model. There are only a few papers and reports in the literature that provide a comparison in hydrologic model performance. From these references, it appears that a simple unit hydrograph model provides as good as or better results than quasi-physically based (or QPB, see the work of Loague and Freeze (1985)) or complex models.

In their paper, Beard and Chang (1979) write that in the case of the unit hydrograph model, "the function of runoff versus rainfall excess is considered to be linear, whereas it usually is not in nature. Also, the variations in shapes of unit hydrographs are not derivable directly from physical factors. However, models of this general nature are usually as representative of physical conditions as can reasonably be validated by available data, and there is little advantage in extending the degree of model sophistication beyond validation capability." It is suggested that "if (data from) 50 - 100 yrs of streamflow were available for a specified condition of watershed development, a frequency curve of flows for that condition can be constructed from a properly selected set of flows."

Schilling and Fuchs (1986) write "that the spatial resolution of rain data input is of paramount importance to the accuracy of the simulated hydrograph" due to "the high spatial variability of storms" and "the amplification of rainfall sampling errors by the nonlinear transformation" of rainfall into runoff. Their recommendations are that a model should employ a simplified surface flow model if there are many subbasins; a simple runoff coefficient loss rate; and a diffusion (zero inertia) or storage channel routing technique. Hornberger et al. (1985) writes that "Even the most physically based models...cannot reflect the true complexity and heterogeneity of the processes occurring in the field. Catchment hydrology is still very much an emperical science."

In attempting to define the modeling processes by the available field data forms, Hornberger et al. find that "Hydrological quantities measured in the field tend to be either integral variables (e.g., stream discharge, which reflects an integrated catchment response) or point estimates of variables that are likely to exhibit marked spatial and/or temporal variation (e.g., soil hydraulic conductivity)." Hence, the precise definition of the physics in a modeling sense becomes a problem that is "poorly posed in the mathematical sense." Typically, the submodel parameters cannot be estimated precisely due to the large associated estimation error. "Such difficulties often indicate that the structural complexity of the model is greater than is warranted on the basis of the calibration data set."

Schilling and Fuchs (1986) note that errors in simulation occur for several reasons, including:

"1. The input data, consisting of rainfall and antecedent conditions, vary throughout the watershed and cannot be precisely measured.

2. The physical laws of fluid motion are simplified.
3. Model parameter estimates may be in error."

By reducing the rainfall data set resolution from a grid of 81 gages to a single catchment-centered gage in an 1,800 acre catchment, variations in runoff volumes and peak flows "is well above 100 percent over the entire range of storms implying that the spatial resolution of rainfall has a dominant influence on the reliability of computed runoff." It is also noted that "errors in the rainfall input are amplified by the rainfall-runoff transformation" so that "a rainfall depth error of 30 percent results in a volume error of 60 percent and a peak flow error of 80 percent."

Schilling and Fuchs (1986) also write that "it is inappropriate to use a sophisticated runoff model to achieve a desired level of modeling accuracy if the spatial resolution of rain input is low" (in their study, the raingage densities considered for the 1,800-acre catchment are 81, 9, and a single centered gage).

Similarly, Beard and Chang (1979) write that in their study of 14 urban catchments, complex models such as continuous simulation typically have 20 to 40 parameters and functions that must be derived from recorded rainfall-runoff data. "Inasmuch as rainfall data are for scattered point locations and storm rainfall is highly variable in time and space, available data are generally inadequate...in this region for reliably calibrating the various interrelated functions of these complex models." Additionally, "changes in the model that would result from urbanization could not be reliably determined." They write that the application "of these complex models to evaluating changes in flood frequencies usually requires simulation of about 50 years of streamflow at each location under each alternative watershed condition."

Garen and Burges (1981) noted the difficulties in rainfall measurement for use in the Stanford Watershed Model, because the K1 parameter (rainfall adjustment factor) and UZSN parameter (upper level storage) had the dominant impact on the model sensitivity. This is especially noteworthy because Dawdy and O'Donnell (1965) concluded that insensitive model coefficients could not be calibrated accurately. Thus, they could not be used to measure physical effects of watershed changes.

In the extensive study by Loague and Freeze (1985), three event-based rainfall-runoff models (a regression model, a unit hydrograph model, and a kinematic wave quasi-physically based model) were used on three data sets of 269 storm events from three small upland catchments. In that paper, the term "quasi-physically based" or QPB is used for the kinematic wave model. The three catchments were 25 acres, 2.8 square-miles, and 35 acres in size, and were extensively monitored with rain gage, stream gage, neutron probe, and soil parameter site testing.

For example, the 25 acre site instrumentation (Fig. 1.3) contained 35 neutron probe access sites, 26 soil parameter sites (all equally spaced), an on-site rain gage, and a stream gage. The QPB model (Fig. 1.4) utilized 22 overland flow planes and four channel segments. In comparative tests between the three modeling approaches to measured rainfall-runoff data it was concluded that all models performed poorly and that the QPB performance was only slightly improved by calibration of its most sensitive parameter, hydraulic conductivity. They write that the "conclusion one is forced to draw...is that the QPB model does not represent reality very well; in other words, there is considerable model error present. We suspect this is the case with most, if not all conceptual models currently in use." Additionally, "the fact that simpler, less data intensive models provided as good or better predictions than a QPB is food for thought."

Based on the literature, the main difficulty in the use, calibration, and development of complex models appears to be the lack of precise rainfall data and the high model sensitivity to (and magnification of) rainfall measurement errors. Nash and Sutcliffe (1970) write that "As there is little point in applying exact laws to approximate boundary conditions, this, and the limited ranges of the variables encountered, suggest the use of simplified empirical relations."

It is noteworthy to consider the HEC Research Note No. 6 (1979) where the Hydrocomp HSP continuous simulation model was applied to the West Branch DuPage River in Illinois. Personnel from Hydrocomp, HEC and COE participated in this study which started with a nearly complete hydrologic/meteorologic data base. The report stated that "It took one person six months to assemble and analyze additional data, and to learn how to use the model. Another six months were spent in calibration and long-record simulation." This time allocation applies to only a 28.5 square-mile basin. The quality of the final model is indicated by the average absolute monthly volume error of 32.1 and 28.1 percent for calibration and verification periods, respectively. Figure 1.5 shows a typical comparison of modeled and measured results. Peak flow rate average absolute errors were 26 and 36 percent for calibration and verification periods, respectively. It was concluded that "Discharge frequency under changing urban conditions is a problem that could be handled by simpler, quicker, less costly approaches requiring much less data; e.g., design storms or several historical events used as input to a single-event model, or a continuous model with a less complex soil-moisture accounting algorithm."

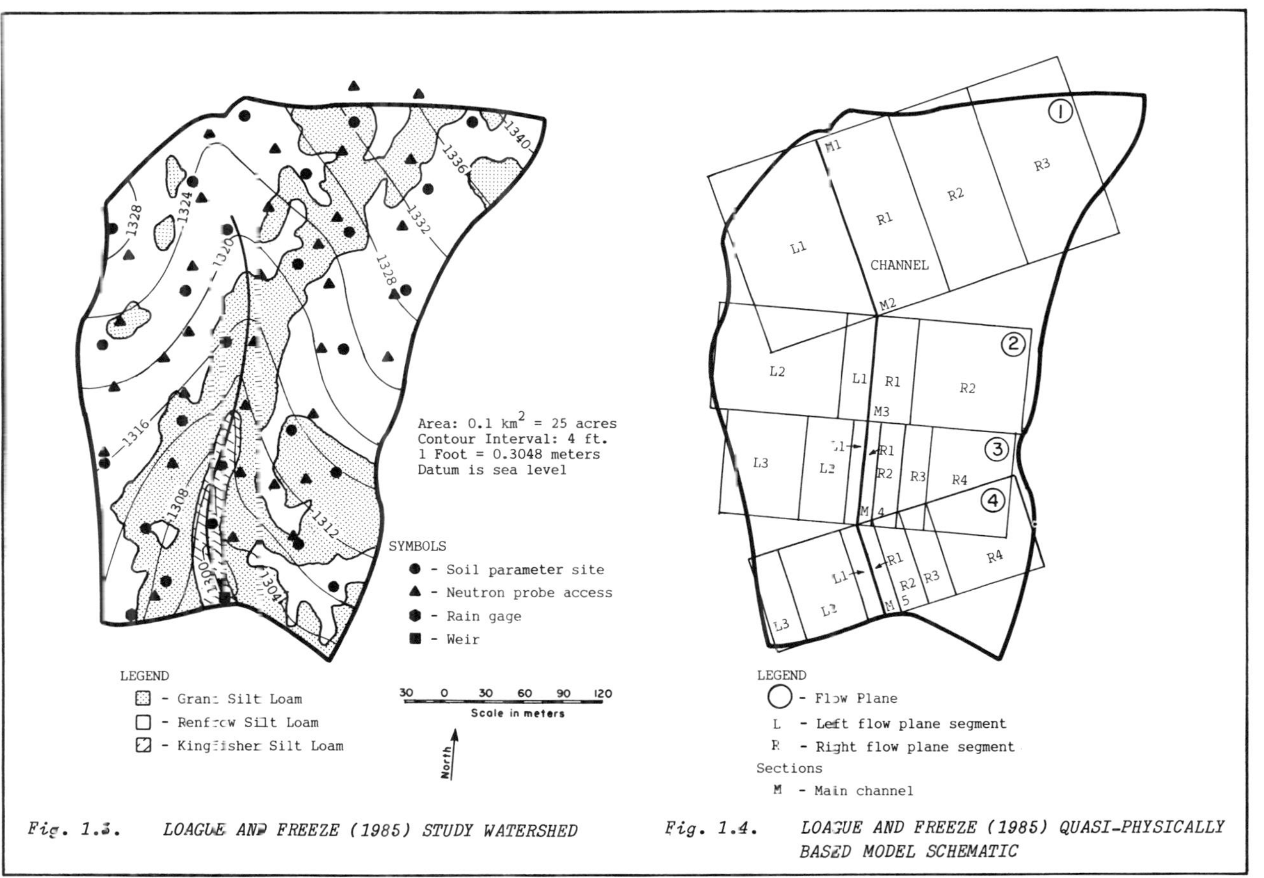

Fig. 1.3. LOAGUE AND FREEZE (1985) STUDY WATERSHED

Fig. 1.4. LOAGUE AND FREEZE (1985) QUASI-PHYSICALLY BASED MODEL SCHEMATIC

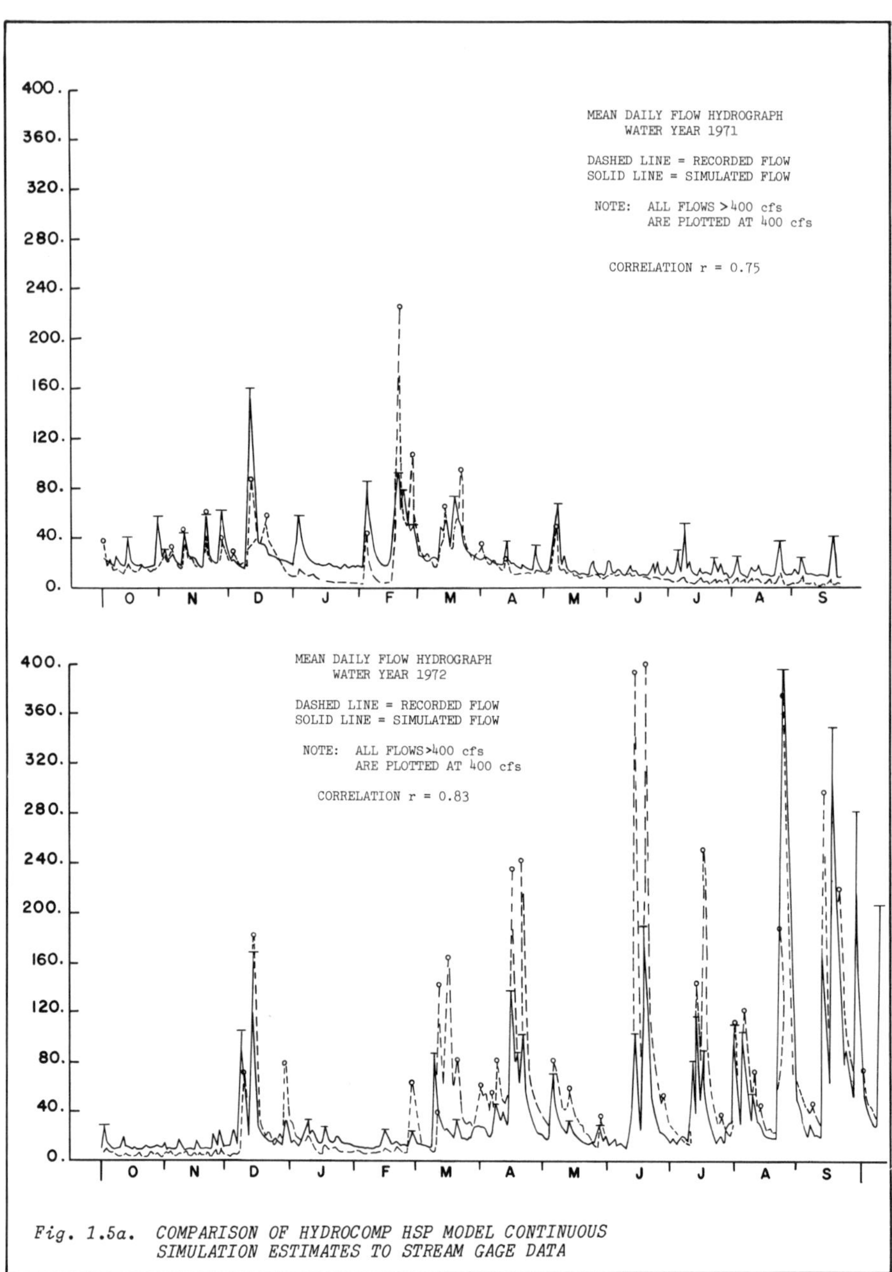

Fig. 1.5a. COMPARISON OF HYDROCOMP HSP MODEL CONTINUOUS SIMULATION ESTIMATES TO STREAM GAGE DATA

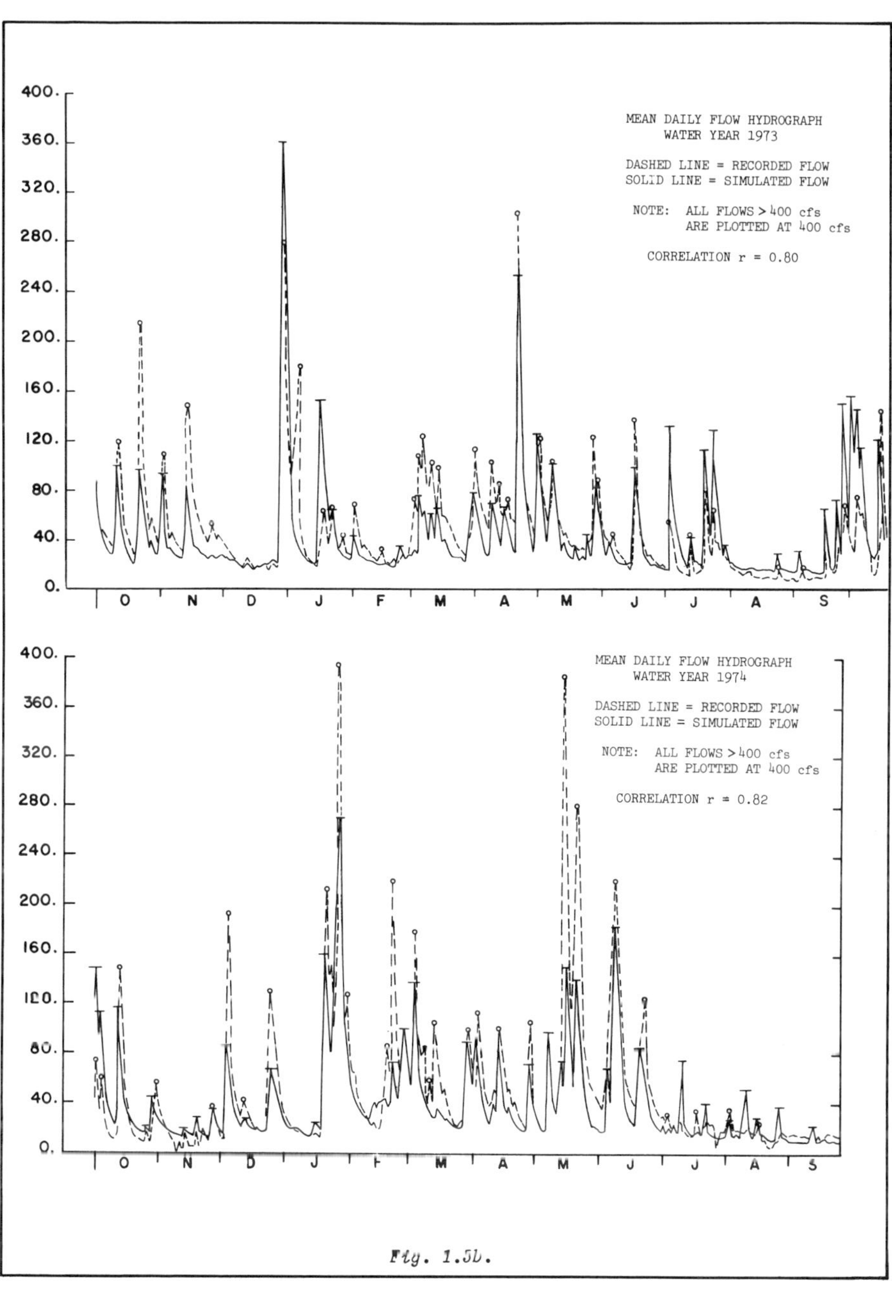
400.
360.
320.
280.
240.
200.
160.
120.
80.
40.
0.
O
N
D
J
F
M
A
M
J
J
A
S
MEAN DAILY FLOW HYDROGRAPH
WATER YEAR 1973
DASHED LINE = RECORDED FLOW
SOLID LINE = SIMULATED FLOW
NOTE: ALL FLOWS > 400 cfs
ARE PLOTTED AT 400 cfs
CORRELATION r = 0.80
MEAN DAILY FLOW HYDROGRAPH
WATER YEAR 1974
DASHED LINE = RECORDED FLOW
SOLID LINE = SIMULATED FLOW
NOTE: ALL FLOWS > 400 cfs
ARE PLOTTED AT 400 cfs
CORRELATION r = 0.82

Fig. 1.5b.

In another study, HEC Technical Paper No. 59 (Abbott, 1978) compared six hydrologic models, plus two variants of one and a variant of another, in a preliminary evaluation of their relative capabilities, accuracy and ease of application on a 5.5 square-mile urban watershed near Oakland, California. Four continuous simulation models were tested: Storage Treatment Overflow Runoff Model (STORM), Hydrocomp Simulation Program (HSP), Streamflow Synthesis and Reservoir Regulation (SSARR), and Continuous Flood Hydrographs (HEC-IC). Single-storm event comparisons were made using STORM, HSP, SSARR, Storm Water Managerment Model (SWMM), Massachusetts Institute of Technology Catchment Model (MITCAT), and the HEC-1 unit hydrograph model (single area analysis). Each model was calibrated with the first 40 percent of a 42 month record, and the resulting calibration coefficients were used in simulating the remaining record. The study results showed that the more complex models did not produce better results in developing watershed runoff quantities than the simple models for this test watershed (see Fig. 1.6).

In the absence of more encouraging results in the use of complex hydrology models, the widespread use and continued acceptance of design storm unit hydrograph methods for the estimation of watershed runoff quantities is understandable. For a new rainfall-runoff modeling approach to achieve widespread acceptance, it must clearly demonstrate a superiority in performance. For example, Hall (1984) writes that some predetermined criterion of "goodness-of-fit" is typically used to assess a new model's capability in reproducing historic storm event runoff quantities. The new model is first calibrated to observed rainfall-runoff data and then "verified" using storm events excluded from the calibration storm event data set. This type of split-sample testing (for example, TP-59, 1978; Loague and Freeze, 1985) has been found to be a standard in comparing rainfall-runoff model performance.

A second set of criteria must be evaluated when using a new rainfall-runoff model for design storm flood estimation. Model parameters must be correlated to watershed characteristics, or regional values of the parameters must be established. More specifically, the model parameters used as the dependent variables must provide a relationship between the return frequency of runoff and the return frequency of the input rainfall. Acceptance of any new modeling technique typically depends upon the models ease of use and reproducibility of the results by different engineers and hydrologists. Hall (1984) concludes that "until the additional steps required to develop a rainfall-runoff model into a flood estimation method are more widely appreciated, this apparent reluctance to accept innovation is liable to remain a feature of design practice."

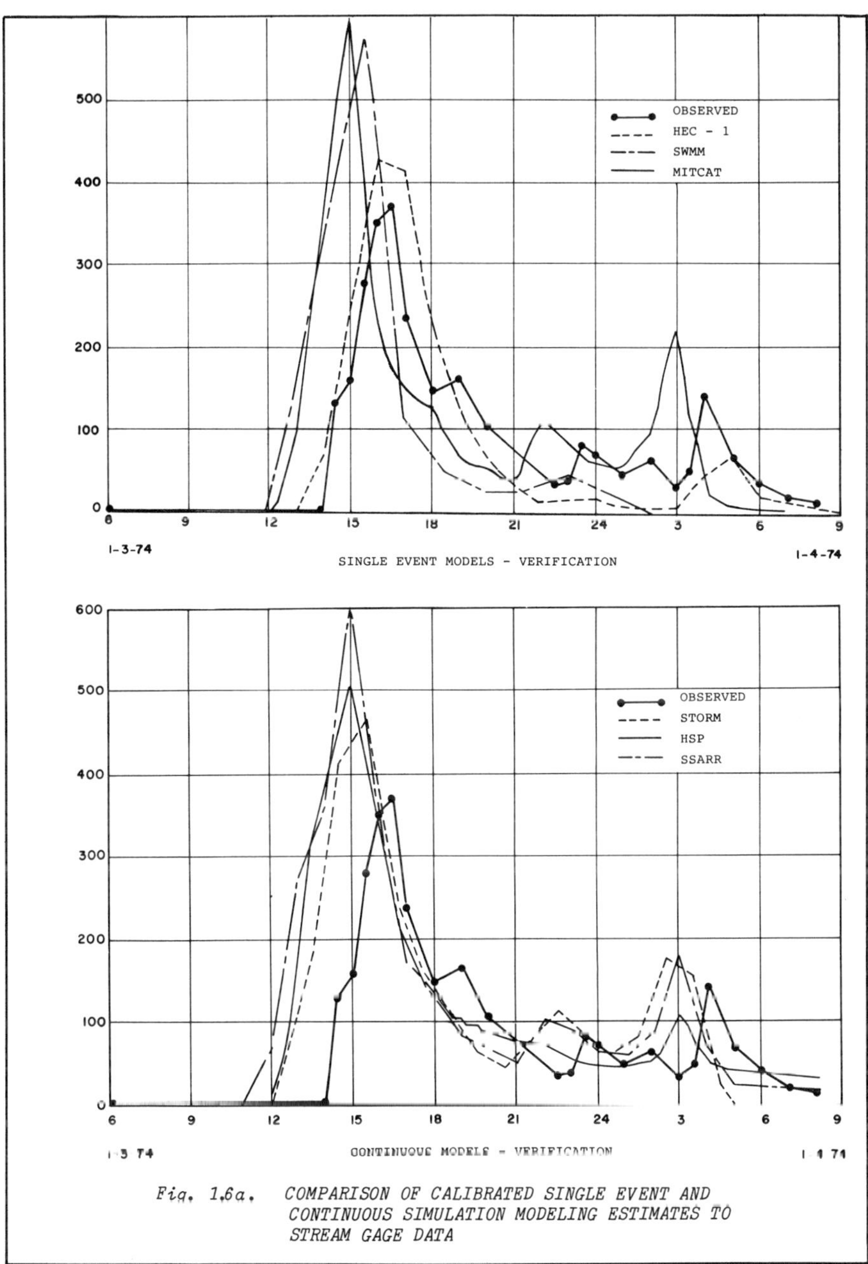

Fig. 1.6a. COMPARISON OF CALIBRATED SINGLE EVENT AND CONTINUOUS SIMULATION MODELING ESTIMATES TO STREAM GAGE DATA

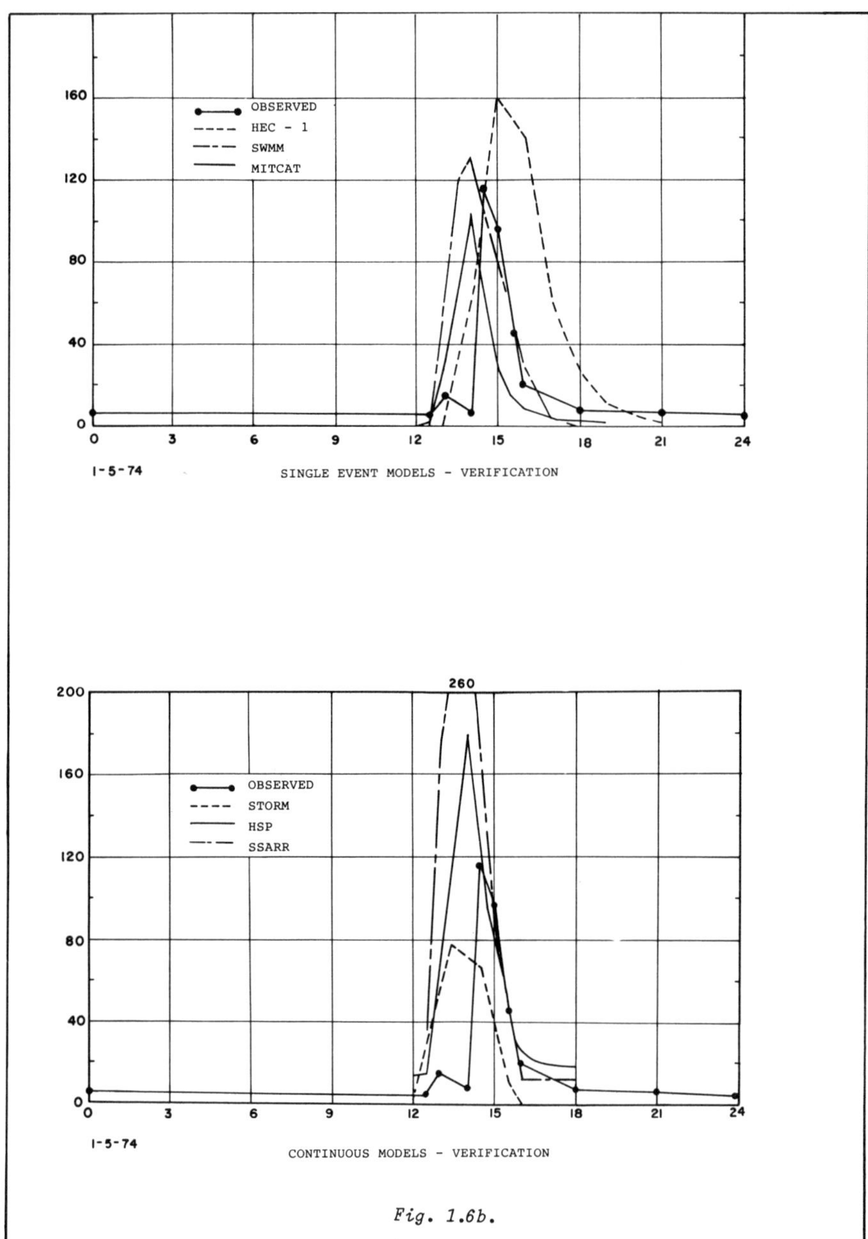

160
120
80
40
0
OBSERVED
HEC - 1
SWMM
MITCAT
0
3
6
9
12
15
18
21
24
1-5-74
SINGLE EVENT MODELS - VERIFICATION
260
200
160
120
80
40
0
OBSERVED
STORM
HSP
SSARR
0
3
6
9
12
15
18
21
24
1-5-74
CONTINUOUS MODELS - VERIFICATION

Fig. 1.6b.

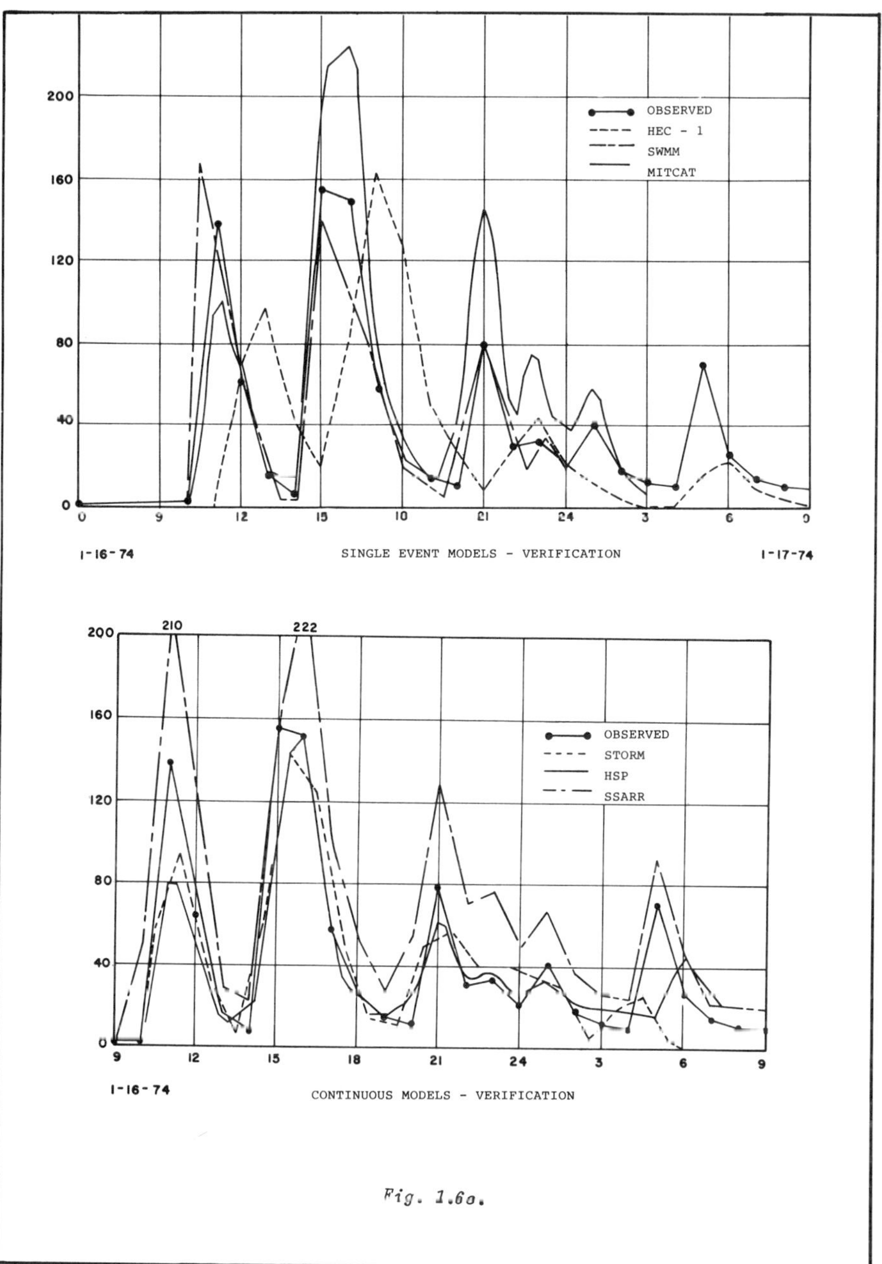

200
160
120
80
40
0
OBSERVED
HEC - 1
SWMM
MITCAT
0
9
12
15
18
21
24
3
6
9
1-16-74
SINGLE EVENT MODELS - VERIFICATION
1-17-74
210
222
OBSERVED
STORM
HSP
SSARR
1-16-74
CONTINUOUS MODELS - VERIFICATION

Fig. 1.6a.

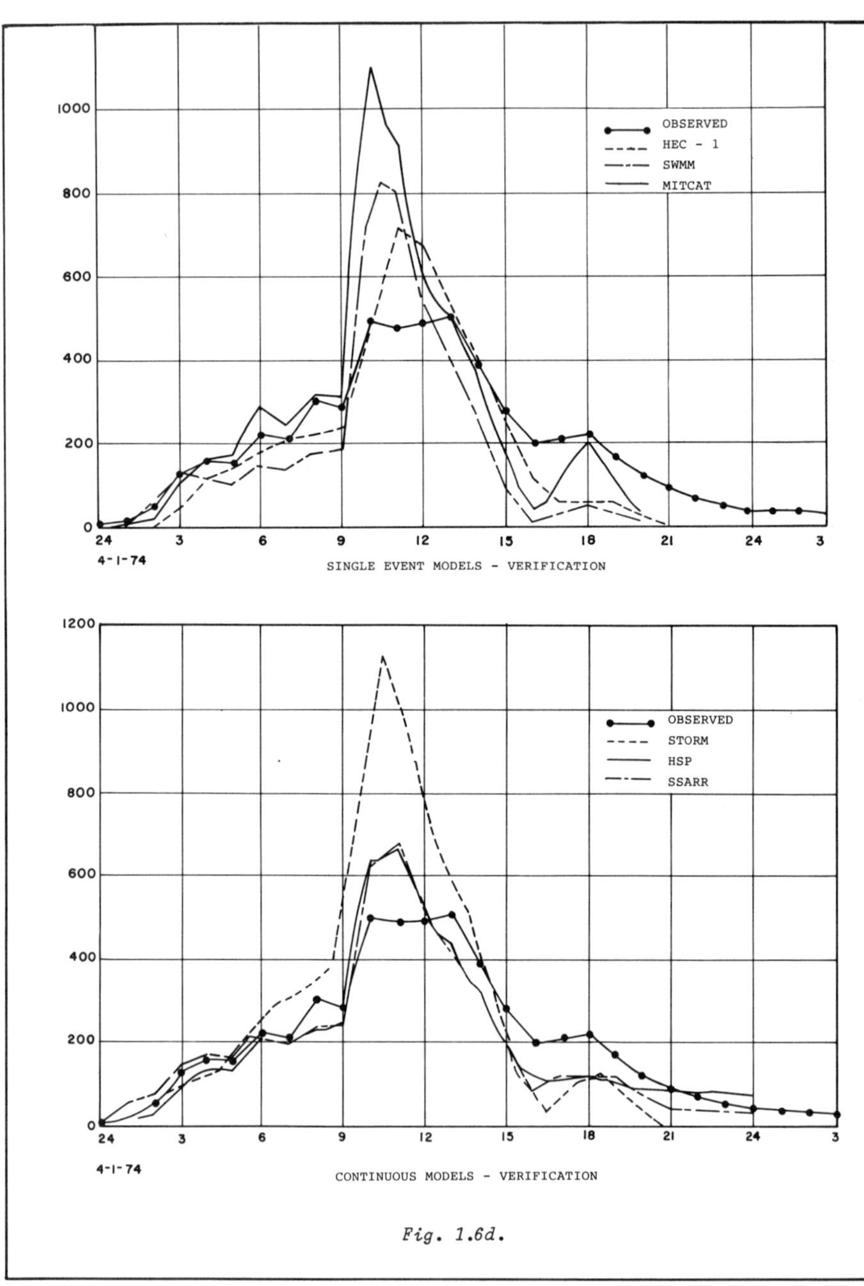
1000
800
600
400
200
0
OBSERVED
HEC - 1
SWMM
MITCAT
24
3
6
9
12
15
18
21
24
3
4-1-74
SINGLE EVENT MODELS - VERIFICATION
1200
1000
800
600
400
200
0
OBSERVED
STORM
HSP
SSARR
24
3
6
9
12
15
18
21
24
3
4-1-74
CONTINUOUS MODELS - VERIFICATION

Fig. 1.6d.

The issue as to whether complex models provided better results than simpler models in developing watershed runoff quantities has motivated the proliferation of dozens of complex, conceptual or so-called physically-based models. However, based upon the available literature, the weight of evidence indicates that use of simpler models such as the well-known unit hydrograph approach will continue to be the most widely used modeling technique. Consequently it is clear that so far, complex watershed models have not proven themselves to provide better results than the standard,simple unit-hydrograph models. In a study of stochastic hydrologic methods, Klemes and Bulu (1979) write that often modelers "sidestep the real problem of modeling—the problem of how well a model is likely to reflect the future events--and divert the user to a more tractable, though less useful, problem of how to construct a model that will reproduce the past events. In so doing they expect, and perhaps rightly so, that by the time the prospective modeler has dug himself out of the heaps of technicalities, he either will have forgotten what the true purpose of modeling is or will have invested so much effort into the modeling game that he would prefer to avoid questions about its relevance." According to Gburek (1971), "...a model system is merely a researcher's idea of how a physical system interacts and behaves, and in the case of watershed research, watershed models are usually extremely simplified mathematical descriptions of a complex situation...until each internal submodel of the overall model can be independently verified, the model remains strictly a hypothesis with respect to its internal locations and transformations...".

Conceptual rainfall-runoff models are typically advertised to be physically based in that they model all the several components of the hydrologic cycle in rainfall-runoff processes. However in reality, most existing conceptual models "...are simplified nonlinear, lumped parameter, time-invariant, discontinuous representaions of a complex nonlinear, distributed parameter, time-variant, and continuous system" (Sorooshian and Gupta, 1983). The use of a lumped parameter approach means that a characteristic or representative value of a parameter is assumed to apply for the entire watershed, for each parameter used in the model. The invariant parameter assumption assumes that all parameters are constant with respect to seasonal moisture changes. Rain gage data are also lumped by some selected procedure which ignores the time and spatial variations of rainfall over the watershed, and between storm events. Watt and Kidd (1975) write that the differences between physically based and so-called "black box" models, (e.g., unit hydrograph models), become less obvious when applied to a field situation. The authors conclude that the considerations of whether the model is physically based or is a black box model "should carry very little weight in the selection process."

Another major issue involving continuous simulation models is that each of these conceptual models requires a calibration of the model parameters be performed in order to obtain an optimum parameter set. However, currently there is no proven technique to obtain this true optimum parameter set.

In Sorooshian and Gupta (1983), several papers are cited which report problems such as spending "over two years of full-time effort, to find a 'true optimum' parameter set for a nine-parameter version of the Boughton model" (see Johnston and Pilgrim, 1976). Other conceptual models are cited (e.g., the Stanford watershed model) which also have a similar level of difficulty in parameter optimization. Sorooshian and Gupta (1983) discuss in detail several reasons for this difficulty, one of which is the conceptual model's parameter interdependence (i.e., more than one parameter simulates an identical modeling output behavior) and another being problems with the so-called model's response surface (i.e., model structural representation, data measurement errors, and imperfect representation of the hydrologic process by the model). In another paper, Gupta and Sorooshian (1983) wrote that "one of the most often-reported problems has been the inability to identify the parameters associated with the infiltration or percolation process."

The study of Johnston and Pilgrim (1976) highlighted the complexities associated to determining the optimum parameter set for a conceptual model, and although the Boughton model was used, it was concluded that "most of the findings are applicable to all rainfall-runoff models." Their study identified nine levels of difficulty in optimizing a parameter set, most of which are related to parameter interdependence and the use of a specific objective function to optimize the parameters. They conclude that "until more confidence can be placed in the derivation of truly optimum values, some doubt must remain on the potential usefulness of rainfall-runoff models." When attempting to calibrate a simulation model to model-produced runoff data, Gupta and Sorooshian (1983) reported that "even when calibrated under ideal conditions, it is often impossible to obtain unique estimates for the parameters."

In another examination of the 13-parameter Boughton model, Mein and Brown (1978) examine the conceptual rainfall-runoff model's sensitivity to variations in each parameter of the 'optimized' parameter set. They conclude that "relationships derived between any given parameter value and measureable watershed characteristics would be imprecise, i.e., they would have wide confidence limits" and that "one could not be confident therefore in changing a particular parameter value of this model and then claiming that this alteration represented the effect of some proposed land use change. On the other hand, the model performed quite well in predicting flows with these insensitive parameters, showing that individual parameter precision is not a prerequisite to study output performance."

Dawdy and Bergmann (1969) identify two categories of error which impact complex models, namely, errors in the estimation of an optimum parameter set and errors resulting due to the unknown variability and intensity of rainfall and storm volume over the watershed. The second error category "places a limit of accuracy upon simulation results," even given the true long-term parameter set. The study concluded that for the test 9.7 square-mile California watershed, using data from a single rain gage whose data had been adjusted to represent mean basin conditions, the prediction of flood peaks could not be made better than about 20 to 25% using a rainfall-runoff simulation model.

Ideally, a dense network of rain gages within the watershed should be used to determine the spatial and temporal variation in storm rainfalls for each storm event. However, usually only one or two gages are available, and often not within the watershed. "Even if measurements from a single gage may be assumed to be representative of overall basin precipitation in an expected value sense, other statistical properties of point rainfall, mainly variability, will differ considerably from the corresponding properties of average basin rainfall. The result can be serious errors in runoff prediction and large biases in parameter estimates obtained by calibration of the model" (Troutman, 1982).

Indeed, rainfall measurement errors at the rain gages themselves provide a source of concern (see for example, Kelway, 1975). "For single rainfall events, where the total catch exceeded 12mm (0.5 inch), the error ranged between 0 and 75%, depending on wind characteristics during the storm," (Neff, 1977).

Another source of difficulty in the determination of the true optimum parameter set is the optimization procedure used during the calibration process, that is, the so-called objective function which is to be minimized. "The choice of the set of data and of the objective function to be used for any given model is a subjective decision which influences the values of the model parameters and the performance of the model," (Dickin and Simon, 1977).

1.6.2. Runoff Hydrograph Generation Techniques

A common critique of the unit hydrograph method for producing runoff is that this approach results in a linear model whereas the watershed response is nonlinear. A popular technique for providing a nonlinear response is the kinematic wave modeling approach which involves the use of overland flowplanes and kinematic routing for channel flow. HEC TD#15 (1982) provides a description and comparison of both the unit hydrograph (UH) and kinematic wave (KW) runoff hydrograph generation techniques. The relative usage of KW by 1983 is indicated in Cermak and Feldman (1983) who write that "actual applications by Corps field offices have been few to nonexistent. Even at HEC the KW approach has not been utilized in any special assistance projects." The relatively small usage of KW was then explained as being due to the slack in hydrologic studies and due to unfamiliarity with the technique.

Watt and Kidd (1975) write that in the comparison of so-called 'physically-based' or 'black-box' modeling types (e.g., UH or n-linear reservoirs) the differences are not clear. For example, "except for certain 'ideal' laboratory catchments, the flow does not conform to the sheet-flow model but instead occurs in many small rivulets.... The choice is then between a 'black-box' model and a 'physically-based' model which is based on a physical situation quite different than the actual field situation, i.e., a 'black-box' model."

However, use of KW implies a non-linear response whereas the UH implies a linear response. Nash and Sutcliffe (1970) write that "the UH assumption of a linear time invariant relationship cannot be tested because neither the input (effective rainfall) nor output (storm runoff) are unequivocally defined." Although watershed response is often considered to be mathematically nonlinear, the nonlinearity of the total watershed response has not been shown to be exactly described as a KW. Indeed, a diffusion hydrodynamic model, DHM (Hromadka and Yen, 1986), provides another nonlinear watershed response that includes an additional term in the governing St. Venant flow equations and that may differ significantly in response from a KW model (e.g., overland flow planes with KW channel routing). There are an infinity of nonlinear mathematical representations possible as a combination of surface runoff and channel routing analogs, therefore, merely claiming that the response of a watershed model can be classified as 'nonlinear' is not proof that the model represents the true nonlinear response of the catchment.

Given that the KW analog is only used to obtain an approximation to catchment response, the KW approach does not appear to provide significantly better computational results (for floods of interest in flood control design and planning) than the commonly used UH method. Dickinson et al. (1967) noted that "in the range of discharges normally considered as flood hydrographs, the time [of concentration] remained virtually constant. In other words, in the range of flood interest, the nonlinear effect approached linearity." An explanation was advanced that "at low discharges, the mean velocity may vary considerably with discharge. However, for higher discharges contained within banks, the mean velocity in the channel remains approximately constant."

In actual travel time measurements of flows in a 96-acre catchment using a radioactive tracing technique, Pilgrim (1976) noted that although the flood runoff process "is grossly nonlinear at low flows, linearity is approximated at high flows." Pilgrim also writes that "simple nonlinear models fitted by data from events covering the whole range of flow may give gross errors when used to estimate large events." It is noted that overbank flow was one of the factors for linearity in this study.

Beven (1979) proposed to place limits on the nonlinearity associated to KW by the specification of a constant flow velocity for catchment runoff for large floods. He proposes "a nonlinear channel system at low flows and a linear system at high flows into a single model." Hence for flood flows of interest in flood control planning and design, Beven's model would reduce to a linear representation of the catchment hydraulics.

A physical test of the KW concept was provided by Hjelmfelt and Burwell (1984), who studied a set of 40 similar erosion plots (which satisfied the definition of being kinematic overland flow planes) and the net response to storm events. Due to the large variability in measured runoff quantities from the plots, however, it was concluded that a criterion for a valid rainfall-runoff model "is that it predicts the mean runoff for each event." The KW method did not fulfill this objective.

In HEC Technical Paper No. 59 (1978), six models, plus two variants of one of these models and a variant of another, were calibrated and tested on a 5.5 square-mile urban catchment in Castro Valley near Oakland, California. Both single event and continuous simulation models based on both UH and KW techniques were used in the text. The study concluded that for this watershed "the more complex models did not produce better results than the simple models..." An examination of the test results between the KW and HEC-1 UH models did not show a clear difference between the methods (see Fig. 1.6).

It is of interest that Singh (1977) concluded "if one is not very confident in estimates of watershed infiltration then in some circumstances linear models may have an advantage over nonlinear models in runoff peak predictions because they do not amplify the input errors." That is, the uncertainty in effective rainfall quantities may be magnified by a nonlinear model; consequently, there is an advantage in using a linear model when there are errors in loss rate and precipitation estimates.

Because it is not clear whether the nonlinear KW method for modeling surface runoff provides an improvement in accuracy over the linear UH based hydrologic models, the UH model will probably continue to be used in hydrologic design and planning modeling studies. The UH approach is simpler to apply, and there is less chance that the UH approach will be incorrectly applied.

1.6.3. The Choice of Design Storm

A final issue is the choice of the design storm to be used with the single event unit hydrograph model. Beard (The HEC publication Hydrologic Engineering Methods for Water Resources: Volume 5- Hypothetical Floods; March, 1975) provides an in-depth study of the use of design runoff hydrographs for flood control studies. "Hypothetical floods consists of hydrographs of artificial flood flows...that can be used as a basis for flood-control planning, design and operation decisions or evaluations. These floods represent classes of floods of a specified or implied range of severity." Such "floods are ordinarily derived from rainfall or snowmelt or both, with ground conditions that are appropriate to the objectives of the study, but they can be derived from runoff data alone, usually on the basis of runoff volume and peak-flow frequency studies and representative time sequences of runoff."

In complex watershed systems that include catchment subareas, and channel and basin routing components, Beard (1975) writes that "it is usually necessary to simulate the effects of each reservoir on downstream flows for all relevant magnitudes of peaks and volumes of inflows. Here it is particularly important that each hypothetical flood has a peak flow and volumes for all pertinent durations that are commensurate in severity, so that each computed regulated flow will have a probability or frequency that is comparable to that of the corresponding unregulated flow.... In the planning of a flood control project involving storage or in the development of reservoir operation rules, it is not ordinarily known what the critical duration will be, because this depends on the amounts of reservoir space and release in relation to flow magnitude. When alternate types of projects are considered, critical durations will be different, and a design flood should reflect a degree of protection that is comparable for the various types of projects."

Beard (1975) notes that the balanced storm concept is an important argument for not using a historic storm pattern or sequence of storm patterns as "No one historical flood would ordinarily be representative of the same severity of peak flow and runoff volumes for all durations of interest." Indeed, should a continuous simulation study be proposed such that the "project is designed to regulate all floods of record, it is likely that one flood will dictate the type of project and its general features, because the largest flood for peak flows is also usually the largest-volume flood." Hence, a continuous simulation model of say 40 years of data can be thought of as a 40 year duration design storm with its own probability of re-occurrence, which typically reduces for modeling purposes to simply a single or double day storm pattern.

Beard and Chang (1979) write that for design storm construction, "it is generally considered that a satisfactory procedure is to construct an approximately symmetrical pattern of rainfall with uniform areal distribution having intensities for all durations corresponding to the same recurrence interval and for that location and size of area" (i.e., depth-area effects). The nested design storm concept is developed in detail in HEC TD#15 (1982), including the use of depth-area adjustments.

1.6.4. The Design Storm/Unit Hydrograph Approach

Of the over 100 models reported in the literature, a design storm/unit hydrograph model (i.e.,"model") continues to be widely used in engineering practice. Some of the reasons are as follows: (1) the design storm approach--the multiple discrete event and continuous simulation categories of models have not been clearly established to provide better predictions of flood flow frequency estimates for evaluating the impact of urbanization and for design of flood control systems than a calibrated design storm model; (2) the unit hydrograph method--it has not been shown that the kinematic wave modeling technique provides a significantly better representation of watershed hydrologic response than a model based on unit hydrographs (locally calibrated or regionally calibrated) that represent free-draining catchments; (3) model usage--the "model" has been used extensively nationwide and has proved generally acceptable and reliable; (4) parameter calibration--the "model" is based on a minimal number of parameters, giving higher accuracy in calibration of model parameters to rainfall-runoff data, and the design storm to local flood flow frequency tendencies; (5) calibration effort--the "model" does not require large data or time requirements for calibration; (6) application effort--the "model" does not require excessive computation for application; (7) acceptability—the "model" uses algorithms that are accepted in engineering practice; (8) model flexibility for planning—data handling and computational submodels can be coupled to the "model" (e.g., channel and basin routing) resulting in a highly flexible modeling capability; (9) model certainty evaluation--the certainty of modeling results can be readily evaluated as a distribution of possible outcomes over the probabilistic distribution of parameter values.

Based on the literature, a major obstacle to successfully calibrating and operating any hydrologic model is the knowledge of the exact rainfall distribution for every storm, for every portion of the catchment. Obviously, the incorporation of detailed hydraulic network approximations and soil-moisture accounting budgets in a model does not substitute for the knowledge of rainfall over the catchment, and the incorporation of such detail does not appear to improve computational results over that obtained by simpler models.

1.7. Hypothetical Floods, Balanced Floods, and Design Storm Methods

The use of the design storm approach for generating hypothetical floods (e.g., HEC TD#15) are by far the most commonly used technique for floodflow estimation in the design and planning of flood control facilities. The design storm approach used in the HEC TD#15 results in a balanced hypothetical runoff hydrograph for use in testing both simple and complex watershed flood control systems. The use of hypothetical floods, therefore, is applicable for catchments where stream gage data is insufficient to conclude a design level of flood protection. The nature, need, application and development of hypothetical floods and balanced floods is discussed in detail in the HEC publication Hydrologic Engineering Methods for Water Resources Development: Volume 5 - Hypothetical Floods (March, 1975) and is paraphrased in the following text.

Hypothetical floods consist of runoff hydrographs of artificial flood flows that can be used as a basis for flood-control planning and design. These synthetic floods represent classes of floods of a specified or implied range of severity. They can be used as a direct basis of design or as representative flood magnitudes for use in evaluating the effectiveness of a flood control project. Hypothetical floods are typically derived from rainfall with soil wetting conditions that are appropriate to the study objectives, but can also be derived from runoff data using runoff volume and peak-flow frequency studies and a representative runoff hydrograph shape. A balanced flood is one that is of equal severity for all possible catchment critical durations, and is expressed in terms of exceedence frequency. In the planning of a flood control project, especially a project involving detention basins, it is not known what the critical duration will be due to the relationship between detention basin storage release and flood magnitude. When alternative types of projects are considered, critical durations will be different, and a design flood should reflect a degree of protection that is comparable for the various types of projects.

Time patterns of floods that occur at any study side are highly variable. No one historical flood would be representative of the same severity of peak flow and runoff volumes for all durations of interest. If a project is designed to control all floods of record, it is likely that one flood will dictate the type of project and its general features, because the largest flood for peak flows is also usually the largest-volume flood. Yet, there is information in the other floods that can be used to obtain a better probabilistic balance of the peak flow and volumes for various durations of a design flood.

The varying characteristics of runoff volume vs. duration can be represented by a set of runoff peak and volume frequency curves such as shown in Fig. 1.7. Floods that dominate the peak-flow and short-duration-volume curves are not necessarily the same ones that dominate the long-duration volume curves. However, the curves of Fig. 1.7 for peak flow and each duration that have the same exceedence probability would represent the same degree of severity, and can be used to construct a balanced hypothetical flood.

In order to develop a balanced flood shape, an analysis can be prepared of the relative timing of maximum volumes for specific durations in relations to the time of peak flows for the largest floods of record. A representative flood hydrograph shape can then be synthesized from among recorded floods. This would be representative only with respect to flow sequences and not with respect to absolute or relative magnitudes of flow for various durations.

In the cited HEC document (March, 1975), the steps to develop the balanced flood are given as follows:

"a. Select an exceedence frequency suited to the needs of a particular problem. This may be a design frequency or a frequency that represents a range of frequencies particularly pertinent to flood damage evaluation.

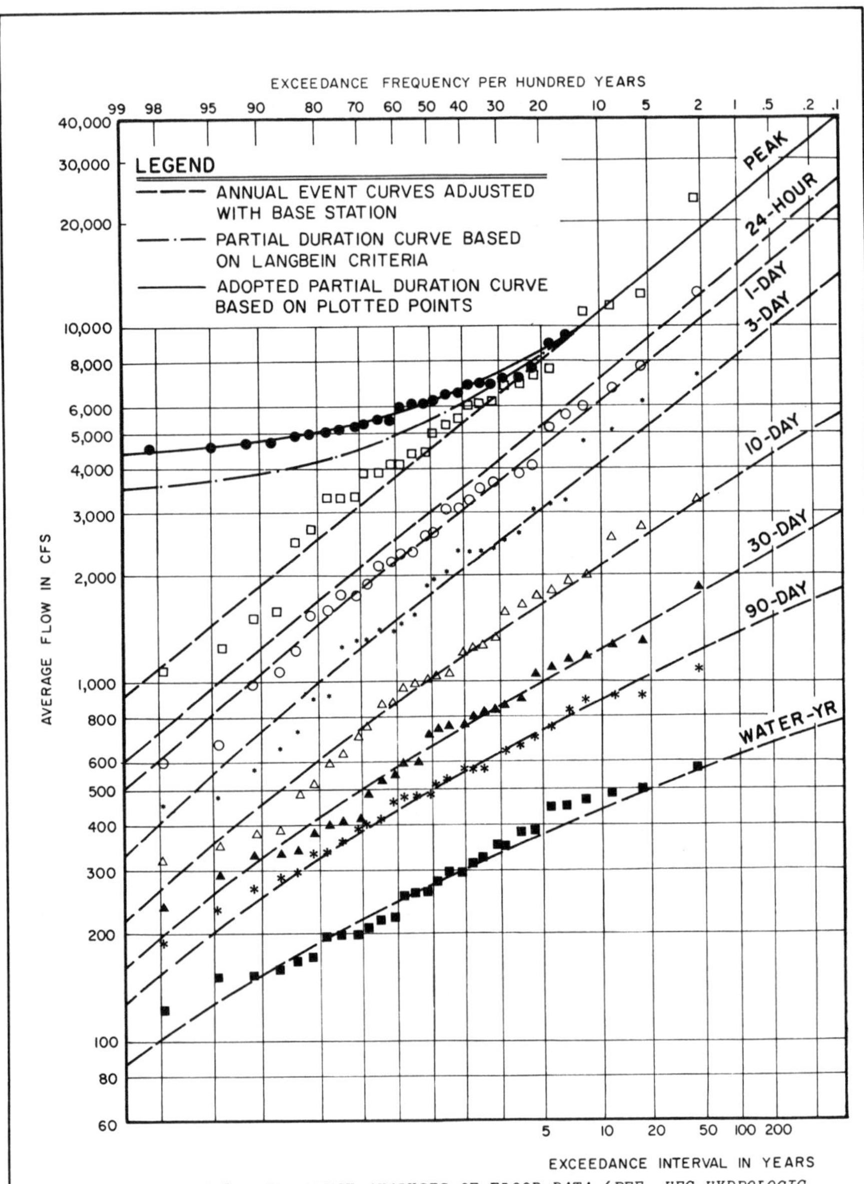

Fig. 1.7. FREQUENCY ANALYSIS OF FLOOD DATA (REF: HEC HYDROLOGIC ENGINEERING METHODS FOR WATER RESOURCES: VOLUME 5-HYPOTHETICAL FLOODS; MARCH, 1975)

b. Obtain target flow peak and volumes corresponding to the selection exceedence frequency, using a set of frequency curves such as that illustrated in Fig. 1.7.

c. Change the maximum ordinate of the representative hydrograph to the target value for the duration equal to the hydrograph tabulation interval. This is done ordinarily, because each ordinate of a hydrograph is considered to represent a volume of flow lasting one computation interval of time. In most cases, it may be desirable to use the target peak flow for this and to ignore the small error in volume thus incurred. This is because it is important to represent the maximum instantaneous flow, and the computation interval should be short enough that instantaneous flows effectively represent the average of flows for a computation interval.

d. When some balanced hydrograph values have been adjusted for any specific duration, the segment of the unadjusted representative hydrograph that has maximum volume for the next longer duration is determined. All adjusted flows within this segment are expressed as a total runoff volume, which is then subtracted from the target volume for the next longer duration. All of the unadjusted flows within this duration are then adjusted so that the total volume they represent will correspond to the target value. This step is repeated until the last duration is reached.

e. It is possible, although a rare occurrence, that the representative hydrograph will not have maximum volumes for all durations within the time period of maximum volumes for longer durations. It is also possible that successive adjustments as described in step d will substantially modify the shape of the representative hydrograph. Either of these circumstances could result in a hydrograph whose volume-duration characteristics do not conform to target values. Consequently, it is necessary to check this at the end of each adjustment sequence and, if substantial differences exist, steps c and d should be repeated using the adjusted hydrograph as the new representative hydrograph. A number of such repetitions might be necessary in some cases.

An example of a balanced hydrograph is shown in Fig. 1.8.

A balanced flood can be used as a design flood to assure that the project will provide protection against a flood having an

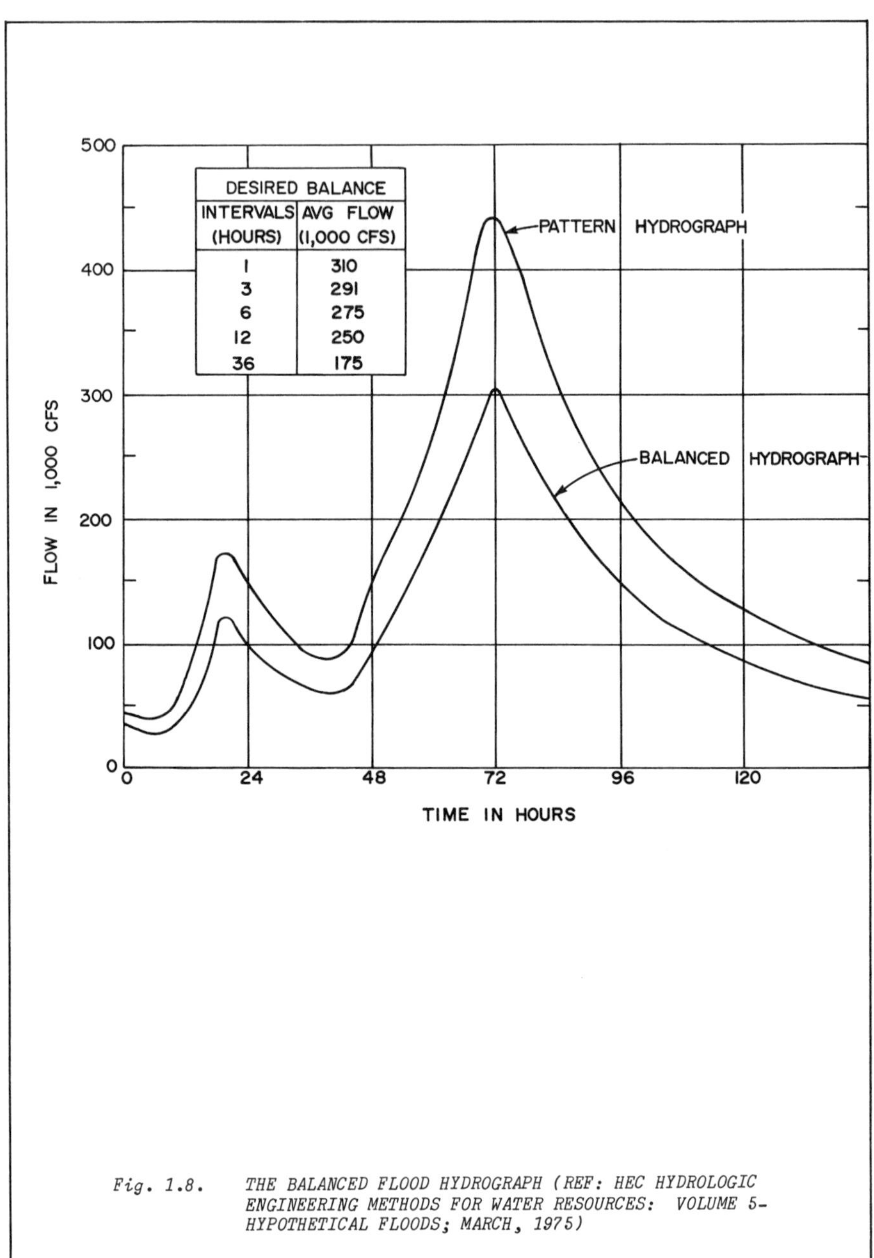

Fig. 1.8. THE BALANCED FLOOD HYDROGRAPH (REF: HEC HYDROLOGIC ENGINEERING METHODS FOR WATER RESOURCES: VOLUME 5-HYPOTHETICAL FLOODS; MARCH, 1975)

exceedence frequency equal to that selected for the balanced flood. If a project design is sensitive to peak flows or short-duration volumes, it will not be sensitive to long-duration volumes, and vice versa. Accordingly, there is no significant escalation of severity by including maximum volumes for all durations in the same design hydrograph."

When stream gage data is inadequate, the runoff volume and peak flow data can be correlated to rain gage data where available. To insure a critical volume for each critical duration, the rainfall depths are defined according to identical derived return frequencies of rainfall (e.g., HEC TD-15). The resulting nested storm pattern, when adjusted for depth-area effects (see Chapter 2), distributes the appropriate depth of rainfall for all durations. The resulting synthetic runoff hydrograph provides an analog to the balanced flood approach.

precipitation

2

2.1. Introduction

Precipitation is an important input to most hydrologic models. The physical processes that control the temporal and spatial variation of precipitation produce a highly variable design input. The variation is represented by a statistical model, with the rainfall depth and intensity statistically related to the storm duration and frequency of occurrence. To properly use precipitation as an input to hydrologic designs it is important to understand the analysis of both actual and synthetic precipitation data. After discussing the source and characteristics of measured rainfall data, methods of spatially averaging measured point data are discussed. While most hydrologic designs only require point rainfall characteristics, some design problems for large watershed areas require rainfall depths that are characteristic of large watersheds; the depth-area adjustment can be applied to point rainfall depths to ensure that the values are representative of the volume of rain that can be expected over large areas. Recognizing that rainfall characteristics vary spatially, as well as with time, methods for deriving isohyetal maps of the spatial variation of rainfall are discussed. While point rainfall depths serve as input to many hydrologic models, many design problems require rainfall input that shows the temporal variation of rainfall characteristics. Therefore, the final topic discussed is the development of synthetic rainfall patterns, which are usually referred to as design storm hyetographs.

2.2. Classification of Precipitation

The classification of storm precipitation is generally based on the mechanism that elevates the moisture-laden air mass to be subsequently cooled. Frontal precipitation, for example, occurs when moisture laden warm air is elevated along a frontal surface over a region of cooler air. Likewise, nonfrontal precipitation implies that the storm rainfall is not attributed to fronts. If the movement of air masses results in warm air replacing cooler air, a "warm front" is identified. In contrast, if warm air is replaced by cooler air, then a "cold front" is identified. A "stationary front" indicates that the front is not in motion. Convective precipitation is generally caused by the upward migration of warm air (usually heated at the ground surface) through denser cold air with resulting rainfall intensities ranging from light showers to catastrophic cloudbursts such as associated with severe thunderstorms. This type of precipitation is generally found in tropical regions. Orographic precipitation describes the process of moisture-laden air being cooled due to the elevation of the air mass over mountain barriers. Cyclonic precipitation is identified with the movement of air masses from high pressure to low pressure regions, caused by the uneven heating of the atmosphere by the sun.

2.3. Mechanisms for Cooling

Each of the described elevation mechanisms involves a cooling process which is associated with the resulting precipitation. Cyclonic cooling, for example, is generally classified according to the association with frontal and nonfrontal effects. The nonfrontal grouping is associated with the convergence and resulting elevation of air accompanied with a low pressure area. Nonfrontal cyclonic precipitation of extratropical origin is associated with rainfall (or snows) of moderate intensities and of long duration. This type of storm may last for a few days and deliver from one to six inches of rainfall. In contrast nonfrontal cyclonic precipitation of tropical origin may result in more than double the rainfall mass in half the storm duration.

Orographic cooling occurs when the moisture laden air is mechanically elevated along a mountain barrier, allowing the air to expand and cool at the lower pressure upper elevation. Because of the mountain barrier, a rain shadow may occur on the lee side of the barrier. Generally, the windward side of the mountains are more cloudy, have more precipitation, and have smaller temperature ranges than the leeward side.

Convective cooling is associated with the case when a vertical instability of moisture laden air is produced by surface heating. Convective precipitation is usually associated with short duration events of high rainfall intensities and can result in catastrophic flooding.

2.4. Measurement and Synthesis of Precipitation Data

The primary source of rainfall data in the United States is the U.S. Weather Service, which maintains rain gage networks throughout the country and provides record information for maximum precipitation for durations of 5 minutes and longer. Other sources of rainfall data include the U.S. Geological Survey, the U.S. Department of Agriculture, and state and local governmental and private cooperative weather observers. Of prime importance are the publications from the National Weather Service including the NOAA Atlas 2 (Miller et al., 1973). These reports provide point precipitation isohyetal maps for 6- and 24-hour durations with return frequencies of 2 and 100 years. Also included are procedures for adjusting the rainfall data for other return frequencies and durations.

2.5. Comparison of Point Rainfall Information

In addition to point rainfall data prepared by the U.S. Weather Service, other analyses are often available from other sources. Typically, synthesis results differ between the sources of information due principally to the different lengths of data records used, differences in the assumed regionalization procedures and differences in the statistical analysis procedures. Consequently, the decision to use a 100-year rainfall input into a rainfall-runoff model also entails the decision to use a specified data bank for point rainfall values.

As an example, Figs. 2.1 (a,b) compare NOAA Atlas 2 (1973) 6- and 24-hours, 100-year return frequency point rainfall values to corresponding rain gage estimates developed by an independent statistical analysis prepared and published by the State of California Department of Water Resources (1982). In the figures, the solid lines represent the NOAA Atlas 2 (1973) isohyetals, whereas the numbers in parenthesis represent the State of California estimated point rainfall values. The study area shown is the coastal county of Orange, California. As another example, Figs. 2.2a,b,c illustrate a considerable discrepancy between the two cited rainfall data base analyses for the adjacent county of San Bernardino, California.

2.6. Precipitation Depth-Duration-Frequency

Due to the apparent randomness of precipitation patterns and intensities, a strictly deterministic analysis of precipitation quantities is not possible and, consequently, a statistical evaluation is used. In the statistical analysis, the following definitions of precipitation depth, duration, and frequency are used:

> Precipitation depth: the amount of precipitation occurring during a specified duration of storm time. Precipitation depth is usually expressed in units of inches.
>
> Duration: the specified length of storm time under study. Duration may be expressed in any time unit such as seconds, hours, days, or season.

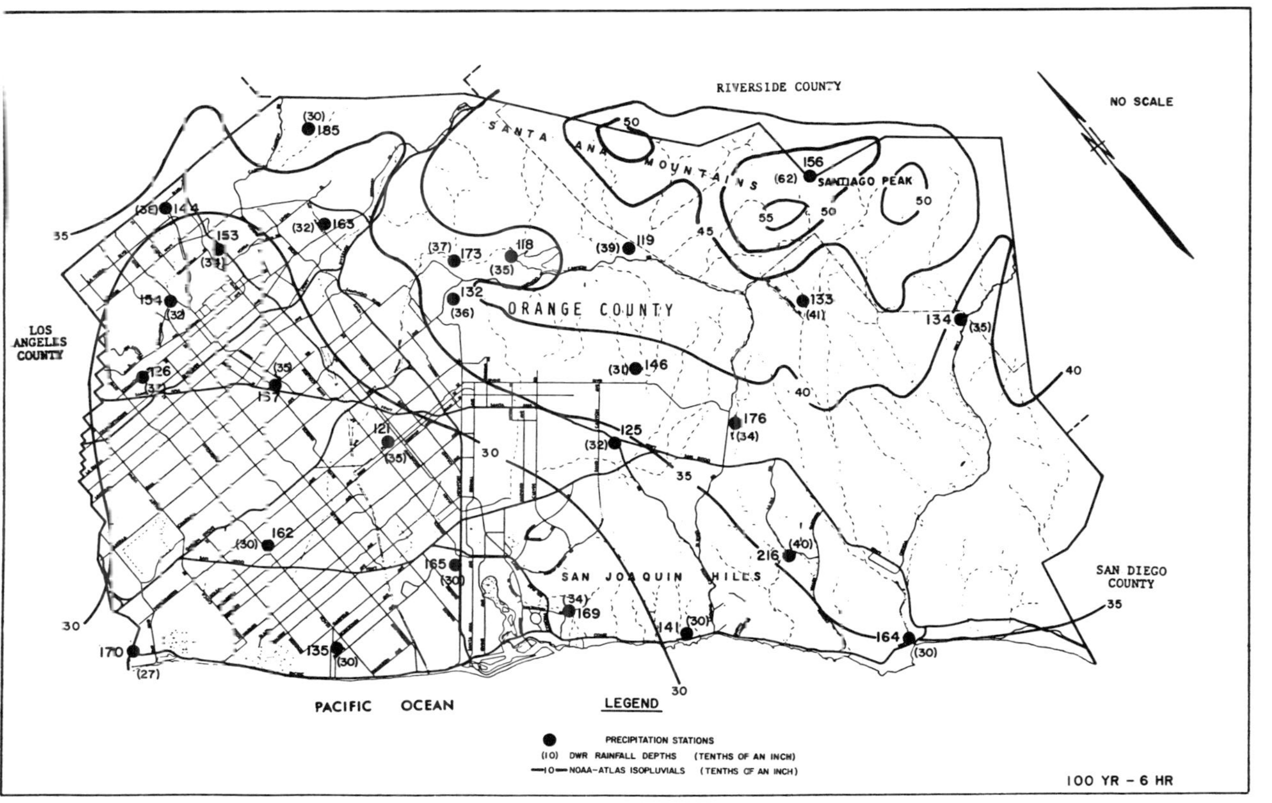

Fig. 2.1a. COMPARISON OF NOAA ATLAS 2 (1973) AND STATE OF CALIFORNIA DWR (1983) POINT RAINFALL DATA

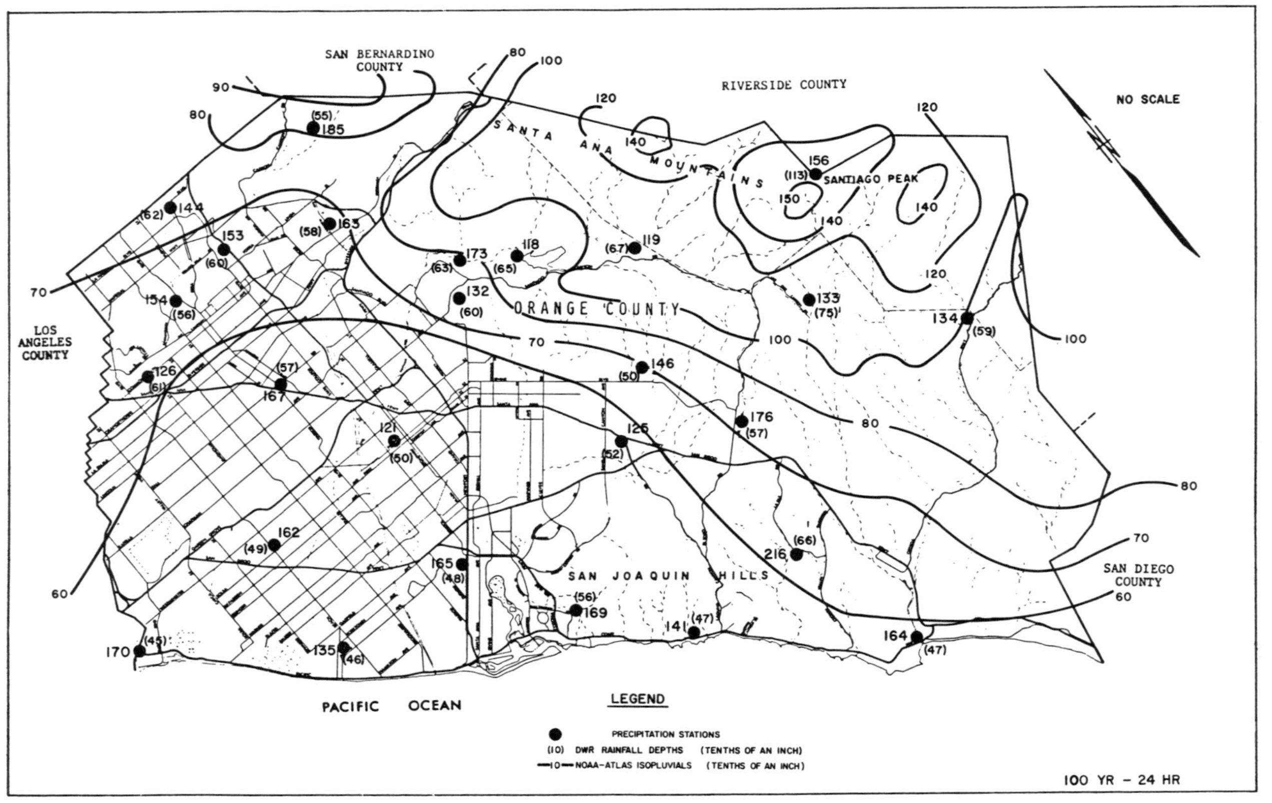

SAN BERNARDINO COUNTY
RIVERSIDE COUNTY
NO SCALE
LOS ANGELES COUNTY
SANTA ANA MOUNTAINS
SANTIAGO PEAK
ORANGE COUNTY
SAN JOAQUIN HILLS
SAN DIEGO COUNTY
PACIFIC OCEAN
LEGEND
PRECIPITATION STATIONS
(10) DWR RAINFALL DEPTHS (TENTHS OF AN INCH)
—10— NOAA-ATLAS ISOPLUVIALS (TENTHS OF AN INCH)
100 YR - 24 HR

Fig. 2.1b.

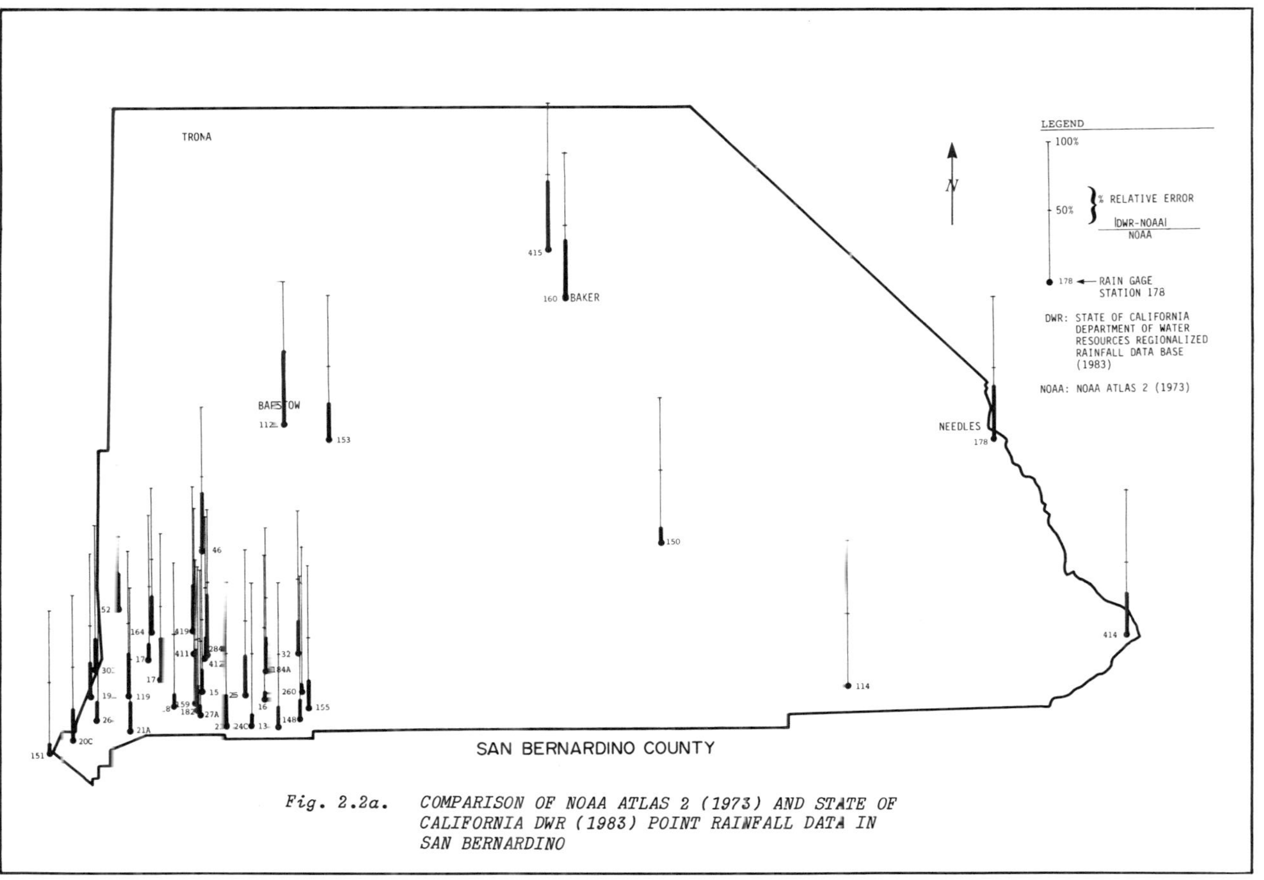

Fig. 2.2a. COMPARISON OF NOAA ATLAS 2 (1973) AND STATE OF CALIFORNIA DWR (1983) POINT RAINFALL DATA IN SAN BERNARDINO

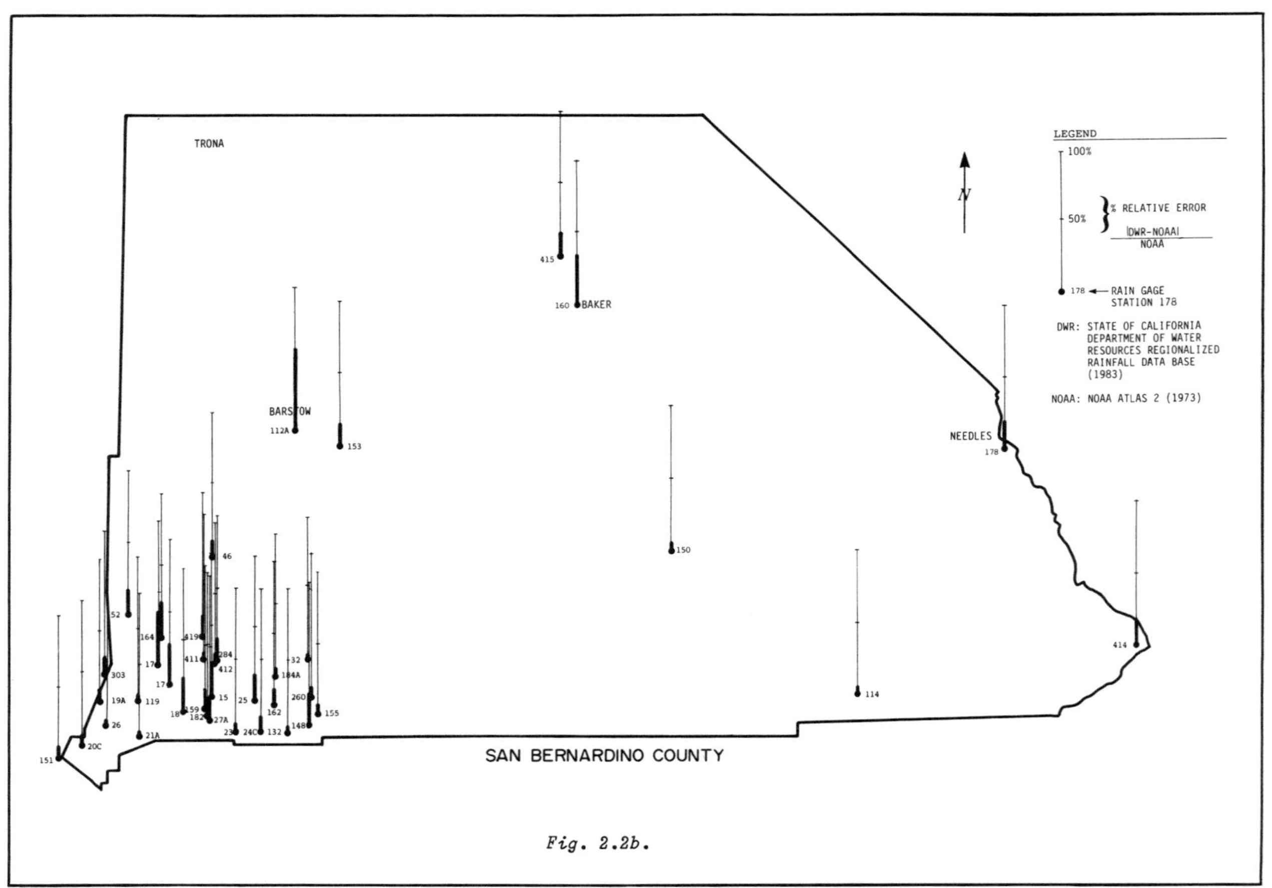
LEGEND
100%
50%
% RELATIVE ERROR
|DWR-NOAA| / NOAA
178 ← RAIN GAGE STATION 178
DWR: STATE OF CALIFORNIA DEPARTMENT OF WATER RESOURCES REGIONALIZED RAINFALL DATA BASE (1983)
NOAA: NOAA ATLAS 2 (1973)
N
TRONA
BAKER
BARSTOW
NEEDLES
SAN BERNARDINO COUNTY

Fig. 2.2b.

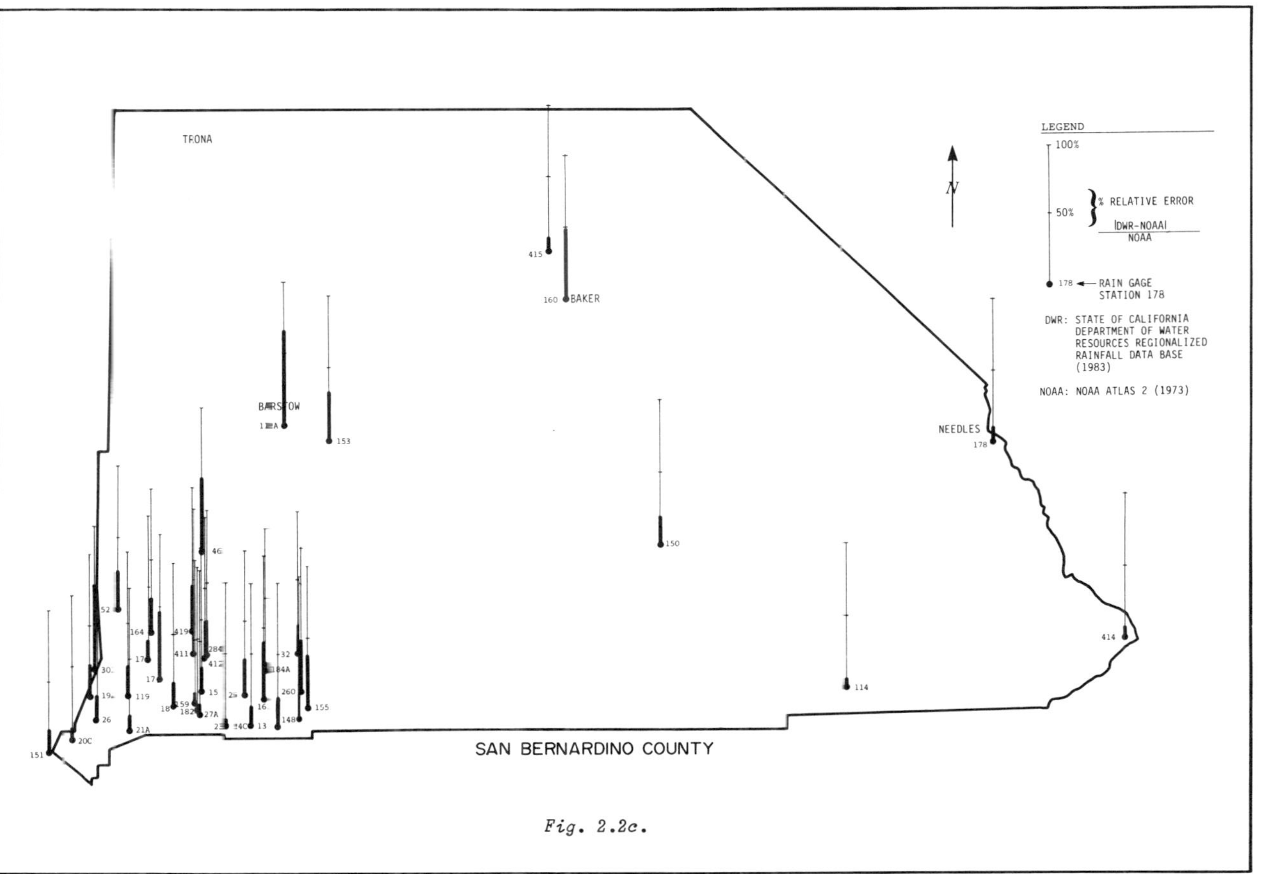
LEGEND
100%
50%
% RELATIVE ERROR
|DWR-NOAA| / NOAA
178 ← RAIN GAGE STATION 178
DWR: STATE OF CALIFORNIA DEPARTMENT OF WATER RESOURCES REGIONALIZED RAINFALL DATA BASE (1983)
NOAA: NOAA ATLAS 2 (1973)
N
TRONA
BARSTOW
BAKER
NEEDLES
415
160
153
150
114
178
414
151
SAN BERNARDINO COUNTY

Fig. 2.2c.

Frequency: the return period (or frequency) is the average period of time, expressed in years, in which the precipitation depth (or intensity) associated with a specified duration is equaled or exceeded.

Intensity: dividing precipitation depth by duration, an average intensity for a specified duration is obtained.

In hydrologic analysis, the rainfall intensity is usually the most important parameter. This parameter relates both precipitation volume to storm duration and also relates storm runoff to storm precipitation with the intensity being an upper bound to the watershed runoff rate. However, in order to provide a reasonable level of flood protection, the statistical concept of return frequency is used which aids in assigning a probabilistic meaning to a precipitation event and the corresponding risk to the public protection. The following definitions are often used for this statistical interpretation:

Exceedance (cumulative) probability: the probability that a precipitation event of a specified depth and duration will be equaled or exceeded in one year.

Return period: the long term average number of years within which a given depth and duration will be equaled or exceeded.

The exceedance probability (P) and return period (T) maybe related by

$$P = 1/T \tag{2.1}$$

From the above definitions it can be argued that a 100-year precipitation event, for example, will not necessarily occur once in every 100 years but actually has a finite probability of occurring in several consecutive years.

2.7. Event Depth-Duration

Of interest for hydrologic studies is the maximum intensities of any type of precipitation event possible within a watershed. Given a long history of such maximum rainfall intensity-duration data, a statistical interpretation can be made to determine estimates of maximum rainfall intensities or depths as a function of storm duration and return frequency.

For most hydrologic study purposes, the most important relationship is that of precipitation depth for any rainfall event (regardless of storm origin) with respect to a specified duration. Such information should include total precipitation from storms of a given duration as well as depths from independent continuous partial storm durations (e.g., the NOAA Atlas 2). This information can be represented by event depth-duration curves that are constructed by ranking in the order of increasing rainfall depth all storm events of some specified duration. From the position of a precipitation depth, an estimate can be made of the number of years during which the subject event will be equaled or exceeded.

2.8. Point Precipitation Data and the Design Storm

When rain gage records are examined to identify tendencies and patterns, an extremely wide range of variations are found. These random variations generally are so great that they obscure any long term pattern or periodicity which may exist. Therefore, to utilize point precipitation data in a design storm method, a combination of probabilistic and deterministic methods are needed. The duration and magnitude of individual storm events is assumed to be probabilistic, while the internal assemblage of the design storm is often assumed deterministic. Additionally, the origin of severe storm events, such as whether the storm is classified as being convective, orographic, or of some other category adds to the difficulty of developing a comprehensive analysis and concluding the actual probabilistic distribution of severe storm rainfall quantities.

2.9. Intensity-Duration-Frequency Curves

Intensity-duration data may be presented as curves relating rainfall intensity in inches per hour versus duration in minutes. Intensity-duration data for durations less than 3 hours tend to plot as a straight line on standard log-log paper, and the curves for various return frequencies tend to plot parallel to one another (see Fig. 2.3).

For each rain gage, the collected precipitation records are analyzed to determine the annual maximum rainfall depth for several durations of interest (e.g., 5 minutes, 1 hour, 3 hours, etc.). This information can then be arranged in an increasing order of magnitude for each storm duration for the history of the gage and plotted on some type of logarithmic scaled paper. From this accumulation of rainfall depth-duration data, various statistical models can be applied to assign return frequencies. The resulting synthesized data are termed point rainfall values to distinguish them from area-averaged values for large watersheds.

REGRESSION EQUATIONS: $D(t) = at^b$

D = DEPTH IN INCHES, t = DURATION IN MINUTES

RETURN FREQUENCY (YEARS)	a	b
2	0.095	0.426
5	0.131	0.438
10	0.170	0.427
25	0.200	0.434
50	0.225	0.434
100	0.259	0.427

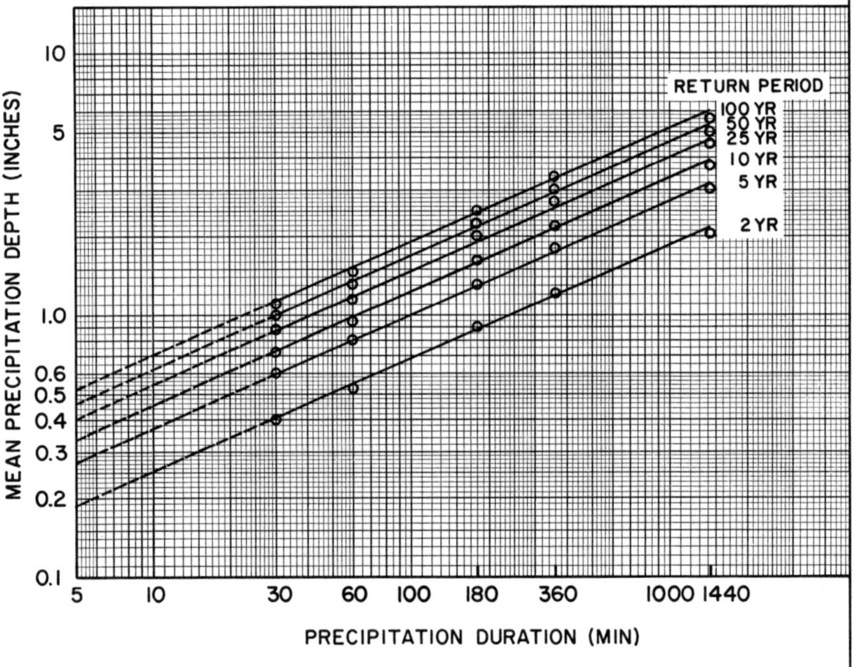

Fig. 2.3. PRECIPITATION INTENSITY-DURATION CURVES (2-100 YEAR RETURN FREQUENCY)

Because peak intensities of storm events are seldom located directly over rain gages and because the rain gage network is usually of low density, the rain gage data often may not represent the true maximum rainfall intensities possible within the locality. Consequently, care must be taken to thoroughly evaluate all available rainfall data, including anecdotal accounts, prior to deriving a depth-duration-frequency relationship.

2.10. Depth-Duration-Frequency Relationships

In rainfall-based hydrologic modeling methods, the depth-duration-frequency information is usually the keystone in formulating a critical flood control design objective. Based on an acceptable level of public protection, the design objective can be implemented for designing flood control facilities and the corresponding costs distributed by a cost-to-benefit formula. Therefore, the engineer may be interested in the costs for protection from several return frequency flood events and must use extrapolated rainfall depth-duration-frequency data in order to model the higher return frequency events. To aid in estimating this rainfall data, various empirical formulas have been proposed. One equation relates rainfall intensity to storm duration by:

$$I(d,T) = C_o(d + C_1)^E \tag{2.2}$$

where I is the rainfall intensity; C_o a constant which is a function of the return period, T; and C_1 and E are constants; and d is the storm duration. The coefficients are usually determined by curve fitting to known precipitation data, and then the formula is assumed to apply for the intermediate rainfall data values. Other equations can be used which reduce to straight lines on specially prepared graphs such as used in the NOAA Atlas 2 (see Chapter 4).

2.11. Area-Averaged Point Rainfall

Due to the potential variation of precipitation depth (for a given duration) throughout a watershed, approximate methods are required in order to estimate an area-averaged point rainfall for the entire watershed. The three most often used methods are the arithmetic method, the Thiessen polygon method, and the isohyetal method. The application of these methods is shown in Fig. 2.4.

The arithmetic method is a simple average of point rainfall values. It is usually considered to be the least accurate of the three approaches for area-averaging.

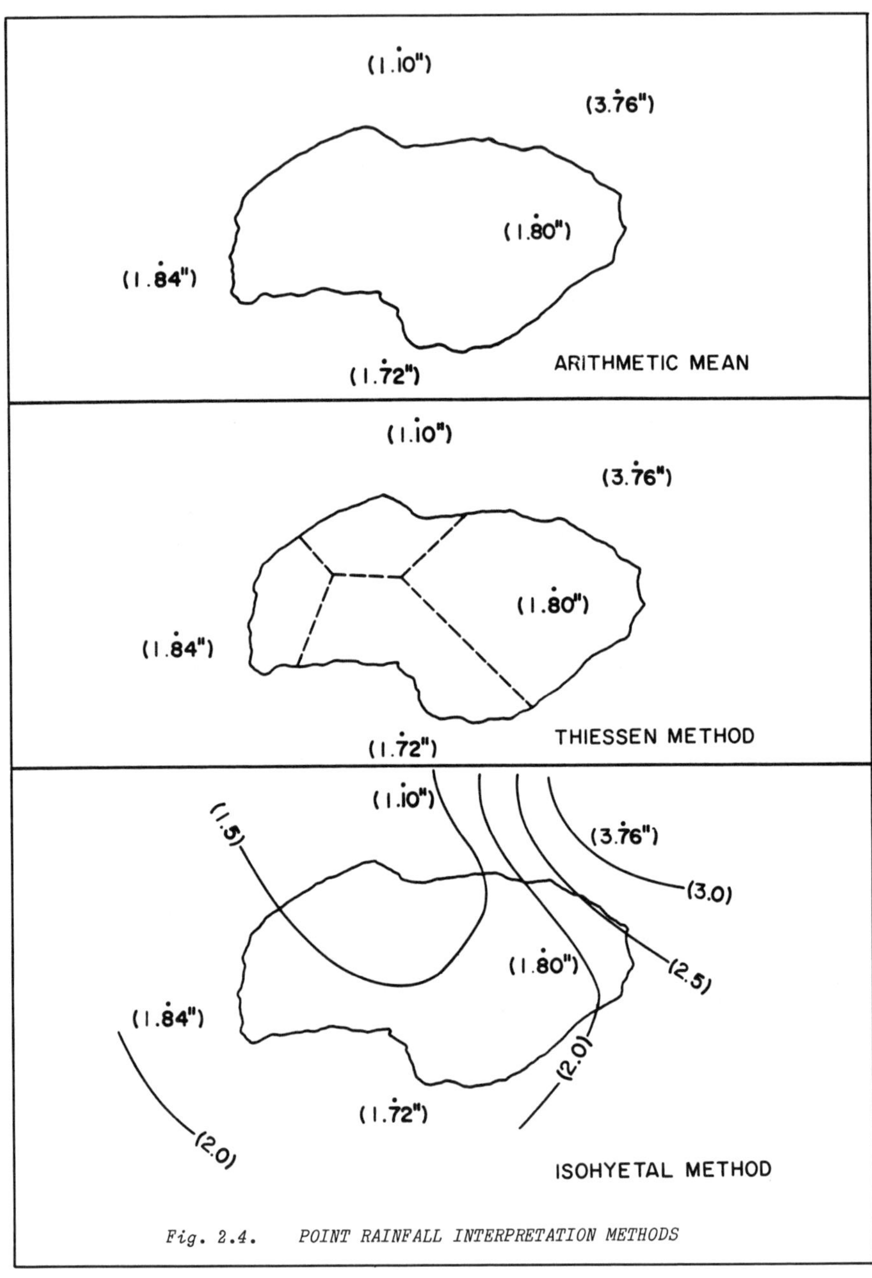

Fig. 2.4. POINT RAINFALL INTERPRETATION METHODS

The Thiessen polygon method utilizes an area-averaged value based on polygons associated to each rain gage in the vicinity of the watershed. The polygons are determined by constructing perpendicular bisectors to lines connecting the several rain gages.

Finally, the isohyetal method is a more precise approach in which a map of the interpolated point rainfall isohyetals are plotted on the watershed and the area-averaged point rainfall is directly estimated by contour integration.

2.12. Depth-Area Relationships

A design storm pattern can be directly applied towards estimating peak flow rates and runoff volumes for small watersheds where stream gage records are inadequate for statistical frequency analysis. As the watershed area increases in size, however, studies indicate (Miller et al., 1973; Myers and Zehr, 1980) that the area-averaged point rainfall values used in the design storm pattern should be adjusted so that the volume of rainfall over the area is accurately represented by the point rainfall characteristics. This adjustment is called the depth-area adjustment. For example, the NOAA Atlas 2 includes point rainfall depth-area relationships (Fig. 2.5) for 30-minute, 1-, 3-, 6-, and 24-hour point rainfall values as a function of watershed size. These relationships are generalized for all locations. Such depth-area curves are used to modify local rainfall data when precipitation records are inadequate to develop localized adjustment factors. If they could be developed, localized depth-area relationships would be more accurate than the generalized relationships of Fig. 2.5. For example, Fig. 2.6 shows depth-area curves developed from severe storm analysis in the Los Angeles basin, California, and also includes curves from neighboring regions.

2.13. Synthetic Critical Storm Patterns

The use of computer watershed modeling techniques for hydrologic analysis has become widespread with the advent of inexpensive digital computer availability. Generally speaking, the basis of the hydrologic modeling approach is to study the effects of a severe storm event (or design storm) upon the watershed. Consequently, the decision as to what critical storm to use for study purposes has a significant impact upon the ultimate design objective in providing flood control.

The statistical relationships mentioned above (time variation in precipitation and the observation of severe rainfall storm

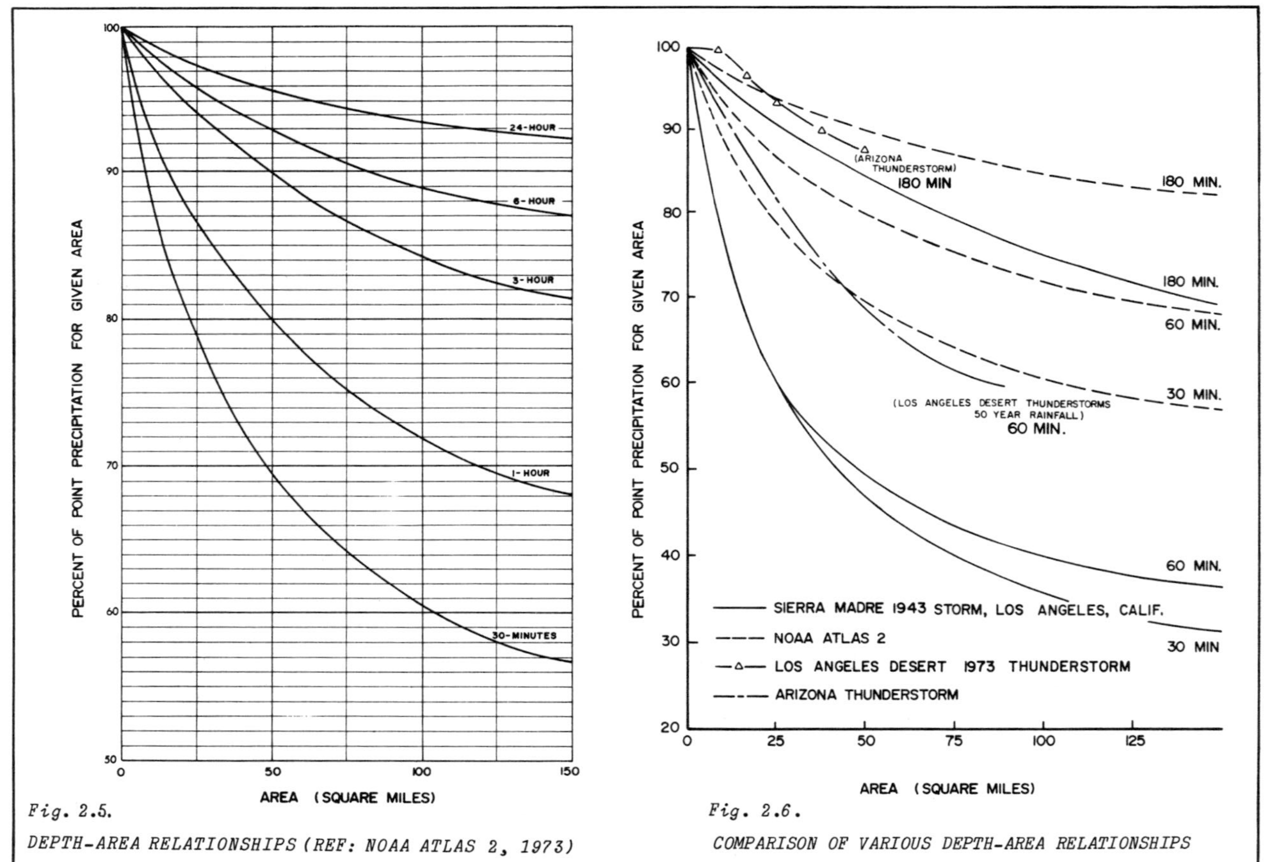

Fig. 2.5.
DEPTH-AREA RELATIONSHIPS (REF: NOAA ATLAS 2, 1973)

Fig. 2.6.
COMPARISON OF VARIOUS DEPTH-AREA RELATIONSHIPS

events covering various watershed areas) allow for the generalization of three important characteristics of storm rainfall:

(1) The average precipitation intensity decreases as the storm duration increases,

(2) The average precipitation intensity increases as the return frequency increases,

(3) The area-averaged precipitation intensity decreases as the watershed area covered by a storm increases.

Therefore, the selection of the design storm needs to accommodate these generalizations in order to rationally develop the design storm runoff hydrograph. In the following, the information and procedures needed to construct a typical design storm pattern will be presented. The procedures and assumptions used to construct the various design storms used in the provided computer software code closely follows several approaches, including the development of the U.S. Army Corps of Engineers TD-15 (1982).

The critical storm pattern approach falls into two categories: (1) a severe storm pattern of record; and (2) a synthetic critical storm pattern. The critical storm pattern of record is a historical rainfall event which is assumed to be associated with the severe flooding event. The storm may have occurred within the vicinity of the watershed or may have occurred elsewhere but is assumed to be transposable to the study watershed without violating the assumptions of uniformity of hydrologic or geographic characteristics.

For example, the U.S. Army Corps of Engineers (Los Angeles District Office, California) utilize for study purposes a critical storm pattern based on a 1943 thunderstorm (Fig. 2.7). From Fig. 2.7, the 3 hour duration storm is composed of twelve 15 minute incremental precipitation depths. These incremental values are estimated from peak depth-duration data for 15-, 30-, 60-, and 180- minute values as a function of watershed area in square miles. This specific historic storm pattern is based upon a 3.3 inch total storm rainfall. In transposing the storm to another watershed, the local 3- hour duration precipitation-depth is estimated (for the desired return frequency) and the adjusted storm pattern is obtained by a simple scaling ratio. For a depth of 3.3 inches, the resulting storm pattern is shown in Fig. 2.8. A difficulty with this storm pattern is that the precipitation-depth ratio of interior durations to the total 3-hour duration may not be appropriate for the local region of interest, and an unacceptably high (or low) precipitation depth may be assigned to the interior unit rainfalls. For example, the design objective

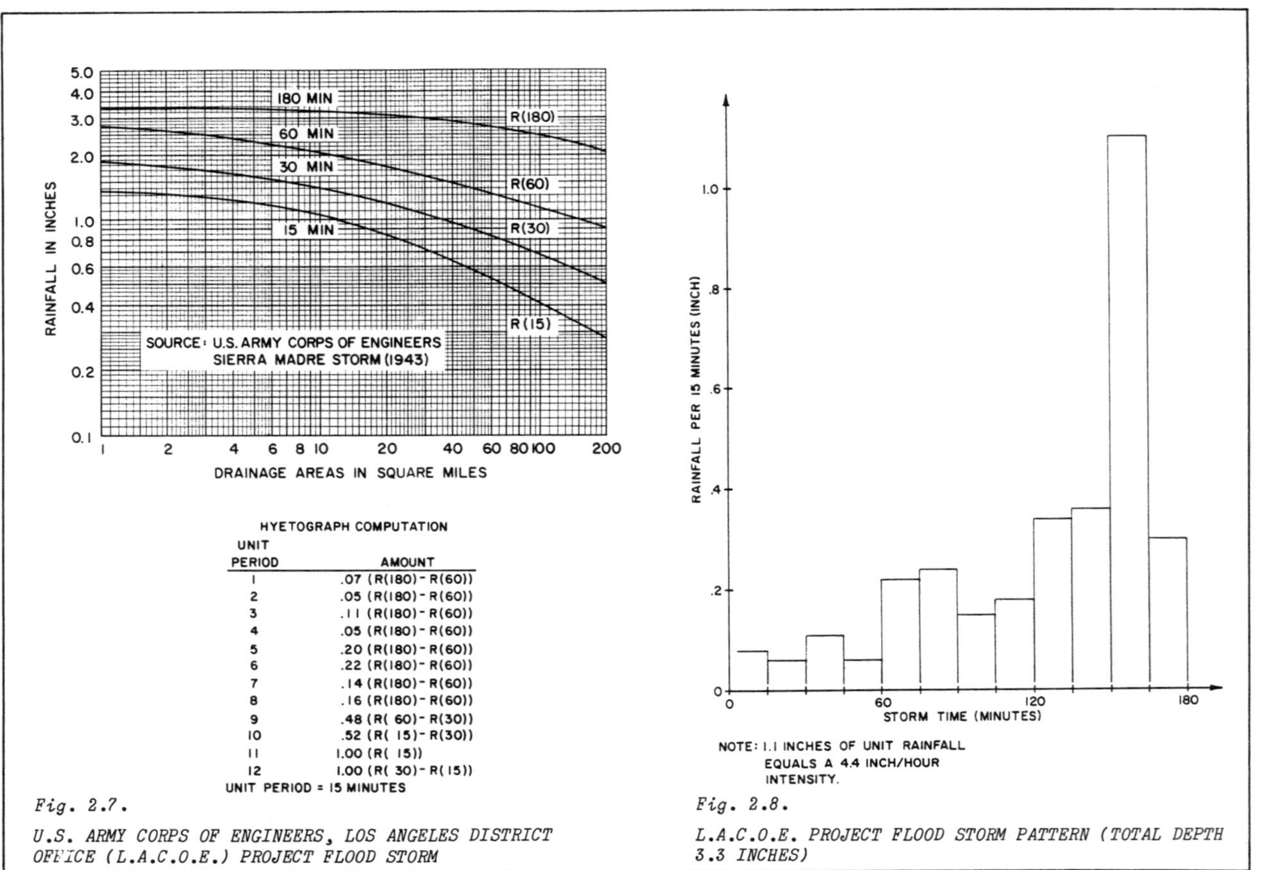

HYETOGRAPH COMPUTATION

UNIT PERIOD	AMOUNT
1	.07 (R(180) - R(60))
2	.05 (R(180) - R(60))
3	.11 (R(180) - R(60))
4	.05 (R(180) - R(60))
5	.20 (R(180) - R(60))
6	.22 (R(180) - R(60))
7	.14 (R(180) - R(60))
8	.16 (R(180) - R(60))
9	.48 (R(60) - R(30))
10	.52 (R(15) - R(30))
11	1.00 (R(15))
12	1.00 (R(30) - R(15))

UNIT PERIOD = 15 MINUTES

Fig. 2.7.

U.S. ARMY CORPS OF ENGINEERS, LOS ANGELES DISTRICT OFFICE (L.A.C.O.E.) PROJECT FLOOD STORM

Fig. 2.8.

L.A.C.O.E. PROJECT FLOOD STORM PATTERN (TOTAL DEPTH 3.3 INCHES)

may be to provide flood protection for a 100- year return frequency rainfall for the critical duration of every hydraulic structure within a watershed. Then using a 100- year 3-hour depth with the given storm pattern may possibly result in a 150- year (or 50- year) return frequency rainfall depth to be assigned for some critical duration. Designing the structure for such a runoff rate would result in an overdesign (or underdesign). Note in Fig. 2.7 that depth-area effects are automatically included in the specified design storm depth-duration relationships.

The second category of critical storm patterns is the synthetic critical storm pattern. This approach is an attempt to accommodate some of the concerns which are associated with using critical storm patterns of record. Several synthetic storm patterns have been proposed, such as Keifer and Chu (1957), Huff (1967), Terstriep and Stall (1974), Pilgrim and Cordery (1975), SCS (1973), TD-15 (1982), among others. For example, Hershfield (1962) developed a dimensionless time distribution of total mass rainfall versus storm duration. This generalized duration is given in Table 2.1. To use a synthetic storm pattern such as given in Table 2.1, the duration of the synthetic storm is used to scale the dimensionless time scale of column 1 and the total storm rainfall (inches) is used to scale the dimensionless mass rainfall ordinates of column 2.

TABLE 2.1
HERSHFIELD MASS RAINFALL DISTRIBUTION

Dimensionless Storm Duration	Cumulative Mass Rainfall
0.00	0.00
0.10	0.06
0.20	0.12
0.30	0.20
0.40	0.29
0.45	0.34
0.50	0.45
0.55	0.63
0.60	0.73
0.65	0.81
0.70	0.86
0.80	0.94
0.90	0.99
1.00	1.00

A widely used set of critical storm patterns were developed by the U.S. Department of Agriculture, Soil Conservation Service or SCS. The synthetic storm patterns are presented as dimensionless mass rainfall curves for several storm durations, (e.g., 6 and 24-hour storms). For example, the 24-hour storm pattern is further classified as one of the following types: I, IA, II, or III (see Fig. 2.9). This delineation was necessary in order to better represent different precipitation characteristics as averaged over broad regions of the United States. The SCS developed the dimensionless synthetic storm patterns using the U.S. Weather Service rainfall frequency reports. The patterns are based upon regionally averaged measured rainfall data for watershed areas less than approximately 400 square miles, with durations to 24 hours, and return frequencies from 1 year to 100 years. These critical storm patterns represent an average of generalized precipitation depth-duration-frequency relationships. A tabulation of SCS Type I, 24-hour storm pattern is provided in Table 2.2.

TABLE 2.2
SCS SYNTHETIC STORM PATTERN
(TYPE I, 24 HOUR STORM)

Storm Hour	Mass Rainfall
0.0	0.000
2.0	0.035
4.0	0.076
6.0	0.125
6.0	0.156
8.0	0.194
8.5	0.219
9.0	0.254
9.5	0.303
9.75	3.362
10.0	0.515
11.0	0.624
12.0	0.682
13.0	0.727
14.0	0.767
16.0	0.830
20.0	0.926
24.0	1.000

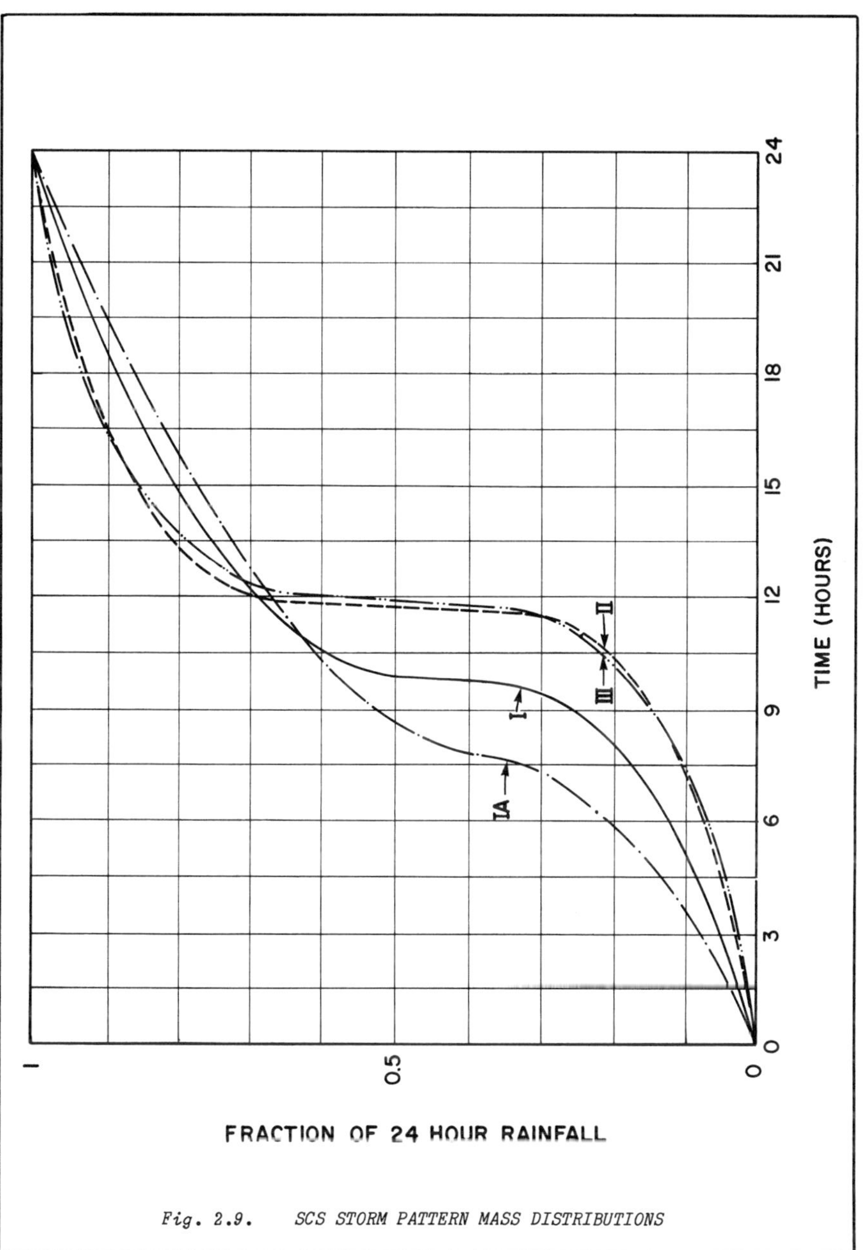

Fig. 2.9. SCS STORM PATTERN MASS DISTRIBUTIONS

The storm patterns are constructed by subtracting the peak 30-minute depth from the peak 1- hour depth, and subtracting the peak 1- hour depth from the peak 1.5- hour depth, and so forth until a set of 30-minute unit rainfalls result. The unit rainfalls are arranged such that a nesting of the peak unit rainfalls are formed where the peak 30 minutes occurs within the peak 1 hour, the peak hour occurs within the peak 1.5 hours, and so forth until the 24-hour storm pattern is defined. Because all of the critical precipitation depths are contained within the storm pattern, the synthetic storm patterns are generally assumed appropriate for design study purposes for both small and large watersheds.

2.14. Modified Composite Storm Pattern

An examination of both the Hershfield and the SCS synthetic storm patterns indicate that neither pattern guarantees a critical precipitation depth for an arbitrary critical duration. That is, both patterns define each interior storm duration as a fixed percentage of total storm mass rainfall. The Composite Method for developing a synthetic storm pattern is analogous to the approach used by the SCS except that the local rain gage data is used to determine the unit rainfalls. This method simply uses the local rainfall intensity-duration curves with the desired return frequency, and incremental unit rainfalls are determined by successive subtractions. The unit rainfalls are then arranged in some nested pattern with the peak intensities usually defined to occur at 0.25, 0.50, 0.67, or some percentage of the total storm duration. Figures 2.10 a,b,c show a typical composite storm pattern which is used for design hydrology purposes in southern California (Hromadka and McCuen, 1986a,b).

The composite storm pattern approach can be modified to include the effects of depth-area adjustment. However, due to the adjustment factors being available for only a few durations, an interpolation procedure is required for the remaining durations within the storm pattern. One approach (which follows the procedures of TD-15) is to plot the point rainfall values, each adjusted for depth-area effects, on standard log-log paper and assume a straight line relationship (see Fig. 2.11). Other relationships can be used, but the log-log plot is convenient in that the resulting mass rainfall plot is simple to approximate numerically. Unit rainfalls are determined for the modified composite storm pattern by successive subtractions of unit durations along the adjusted mass rainfall plot. The unit rainfalls are then arranged as a nested sequence of peak rainfall

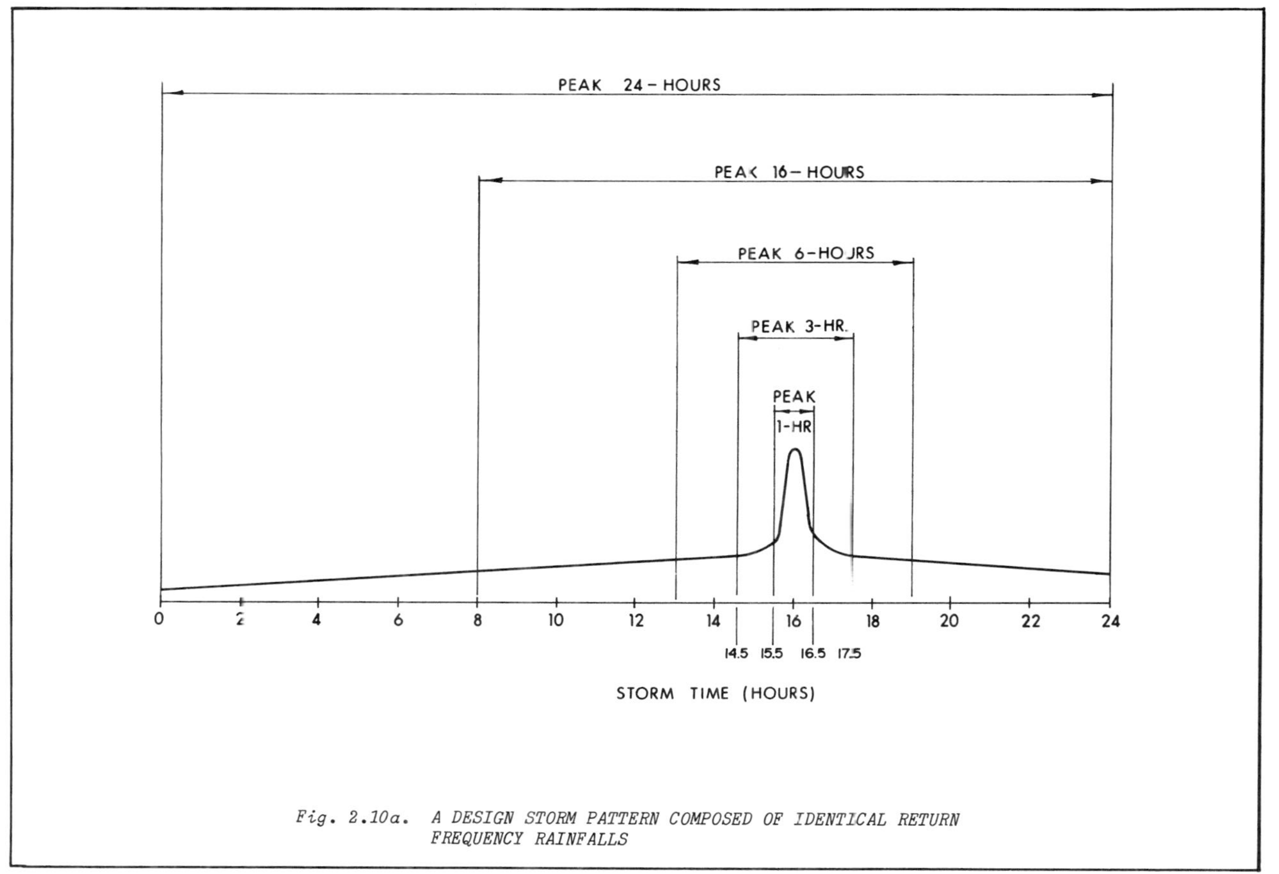

Fig. 2.10a. A DESIGN STORM PATTERN COMPOSED OF IDENTICAL RETURN FREQUENCY RAINFALLS

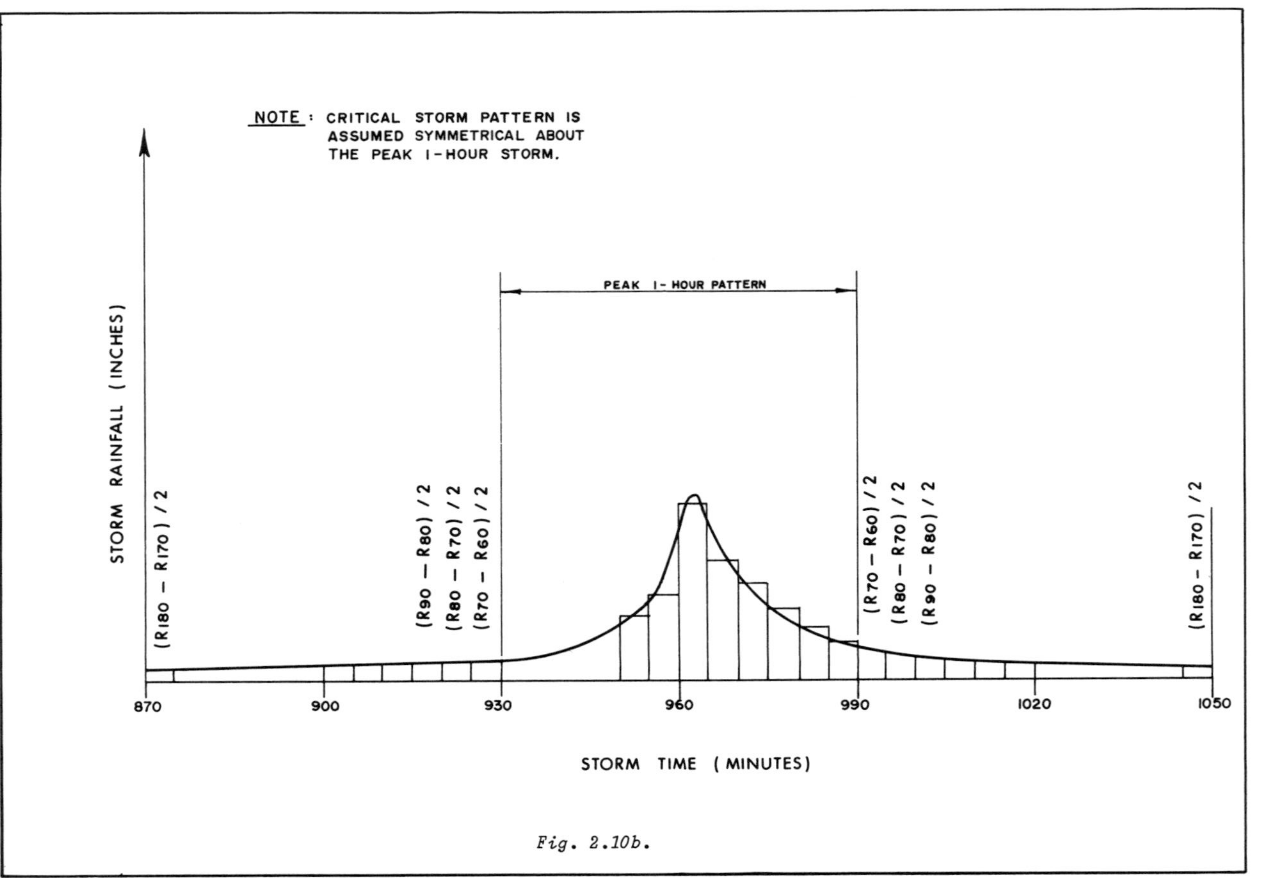
NOTE : CRITICAL STORM PATTERN IS ASSUMED SYMMETRICAL ABOUT THE PEAK 1-HOUR STORM.
PEAK 1-HOUR PATTERN
STORM RAINFALL (INCHES)
(R180 − R170) / 2
(R90 − R80) / 2
(R80 − R70) / 2
(R70 − R60) / 2
(R70 − R60) / 2
(R80 − R70) / 2
(R90 − R80) / 2
(R180 − R170) / 2
870
900
930
960
990
1020
1050
STORM TIME (MINUTES)

Fig. 2.10b.

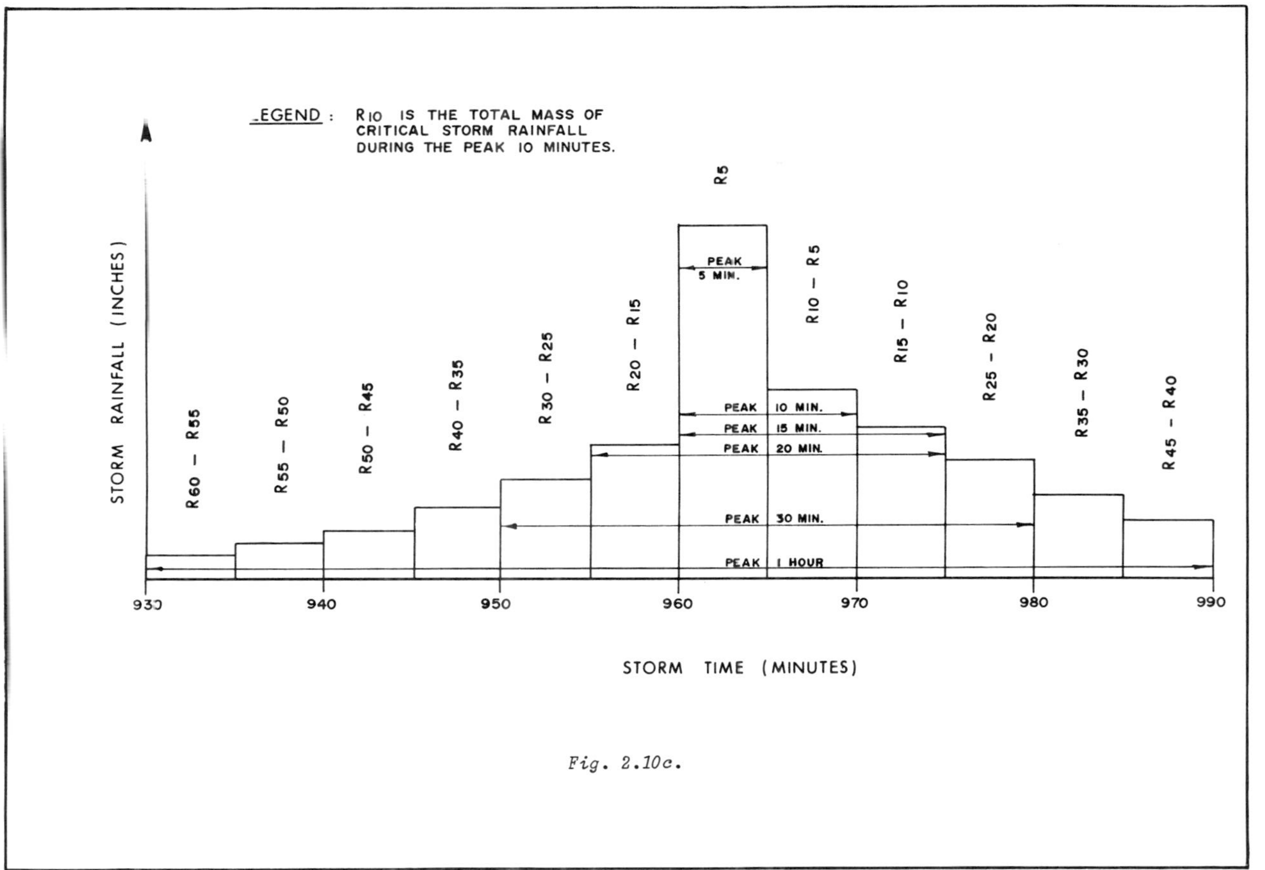

LEGEND : R10 IS THE TOTAL MASS OF CRITICAL STORM RAINFALL DURING THE PEAK 10 MINUTES.
STORM RAINFALL (INCHES)
R5
PEAK 5 MIN.
R60 − R55
R55 − R50
R50 − R45
R40 − R35
R30 − R25
R20 − R15
R10 − R5
R15 − R10
R25 − R20
R35 − R30
R45 − R40
PEAK 10 MIN.
PEAK 15 MIN.
PEAK 20 MIN.
PEAK 30 MIN.
PEAK 1 HOUR
930
940
950
960
970
980
990
STORM TIME (MINUTES)

Fig. 2.10c.

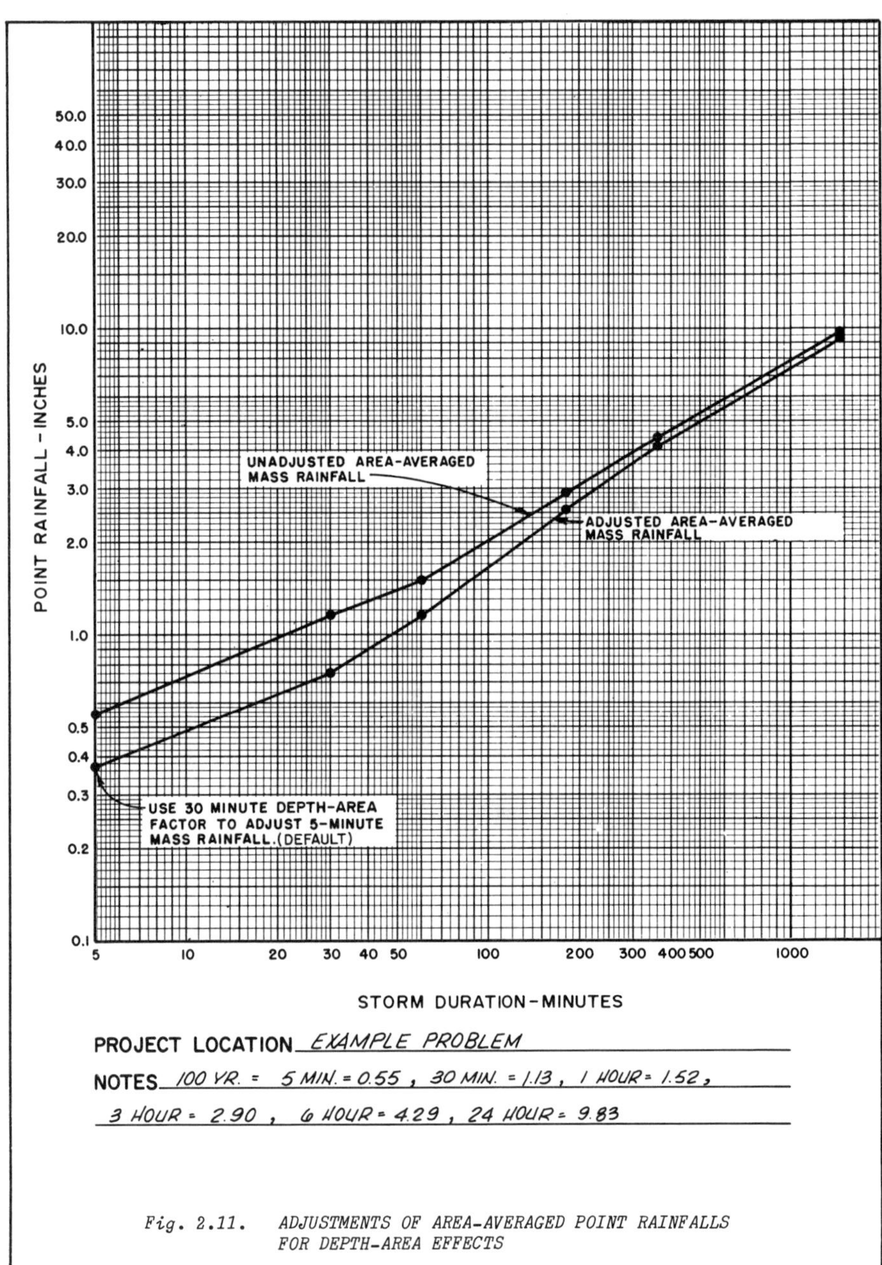

Fig. 2.11. ADJUSTMENTS OF AREA-AVERAGED POINT RAINFALLS FOR DEPTH-AREA EFFECTS

events with the highest intensity defined to occur at a specified time within the 24-hour storm pattern. For example, TD-15 uses hour 12; however, for another storm pattern it is possible to assign the peak unit rainfall to occur at any hour of the 24 storm pattern (Fig. 2.12). The resulting synthetic storm pattern (1) provides a critical precipitation depth for every critical duration; (2) does not exceed any maximum rainfall intensity with a specified return frequency; (3) provides each critical depth with a depth-area adjustment; (4) can be used for both small and large urban watersheds with consistency in results; (5) develops a synthetic severe critical storm pattern to be used for study purposes in urban watershed studies; (6) is easily extended to multiple day storms by extending the log-log mass rainfall plot.

As discussed previously, other design storm patterns can be developed for solving design problems. However, the storm profile is but one of the input variables used in the design storm approach; the other variables of rainfall frequency, the antecedent moisture condition (or wetness of the soil, or availability of losses), and depth-area adjustment curves all need to be chosen due to their impact on the relationship between the flood flow frequency curve estimates and the design storm rainfall frequency input into the model.

The modified composite storm pattern is further discussed in Chapter 5 where a total synthetic runoff hydrograph model is developed, and a corresponding FORTRAN computer program presented. A link-node watershed model is developed in Chapter 6 which includes a generalization of the modified composite storm pattern for developing synthetic runoff hydrographs within the schematic of a watershed.

2.15. PROGRAM 2.1: Subroutine RAIN

This subroutine develops seven 24-hour nested design storm patterns, each continuously adjusted for depth-area effects. Also available are the SCS type I, II, III, and IA storm patterns. Other than the user-specified storm patterns defined in program CONTR (see Chapter 8), two other storm patterns are available, namely, a rational method storm (i.e., a constant rainfall intensity for a specified duration) and a linear interpolation method (i.e., an arbitrary storm pattern defined by use of time versus rainfall intensity data pairs). The three-minute unit interval is used for all the storm patterns.

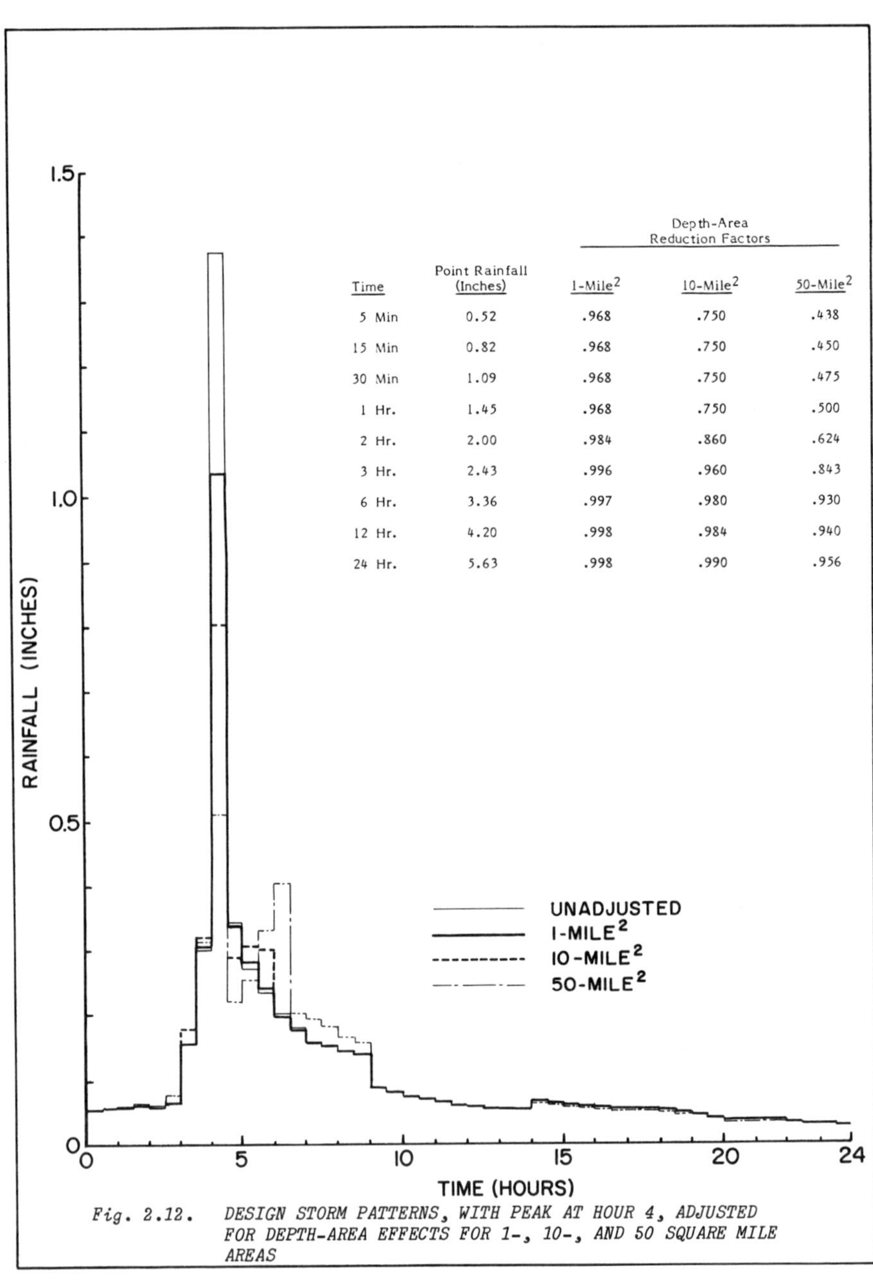

Time	Point Rainfall (Inches)	Depth-Area Reduction Factors 1-Mile²	10-Mile²	50-Mile²
5 Min	0.52	.968	.750	.438
15 Min	0.82	.968	.750	.450
30 Min	1.09	.968	.750	.475
1 Hr.	1.45	.968	.750	.500
2 Hr.	2.00	.984	.860	.624
3 Hr.	2.43	.996	.960	.843
6 Hr.	3.36	.997	.980	.930
12 Hr.	4.20	.998	.984	.940
24 Hr.	5.63	.998	.990	.956

Fig. 2.12. DESIGN STORM PATTERNS, WITH PEAK AT HOUR 4, ADJUSTED FOR DEPTH-AREA EFFECTS FOR 1-, 10-, AND 50 SQUARE MILE AREAS

2.15.1. The 24-Hour Nested Design Storm Patterns

The peak 5-minute, 15-minute, 30-minute, 1-hour, 2-hour, 3-hour, 6-hour, 12-hour, and 24-hour area-averaged point precipitation values are calculated by the user and entered into the program which subsequently constructs a 24-hour nested design storm pattern. Straight line logarithmic interpolation is used to estimate intermediate duration point precipitation values, based on 3-minute unit intervals. The available seven 24-hour nested design storm patterns are shown in Figures 2.13a through 2.13g.

2.15.2. Depth-Area Adjustment

Figures 2.5 and 2.6 depict two a depth-area adjustment curve sets. Table 2.3 and Fig. 2.11 show the watershed area-averaged point rainfalls and depth-area adjustment factors for a watershed area of 10 square miles. Figure 2.14 depicts the 24-hour nested storm pattern with the peak rainfalls selected to occur at storm hour 4, with (shaded) and without the depth-area adjustment, respectively, for an unit interval of 30-minutes. Figure 2.6 is used for adjustment of rainfalls for durations less than 3-hours.

TABLE 2.3

EXAMPLE DEPTH-AREA ADJUSTMENTS FOR 10 SQUARE-MILE AREA

Watershed Area-Averaged Rainfall	(inches)	Depth-Area Adjustment Factors
5 min.	0.52	0.75[1]
15 min.	0.82	0.75[1]
30 min.	1.09	0.75[2]
1 hr.	1.45	0.75[2]
2 hr.	2.00	0.86[2]
3 hr.	2.43	0.96[2]
6 hr.	3.36	0.984[3]
12 hr.	4.20	0.986[3]
24 hr.	5.63	0.989[3]

notes 1: 30-minute depth-area adjustment used as default
2: Fig. 2.6 used.
3: Fig. 2.5 used.

2.15.3. SCS Storm Patterns

Figure 2.15a shows the approximate geographic regions designated for specific SCS rainfall distributions, Fig. 2.15b

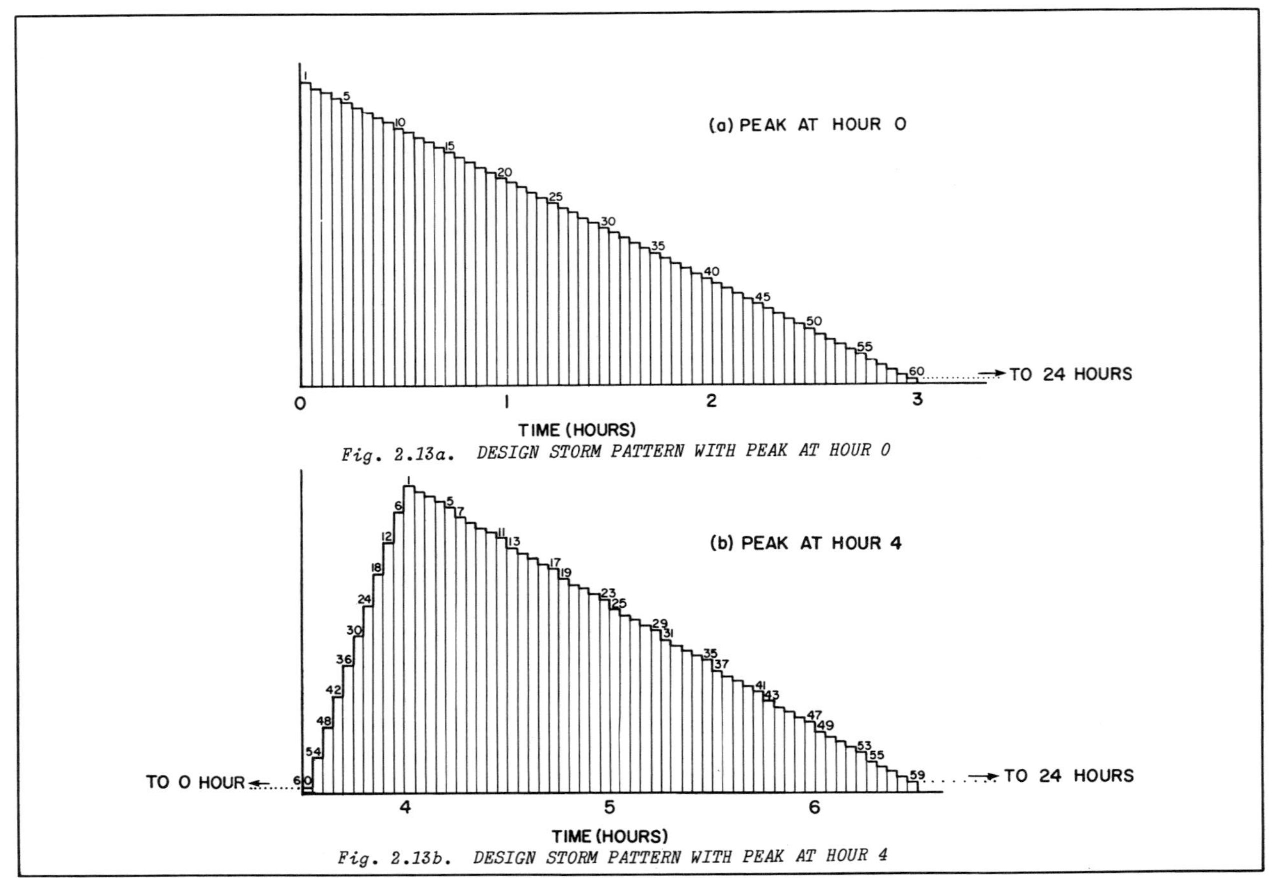

Fig. 2.13a. DESIGN STORM PATTERN WITH PEAK AT HOUR 0

Fig. 2.13b. DESIGN STORM PATTERN WITH PEAK AT HOUR 4

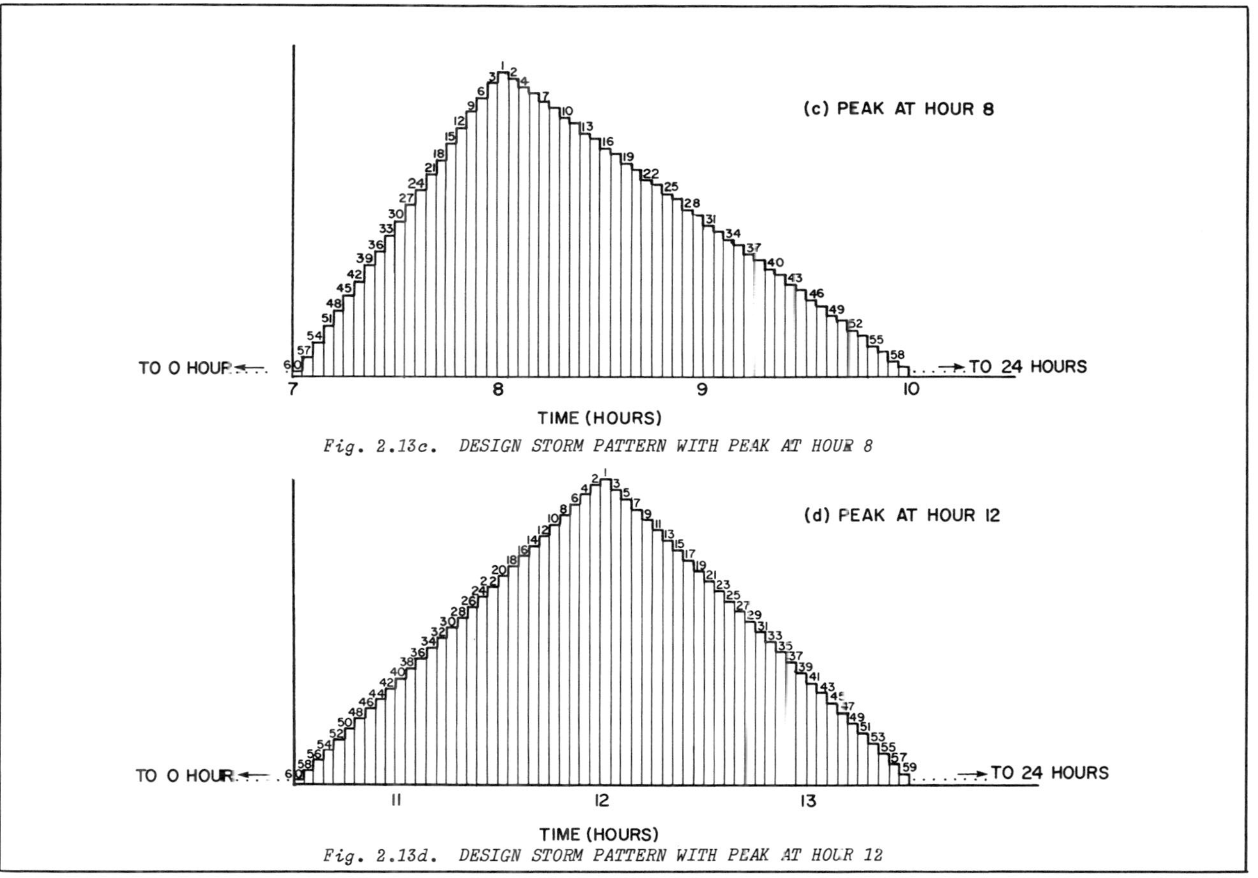

Fig. 2.13c. DESIGN STORM PATTERN WITH PEAK AT HOUR 8

Fig. 2.13d. DESIGN STORM PATTERN WITH PEAK AT HOUR 12

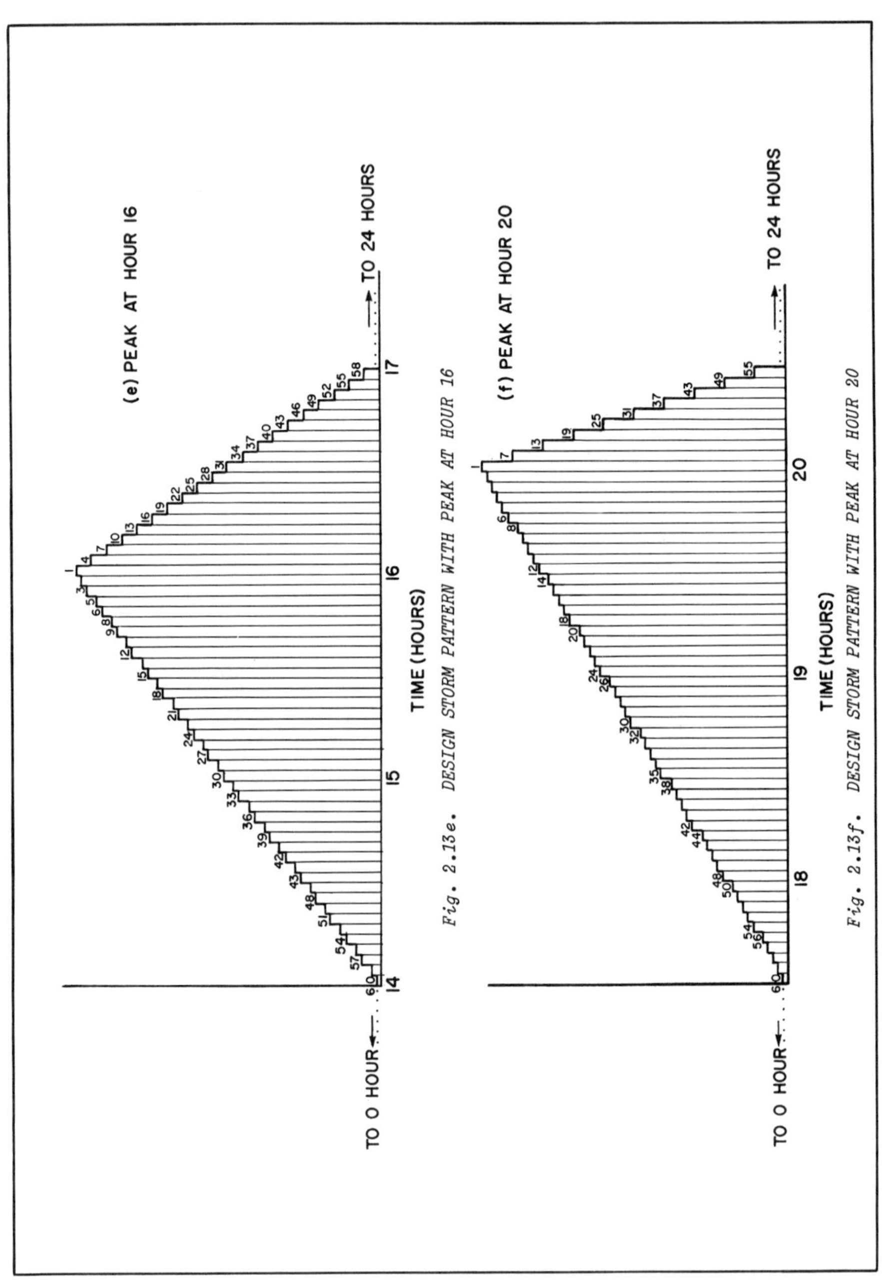

Fig. 2.13e. DESIGN STORM PATTERN WITH PEAK AT HOUR 16

Fig. 2.13f. DESIGN STORM PATTERN WITH PEAK AT HOUR 20

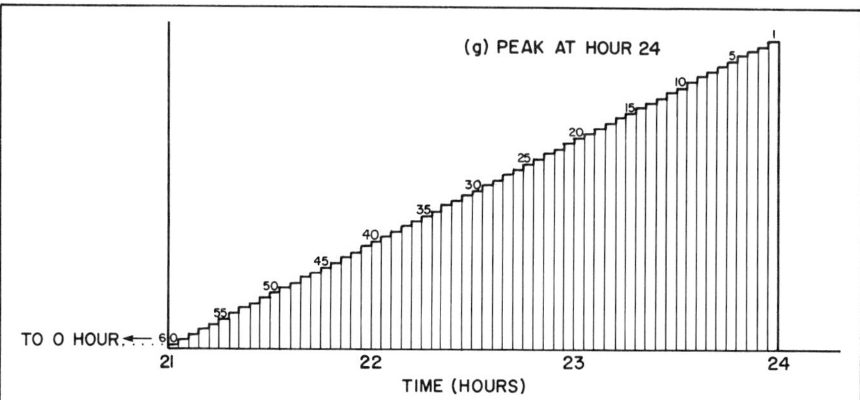

RAIN (1) = RR(0.05)
RAIN (2) = RR(0.1) - RR(0.05)
RAIN (3) = RR(0.15) - RR(0.1)

WHERE RR(TIME) = EXP [A*LOG(TIME)+B]
* RAINFALL IS ASSUMED TO BE LINEAR BETWEEN POINT AVERAGED RAINFALL ON LOG LOG SCALE

Fig. 2.13g. DESIGN STORM PATTERN WITH PEAK AT HOUR 24

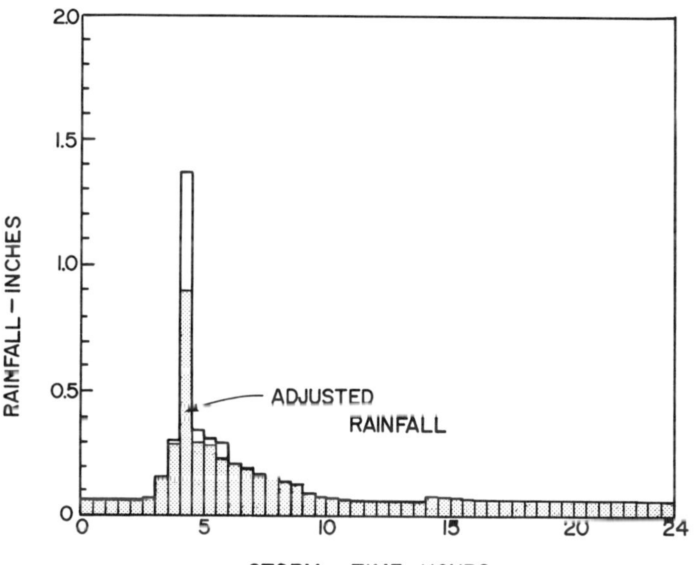

Fig. 2.14. 24-HOUR NESTED STORM PATTERN WITH PEAK AT HOUR 4

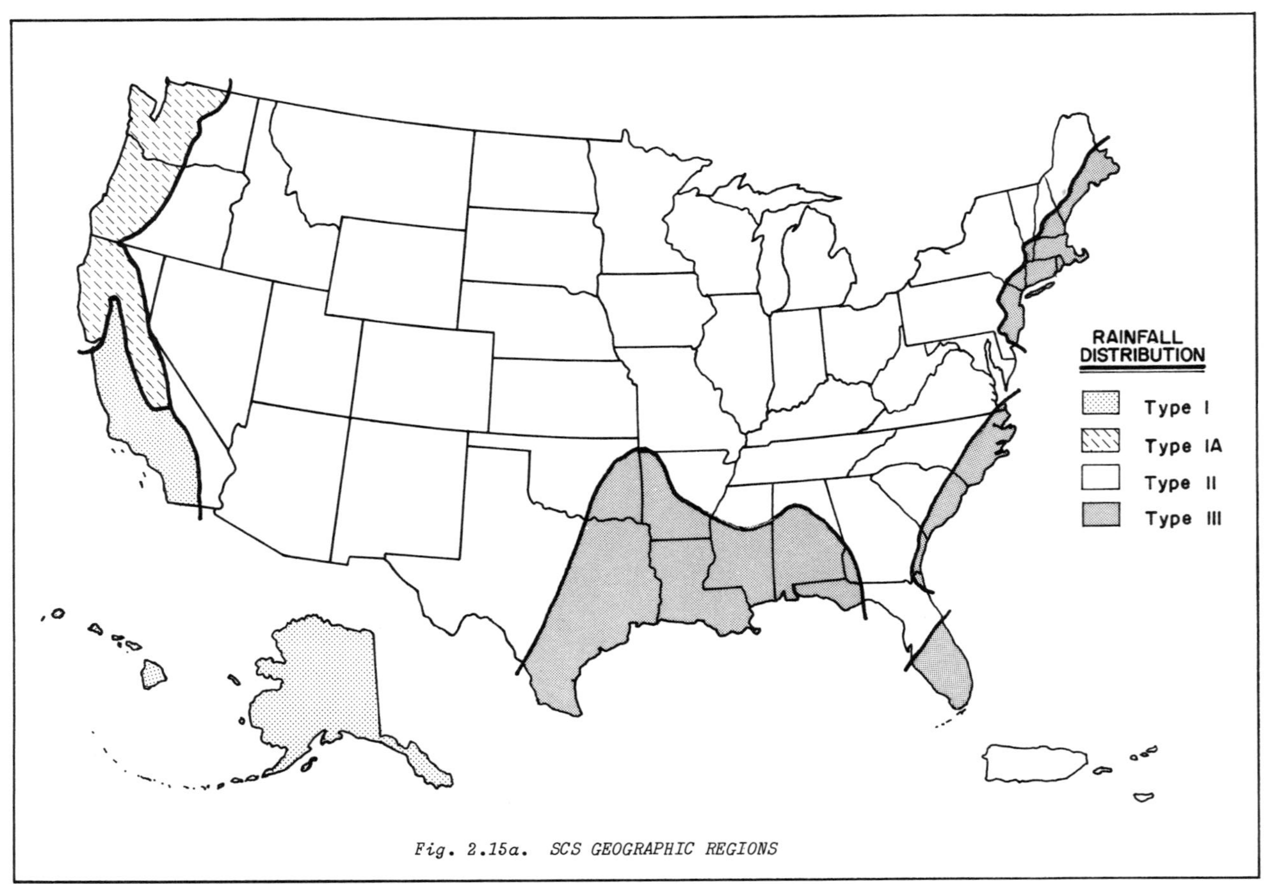

Fig. 2.15a. SCS GEOGRAPHIC REGIONS

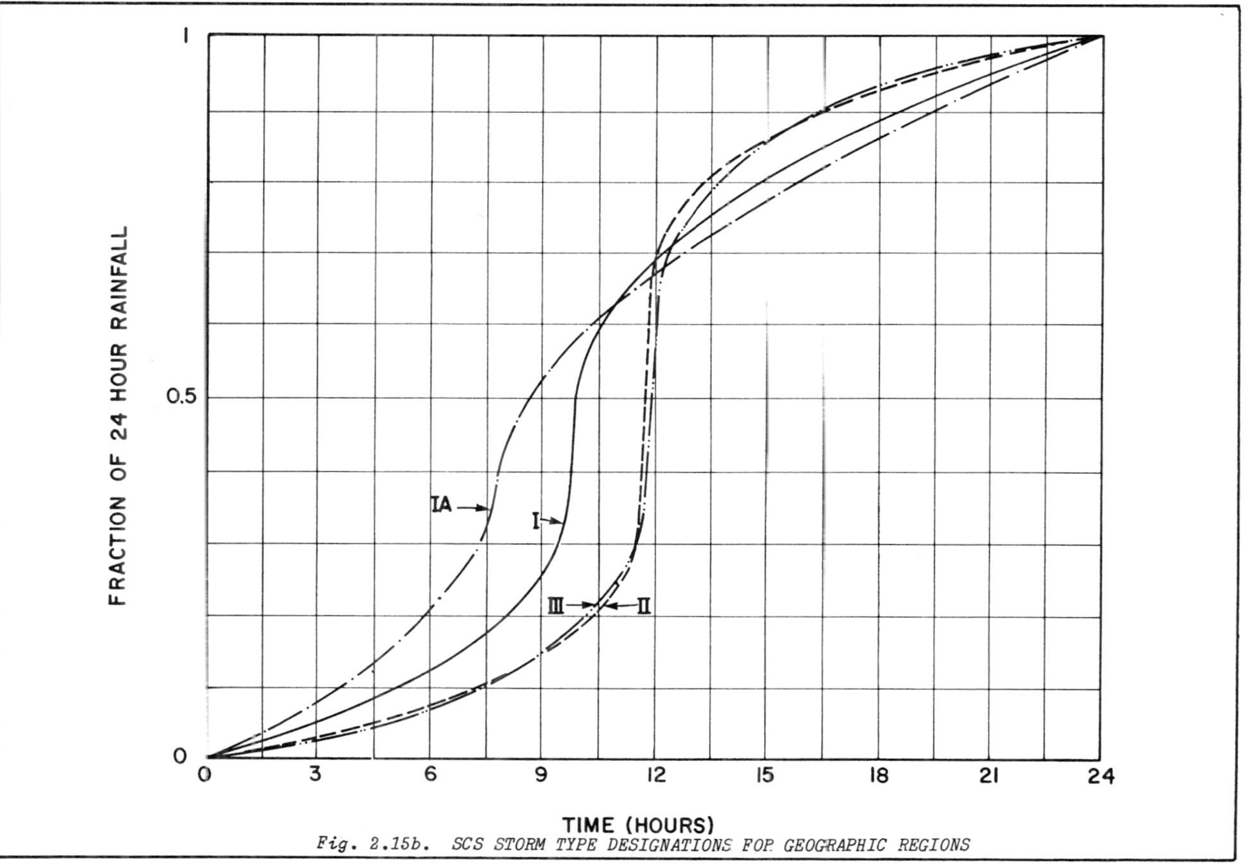

Fig. 2.15b. SCS STORM TYPE DESIGNATIONS FOR GEOGRAPHIC REGIONS

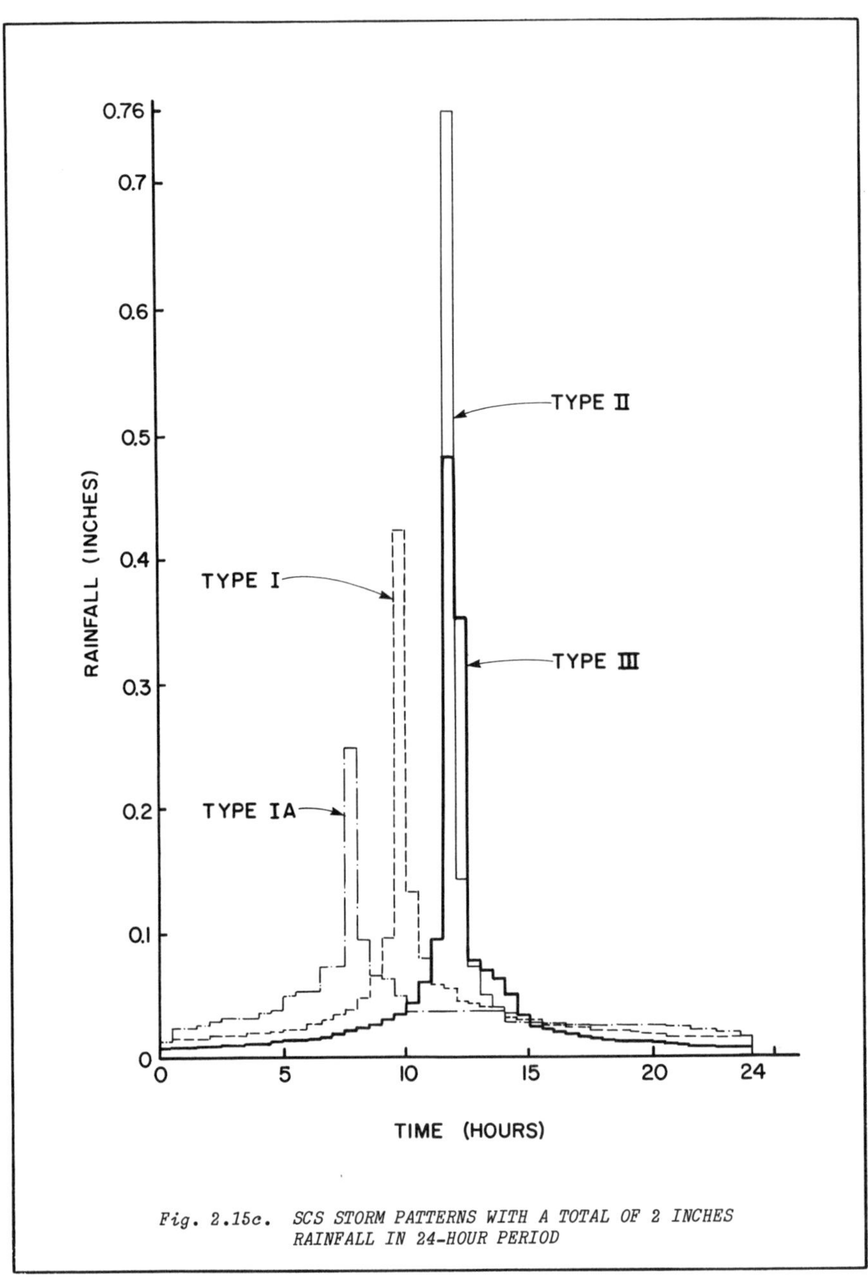

Fig. 2.15c. SCS STORM PATTERNS WITH A TOTAL OF 2 INCHES RAINFALL IN 24-HOUR PERIOD

depicts the SCS 24-hour rainfall distribution curves available in the program, and Fig. 2.15c illustrates the SCS storm patterns with a total rainfall of 2 inches in the 24-hour period.

2.15.4. Constant Rational Method

The rational method provides a constant rainfall intensity (inches/hour) uniformly distributed over a user specified duration (in hours). Figure 2.16 shows a storm pattern of a 2 inches/hour rainfall intensity with a duration equal to 2-hours.

2.15.5. Linear Interpolation Method

An arbitrary storm pattern can be developed in the program by entering data pairs of increasing storm time (hours) versus rainfall intensity (inches/hour). The data pairs are then linearly interpolated to estimate the rainfall intensities occurring between the supplied data values. Figure 2.17 depicts the 3-minute unit interval rainfall depth (inches) being linearly interpolated between two adjacent 3-minute interval numbers.

2.15.6. Program Data Entry Sequences

The watershed area has to be entered in acres, then one of the fifteen storm patterns can be selected by the program user.

```
ENTER WATERSHED AREA (ACRES)
=====> 'AREA'
```

```
              *** PRECIPITATION MODEL ***
         *** SELECT STORM PATTERN OPTION NUMBER ***

** 24-HOUR NESTED DESIGN STORM PATTERN OPTIONS:
   (INCLUDES CONTINUOUS DEPTH-AREA ADJUSTMENT)
 1 = PEAK AT HOUR 0              2 = PEAK AT HOUR 4
 3 = PEAK AT HOUR 8              4 = PEAK AT HOUR 12
 5 = PEAK AT HOUR 16             6 = PEAK AT HOUR 20
 7 = PEAK AT HOUR 24

** S.C.S. STORM PATTERN OPTIONS:
 8 = TYPE I                      9 = TYPE II
10 = TYPE III                   11 = TYPE IA

**USER ENTERED STORM PATTERN OPTIONS:
12 = ENTER CONSTANT RAINFALL INTENSITY FOR A GIVEN DURATION


13 = ENTER A NEW STORM PATTERN (LINEAR INTERPOLATION METHOD)
** USER SPECIFIED INFORMATION OPTIONS FROM DATA BANK:
14 = USE A PRESET STORM PATTERN
15 = USE THE DESIGN STORM PATTERN WITH CONTINUOUS DEPTH-AREA
     ADJUSTMENT
=====> 'KODE'
```

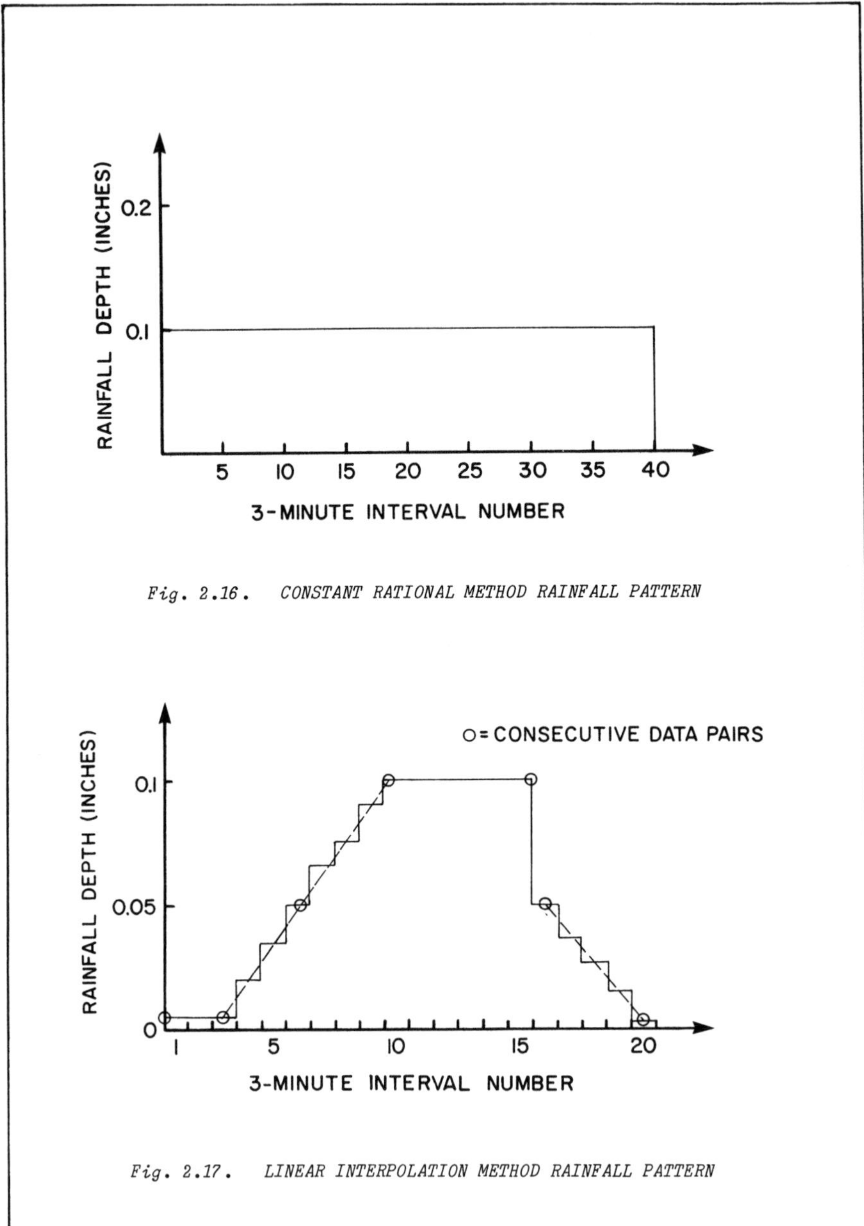

Fig. 2.16. CONSTANT RATIONAL METHOD RAINFALL PATTERN

Fig. 2.17. LINEAR INTERPOLATION METHOD RAINFALL PATTERN

2.15.6.1. <u>24-Hour Nested Design Storm Pattern</u>

If one of the 24-hour nested design storm patterns is chosen, then the watershed area-averaged point rainfall in inches has to be entered as follows:

```
ENTER WATERSHED AREA-AVERAGED POINT RAINFALL (INCHES)
   5-MINUTE VALUE
=====> 'R(1)'

  15-MINUTE VALUE
=====> 'R(2)'

  30-MINUTE VALUE
=====> 'R(3)'

  1-HOUR VALUE
=====> 'R(4)'

  2-HOUR VALUE
=====> 'R(5)'

  3-HOUR VALUE
=====> 'R(6)'

  6-HOUR VALUE
=====> 'R(7)'

 12-HOUR VALUE
=====> 'R(8)'

 24-HOUR VALUE
=====> 'R(9)'
```

A depth-area curve set has to be selected from the data bank after the watershed area-averaged point rainfalls are entered.

2.15.6.2. <u>SCS Storm Pattern</u>

If one of the SCS storm patterns is chosen, then the total rainfall (inches) during the 24-hour period is entered.

2.15.6.3. <u>Constant Rainfall Intensity for A Given Duration</u>

Data requirements are: (1) A constant rainfall intensity (inches/hour), and (2) a specified storm duration (hours).

2.15.6.4. Linear Interpolation Method

Data entries are:

```
ENTER NUMBER OF DATA SET (> 2)
=====> 'NSET'

ENTER THE INTERVAL NUMBER FOR DATA SET [1]
=====> 'N1'

ENTER RAINFALL DEPTH (INCHES)
=====> 'RI1'
```

2.15.6.5. User Pre-Set Storm Pattern

The total rainfall (inches/hour) defined for the 24-hour period, and the pre-set storm pattern number (program CONTR) is entered as requested.

2.15.6.6. Design Storm Pattern with Continuous Depth-Area Adjustment

The watershed area-averaged point rainfall in inches has to be entered as discussed in Section 2.15.2. After one storm pattern is chosen, then the user can print or suppress the storm pattern as follows:

```
ENTER A [1] TO SUPPRESS THE STORM PATTERN
   OR A [2] TO OUTPUT THE STORM PATTERN
=====>   'KKST'
```

2.16. PROGRAM 2.1

```
C
C-----------------------------------------------------------------
C     THIS PROGRAM DETERMINES THE STORM PATTERN
C-----------------------------------------------------------------
C
      SUBROUTINE RAIN(NS,KKK,KRUN,KXY,KRD,KWT)
C
      COMMON/BLK 1/ PRECIP(480)
      COMMON/BLK 2/ DARF(4,110),NDARF(4)
      COMMON/BLK 3/ XF(9)
      COMMON/BLK 4/ NEW(480)
      COMMON/BLK 5/ SMASS(49)
      COMMON/BLK 6/ R(9)
      COMMON/BLK 7/ ST(4,50,2),NST(4)
      COMMON/BLK 10/ AREA
C
C  ENTER WATERSHED AREA
C
      IF(KRUN.EQ.2)GO TO 20
      OPEN(UNIT=10,FILE='TEMP.DAT',STATUS='NEW')
      WRITE(10,*)KXY
 15   WRITE(KWT,17)
 17   FORMAT(24(/),2X,'ENTER WATERSHED AREA (ACRES)')
      WRITE(KWT,99)
 20   READ(KRD,*)AREA
      IF(AREA.LT.1.)GO TO 15
      IF(KRUN.EQ.1)WRITE(10,*)AREA
      IF(KRUN.EQ.2)WRITE(KWT,22)AREA
 22   FORMAT(/,76('*'),//,2X,'WATERSHED AREA = ',F11.3,' ACRES')
      DO 110 I=1,480
 110  PRECIP(I)=0.
C
C  DISPLAY STORM OPTIONS
C
      IF(KRUN.EQ.2)GO TO 90
 100  WRITE(KWT,500)
 500  FORMAT(24(/),12X,'*** PRECIPITATION MODEL ***',/,
     1 '      *** SELECT STORM PATTERN OPTION NUMBER ***',//,
     1 '  ** 24-HOUR NESTED DESIGN STORM PATTERN OPTIONS:',/,
     1 '     (INCLUDES CONTINUOUS DEPTH-AREA ADJUSTMENT)',/,
     2 '   1 = PEAK AT HOUR 0',10X,'   2 = PEAK AT HOUR 4',/,
     3 '   3 = PEAK AT HOUR 8',10X,'   4 = PEAK AT HOUR 12',/,
     5 '   5 = PEAK AT HOUR 16',10X,'  6 = PEAK AT HOUR 20',/,
     5 '   7 = PEAK AT HOUR 24',//,'  ** S.C.S. STORM PATTERN ',
     5 'OPTIONS:',/,'   8 = TYPE I',10X,
     6 '     9 = TYPE II',/,'  10 = TYPE III',9X,'  11 = TYPE IA',
     7 //,'  ** USER ENTERED STORM PATTERN OPTIONS:',/,'  12 = ',
     8 'ENTER CONSTANT RAINFALL INTENSITY FOR A GIVEN DURATION',/,
     8 '  13 = ENTER A NEW STORM PATTERN (LINEAR INTERPOLATION',
```

```
    9 ' METHOD)',/)
      WRITE(KWT,501)
 501  FORMAT('  ** USER SPECIFIED INFORMATION OPTIONS FROM DATA',
    1 ' BANK:',
    9 /,'  14 = USE A PRESET STORM PATTERN'
    * ,/,'  15 = USE THE DESIGN STORM PATTERN WITH',
    * ' CONTINUOUS DEPTH-AREA ADJUSTMENT'/)
      WRITE(KWT,99)
 99   FORMAT(2X,'=====> ',$)
 90   READ(KRD,*)KODE
      IF(KODE.LT.1 .OR. KODE.GT.15)GO TO 100
      IF(KRUN.EQ.1)WRITE(10,*)KODE
      IF(KODE.LE.7)CALL STORMA(KODE,KKK,KRUN,KRD,KWT)
      IF(KODE.GE.8 .AND. KODE.LE.11)CALL STORMB(KODE,KRUN,KRD,KWT)
      IF(KODE.EQ.12)CALL STORMC(KRUN,KRD,KWT)
      IF(KODE.EQ.13)CALL STORMD(KRUN,KRD,KWT)
      IF(KODE.EQ.14)CALL STORME(NS,KRUN,KRD,KWT)
      IF(KODE.EQ.15 .AND. NEW(1).EQ.0)GO TO 100
      IF(KODE.EQ.15 .AND. NEW(1).NE.0)CALL STORMA(KODE,KKK,KRUN,KRD,KWT)
C
      RETURN
      END
C
C-------------------------------------------------------------------
C     24-HOUR NESTED DESIGN STORM PATTERNS
C-------------------------------------------------------------------
C
      SUBROUTINE STORMA(KTYPE,KKK,KRUN,KRD,KWT)
C
      COMMON/BLK 1/ PRECIP(480)
      COMMON/BLK 2/ DARF(4,110),NDARF(4)
      COMMON/BLK 3/ XF(9)
      COMMON/BLK 4/ NEW(480)
      COMMON/BLK 6/ R(9)
      COMMON/BLK 10/ AREA
C
      DIMENSION TEMP(480)
C
      RR(T)=EXP(A*ALOG(T)+B)
C
      AREA=AREA/640.
C
C  INPUT WATERSHED AREA-AVERAGED POINT RAINFALL
C
 512  IF(KRUN.EQ.1)WRITE(KWT,5000)
 5000 FORMAT(24(/),2X,'ENTER WATERSHED AREA-AVERAGED POINT RAINFALL',
     1 '(INCHES)',/,2X,'5-MINUTE VALUE')
      IF(KRUN.EQ.1)WRITE(KWT,99)
 99   FORMAT(2X,'=====> ',$)
      READ(KRD,*)R(1)
      IF(R(1).LT..1)GO TO 512
      IF(KRUN.EQ.1)WRITE(10,*)R(1)
 511  IF(KRUN.EQ.1)WRITE(KWT,5001)
 5001 FORMAT(/,2X,'15-MINUTE VALUE')
      IF(KRUN.EQ.1)WRITE(KWT,99)
```

```
      READ(KRD,*)R(2)
      IF(R(2).LT.R(1))GO TO 511
      IF(KRUN.EQ.1)WRITE(10,*)R(2)
 521  IF(KRUN.EQ.1)WRITE(KWT,5002)
 5002 FORMAT(/,2X,'30-MINUTE VALUE')
      IF(KRUN.EQ.1)WRITE(KWT,99)
      READ(KRD,*)R(3)
      IF(R(3).LT.R(2))GO TO 521
      IF(KRUN.EQ.1)WRITE(10,*)R(3)
 530  IF(KRUN.EQ.1)WRITE(KWT,5003)
 5003 FORMAT(/,2X,'1-HOUR VALUE')
      IF(KRUN.EQ.1)WRITE(KWT,99)
      READ(KRD,*)R(4)
      IF(R(4).LT.R(3))GO TO 530
      IF(KRUN.EQ.1)WRITE(10,*)R(4)
 540  IF(KRUN.EQ.1)WRITE(KWT,5004)
 5004 FORMAT(/,2X,'2-HOUR VALUE')
      IF(KRUN.EQ.1)WRITE(KWT,99)
      READ(KRD,*)R(5)
      IF(R(5).LT.R(4))GO TO 540
      IF(KRUN.EQ.1)WRITE(10,*)R(5)
 550  IF(KRUN.EQ.1)WRITE(KWT,5005)
 5005 FORMAT(/,2X,'3-HOUR VALUE')
      IF(KRUN.EQ.1)WRITE(KWT,99)
      READ(KRD,*)R(6)
      IF(R(6).LT.R(5))GO TO 550
      IF(KRUN.EQ.1)WRITE(10,*)R(6)
 560  IF(KRUN.EQ.1)WRITE(KWT,5006)
 5006 FORMAT(/,2X,'6-HOUR VALUE')
      IF(KRUN.EQ.1)WRITE(KWT,99)
      READ(KRD,*)R(7)
      IF(R(7).LT.R(6))GO TO 560
      IF(KRUN.EQ.1)WRITE(10,*)R(7)
 570  IF(KRUN.EQ.1)WRITE(KWT,5007)
 5007 FORMAT(/,2X,'12-HOUR VALUE')
      IF(KRUN.EQ.1)WRITE(KWT,99)
      READ(KRD,*)R(8)
      IF(R(8).LT.R(7))GO TO 570
      IF(KRUN.EQ.1)WRITE(10,*)R(8)
 580  IF(KRUN.EQ.1)WRITE(KWT,5008)
 5008 FORMAT(/,2X,'24-HOUR VALUE')
      IF(KRUN.EQ.1)WRITE(KWT,99)
      READ(KRD,*)R(9)
      IF(R(9).LT.R(8))GO TO 580
      IF(KRUN.EQ.1)WRITE(10,*)R(9)
C
C  SELECT A DEPTH-AREA CURVE SET
C
      IF(KKK.EQ.1)KCU=NDARF(1)
      IF(KKK.EQ.1)GO TO 121
      IF(KRUN.EQ.2)GO TO 123
 120  WRITE(KWT,52)KKK,(NDARF(I),I=1,KKK)
 52   FORMAT(/,2X,' [',I2,' ] DEPTH-AREA CURVE SETS EXIST',/,
     1 5X,'THE FOLLOWING DEPTH-AREA CURVE SETS ARE DEFINED :',/,
     2 7X,I2,:,I2,:,I2,:,I2)
```

```
      WRITE(KWT,53)
 53   FORMAT('  ENTER CURVE SET NUMBER')
      WRITE(KWT,99)
 123  READ(KRD,*)KCU
      DO 125 I=1,KKK
      IF(KCU.EQ.NDARF(I))GO TO 127
 125  CONTINUE
      GO TO 120
 127  IF(KRUN.EQ.1)WRITE(10,*)KCU
C
C  DEFINE VALUES AT CUTOFF POINTS
C
 121  XF(1)=DARF(KCU,22)
      XF(2)=DARF(KCU,33)
      XF(3)=DARF(KCU,44)
      XF(4)=DARF(KCU,55)
      XF(5)=DARF(KCU,66)
      XF(6)=DARF(KCU,77)
      XF(7)=DARF(KCU,88)
      XF(8)=DARF(KCU,99)
      XF(9)=DARF(KCU,110)
      IF(AREA.GE.150.)GOTO 180
      DO 101 I=1,11
      IF(AREA.LE.DARF(KCU,I))GOTO 110
 101  CONTINUE
 110  I=I-1
      DX=DARF(KCU,I+1)-DARF(KCU,I)
      FACT=AREA-DARF(KCU,I)
      XF(1)=(DARF(KCU,11+I)-FACT*(DARF(KCU,11+I)-DARF(KCU,12+I))/DX)
      XF(2)=(DARF(KCU,22+I)-FACT*(DARF(KCU,22+I)-DARF(KCU,23+I))/DX)
      XF(3)=(DARF(KCU,33+I)-FACT*(DARF(KCU,33+I)-DARF(KCU,34+I))/DX)
      XF(4)=(DARF(KCU,44+I)-FACT*(DARF(KCU,44+I)-DARF(KCU,45+I))/DX)
      XF(5)=(DARF(KCU,55+I)-FACT*(DARF(KCU,55+I)-DARF(KCU,56+I))/DX)
      XF(6)=(DARF(KCU,66+I)-FACT*(DARF(KCU,66+I)-DARF(KCU,67+I))/DX)
      XF(7)=(DARF(KCU,77+I)-FACT*(DARF(KCU,77+I)-DARF(KCU,78+I))/DX)
      XF(8)=(DARF(KCU,88+I)-FACT*(DARF(KCU,88+I)-DARF(KCU,89+I))/DX)
      XF(9)=(DARF(KCU,99+I)-FACT*(DARF(KCU,99+I)-DARF(KCU,100+I))/DX)
 180  CONTINUE
C
      AREA=AREA*640.
      IF(KRUN.EQ.1)GO TO 5
      WRITE(KWT,5009)KTYPE
 5009 FORMAT(/,2X,'24-HOUR NESTED DESIGN STORM PATTERN #',I2,
     C /,2X,'WATERSHED AREA-AVERAGED POINT RAINFALL')
      WRITE(KWT,5010)R(1)
 5010 FORMAT(5X,' 5-MINUTE VALUE = ',F5.2,' INCHES')
      WRITE(KWT,5011)R(2)
 5011 FORMAT(5X,'15-MINUTE VALUE = ',F5.2,' INCHES')
      WRITE(KWT,5012)R(3)
 5012 FORMAT(5X,'30-MINUTE VALUE = ',F5.2,' INCHES')
      WRITE(KWT,5013)R(4)
 5013 FORMAT(5X,' 1-HOUR   VALUE = ',F5.2,' INCHES')
      WRITE(KWT,5014)R(5)
 5014 FORMAT(5X,' 2-HOUR   VALUE = ',F5.2,' INCHES')
      WRITE(KWT,5015)R(6)
```

```
 5015 FORMAT(5X,' 3-HOUR   VALUE = ',F5.2,' INCHES')
      WRITE(KWT,5016)R(7)
 5016 FORMAT(5X,' 6-HOUR   VALUE = ',F5.2,' INCHES')
      WRITE(KWT,5017)R(8)
 5017 FORMAT(5X,'12-HOUR   VALUE = ',F5.2,' INCHES')
      WRITE(KWT,5018)R(9)
 5018 FORMAT(5X,'24-HOUR   VALUE = ',F5.2,' INCHES')
      WRITE(KWT,13)(XF(I),I=1,9)
 13   FORMAT(/,2X,'*** DEPTH-AREA REDUCTION FACTORS :',/,
     C 5X,' 5-MINUTE FACTOR = ',F5.3,/,5X,'15-MINUTE FACTOR = ',F5.3,/
     C 5X,'30-MINUTE FACTOR = ',F5.3,/,5X,' 1-HOUR   FACTOR = ',F5.3,/
     C 5X,' 2-HOUR   FACTOR = ',F5.3,/,5X,' 3-HOUR   FACTOR = ',F5.3,/
     C 5X,' 6-HOUR   FACTOR = ',F5.3,/,5X,'12-HOUR   FACTOR = ',F5.3,/
     C 5X,'24-HOUR   FACTOR = ',F5.3,/)
C
C  DEPTH-AREA ADJUSTMENT
C
 5    DO 15 I=1,9
      R(I)=R(I)*XF(I)
 15   CONTINUE
      IF(R(1).LT.0.01)R(1)=0.01
      DO 25 I=1,8
      IF(R(I).EQ.0)GO TO 35
      DIFF=R(I+1)-R(I)
      IF(DIFF.LT.0.01)R(I+1)=R(I)+0.01
 25   CONTINUE
 35   CONTINUE
C
      IF(KTYPE.EQ.15)ITYPE=15
      IF(KTYPE.EQ.15)KTYPE=1
      IF(KTYPE.EQ.1)GO TO 10
      IF(KTYPE.EQ.2)GO TO 20
      IF(KTYPE.EQ.3)GO TO 30
      IF(KTYPE.EQ.4)GO TO 40
      IF(KTYPE.EQ.5)GO TO 50
      IF(KTYPE.EQ.6)GO TO 60
      IF(KTYPE.EQ.7)GO TO 70
C
C  STORM PATTERN #1 -- PEAK AT 0-HOUR
C
 10   NC=1
      NF=1
      NB=1
      NCH=500
      GO TO 100
C
C  STORM PATTERN #2 -- PEAK AT 4-HOUR
C
 20   NC=81
      NF=81
      NB=81
      NCH=6
      GO TO 100
C
C  STORM PATTERN #3 -- PEAK AT 8-HOUR
```

```
C
 30   NC=161
      NF=161
      NB=161
      NCH=3
      GO TO 100
C
C  STORM PATTERN #4 -- PEAK AT 12-HOUR
C
 40   NC=241
      NF=241
      NB=241
      NCH=2
      GO TO 100
C
C  STORM PATTERN #5 -- PEAK AT 16-HOUR
C
 50   NC=321
      NF=321
      NB=321
      NCH=3
      GO TO 100
C
C  STORM PATTERN #6 -- PEAK AT 20-HOUR
C
 60   NC=401
      NF=401
      NB=401
      NCH=6
      GO TO 100
C
C  STORM PATTERN #7 -- PEAK AT 24-HOUR
C
 70   NC=480
      NF=480
      NB=480
      NCH=1
C
 100  CONTINUE
C
C  PEAK 5-MINUTE STORM
C
      A=(ALOG(R(2))-ALOG(R(1)))/(ALOG(.25)-ALOG(.0833))
      B=ALOG(R(1))-A*ALOG(.0833)
      PRECIP(NC)=RR(.05)
      K=0
      IF(KTYPE.LE.3)K=1
      XX=.1
      YY=.05
C
C  PEAK 15-MINUTE STORM
C
      DO 200 I=1,4
      K=K+1
      IF(K.EQ.NCH)K=0
```

```
      IF(KTYPE.GE.4 .AND. KTYPE.NE.7)GO TO 210
      IF(K.NE.0)NF=NF+1
      IF(K.NE.0)NN=NF
      IF(K.EQ.0)NB=NB-1
      IF(K.EQ.0)NN=NB
      GO TO 220
 210  IF(K.NE.0)NB=NB-1
      IF(K.NE.0)NN=NB
      IF(K.EQ.0)NF=NF+1
      IF(K.EQ.0)NN=NF
 220  IF(NB.LT.1)NN=NF+1
      IF(NF.GT.480)NN=NB-1
      PRECIP(NN)=RR(XX)-RR(YY)
      XX=XX+.05
      YY=YY+.05
 200  CONTINUE
C
C  PEAK 30-MINUTE STORM
C
      IF(R(3).EQ.0.)GO TO 3000
      A=(ALOG(R(3))-ALOG(R(2)))/(ALOG(.5)-ALOG(.25))
      B=ALOG(R(2))-A*ALOG(.25)
      DO 300 I=1,5
      K=K+1
      IF(K.EQ.NCH)K=0
      IF(KTYPE.GE.4 .AND. KTYPE.NE.7)GO TO 310
      IF(K.NE.0)NF=NF+1
      IF(K.NE.0)NN=NF
      IF(K.EQ.0)NB=NB-1
      IF(K.EQ.0)NN=NB
      GO TO 320
 310  IF(K.NE.0)NB=NB-1
      IF(K.NE.0)NN=NB
      IF(K.EQ.0)NF=NF+1
      IF(K.EQ.0)NN=NF
 320  IF(NB.LT.1)NN=NF+1
      IF(NF.GT.480)NN=NB-1
      PRECIP(NN)=RR(XX)-RR(YY)
      XX=XX+.05
      YY=YY+.05
 300  CONTINUE
C
C  PEAK 1-HOUR STORM
C
      IF(R(4).EQ.0.)GO TO 3000
      A=(ALOG(R(4))-ALOG(R(3)))/(ALOG(1.)-ALOG(.5))
      B=ALOG(R(3))-A*ALOG(.5)
      DO 400 I=1,10
      K=K+1
      IF(K.EQ.NCH)K=0
      IF(KTYPE.GE.4 .AND. KTYPE.NE.7)GO TO 410
      IF(K.NE.0)NF=NF+1
      IF(K.NE.0)NN=NF
      IF(K.EQ.0)NB=NB-1
      IF(K.EQ.0)NN=NB
```

```
      GO TO 420
 410  IF(K.NE.0)NB=NB-1
      IF(K.NE.0)NN=NB
      IF(K.EQ.0)NF=NF+1
      IF(K.EQ.0)NN=NF
 420  IF(NB.LT.1)NN=NF+1
      IF(NF.GT.480)NN=NB-1
      PRECIP(NN)=RR(XX)-RR(YY)
      XX=XX+.05
      YY=YY+.05
 400  CONTINUE
C
C  PEAK 2-HOUR STORM
C
      IF(R(5).EQ.0.)GO TO 3000
      A=(ALOG(R(5))-ALOG(R(4)))/(ALOG(2.)-ALOG(1.))
      B=ALOG(R(4))-A*ALOG(1.)
      DO 500 I=1,20
      K=K+1
      IF(K.EQ.NCH)K=0
      IF(KTYPE.GE.4 .AND. KTYPE.NE.7)GO TO 510
      IF(K.NE.0)NF=NF+1
      IF(K.NE.0)NN=NF
      IF(K.EQ.0)NB=NB-1
      IF(K.EQ.0)NN=NB
      GO TO 520
 510  IF(K.NE.0)NB=NB-1
      IF(K.NE.0)NN=NB
      IF(K.EQ.0)NF=NF+1
      IF(K.EQ.0)NN=NF
 520  IF(NB.LT.1)NN=NF+1
      IF(NF.GT.480)NN=NB-1
      PRECIP(NN)=RR(XX)-RR(YY)
      XX=XX+.05
      YY=YY+.05
 500  CONTINUE
C
C  PEAK 3-HOUR STORM
C
      IF(R(6).EQ.0.)GO TO 3000
      A=(ALOG(R(6))-ALOG(R(5)))/(ALOG(3.)-ALOG(2.))
      B=ALOG(R(5))-A*ALOG(2.)
      DO 600 I=1,20
      K=K+1
      IF(K.EQ.NCH)K=0
      IF(KTYPE.GE.4 .AND. KTYPE.NE.7)GO TO 610
      IF(K.NE.0)NF=NF+1
      IF(K.NE.0)NN=NF
      IF(K.EQ.0)NB=NB-1
      IF(K.EQ.0)NN=NB
      GO TO 620
 610  IF(K.NE.0)NB=NB-1
      IF(K.NE.0)NN=NB
      IF(K.EQ.0)NF=NF+1
      IF(K.EQ.0)NN=NF
```

```
 620  IF(NB.LT.1)NN=NF+1
      IF(NF.GT.480)NN=NB-1
      PRECIP(NN)=RR(XX)-RR(YY)
      XX=XX+.05
      YY=YY+.05
 600  CONTINUE
C
C  PEAK 6-HOUR STORM
C
      IF(R(7).EQ.0.)GO TO 3000
      A=(ALOG(R(7))-ALOG(R(6)))/(ALOG(6.)-ALOG(3.))
      B=ALOG(R(6))-A*ALOG(3.)
      DO 700 I=1,60
      K=K+1
      IF(K.EQ.NCH)K=0
      IF(KTYPE.GE.4 .AND. KTYPE.NE.7)GO TO 710
      IF(K.NE.0)NF=NF+1
      IF(K.NE.0)NN=NF
      IF(K.EQ.0)NB=NB-1
      IF(K.EQ.0)NN=NB
      GO TO 720
 710  IF(K.NE.0)NB=NB-1
      IF(K.NE.0)NN=NB
      IF(K.EQ.0)NF=NF+1
      IF(K.EQ.0)NN=NF
 720  IF(NB.LT.1)NN=NF+1
      IF(NF.GT.480)NN=NB-1
      PRECIP(NN)=RR(XX)-RR(YY)
      XX=XX+.05
      YY=YY+.05
 700  CONTINUE
C
C  PEAK 12-HOUR STORM
C
      IF(R(8).EQ.0.)GO TO 3000
      A=(ALOG(R(8))-ALOG(R(7)))/(ALOG(12.)-ALOG(6.))
      B=ALOG(R(7))-A*ALOG(6.)
      DO 800 I=1,120
      K=K+1
      IF(K.EQ.NCH)K=0
      IF(KTYPE.GE.4 .AND. KTYPE.NE.7)GO TO 810
      IF(K.NE.0)NF=NF+1
      IF(K.NE.0)NN=NF
      IF(K.EQ.0)NB=NB-1
      IF(K.EQ.0)NN=NB
      GO TO 820
 810  IF(K.NE.0)NB=NB-1
      IF(K.NE.0)NN=NB
      IF(K.EQ.0)NF=NF+1
      IF(K.EQ.0)NN=NF
 820  IF(NB.LT.1)NN=NF+1
      IF(NF.GT.480)NN=NB-1
      PRECIP(NN)=RR(XX)-RR(YY)
      XX=XX+.05
      YY=YY+.05
```

```
 800  CONTINUE
C
C  PEAK 24-HOUR STORM
C
      IF(R(9).EQ.0.)GO TO 3000
      A=(ALOG(R(9))-ALOG(R(8)))/(ALOG(24.)-ALOG(12.))
      B=ALOG(R(8))-A*ALOG(12.)
      DO 900 I=1,240
      K=K+1
      IF(K.EQ.NCH)K=0
      IF(KTYPE.GE.4 .AND. KTYPE.NE.7)GO TO 910
      IF(K.NE.0)NF=NF+1
      IF(K.NE.0)NN=NF
      IF(K.EQ.0)NB=NB-1
      IF(K.EQ.0)NN=NB
      GO TO 920
 910  IF(K.NE.0)NB=NB-1
      IF(K.NE.0)NN=NB
      IF(K.EQ.0)NF=NF+1
      IF(K.EQ.0)NN=NF
 920  IF(NB.LT.1)NN=NF+1
      IF(NF.GT.480)NN=NB-1
      PRECIP(NN)=RR(XX)-RR(YY)
      XX=XX+.05
      YY=YY+.05
 900  CONTINUE
 3000 CONTINUE
C
C  REARRANGE THE PRECIPITATION INTERVALS - STORM #15
C
      IF(ITYPE.NE.15)GO TO 940
      DO 950 I=1,480
      TEMP(I)=PRECIP(I)
      PRECIP(I)=0.
 950  CONTINUE
      DO 960 I=1,480
      II=NEW(I)
      IF(II.EQ.0)GO TO 940
      PRECIP(I)=TEMP(II)
 960  CONTINUE
C
 940  CONTINUE
C
      RETURN
      END
C
C---------------------------------------------------------
C     S.C.S. 24-HOUR STORM PATTERN
C---------------------------------------------------------
C
      SUBROUTINE STORMB(KTYPE,KRUN,KRD,KWT)
C
      COMMON/BLK 1/ PRECIP(480)
      COMMON/BLK 5/ SMASS(49)
C
```

```
C  INPUT TOTAL RAINFALL (INCHES) IN 24-HOUR PERIOD
C
 40   IF(KRUN.EQ.1)WRITE(KWT,50)
 50   FORMAT(24(/),'  ENTER TOTAL RAINFALL(INCHES) IN 24-HOUR PERIOD')
      IF(KRUN.EQ.1)WRITE(KWT,99)
 99   FORMAT(2X,'=====> ',$)
      READ(KRD,*)RTOTL
      IF(RTOTL.LT..1)GO TO 40
      IF(KRUN.EQ.1)WRITE(10,*)RTOTL
C
      KK=KTYPE-7
C
C  ENTER SCS 24-HOUR RAINFALL DISTRIBUTION CURVE
C
      OPEN(UNIT=3,FILE='SCSMASS.DAT',STATUS='OLD')
      READ(3,*)((SMASS(K),K=1,49),I=1,KK)
      CLOSE(UNIT=3)
C
      K=0
      SUM=0.
      DO 200 I=1,48
      DEL=SMASS(I+1)-SMASS(I)
      RI=DEL*0.1*RTOTL
      DO 200 J=1,10
      SUM=SUM+RI
      K=K+1
      PRECIP(K)=RI
 200  CONTINUE
C
      IF(KRUN.EQ.2)WRITE(KWT,6)KK,SUM
 6    FORMAT(/,2X,'S.C.S. 24-HOUR STORM PATTERN #',I2,/,
     1 2X,'WITH TOTAL RAINFALL = ',F6.3,' INCHES IN 24-HOUR')
C
      RETURN
      END
C
C-------------------------------------------------------------
C     CONSTANT RATIONAL METHOD
C-------------------------------------------------------------
C
      SUBROUTINE STORMC(KRUN,KRD,KWT)
C
      COMMON/BLK 1/ PRECIP(480)
C
C  READ RAINFALL INTENSITY (INCHES/HOUR)
C
 40   IF(KRUN.EQ.1)WRITE(KWT,50)
 50   FORMAT(24(/),2X,'ENTER RAINFALL INTENSITY (INCHES/HOUR)')
      IF(KRUN.EQ.1)WRITE(KWT,99)
 99   FORMAT(2X,'=====> ',$)
      READ(KRD,*)RI
      IF(RI.LT..1)GO TO 40
      IF(KRUN.EQ.1)WRITE(10,*)RI
C
C  READ STORM DURATION (HOUR)
```

```
C
 30    IF(KRUN.EQ.1)WRITE(KWT,51)
 51    FORMAT(/,2X,'ENTER STORM DURATION (HOURS)')
       IF(KRUN.EQ.1)WRITE(KWT,99)
       READ(KRD,*)DUR
       IF(DUR.LT..1)GO TO 30
       IF(KRUN.EQ.1)WRITE(10,*)DUR
       RI=RI/20.
       NDUR=DUR*20.
       DO 100 I=1,NDUR
 100   PRECIP(I)=RI
C
       RI=RI*20
       IF(KRUN.EQ.2)WRITE(KWT,10)RI,DUR
 10    FORMAT(/,2X,'CONSTANT RAINFALL INTENSITY = ',F6.3,' INCHES/HOUR',/,
      1 2X,'FOR DURATION = ',F6.3,' HOURS')
C
       RETURN
       END
C
C------------------------------------------------------------------
C      LINEAR INTERPOLATION FOR RAINFALL INTENSITY
C------------------------------------------------------------------
C
       SUBROUTINE STORMD(KRUN,KRD,KWT)
C
       COMMON/BLK 1/ PRECIP(480)
C
C  INPUT NUMBER OF DATA SET (>=2)
C
 400   IF(KRUN.EQ.1)WRITE(KWT,50)
 50    FORMAT(24(/),2X,'ENTER NUMBER OF DATA SET (>=2)')
       IF(KRUN.EQ.1)WRITE(KWT,99)
 99    FORMAT(2X,'=====> ',$)
       READ(KRD,*)NSET
       IF(NSET.LT.2)GO TO 400
       IF(KRUN.EQ.1)WRITE(10,*)NSET
C
C  INPUT INTERVAL NUMBER (N1>0)
C
       KK=1
       IF(KRUN.EQ.1)WRITE(KWT,51)KK
 51    FORMAT(/,2X,'ENTER THE INTERVAL NUMBER FOR DATA SET [',I3,']')
       IF(KRUN.EQ.1)WRITE(KWT,99)
       READ(KRD,*)N1
       IF(KRUN.EQ.1)WRITE(10,*)N1
C
C  INPUT RAINFALL DEPTH (INCHES)
C
 30    IF(KRUN.EQ.1)WRITE(KWT,52)
 52    FORMAT(/,2X,'ENTER RAINFALL DEPTH (INCHES)')
       IF(KRUN.EQ.1)WRITE(KWT,99)
       READ(KRD,*)RI1
       IF(RI1.LT.0.)GO TO 30
       IF(KRUN.EQ.1)WRITE(10,*)RI1
```

```
 200  KK=KK+1
C
C  INPUT NEXT INTERVAL NUMBER (N2>N1)
C
 55   IF(KRUN.EQ.1)WRITE(KWT,51)KK
      IF(KRUN.EQ.1)WRITE(KWT,99)
      READ(KRD,*)N2
      IF(N2.LE.N1)GO TO 55
      IF(KRUN.EQ.1)WRITE(10,*)N2
C
C  INPUT NEXT RAINFALL DEPTH (INCHES)
C
 40   IF(KRUN.EQ.1)WRITE(KWT,52)
      IF(KRUN.EQ.1)WRITE(KWT,99)
      READ(KRD,*)RI2
      IF(RI2.LT.0.)GO TO 40
      IF(KRUN.EQ.1)WRITE(10,*)RI2
      IF(N2-N1 .EQ. 1)GO TO 150
      K=0
      B=N2-N1
      A=RI2-RI1
      DO 100 I=N1,N2
      PRECIP(I)=RI1+K*A/B
      K=K+1
 100  CONTINUE
 150  N1=N2
      RI1=RI2
      IF(KK.GE.NSET)GO TO 300
      GO TO 200
 300  CONTINUE
C
      IF(KRUN.EQ.2)WRITE(KWT,6)
 6    FORMAT(/,2X,'LINEAR INTERPOLATION BETWEEN RAINFALL DATA SET')
C
      RETURN
      END
C
C-----------------------------------------------------------------
C     USER SPECIFIED STORM PATTERNS
C                                   -----------------------------
C
      SUBROUTINE STORME(N3,KRUN,KRD,KWT)
C
      COMMON/BLK 1/ PRECIP(480)
      COMMON/BLK 7/ ST(4,50,2),NST(4)
C
 40   IF(KRUN.EQ.1)WRITE(KWT,50)
 50   FORMAT(24(/),'  ENTER TOTAL RAINFALL(INCHES) IN 24-HOUR PERIOD'
     1 ,/,'  =====> ',$)
      READ(KRD,*)RTOTL
      IF(RTOTL.LT..1)GO TO 40
      IF(KRUN.EQ.1)WRITE(10,*)RTOTL
      KST=1
      KIN=0
      SUM=0.
```

```
      IF(NS.EQ.1)NNS=NST(1)
      IF(NS.EQ.1)GO TO 110
100   IF(KRUN.EQ.1)WRITE(KWT,51)NS,(NST(I),I=1,NS)
51    FORMAT(/,2X,' [',I2,' ] 24-HOUR STORM PATTERNS EXIST',/,
     1 5X,'THE FOLLOWING 24-HOUR STORM PATTERNS ARE DEFINED :',/,
     2 7X,I2,:,I2,:,I2,:,I2)
      IF(KRUN.EQ.1)WRITE(KWT,52)
52    FORMAT('  ENTER STORM NUMBER',/,'  =====> ',$)
      READ(KRD,*)NNS
      DO 105 I=1,NS
      IF(NNS.EQ.NST(I))GO TO 107
105   CONTINUE
      GO TO 100
107   IF(KRUN.EQ.1)WRITE(10,*)NNS
110   DO 600 I=1,49
      NO=(ST(NNS,I+1,1)-ST(NNS,I,1))*20.
      KIN=KIN+NO
      RI=RTOTL*(ST(NNS,I+1,2)-ST(NNS,I,2))/NO
      DO 720 J=KST,KIN
720   PRECIP(J)=RI
      KST=KIN+1
      SUM=SUM+RI*NO
      IF(ST(NNS,I+1,1).EQ.24.)GO TO 620
600   CONTINUE
620   IF(KRUN.EQ.2)WRITE(KWT,61)NNS,SUM
61    FORMAT(2X,'USER SPECIFIED STORM PATTERN #',I2,/,
     1 '  WITH TOTAL RAINFALL (INCHES) = ',F6.3,' IN 24-HOUR PEROID')
C
      RETURN
      END
```

estimation of effective rainfall

3

3.1. Introduction

Rainfall, which was discussed in Chapter 2, represents the maximum amount of water that can appear as runoff. But given that the total volume of rainfall does not appear immediately at the watershed outlet, it is only necessary to design for part of the total volume of rainfall. Part of the total rainfall, the rainfall excess, is converted into immediate runoff, usually referred to as the direct runoff, and the remainder of the rainfall is assumed to be losses. The calculation of losses is the subject of this chapter.

Watershed losses consist of infiltration, depression storage, vegetation and man-made interception, and minor amounts of evaporation. In urban hydrology studies, infiltration and depression storage losses are usually considered the most important watershed loss processes that occur during a severe design storm event and are often assumed for study purposes to equal the total of the watershed losses.

While a number of methods have been proposed for estimating losses, the two methods used most frequently are the phi-index and SCS methods. After briefly describing the physical processes involved, these two methods of computing losses are described. Other methods are also discussed.

3.2. Physical Processes: An Overview

Infiltration is the process by which water enters the soil surface under the influences of gravity, viscosity, and capillarity. The infiltration process also includes the percolation of soil water into the soil where it is stored in the soil pores or migrates further downward until the groundwater table is reached. The stored water may be utilized by vegetation, remain in groundwater storage, or leave the soil system through seeps and springs. Seepage from groundwater storage is the primary source of watershed baseflow, which is derived from previous storm events. When making estimates of stormwater runoff quantities, it is often assumed that the infiltration process is a watershed loss for the storm event. Infiltration rates are usually highest at the beginning of a storm and become less at the end of the storm. However, there are several infiltration process approximations available for hydrologic modeling purposes, including assuming infiltration to be a constant during the severe design storm situation.

There are many factors that influence infiltration rates: soil type, soil depth, soil moisture amount and distribution, vegetation cover, cultural practices, and the availability of water at the ground surface. When the availability of water exceeds the maximum possible infiltration rate, which corresponds to in-situ conditions, then the infiltration rate equals the infiltration capacity. Infiltration loss rates can be examined by analysis of rainfall-runoff data and by relating the study results to soil type, vegetation cover, and other factors associated with the watershed. Because such data are usually unavailable for local watersheds, approximation methods for estimating the infiltration loss rates are required.

For design storm conditions on urbanized watersheds where the goal is to provide public protection against flood damage from a selected return frequency storm, the watershed depression storage and interception losses are usually small in comparison to infiltration losses or are considered satisfied by a previous storm event and the watershed losses assumed to approximately equal the infiltration losses. Consequently for this severe design storm condition, the process of infiltration is a major component in the estimation of runoff quantities.

In studies by Horton (1933), it was shown that the infiltration rate of a soil decreases with the total storm duration. Additionally, given a rainfall rate which exceeds the infiltration rate of the soil, then a limiting value of infiltration exists, which is called the infiltration capacity of the soil, and defines a maximum rate of infiltration. The infiltration capacity curve decreases with storm duration (see Fig. 3.1) given an ample availability of water. The reduction in the infiltration capacity is caused by the filling of the unsaturated soil pores and the corresponding reduction in the soil water total energy head gradient (the gradient of the sum of soil water pressure head and gravity head). From Fig. 3.1, the infiltration capacity curve appears to approach a constant lower value which is a function of the native soil type and condition. For example, a loose grained sandy soil has a higher limiting value of the infiltration capacity curve than does a compacted, fine grained clay soil.

Rubin et al., (1963, 1964) proposed that the limiting infiltration capacity is essentially the hydraulic conductivity of the soil with an associated total energy head gradient of unity. Using the fundamental equation of continuity for both a saturated and unsaturated soil, Freeze (1969) numerically modeled the rainfall soil infiltration process as a mathematical boundary

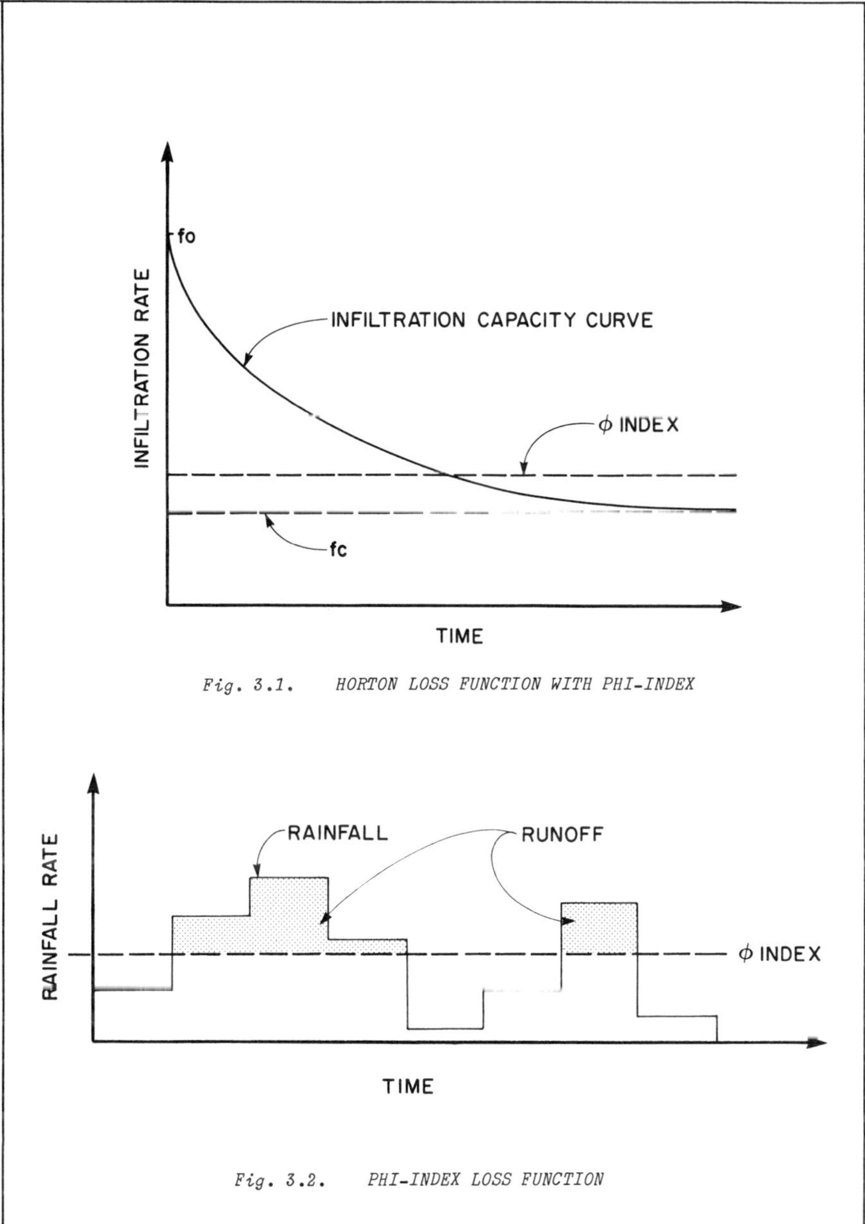

Fig. 3.1. HORTON LOSS FUNCTION WITH PHI-INDEX

Fig. 3.2. PHI-INDEX LOSS FUNCTION

value problem. That simulation study showed the interdependence of soil moisture content, soil water pore pressure head, and the soil water total energy head with the assumed rainfall rate and subsequent runoff rate. Additionally it was shown that the decrease in the infiltration capacity results from a decrease in the product of the total energy head gradient (due to the saturation of the soil) and the hydraulic conductivity of the soil.

For severe design storm conditions, it can be argued that a large portion of the storm contains rainfall intensities that exceed the infiltration capacity of the soil and that the soil infiltration rate may be at a minimum during the storm's peak rainfall intensities. For example, Viessman et al., (1977, pg. 72, Fig. 11a) shows that the infiltration capacity essentially reaches equilibrium approximately 20 minutes to 30 minutes after the beginning of rainfall. Consequently, for many cases in urban hydrologic studies, the equilibrium infiltration capacity value may be an appropriate value for the minimum watershed infiltration loss rate.

Several other infiltration models have been reported in the literature such as Green and Ampt (1911), the classic infiltration studies by Philip (1957 a,b,c,d,e; 1958), and Holtan (1961). Freeze (1969) provides a detailed tabulation of the numerical modeling literature based on solving the fundamental infiltration equations. Due to the difficulties in applying these mathematical models, many hydrologic studies utilize a simplified effective rainfall estimation method such as assuming a constant or average value of infiltration for the entire storm. This concept is analogous to the phi index approach (see Fig.3.1) of representing the watershed losses, and has the added advantage of being simple to apply (Viessman et al., 1977, pg. 80).

Depression storage is associated with the rainfall volume which is permanently stored due to the natural or man-made obstacles to runoff flow. Estimates of this parameter range from 0.01 to over 0.50 inches of rainfall depending on development type, soil type, and other factors. Table 3.1 provides some estimates for depression storage; from the table, it is evident that a wide range of values are available, which reflects the uncertainty of this parameter. In comparison, the ASCE (1970) estimates for this parameter are simply 0.0625 inch for impervious surfaces and 0.250 inch for pervious surfaces. For design storm purposes where a significant rainfall event may be assumed to occur prior to the severe design storm (i.e., an antecedent moisture condition) or if the design storm is of long duration such that the initial portions of the storm generate significant quantities of runoff, the watershed depression storage may be assumed to negligible and could be omitted from the rainfall-runoff budget.

TABLE 3.1
DEPRESSION STORAGE ESTIMATES
(Wright-McLaughlin Engineers, 1969)

Land Cover	Depression Storage (inch)	Recommended Value (inch)
Paved	0.05 - 0.15	0.10
Roofs (flat)	0.10 - 0.30	0.10
Lawn grass	0.20 - 0.50	0.30
Wooded areas & open fields	0.20 - 0.60	0.00

Vegetation interception is the volume of rainfall retained on the leaves and stems of the plant cover. That portion of the intercepted volume of rainfall which evaporates from the plant cover and does not contribute to runoff is termed the interception loss. The remaining portion contributing to runoff is termed the throughfall. For a long duration storm or with a significant prior precipitation event, the interception loss is considerably reduced and becomes simply that volume of rainfall which is retained on the plant cover. In most models, interception is simply subtracted from the precipitation budget at the beginning of the storm. Example values of interception are given by Horton (1919) which show this parameter as a function of total precipitation and tree cover. Values ranged from 0.10 to 0.50 inches for willow trees, with precipitation values of 0.20 to 1.30 inches, respectively.

3.3. Phi-Index Method

Field studies have shown that the infiltration capacity is greatest at the start of a storm and that it decreases rapidly to a relatively constant rate. The recession time of the infiltration capacity may be as short as 10 to 15 minutes. Therefore, it is not unreasonable to assume that the infiltration capacity is constant over the entire storm duration. When the rainfall rate exceeds the capacity, the loss rate is assumed to equal the constant capacity, which is called the phi (ϕ) index. When the rainfall rate is less than the value of ϕ, the infiltration rate is assumed to equal the rainfall intensity.

Mathematically, the phi index method for modeling losses is described by (Fig. 3.2)

$$f(t) = \begin{cases} I(t), & \text{for } I(t) < \phi \\ \phi, & \text{for } I(t) \geq \phi \end{cases} \tag{3.1}$$

where f(t) is the loss rate, I(t) is storm rainfall intensity; t is time; and ϕ is the calibration constant, called the phi index.

If measured storm event data are available, the value of ϕ can be estimated by separating baseflow from the total runoff volume, computing the volume of direct runoff, and then finding the value of ϕ that results in the volume of rainfall excess being equal to the volume of direct runoff. A watershed mean phi-index can then be computed as the average of storm event phi values. Where measured storm event data are not available, the ultimate capacity of Horton's equation, fc, can be used. The values in Table 3.2 are typical values of the ultimate capacity for bare soils and can be used as estimates of phi. For an individual watershed, actual ultimate capacities can vary considerably from these values. Furthermore, these values represent ultimate capacities and the phi-index value is usually somewhat higher, as evident in Fig. 3.1; however, the difference is small and is probably not significant when compared to the variations in fc that are evident from the tabular values of Table 3.2.

TABLE 3.2
TYPICAL SOIL CAPACITIES

Soil Group	Ultimate Capacity (fc, in./hr.)
Sands	0.5 to 1.0
Loams and silts	0.1 to 0.5
Clays	0.01 to 0.1

3.4. Constant Proportion Loss Rate

The use of a constant proportion loss rate, where the effective rainfall is estimated to be a constant propotion of precipitation, is reported in several studies including Schilling and Fuchs (1986), Williams et al., (1980), and Scully and Bender (1969). Schilling and Fuchs (1986) recommend "simplifying the loss rate model with a correct overall runoff coefficient" as a means of increasing computational efficiency in rainfall-runoff models..."without seriously affecting the modeling accuracy...." Williams et al. (1980) use a unit hydrograph (UH) method to analyze rainfall-runoff data where "effective rainfall was taken to be a constant proportion of total rainfall," and compared the UH modeling results to the complex TRRL model, noting that "the two calculated hydrographs are very similar." Scully and Bender (1969) found that in use of a unit hydrograph method with given rainfall-runoff data, "the infiltration loss rates are primarily a function of the rainfall intensity." Scully and Bender write that

> "these results may at first seem to contradict previous experimental work on infiltration theory. Infiltrometer tests show that the infiltration capacity of the soil is independent of the rainfall intensity as long as there is an excess of rainfall. But it must be remembered that the term infiltration loss includes more than just soil infiltration. Interception, evaporation, and depression storage are also included in the definition of this term. The results indicate that the net effect of all the quantities included in this infiltration loss rate is directly related to the rainfall intensity."

Because "interception storage capacity is usually satisfied early in the storm...some other phenomenon must account for the changes in the infiltration rates." Scully and Bender propose that "this phenomenon might be attributed to surface detention, which acts like depression storage, except that the surface detention is stored only temporarily and increases as the rainfall intensity increases."

The loss rate function f(t) associated to this effective rainfall estimation procedure is given by (Fig. 3.3):

$$f(t) = \bar{Y}I(t) \tag{3.2}$$

where f(t) is the loss rate; I(t) is the storm rainfall intensity; t is time; and $\bar{Y}$ is a calibration constant chosen such that $0<\bar{Y}<1$.

3.5. Coupled Phi-Index and Constant Proportion Loss Rate Function

The phi-index and constant proportion loss rate methods are both single parameter loss functions. A hybridization of the two techniques is to combine both methods by using ϕ as an upper bound to the loss rate determined from the constant proportion loss rate approach. That is, the loss rate f(t) is defined by (Fig. 3.4)

$$f(t) = \begin{cases} \bar{Y}I(t), & \text{for } I(t) < \phi/\bar{Y} \\ \phi, & \text{otherwise} \end{cases} \tag{3.3}$$

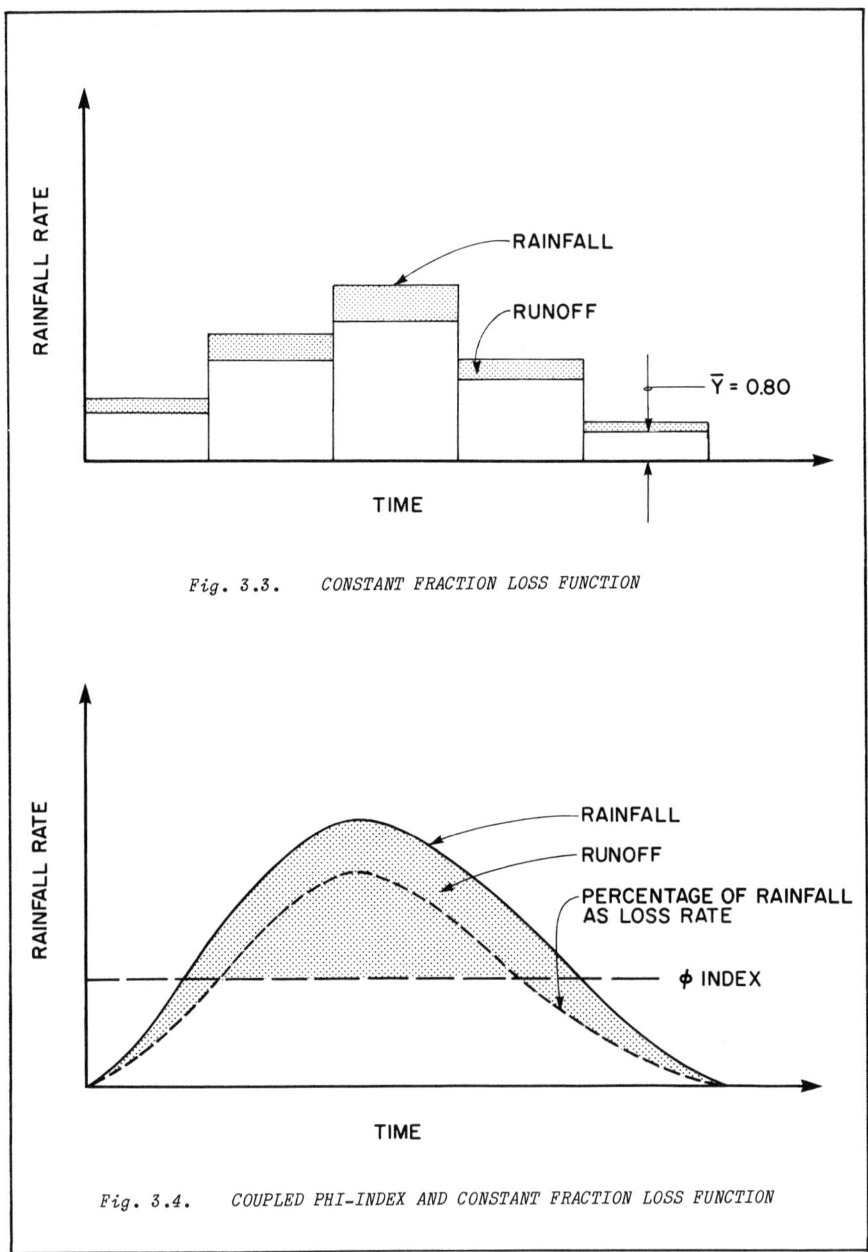

Fig. 3.3. CONSTANT FRACTION LOSS FUNCTION

Fig. 3.4. COUPLED PHI-INDEX AND CONSTANT FRACTION LOSS FUNCTION

where ϕ and $\bar{Y}$ are calibration constants similar to the original definitions given by Eqs. (3.1) and (3.2), respectively. Use of this two parameter model enables a calibration of the loss rate function to both storm peak rainfall intensity loss rates and also the total storm yield. This type of loss function was used in the calibration of the hydrology manuals for Orange and San Bernardino Counties (Hromadka and McCuen, 1986a,b).

3.6. Horton Loss Rate

The Horton loss function of Fig. 3.1 is described by the three parameter model

$$f(t) = fc + (fo - fc)e^{-kt} \tag{3.4}$$

where fo is the initial loss rate; fc is the ultimate loss rate; and k is the third calibration constant and reflects the time rate of decay from fo to fc. In typical applications, the rainfall intensity I(t) serves as an upper bound to the loss function f(t) during low intensity rainfalls by

$$f(t) = \min\left[I(t),\ fc + (fo - fc)e^{-kt}\right] \tag{3.5}$$

3.7. Exponential Loss Rate Function

Another three parameter loss rate function is reported by Ünver and Mays (1984) who suggest that the optimum parameter values of loss rate functions and the optimum unit hydrograph can be determined when using the loss rate function f(t) defined by (Fig. 3.5)

$$f(t) = A[I(t)]^E \tag{3.6}$$

where I(t) is the rainfall intensity at time t; and A and E, are calibration constants.

3.8. Initial Abstraction

The total loss used in the storm rainfall-runoff budget is separated into two parts: the initial abstraction, Ia, and the other losses (and retention). Two models for the initial abstraction estimation which can be coupled to the loss rate functions of Eqs. (3.1) through (3.6) are as follows (Fig. 3.6)

(i) total rainfall up to a total depth of Ia; and

(ii) percentage of total rainfall up to a total depth of Ia.

In the first Ia model, all precipitation is budgeted towards

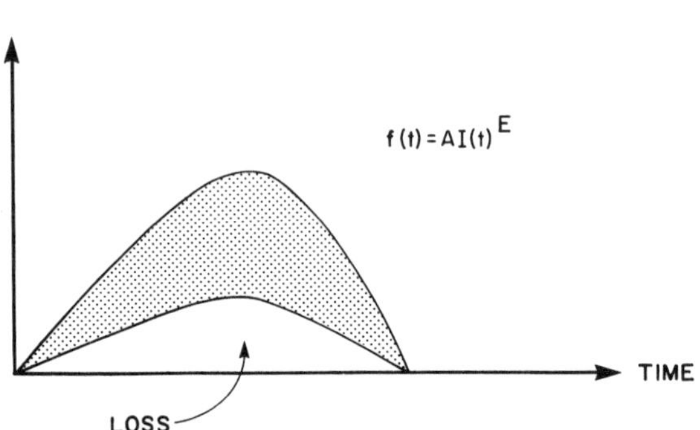

Fig. 3.5. EXPONENTIAL LOSS RATE FUNCTION

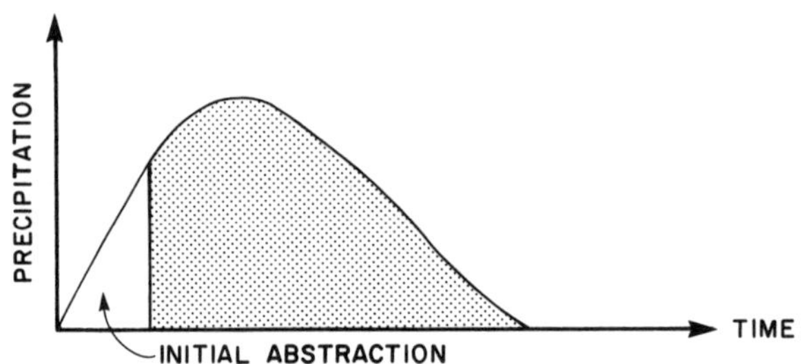

Fig. 3.6a. INITIAL ABSTRACTION MODEL OF PRECIPITATION TO A TOTAL DEPTH

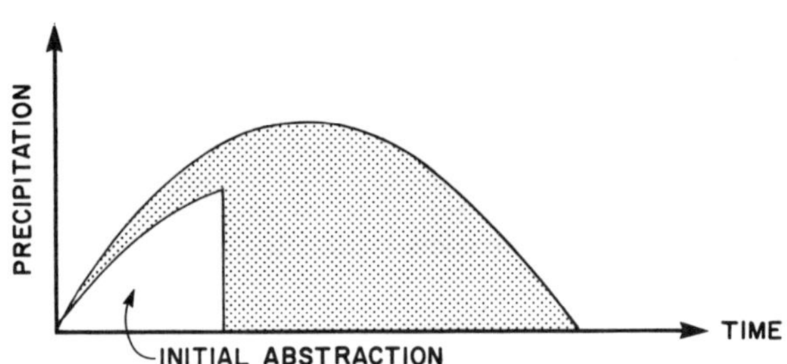

Fig. 3.6b. INITIAL ABSTRACTION MODEL AS A CONSTANT FRACTION OF PRECIPITATION TO A TOTAL DEPTH

the total rainfall depth exceeding Ia. The second Ia model accounts for a fixed percentage of rainfall until the total Ia depth is satisfied. In the second model, however, the loss rate is usually set equal to f(t) or the initial abstraction requirements, whichever is larger.

The first Ia method is a one parameter model whereas the second method is a two parameter model. Use of either Ia method with the previously discussed loss rate models for f(t) can result in up to a 5 parameter effective rainfall estimation model. Other effective rainfall estimation models can be developed, including budgets for accounting groundwater flow, transpiration, and many other effects, resulting in models utilizing a dozen or more calibration parameters. However, the increase in modeling accuracy by using such sophistications is still unclear (see Section 1.6).

It is noted that for the models described, the initial abstraction model can have an important influence in rainfall-runoff studies of storms which have peak rainfall intensities occurring during the early portion of the storm. In such case studies both the loss rate and initial abstraction models are in operation, resulting in the hydrologic model having two similarly acting functions, with up to 5 parameters requiring calibration simultaneously (e.g., for the case of using Eq. (3.4) coupled with the percentage Ia model).

3.9. SCS Loss Separation

The SCS runoff depth equation relates runoff depth to rainfall depth. Specifically, the SCS runoff equation is (see Fig. 3.7)

$$R = (P-Ia)^2/(P-Ia + S) \tag{3.7}$$

in which R and P are the runoff depth (inches) and precipitation depth (inches), respectively; and the maximum watershed retention S is given by:

$$S = (1000/CN) - 10 \tag{3.8}$$

in which CN is a runoff index called the runoff curve number. The difference between the storm event depths of rainfall (P) and runoff (R) represents the total storm event losses.

The total loss is separated into two parts: the initial abstraction Ia and the other losses, or retention. The initial abstraction Ia is related to the CN by the following empirical equation:

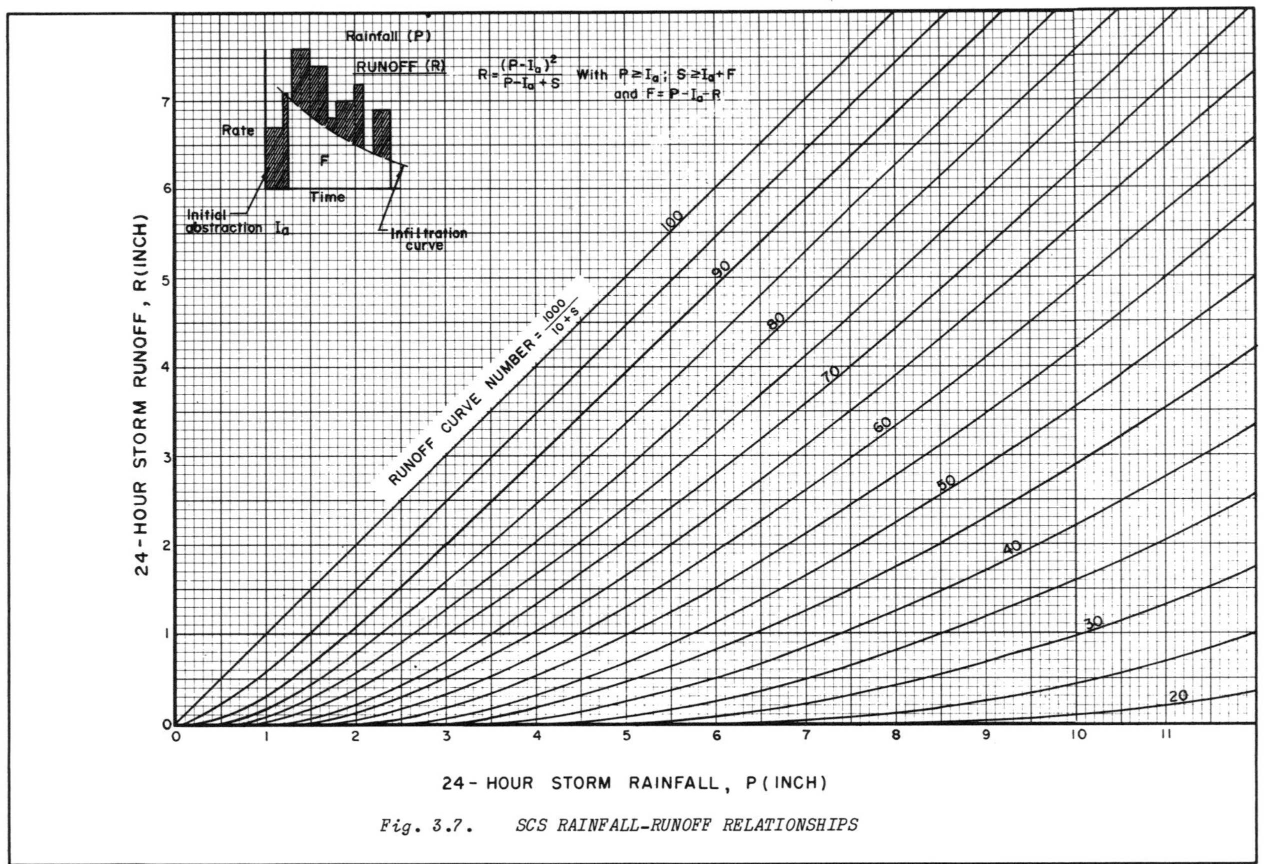

Fig. 3.7. SCS RAINFALL-RUNOFF RELATIONSHIPS

$$Ia = 0.2S \tag{3.9}$$

For large values of S, the initial abstraction can be substantial and may appear to be irrationally large. In spite of this, Eq. 3.9 is inherent in the CNs of Fig. 3.8 and should be used with the CN method and Eq. 3.7.

The SCS runoff equation can be used to estimate incremental runoff quantities dR, as a function of curve number, the accumulated rainfall depth, P, and the incremental rainfall depth, dP, by noting

$$R + dR = (P + dP - Ia)^2/(P + dP - Ia + S) \tag{3.10}$$

where R is the total runoff depth (associated to P) given by Eq. (3.7). Combining Eqs. (3.7) and (3.10) gives the estimate for incremental runoff,

$$dR = (P + dP - Ia)^2/(P + dP - Ia + S) - (P - Ia)^2/(P + Ia + S) \tag{3.11}$$

Given that dP occurs between time t and t+dt, then the associated average rainfall intensity and loss rate are given, respectively, by:

$$\bar{I}(t + dt/2) = dP/dt \tag{3.12a}$$

$$\bar{f}(t + dt/2) = (dP - dR)/dt \tag{3.12b}$$

Where time (t+dt/2) is the midpoint of the timestep, dt, and $\bar{I}$ and $\bar{f}$ are notation for mean values.

3.10. SCS Hydrologic Soil Groups

A major factor affecting infiltration is the condition of the soil itself. The soil surface characteristics, the ability to transmit water through subsurface layers, and the available soil storage capacity are all variables that control the infiltration rate function. The SCS classified more than 4000 soil types into four general categories that provide a general purpose classification of infiltration rates and corresponding runoff rates. These soil groups are defined in Table 3.3.

Soil infiltration rates can be estimated for each of the soil groups by laboratory studies and measurements. Such measurements show that given a sufficient water supply, an initially dry soil will have an associated infiltration rate that essentially decreases with time. Consequently, if the soil is subject to a continuous rainfall rate that exceeds the infiltration capacity, the infiltration loss rate will decrease with increasing storm duration, such as is shown in Fig. 3.1. Depending on the condition and cover of the soil, a relative minimum infiltration

---

TABLE 3.3
SCS HYDROLOGIC SOIL GROUPS

(A): Low runoff potential. Soils having a high infiltration rate even when thoroughly wetted. Consists chiefly of deep, well drained gravels and sands.

(B): Soils having moderate infiltration rates when thoroughly wetted. Consists mainly of moderately deep, well drained soils with moderately fine to moderately coarse texture.

(C): Soils having slow infiltration rates when thoroughly wetted and consisting chiefly of soils with a layer that impedes the downward migration of water, or soils with moderately fine to fine texture.

(D): High runoff potential. Soils having a very low rate of infiltration when thoroughly wetted and consisting chiefly of clay soils with a high swelling potential, soils with a high water table (eliminating soil water storage capacity), soils with a claypan or clay layer at or near the soil surface, or shallow soils over a nearly impervious layer.

TABLE 3.4
MINIMUM INFILTRATION RATES

SCS Soil Group	Infiltration Rate (in./hr.)
A	0.30 - 0.45
B	0.15 - 0.30
C	0.05 - 0.15
D	0.00 - 0.10
Mean value range:	0.125 - 0.25
Mean value:	0.1875

rate for the SCS soil groups are given in Table 3.4 (e.g., U.S Bureau of Reclamation, 1973).

Rallison (1980) gives a detailed account of the origin of the SCS rainfall-runoff equation. The runoff curve numbers were developed by relating daily (24-hour periods) runoff to precipitation records. It is noted that these curves were not designed to be a physical definition of the incremental runoff estimation approach.

In estimating soil infiltration rates, a curve number is determined for each soil cover complex within the watershed. The CN scale has a range of 0 to 100, where a low CN value indicates a low runoff potential and a high CN value indicates a high runoff potential (a CN of 100 indicates total runoff). Selection of the CN value takes into account several major factors affecting the infiltration rate on pervious surfaces including hydrologic soils group, cover type and quality, and the antecedent moisture condition.

3.11. Soil Cover

The type of vegetation or ground cover on a watershed and the quality or density of that cover have a major impact on the infiltration capacity of the soil. A frequently used set of definitions for qualifying the effects of soil cover was developed by the SCS and is based on the three categories given in Table 3.5.

In most cases, cover type and quality can be readily determined by an inspection of the watershed. U.S. Forest Service and SCS personnel may also be helpful in estimation of such information. Cover types, which are listed in Fig. 3.8, relate curve numbers for various cover qualities. General descriptions of cover types are given in Fig. 3.9.

TABLE 3.5
SOIL COVER QUALITY DEFINITIONS

POOR: Heavily grazed or regularly burned ares. Less than 50 percent of the ground surface is protected by plant cover or brush and tree canopy.

FAIR: Moderate cover with 50 percent to 75 percent of the ground surface protected.

GOOD: Heavy or dense cover with more than 75 percent of the ground surface protected.

RUNOFF CURVE NUMBERS FOR URBAN AREAS[1]

Cover Description		Curve Numbers for Hydrologic Soil Group			
Cover Type and Hydrologic Condition	Average Percent Impervious Area[2]	A	B	C	D
Fully developed urban areas (vegetation established)					
Open space (lawns, parks, golf courses, cemeteries, etc.)[3]:					
Poor condition (grass cover 50%)		68	79	86	89
Fair condition (grass cover 50% to 75%)		49	69	79	84
Good condition (grass cover 75%)		39	61	74	80
Impervious areas:					
Paved parking lots, roofs, driveways, etc. (excluding right-of-way)		98	98	98	98
Streets and roads:					
Paved; curbs and storm sewers (excluding right-of-way)		98	98	98	98
Paved; open ditches (including right-of-way)		83	89	92	93
Gravel (including right-of-way)		76	85	89	91
Dirt (including right-of-way)		72	82	87	89
Western desert urban areas:					
Natural desert landscaping (pervious area only)[4]		63	77	85	88
Artificial desert landscaping (impervious weed barrier, desert shrub with 1- to 2-inch sand or gravel mulch and basin borders)		96	96	96	96
Urban districts:					
Commercial and business	85	89	92	94	95
Industrial	72	81	88	91	93
Residential districts by average lot size:					
1/8 acre or less (town houses)	65	77	85	90	92
1/4 acre	38	61	75	83	87
1/3 acre	30	57	72	81	86
1/2 acre	25	54	70	80	85
1 acre	20	51	68	79	84
2 acres	12	46	65	77	82
Developing urban areas					
Newly graded areas (pervious area only, no vegetation)[5]		77	86	91	94
Idle lands (CN's are determined using cover types similar to those in Figure 3.8c)					

1 Average runoff condition, and $I_a = 0.2S$.

2 The average percent impervious area shown was used to develop the composite CN's. Other assumptions are as follows: impervious areas are directly connected to the drainage system, impervious areas have a CN of 98, and pervious areas are considered equivalent to open space in good hydrologic condition. CN's for other combinations of conditions may be computed using Fig. 3.11.

3 CN's shown are equivalent to those of pasture. Composite CN's may be computed for other combinations of open space cover type.

4 Composite CN's for natural desert landscaping should be computed using Fig. 3.11 based on the impervious area percentage (CN = 98) and the pervious area CN. The pervious area CN's are assumed equivalent to desert shrub in poor hydrologic condition.

5 Composite CN's to use for the design of temporary measures during grading and construction should be computed using Fig. 3.11, based on the degree of development (impervious area percentage) and the CN's for the newly graded pervious areas.

Fig. 3.8a. (REF. SCS-TR-55, SECOND ED., JUNE 1986)

RUNOFF CURVE NUMBERS FOR CULTIVATED AGRICULTURAL LANDS[1]

Cover Description			Curve Numbers for Hydrologic Soil Group			
Cover Type	Treatment[2]	Hydrologic Condition[3]	A	B	C	D
Fallow	Bare soil	----	77	86	91	94
	Crop residue cover (CR)	Poor	76	85	90	93
		Good	74	83	88	90
Row Crops	Straight row (SR)	Poor	72	81	88	91
		Good	67	78	85	89
	SR + CR	Poor	71	80	87	90
		Good	64	75	82	85
	Contoured (C)	Poor	70	79	84	88
		Good	65	75	82	86
	C + CR	Poor	69	78	83	87
		Good	64	74	81	85
	Contoured & terraced (C&T)	Poor	66	74	80	82
		Good	62	71	78	81
	C&T + CR	Poor	65	73	79	81
		Good	61	70	77	80
Small Grain	SR	Poor	65	76	84	88
		Good	63	75	83	87
	SR + CR	Poor	64	75	83	86
		Good	60	72	80	84
	C	Poor	63	74	82	85
		Good	61	73	81	84
	C + CR	Poor	62	73	81	84
		Good	60	72	80	83
	C&T	Poor	61	72	79	82
		Good	59	70	78	81
	C&T + CR	Poor	60	71	78	81
		Good	58	69	77	80
Close-Seeded or Broadcast Legumes or Rotation Meadow	SR	Poor	66	77	85	89
		Good	58	72	81	85
	C	Poor	64	75	83	85
		Good	55	69	78	83
	C&T	Poor	63	73	80	83
		Good	51	67	76	80

1 Average runoff condition, and $I_a = 0.2S$.

2 Crop residue cover applies only if residue is on at least 5% of the surface throughout the year.

3 Hydrologic condition is based on combination of factors that affect infiltration and runoff, including (a) density and canopy of vegetative areas, (b) amount of year-round cover, (c) amount of grass or close-seeded legumes in rotations, (d) percent of residue cover on the land surface (good 20%), and (e) degree of surface roughness.

Poor: Factors impair infiltration and tend to increase runoff.

Good: Factors encourage average and better than average infiltration and tend to decrease runoff.

Fig. 3.8b.

RUNOFF CURVE NUMBERS FOR OTHER AGRICULTURAL LANDS[1]

Cover Description		Curve Numbers for Hydrologic Soil Group			
Cover Type	Hydrologic Condition	A	B	C	D
Pasture, grassland, or range -- continuous forage for grazing.[2]	Poor	68	79	86	89
	Fair	49	69	79	84
	Good	39	61	74	80
Meadow -- continuous grass, protected from grazing and generally mowed for hay.	----	30	58	71	78
Brush — brush-weed-grass mixture with brush the major element.[3]	Poor	48	67	77	83
	Fair	35	56	70	77
	Good	30[4]	48	65	73
Woods -- grass combination (orchard or tree farm).[5]	Poor	57	73	82	86
	Fair	43	65	76	82
	Good	32	58	72	79
Woods.[6]	Poor	45	66	77	83
	Fair	36	60	73	79
	Good	30[4]	55	70	77
Farmsteads — buildings, lanes, driveways, and surrounding lots.	----	59	74	82	86

1 Average runoff condition, and $I_a = 0.2S$.

2 Poor: 50% ground cover or heavily grazed with no mulch.
Fair: 50 to 75% ground cover and not heavily grazed.
Good: 75% ground cover and lightly or only occasionally grazed.

3 Poor: 50% ground cover.
Fair: 50 to 75% ground cover.
Good: 75% ground cover.

4 Actual curve number is less than 30; use CN = 30 for runoff computations.

5 CN's shown were computed for areas with 50% woods and 50% grass (pasture) cover. Other combinations of conditions may be computed from the CN's for woods and pasture.

6 Poor: Forest litter, small trees, and brush are destroyed by heavy grazing or regular burning.
Fair: Woods are grazed but not burned, and some forest litter covers the soil.
Good: Woods are protected from grazing, and litter and brush adequately cover the soil.

Fig. 3.8c.

RUNOFF CURVE NUMBERS FOR ARID AND SEMIARID RANGELANDS[1]

Cover Description		Curve Numbers for Hydrologic Soil Group			
Cover Type	Hydrologic Condition[2]	A[3]	B	C	D
Herbaceous -- mixture of grass, weeds, and low-growing brush, with brush the minor element.	Poor		80	87	93
	Fair		71	81	89
	Good		62	74	85
Oak-aspen -- mountain brush mixture of oak brush, aspen, mountain mahogany, bitter brush, maple, and other brush.	Poor		66	74	79
	Fair		48	57	63
	Good		30	41	48
Pinyon-juniper -- pinyon, juniper, or both; grass understory.	Poor		75	85	89
	Fair		58	73	80
	Good		41	61	71
Sagebrush with grass understory.	Poor		67	80	85
	Fair		51	63	70
	Good		35	47	55
Desert shrub -- major plants include saltbush, greasewood, creosotebush, blackbrush, bursage, palo verde, mesquite, and cactus.	Poor	63	77	85	88
	Fair	55	72	81	86
	Good	49	68	79	84

1 Average runoff condition, and $I_a = 0.2S$. For range in humid regions, use Fig. 3.8c.

2 Poor: 30% ground cover (litter, grass, and brush overstory).
Fair: 30 to 70% ground cover.
Good: 70% ground cover.

3 Curve numbers for Group A have been developed only for desert shrub.

Fig. 3.8d.

Residential Landscaping (Lawn, Shrubs, etc.) - The pervious portions of commercial establishments, single and multiple family dwellings, trailer parks and schools where the predominant land cover is lawn, shrubbery and trees.

Row Crops - Lettuce, tomatoes, beets, tulips or any field crop planted in rows far enough apart that most of the soil surface is exposed to rainfall impact throughout the growing season. At plowing, planting and harvest times it is equivalent to fallow.

Small Grain - Wheat, oats, barley, flax, etc. planted in rows close enough that the soil surface is not exposed except during planting and shortly thereafter.

Legumes - Alfalfa, sweetclover, timothy, etc. and combinations are either planted in close rows or broadcast.

Fallow - Fallow land is land plowed but not yet seeded or tilled.

Woodland - grass - Areas with an open cover of broadleaf or coniferous trees usually live oak and pines, with the intervening ground space occupied by annual grasses or weeds. The trees may occur singly or in small clumps. Canopy density, the amount of ground surface shaded at high noon, is from 20 to 50 percent.

Woodland - Areas on which coniferous or broadleaf trees predominate. The canopy density is at least 50 percent. Open areas may have a cover of annual or perennial grasses or of brush. Herbaceous plant cover under the trees is usually sparse because of leaf or needle litter accumulation.

Chaparral - Land on which the principal vegetation consists of evergreen shrubs with broad, hard, stiff leaves such as manzonita, ceanothus and scrub oak. The brush cover is usually dense or moderately dense. Diffusely branched evergreen shrubs with fine needle-like leaves, such as chamise and redchank, with dense high growth are also included in this soil cover.

Annual Grass - Land on which the principal vegetation consists of annual grasses and weeds such as annual bromes, wild barley, soft chess, ryegrass and filaree.

Irrigated Pasture - Irrigated land planted to perennial grasses and legumes for production of forage and which is cultivated only to establish or renew the stand of plants. Dry land pasture is considered as annual grass.

Meadow - Land areas with seasonally high water table, locally called cienegas. Principal vegetation consists of sod-forming grasses interspersed with other plants.

Orchard (Deciduous) - Land planted to such deciduous trees as apples, apricots, pears, walnuts, and almonds.

Orchard (Evergreen) - Land planted to evergreen trees which include citrus and avocados and coniferous plantings.

Turf - Golf courses, parks and similar lands where the predominant cover is irrigated mowed close-grown turf grass. Parks in which trees are dense may be classified as woodland.

Fig. 3.9. SCS COVER TYPE DESCRIPTIONS

3.12. Watershed Development Conditions

Should the flood control facilities be expected to provide the public protection for an extended period, then the maximum urbanization should be assumed in determining the watershed loss rates. All available long range urbanization master plans should be examined to ensure that reasonable land use assumptions are included in the analysis. Particular attention should be directed to local landscape practices. For example, it is common to use ornamental gravels underlain by impervious plastic materials as a substitute for lawns and shrubs, resulting in an increase in the effective impervious percentage.

3.13. Impervious Areas

In the analysis of urban watersheds, the effects of the impervious surfaces on the assumed area-averaged infiltration rate for the entire watershed must be included. Estimated ranges of impervious percentages for various types of land use are given in Fig. 3.8. Values given are for the actual percentage of area covered by an impervious surface; however, the actual impervious area may need to be reduced due to the local drainage practices. For example, runoff from an impervious surface may drain across a pervious surface where infiltration may take place. To account for this infiltration, the actual impervious area may be reduced by a factor such as 10 percent. This type of adjustment is generally included in the computation of runoff coefficients for use in the rational method and in the estimation of watershed loss rates for use in the synthetic unit hydrograph method.

In using the SCS methods, the effects of impervious land covers are handled through adjustments to the CN and the time of concentration. The definition of CNs (Fig. 3.8) includes CNs for both urban and rural/natural land covers; however, the CNs for the urban land covers are based on specific percentages of imperviousness and are accurate only when these stated percentages are representative of actual watershed conditions. Where the stated percentages of imperviousness are not accurate, the CNs can be adjusted using Fig. 3.10. The pervious CN is obtained from the CN table (Fig. 3.8) and used with Fig. 3.10 to obtain the composite CN. Fig. 3.10 is used when all or most of the impervious cover is connected directly to the drainage system. In some localities, drainage policies require parts of the impervious cover to be directed to pervious areas where the water is spread as sheet flow over the pervious area. For example, rooftop runoff can be directed to adjacent grassy areas. Where the impervious

ACTUAL IMPERVIOUS COVER		
Land Use (1)	Range-Percent	Recommended Value For Average Conditions-Percent (2)
Natural or Agriculture	0 - 10	0
Single Family Residential: (3)		
100,000 SF (2.5 acre) Lots	5 - 15	10
40,000 SF (1 acre) Lots	10 - 25	20
20,000 SF (1/2 acre) Lots	30 - 45	40
7,200 - 10,000 SF Lots	45 - 55	50
Multiple Family Residential:		
Condominiums	45 - 70	65
Apartments	65 - 90	80
Mobile Home Park	60 - 85	75
Commercial, Downtown Business or Industrial	80 - 100	90

Notes:

1. Land use should be based on ultimate development of the watershed. Long range master plans for the County and incorporated cities should be reviewed to insure reasonable land use assumptions.

2. Recommended values are based on average conditions which may not apply to a particular study area. The percentage impervious may vary greatly even on comparable sized lots due to differences in dwelling size, improvements, etc. Landscape practices should also be considered as it is common in some areas to use ornamental gravels underlain by impervious plastic materials in place of lawns and shrubs. A field investigation of a study area should always be made, and a review of aerial photos, where available, may assist in estimating the percentage of impervious cover in developed areas.

3. For typical horse ranch subdivisions increase impervious area 5 percent over the values recommended in the table above.

Fig. 3.10. ACTUAL PERCENT IMPERVIOUS

land cover is unconnected, the composite CN can be obtained from Fig. 3.11 using the pervious area CN, the percentage of imperviousness, and the percentage of the imperviousness that is unconnected. Figure 3.11 should only be used when the total imperviousness is less than 30 percent; the figure should not be extrapolated for percentages greater than 30 percent.

3.14. PROGRAM 3.1: Subroutine SLOSS

The effective rainfall distribution is determined by modifying the precipitation values developed in program RAIN according to a selected loss function in program SLOSS.

3.14.1. Initial Abstraction Models

The initial abstraction (Ia) is a function of land use, treatment, and condition; interception; infiltration; depression storage; and antecedent soil moisture. Three options are described in the following text.

3.14.1.1. SCS Model

An empirical analysis was performed for the development of the SCS rainfall-runoff relation (see Section 3.9), and the following relationship is available for estimating Ia:

$$Ia = kS$$

where k is a constant, and S can be estimated by:

$$S = (1000/CN) - 10$$

in which CN is the SCS runoff curve number. The SCS recommends setting k = 0.2.

3.14.1.2. Total Depth of Abstraction

The program user specifies a total depth of initial abstraction (inches) such that no runoff is developed until the Ia budget is satisfied.

3.14.1.3. Percentage of Precipitation to a Maximum Total Depth of Abstraction

The program user specifies a percentage of precipitation as initial abstraction, but the total abstraction depth is bounded by a specified total depth of abstraction.

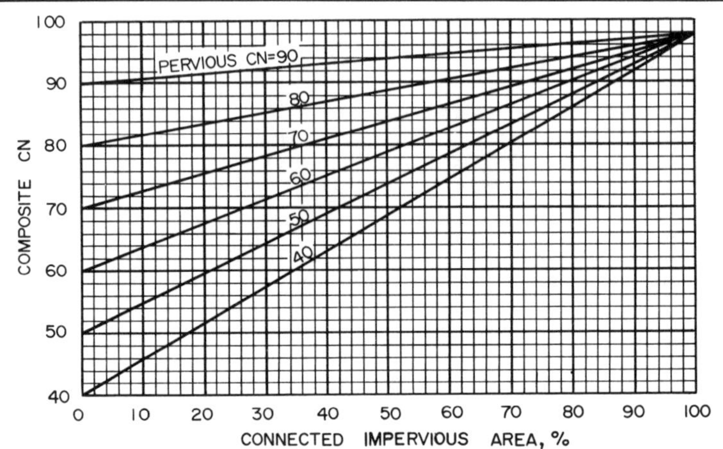

Fig. 3.11a. COMPOSITE CN WITH CONNECTED IMPERVIOUS AREA

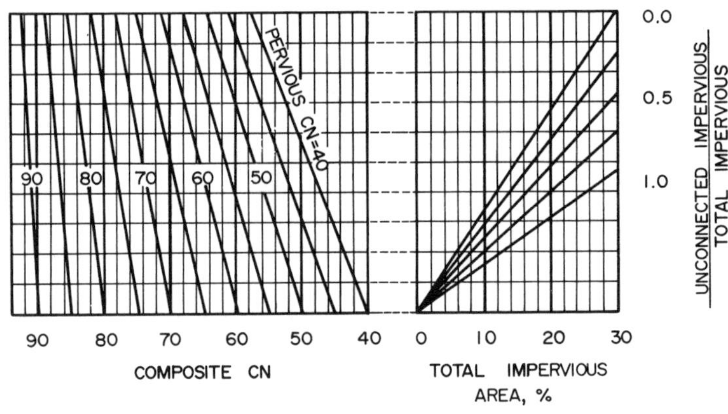

Fig. 3.11b. COMPOSITE CN WITH UNCONNECTED IMPERVIOUS AREAS AND TOTAL IMPERVIOUS AREA LESS THAN 30%

3.14.2. Loss Models

Six loss models are available in program SLOSS. These models can be coupled to the initial abstraction models of the previous section, except that the SCS models are uniquely coupled.

3.14.2.1. SCS Model

The SCS model assumes the following rainfall-runoff relation:

$$R = \begin{cases} (P-Ia)^2/(P-Ia + S), & \text{for } P > Ia \\ 0, & \text{otherwise} \end{cases}$$

where R is the cumulative amount of runoff (inches), P is the cumulative amount of rainfall (inches), Ia is the initial abstraction (inches), and S is the potential maximum retention (inches). The SCS initial abstraction is coupled to the SCS loss model.

3.14.2.2. Phi-Index

The phi-index is defined as the rate of rainfallabove which the rainfall volume equals the runoff volume (Fig. 3.12).

3.14.2.3. Percentage of Rainfall

This loss rate is defined as a simple percentage of the rainfall rate (Fig. 3.13).

3.14.2.4. Percentage of Rainfall Limited by a phi-index

The loss rate is based on the percentage of rainfall rate, but not to exceed the phi-index value (Fig. 3.14).

3.14.2.5. Horton Equation

An often used infiltration equation in soil-water modeling is the Horton equation (Fig. 3.15):

$$f = fc + (fo - fc)e^{-kt}$$

where fo and fc are the initial and ultimate infiltration rates (inches/hour), and k is an exponential delay coefficient to be evaluated by field experiments.

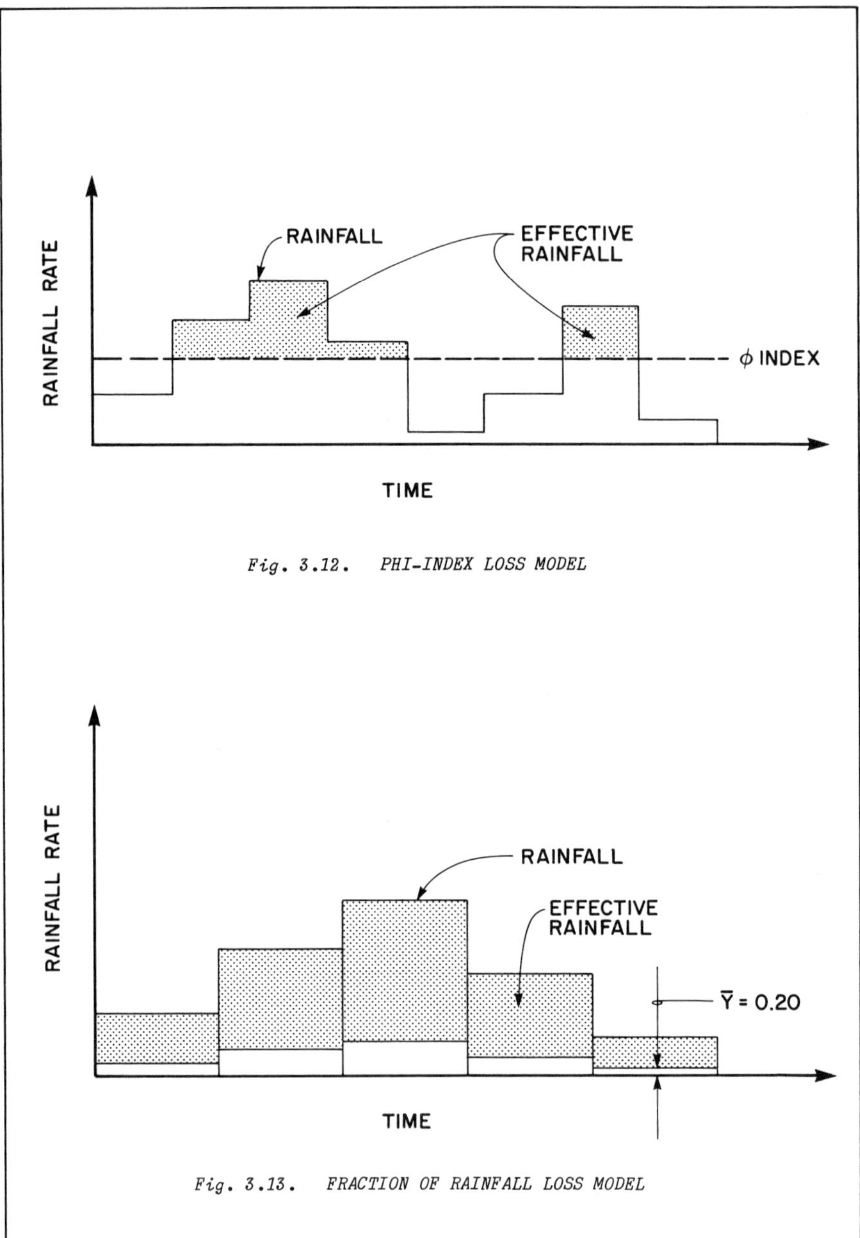

Fig. 3.12. PHI-INDEX LOSS MODEL

Fig. 3.13. FRACTION OF RAINFALL LOSS MODEL

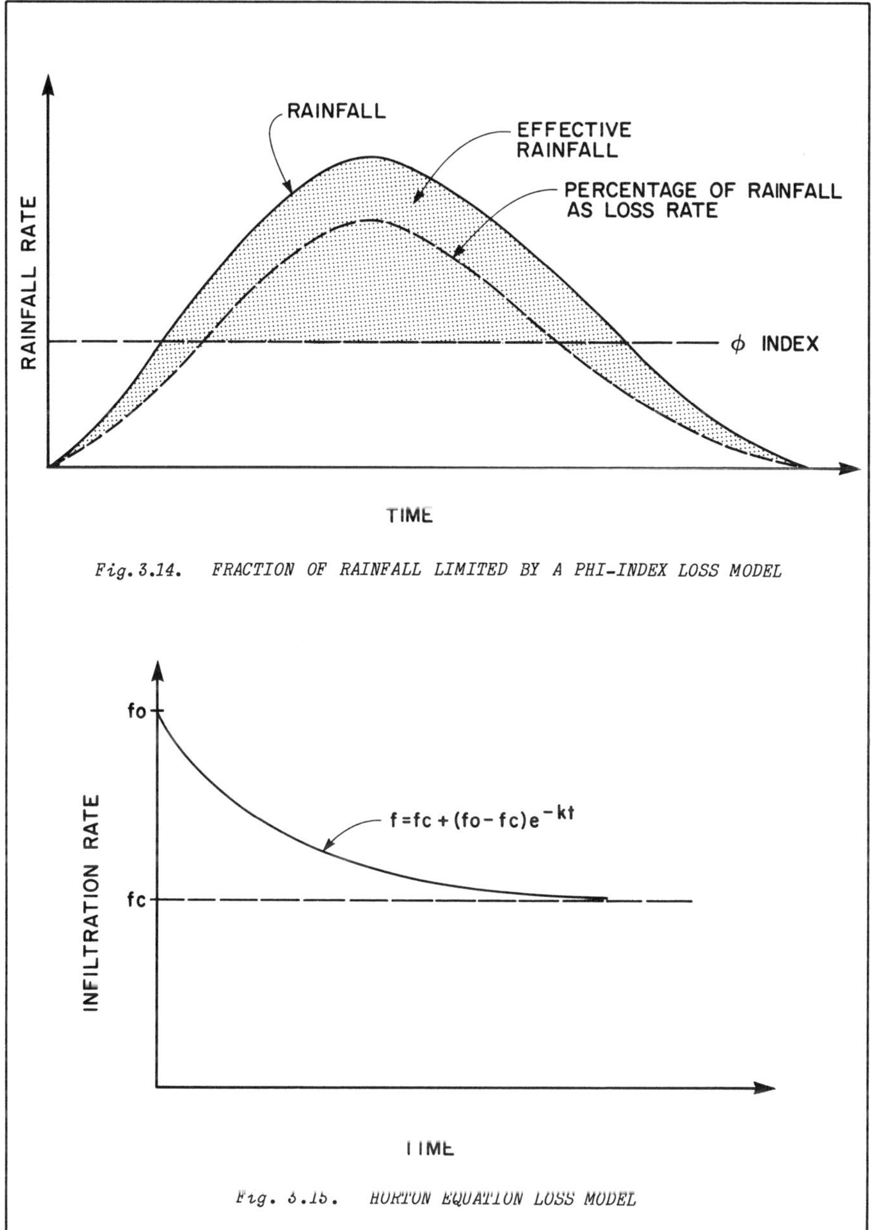

Fig. 3.14. FRACTION OF RAINFALL LIMITED BY A PHI-INDEX LOSS MODEL

Fig. 3.15. HORTON EQUATION LOSS MODEL

3.14.2.6. Exponential of Rainfall Intensity

This loss rate is approximated by the following equation:

$$f = A\,[I(t)]^E$$

where A and E are constants estimated from field experiments, and I(t) is the rainfall intensity at time t.

3.14.3. Applications of Loss Function Models

Figure 3.16 depicts the effective rainfall resulting from an initial abstraction model based on a total depth of 1 inch. Figure 3.17 depicts the effective rainfall resulting from the initial abstraction specified as 50 percent of the rainfall to a maximum of 1 inch depth. Figures 3.18 through 3.22 depict the total rainfall and effective rainfall (30-minute intervals) for five different loss functions: (1). Phi-index (0.1 inches/hour), (2). Percentage of rainfall (10%), (3).Percentage of rainfall (10%) limited by phi-index (0.1 inches/hour), (4). Horton equation (f(t) = 0.1 + (0.4 - 0.1) Exp [-.1*Time (hours)]), and (5). Exponential loss function (f(t) = 0.1* (Intensity)**0.5). Table 3.6 shows the depths of rainfall and soil-loss in 3-minute unit intervals for storm hours 3 through 5. Figure 3.23 depicts the effective rainfall for curve number (CN) of 85 and a coefficient of initial abstraction, k = 0.2. Table 3.7 summarizes the total runoff volume and total soil-loss volume for the different soil-loss functions applied to an example 24-hour storm pattern, and a 10 square-mile watershed.

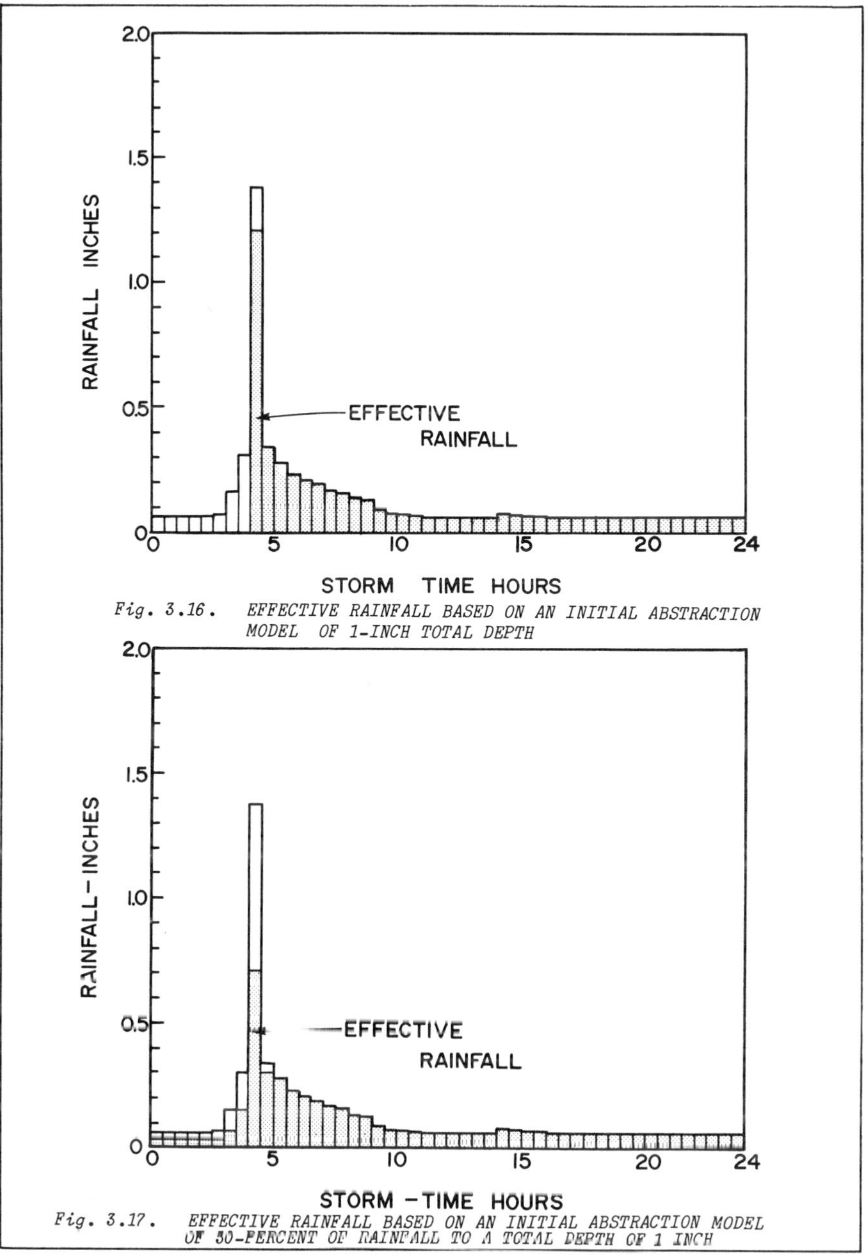

Fig. 3.16. EFFECTIVE RAINFALL BASED ON AN INITIAL ABSTRACTION MODEL OF 1-INCH TOTAL DEPTH

Fig. 3.17. EFFECTIVE RAINFALL BASED ON AN INITIAL ABSTRACTION MODEL OF 50-PERCENT OF RAINFALL TO A TOTAL DEPTH OF 1 INCH

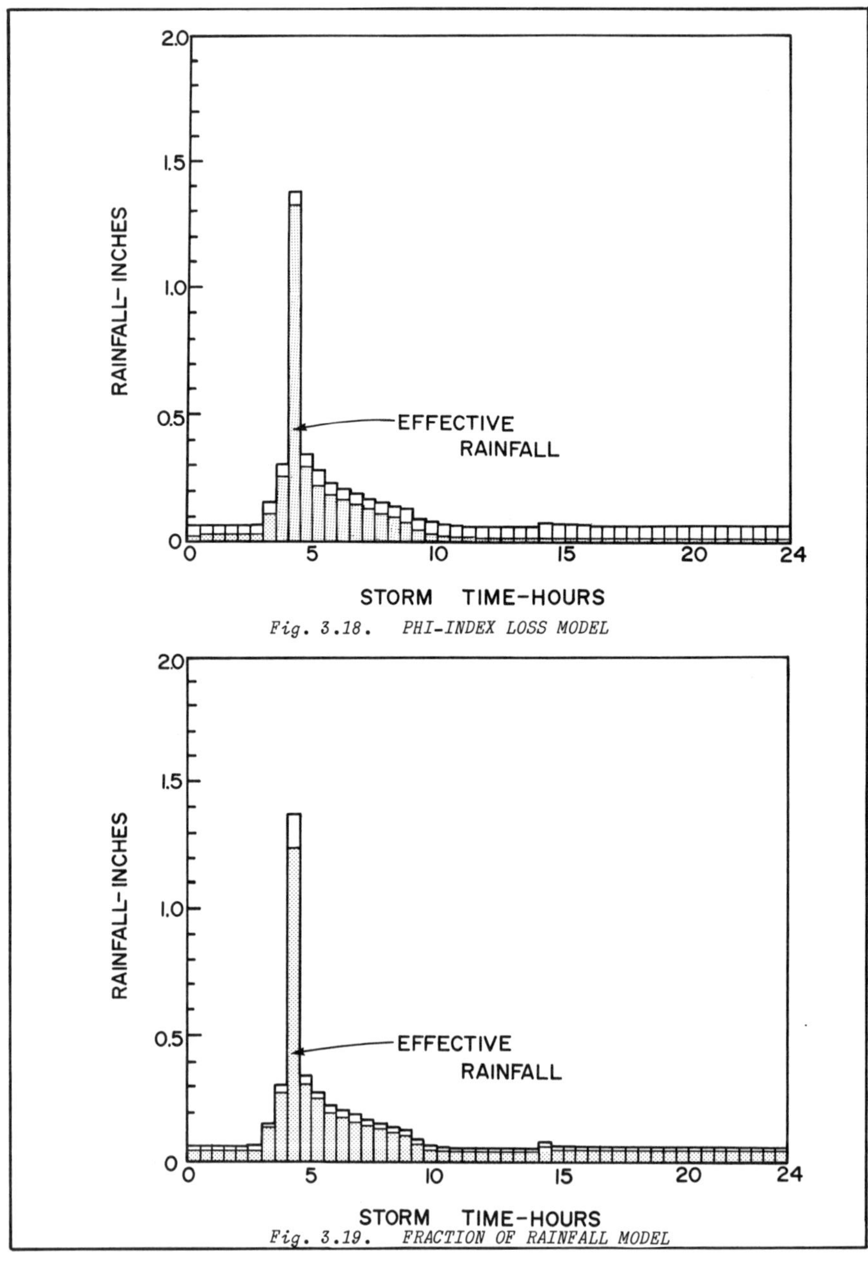

Fig. 3.18. PHI-INDEX LOSS MODEL

Fig. 3.19. FRACTION OF RAINFALL MODEL

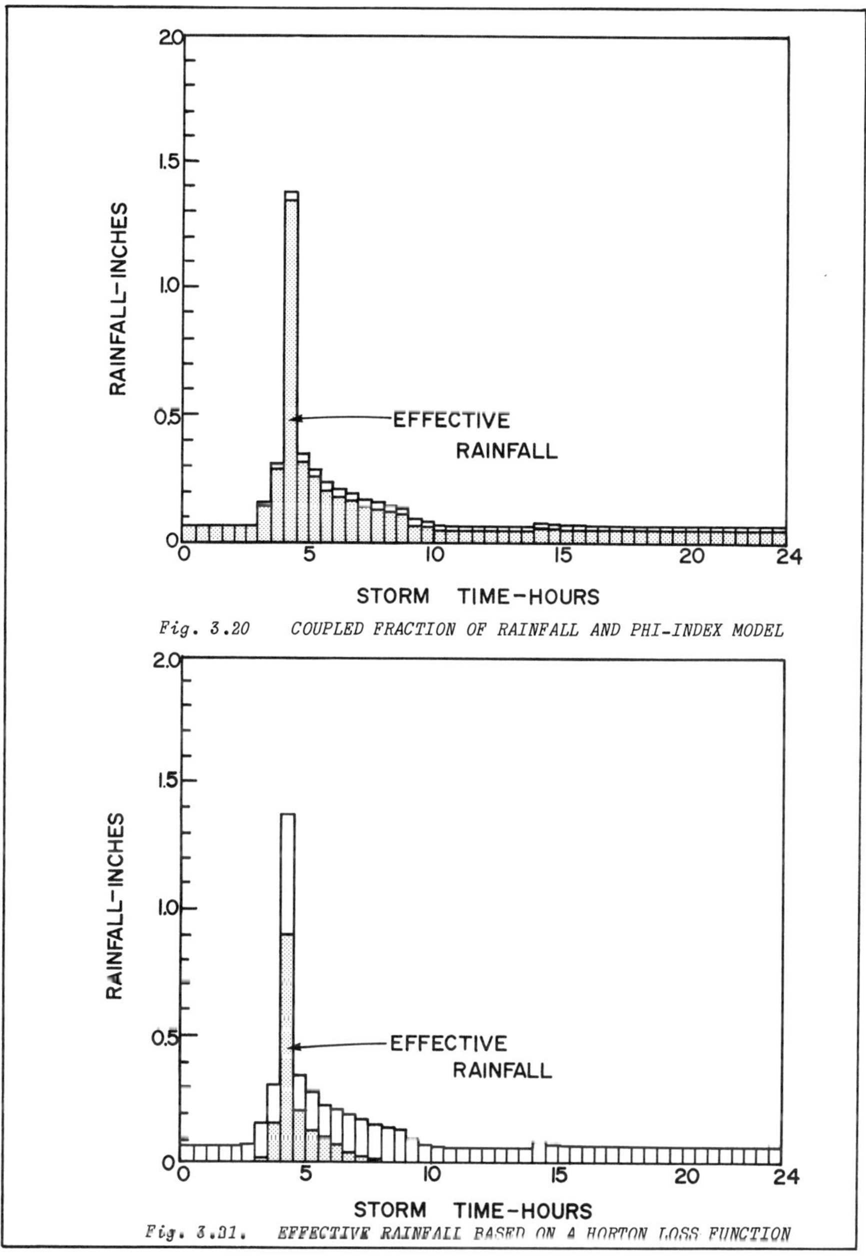

Fig. 3.20 COUPLED FRACTION OF RAINFALL AND PHI-INDEX MODEL

Fig. 3.21. EFFECTIVE RAINFALL BASED ON A HORTON LOSS FUNCTION

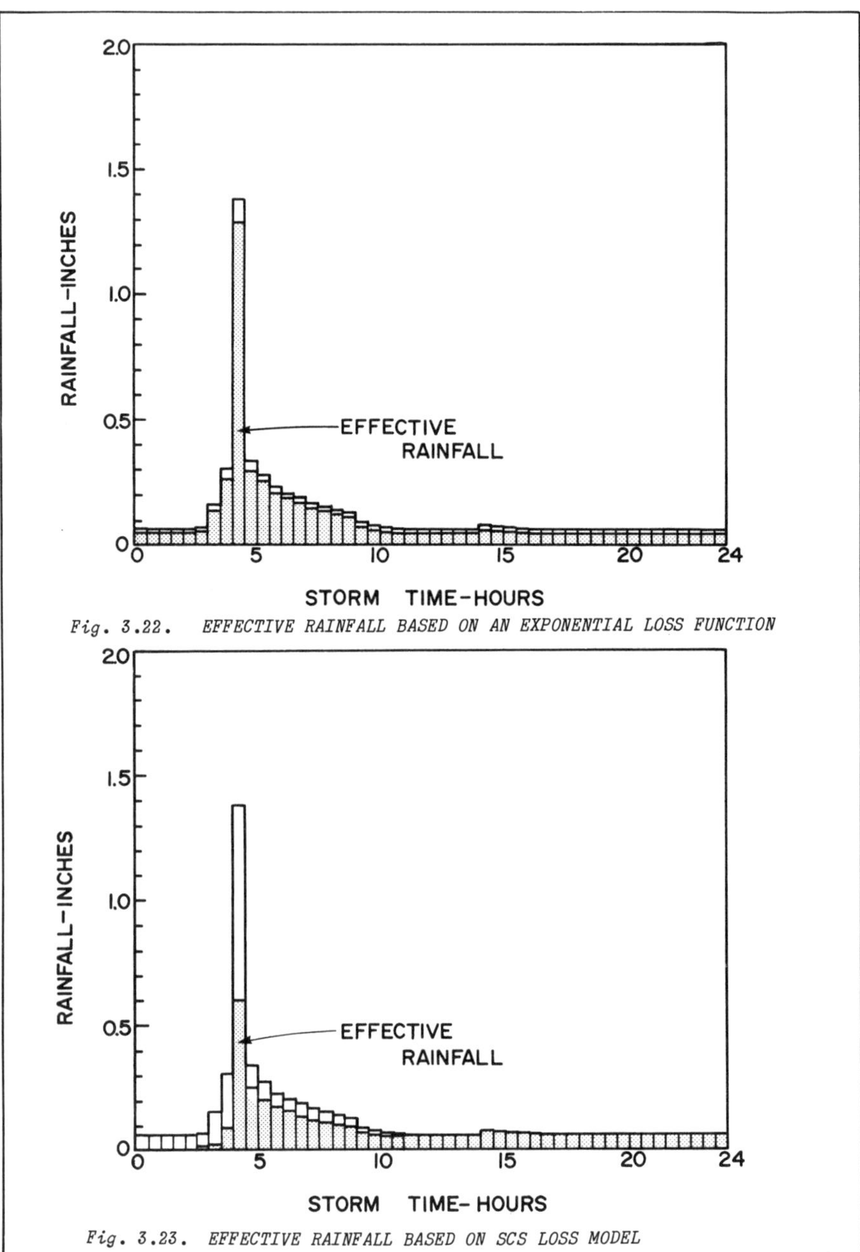

Fig. 3.22. EFFECTIVE RAINFALL BASED ON AN EXPONENTIAL LOSS FUNCTION

Fig. 3.23. EFFECTIVE RAINFALL BASED ON SCS LOSS MODEL

TABLE 3.6
SOIL-LOSS MODELS

		Soil-Loss Model				
(Hour)	(inches/hour)	1	2	3	4	5
3.00	.0087	.065	.0009	.0009	.0087	.0021
3.05	.0131	.005	.0013	.0013	.0131	.0026
3.10	.0135	.005	.0013	.0013	.0135	.0026
3.15	.0139	.005	.0014	.0014	.0139	.0026
3.20	.0143	.005	.0014	.0014	.0143	.0027
3.25	.0148	.005	.0015	.0015	.0148	.0027
3.30	.0153	.005	.0015	.0015	.0153	.0028
3.35	.0159	.005	.0016	.0016	.0157	.0028
3.40	.0165	.005	.0017	.0017	.0157	.0029
3.45	.0172	.005	.0017	.0017	.0156	.0029
3.50	.0181	.005	.0018	.0018	.0156	.0030
3.55	.0195	.005	.0020	.0020	.0155	.0031
3.60	.0206	.005	.0021	.0021	.0155	.0032
3.65	.0220	.005	.0022	.0022	.0154	.0033
3.70	.0236	.005	.0024	.0024	.0154	.0034
3.75	.0247	.005	.0025	.0025	.0153	.0035
3.80	.0273	.005	.0027	.0027	.0153	.0037
3.85	.0309	.005	.0031	.0031	.0152	.0039
3.90	.0323	.005	.0032	.0032	.0152	.0040
3.95	.0413	.005	.0041	.0041	.0151	.0045
4.00	.0637	.005	.0064	.0050	.0151	.0056
4.05	.4209	.005	.0421	.0050	.0150	.0145
4.10	.1401	.005	.0140	.0050	.0150	.0084
4.15	.1026	.005	.0103	.0050	.0149	.0072
4.20	.0840	.005	.0084	.0050	.0149	.0065
4.25	.0724	.005	.0072	.0050	.0148	.0060
4.30	.0577	.005	.0058	.0050	.0148	.0054
4.35	.0531	.005	.0053	.0050	.0147	.0052
4.40	.0493	.005	.0049	.0049	.0147	.0050
4.45	.0462	.005	.0046	.0046	.0146	.0048
4.50	.0436	.005	.0044	.0044	.0146	.0047
4.55	.0394	.005	.0039	.0039	.0145	.0044
4.60	.0376	.005	.0038	.0038	.0145	.0043
4.65	.0361	.005	.0036	.0036	.0144	.0042
4.70	.0347	.005	.0035	.0035	.0144	.0042
4.75	.0034	.005	.0033	.0033	.0143	.0041
4.80	.0313	.005	.0031	.0031	.0143	.0040
4.85	.0303	.005	.0030	.0030	.0142	.0039
4.90	.0332	.005	.0033	.0033	.0142	.0041
4.95	.0324	.005	.0032	.0032	.0141	.0040
5.00	.0316	.005	.0032	.0032	.0141	.0040

Note:
1. ϕ = 0.1 inches/hour
2. 10% of rainfall intensity
3. 10% of rainfall intensity limited by 0.1 inches/hour
4. f(cfs) = 0.1 + (0.4 - 0.1) exp [- .1* TIME (hour)]
5. f(cfs) = 0.1* (Rainfall Intensity)**.5

TABLE 3.7
TOTAL RUNOFF VOLUME FOR DIFFERENT SOIL-LOSS MODELS

Runoff Model	Runoff Volume (A.F.)
. without depth-area adjustment	3000.
. with depth-area adjustment	2970.
Initial Abstraction Model	
. Total depth (1 inch)	2470.
. Percentage of rainfall (50%) limited by total depth (1 inch)	2470.
Soil-loss Model	
. Phi-index (0.1 inch/hour)	1720.
. Percentage of rainfall (10%)	2700.
. Percentage of rainfall (10%) limited by phi-index (0.1 inch/hour)	2732.
. Horton equation: f(t)=0.1+(0.4-0.1)exp[-0.1*Time(hour)]	940.
. Exponential function: f(t)=0.1(Rainfall Intensity)**0.5	2440.
. SCS Model (CN = 85, k = 0.2)	2110.

3.14.4. Data Entry Sequences

Four options for the initial abstraction model are available as follows:

```
*** INITIAL ABSTRACTION MODEL ***
   1 = S.C.S. MODEL (APPLIES TO S.C.S. STORM PATTERN ONLY)
   2 = TOTAL DEPTH OF INITIAL ABSTRACTION
   3 = PERCENT OF PRECIPITATION TO A MAXIMUM TOTAL DEPTH OF
       INITIAL ABSTRACTION
   4 = BYPASS THE PROCESS
=====> 'KOPT'
```

3.14.4.1. SCS Model

```
   1 = S.C.S. MODEL
   2 = MODIFIED S.C.S. MODEL
=====> 'KOPTT'
   ENTER THE CURVE NUMBER BETWEEN [0] AND [100]
=====> 'CN'
   ENTER THE COEFFICIENT k (Ia = kS) BETWEEN [0] AND [.9]
=====> 'XFACT'
```

If (KOPTT.EQ.1), then

```
   ENTER THE MAXIMUM LOSS RATE (INCHES/HOUR)
=====> 'FMAX'
   ENTER THE MINIMUM LOSS RATE (INCHES/HOUR)
   ALLOWABLE VALUE BETWEEN ['FMAX'] AND [0]
=====> 'FMIN'
```

3.14.4.2. Total Depth of Initial Abstraction

The total depth (inches) of rainfall is entered as the initial abstraction.

3.14.4.3. Percent of Precipitation to a Maximum Total Depth of Initial Abstraction

The percentage of the rainfall and a maximum total depth (inches) of rainfall are entered.

After an initial abstraction model is selected, five options for the watershed loss model are available as follows:

```
*** WATERSHED LOSS MODEL ***
   1 = PHI INDEX
   2 = PERCENTAGE OF RAINFALL
   3 = PERCENTAGE OF RAINFALL LIMITED BY A PHI INDEX
   4 = HORTON EQUATION (F=FC+(F0-FC)*EXP(-K*TIME))
   5 = C1*(RAINFALL INTENSITY)**C2
=====>  'KLOSS'
```

3.14.4.4. Phi Index

Data requirements: The phi index (inches/hour).

3.14.4.5. Percentage of Rainfall

Enter the decimal value that the product with the rainfall intensity will be used as the watershed loss rate function.

3.14.4.6. Percentage of Rainfall Limited by a Phi Index

Both the phi index (inches/hour) and the decimal equivalent for the percentage of rainfall are entered.

3.14.4.7. Horton Equation

The initial infiltration rate (inches/hour), the final infiltration rate (inches/hour), and the exponential timing coefficient are entered as requested by the CRT.

3.14.4.8. Exponential Loss Function

The loss rate coefficient and the rainfall intensity exponential coefficient are entered as requested by the CRT.

3.15. PROGRAM 3.1

```
C
C-----------------------------------------------------------------
C     THIS SUBROUTINE CALCULATES THE LOSS RATE
C-----------------------------------------------------------------
C
      SUBROUTINE SLOSS(KKST,KRUN,KRD,KWT)
C
      COMMON/BLK 1/ PRECIP(480)
      COMMON/BLK 10/ AREA
C
 461  IF(KRUN.EQ.1)WRITE(KWT,43)
 43   FORMAT(/,2X,'ENTER A [1] TO SUPPRESS THE STORM PATTERN',/,
     1 5X,'OR A [2] TO OUTPUT THE STORM PATTERN')
      IF(KRUN.EQ.1)WRITE(KWT,99)
      READ(KRD,*)KKST
      IF(KKST.LT.1 .OR. KKST.GT.2)GO TO 461
      IF(KRUN.EQ.1)WRITE(10,*)KKST
C
C  INITIAL ABSTRACTION
C
      ABSD=0.
      ABSP=0.
      XXSUM=0.
 100  IF(KRUN.EQ.1)WRITE(KWT,51)
 51   FORMAT(24(/),2X,'*** INITIAL ABSTRACTION MODEL ***'/,
     1 5X,'1 = S.C.S. MODEL (APPLIES TO S.C.S. STORM PATTERN ONLY)',
     1 /5X,'2 = TOTAL DEPTH OF INITIAL ABSTRACTION',
     2 /,5X,'3 = PERCENT OF PRECIPITATION TO A MAXIMUM TOTAL ',
     3 'DEPTH OF INITIAL ABSTRACTION',/,5X,'4 = BYPASS THE PROCESS')
      IF(KRUN.EQ.1)WRITE(KWT,99)
 99   FORMAT(2X,'=====> ',$)
      READ(KRD,*)KOPT
      IF(KOPT.LT.1 .OR. KOPT.GT.4)GO TO 100
      IF(KRUN.EQ.1)WRITE(10,*)KOPT
      IF(KOPT.EQ.1)GO TO 110
      IF(KOPT.EQ.2)GO TO 120
      IF(KOPT.EQ.3)GO TO 120
      IF(KOPT.EQ.4)GO TO 150
 120  IF(KRUN.EQ.1)WRITE(KWT,53)
 53   FORMAT(/,2X,'ENTER TOTAL DEPTH (INCHES) FOR INITIAL',
     1 ' ABSTRACTION')
      IF(KRUN.EQ.1)WRITE(KWT,99)
      READ(KRD,*)ABSD
      IF(KRUN.EQ.1)WRITE(10,*)ABSD
      IF(KOPT.EQ.2)GO TO 135
 125  IF(KRUN.EQ.1)WRITE(KWT,54)
 54   FORMAT(/'  ENTER THE DECIMAL PROPORTION OF PRECIPITATION TO A ',
     1 'MAXIMUM DEPTH'/,5X,'NOTE: ALLOWABLE VALUE BETWEEN ',
     1 '[0] AND [1]')
```

```
      IF(KRUN.EQ.1)WRITE(KWT,99)
      READ(KRD,*)ABSP
      IF(ABSP.LT.0. .OR. ABSP.GT.1.)GO TO 125
      IF(KRUN.EQ.1)WRITE(10,*)ABSP
C
 135  IF(KRUN.EQ.1)GO TO 150
      WRITE(KWT,401)ABSD
 401  FORMAT(/,2X,'DEPTH OF INITIAL ABSTRACTION = ',F6.3,' INCHES')
      IF(ABSP.NE.0.)WRITE(KWT,402)ABSP
 402  FORMAT(2X,'PERCENT OF PRECIPITATION TO A MAXIMUM DEPTH OF ',
     C 'ABSTRACTION = ',F6.3)
C
C  WATERSHED LOSS MODELS
C
 150  IF(KRUN.EQ.1)WRITE(KWT,56)
 56   FORMAT(24(/),2X,'*** WATERSHED LOSS MODEL ***',/,5X,'1 = ',
     1 'PHI INDEX',/,5X,'2 = PERCENTAGE OF RAINFALL',/,5X,
     2 '3 = PERCENTAGE OF RAINFALL LIMITED BY A PHI INDEX',/,5X,
     3 '4 = HORTON EQUATION ( F=FC+(FO-FC)*EXP(-K*TIME) )',/,
     3 5X,'5 = C1*(RAINFALL INTENSITY)**C2')
      IF(KRUN.EQ.1)WRITE(KWT,99)
      READ(KRD,*)KLOSS
      IF(KLOSS.LT.1 .OR. KLOSS.GT.5)GO TO 150
      IF(KRUN.EQ.1)WRITE(10,*)KLOSS
      IF(KLOSS.EQ.1)GO TO 200
      IF(KLOSS.EQ.2)GO TO 210
      IF(KLOSS.EQ.3)GO TO 200
      IF(KLOSS.EQ.4)GO TO 230
      IF(KLOSS.EQ.5)GO TO 240
 200  IF(KRUN.EQ.1)WRITE(KWT,57)
 57   FORMAT(/,2X,'ENTER THE PHI INDEX (INCHES/HOUR)')
      IF(KRUN.EQ.1)WRITE(KWT,99)
      READ(KRD,*)PHI
      IF(PHI.LT.0.)GO TO 200
      IF(KRUN.EQ.1)WRITE(10,*)PHI
      PERPHI=0.
      IF(KLOSS.EQ.1)GO TO 250
 205  IF(KRUN.EQ.1)WRITE(KWT,58)
 58   FORMAT(/,'  ENTER THE DECIMAL PROPORTION OF RAINFALL TO ',
     1 'BE USED AS THE LOSS RATE')
      IF(KRUN.EQ.1)WRITE(KWT,99)
      READ(KRD,*)PERPHI
      IF(PERPHI.LT.0. .OR. PERPHI.GT.1.)GO TO 205
      IF(KRUN.EQ.1)WRITE(10,*)PERPHI
      GO TO 250
 210  IF(KRUN.EQ.1)WRITE(KWT,58)
      IF(KRUN.EQ.1)WRITE(KWT,99)
      READ(KRD,*)PERPHI
      IF(PERPHI.LT.0. .OR. PERPHI.GT.1.)GO TO 210
      IF(KRUN.EQ.1)WRITE(10,*)PERPHI
      PHI=1000.
      GO TO 250
 230  IF(KRUN.EQ.1)WRITE(KWT,59)
 59   FORMAT(/,2X,'ENTER INITIAL LOSS RATE FO (INCHES/HOUR)')
      IF(KRUN.EQ.1)WRITE(KWT,99)
```

```
      READ(KRD,*)FO
      IF(FO.LT.O.)GO TO 230
      IF(KRUN.EQ.1)WRITE(10,*)FO
      FO1=FO-.01
 215  IF(KRUN.EQ.1)WRITE(KWT,61)FO1
 61   FORMAT(/,2X,'ENTER FINAL LOSS RATE FC (INCHES/HOUR)',
     1 /,2X,'(ALLOWABLE VALUE BETWEEN [',F6.3,'] AND [0])')
      IF(KRUN.EQ.1)WRITE(KWT,99)
      READ(KRD,*)FC
      IF(FC.LT.0. .OR. FC.GT.FO1)GO TO 215
      IF(KRUN.EQ.1)WRITE(10,*)FC
 235  IF(KRUN.EQ.1)WRITE(KWT,62)
 62   FORMAT(/,2X,'ENTER THE EXPONENTIAL DECAY COEFFICIENT K',
     1 /,2X,'NOTE : K (1/HOUR) > 0')
      IF(KRUN.EQ.1)WRITE(KWT,99)
      READ(KRD,*)COEH
      IF(COEH.LT.0.)GO TO 235
      IF(KRUN.EQ.1)WRITE(10,*)COEH
      GO TO 260
 240  IF(KRUN.EQ.1)WRITE(KWT,63)
 63   FORMAT(/,2X,'ENTER THE LOSS RATE COEFFICIENT, C1, BETWEEN'
     1 ' [0] AND [10.]')
      IF(KRUN.EQ.1)WRITE(KWT,99)
      READ(KRD,*)C1
      IF(C1.LT.0. .OR. C1.GT.10.)GO TO 240
      IF(KRUN.EQ.1)WRITE(10,*)C1
 245  IF(KRUN.EQ.1)WRITE(KWT,64)
 64   FORMAT(/,2X,'ENTER RAINFALL INTENSITY EXPONENT, C2, BETWEEN',
     1 ' [0] AND [.99]')
      IF(KRUN.EQ.1)WRITE(KWT,99)
      READ(KRD,*)C2
      IF(C2.LT.0. .OR. C2.GT..99)GO TO 245
      IF(KRUN.EQ.1)WRITE(10,*)C2
      GO TO 270
C
C  PHI INDEX, PERCENTAGE, AND PHI INDEX AND PERCENTAGE
C
 250  REM=ABSD
      PHI=PHI/20.
      IF(KRUN.EQ.1)GO TO 255
      WRITE(KWT,403)
 403  FORMAT(/,2X,'SOIL-LOSS MODEL')
      IF(KLOSS.EQ.1)GO TO 420
      IF(KLOSS.EQ.2)GO TO 430
      IF(KLOSS.EQ.3)GO TO 440
 420  WRITE(KWT,404)PHI
 404  FORMAT(2X,'PHI INDEX = ',F6.3,' INCHES/HOUR')
      GO TO 470
 430  WRITE(KWT,405)PERPHI
 405  FORMAT(2X,'PERCENT OF PRECIPITATION = ',F6.3)
      GO TO 470
 440  WRITE(KWT,404)PHI
      WRITE(KWT,405)PERPHI
 470  IF(KKST.EQ.2)WRITE(KWT,23)
 23   FORMAT(/,76('*'),//,10X,'UNIT',14X,'UNIT',12X,'UNIT',
```

```
     1 14X,'EFFECTIVE',/,9X,'PERIOD',11X,'RAINFALL',7X,
     2 'SOIL-LOSS',12X,'RAINFALL',/,8X,'(NUMBER)',10X,
     3 '(INCHES)',8X,'(INCHES)',12X,'(INCHES)')
      IF(KKST.EQ.2)WRITE(KWT,21)
 21   FORMAT(/,76('-'))
C
 255  DO 300 I=1,480
      ABS=0.
      RAIN=PRECIP(I)
      SOIL=AMIN1(PHI,PERPHI*RAIN)
      IF(PERPHI.EQ.0.)SOIL=PHI
      IF(REM.EQ.0.)GO TO 310
      IF(ABSD.EQ.0.)GO TO 310
      IF(ABSP.NE.0.)RAIN=RAIN*ABSP
      IF(REM.LE.RAIN)GO TO 305
      REM=REM-RAIN
      ABS=RAIN
      IF(KOPT.EQ.3)RAIN=RAIN/ABSP
      GO TO 310
 305  ABS=REM
      REM=0.
      IF(KOPT.EQ.3)RAIN=RAIN/ABSP
 310  XLOSS=AMAX1(SOIL,ABS)
      IF(XLOSS.GT.RAIN)XX=RAIN
      IF(XLOSS.LE.RAIN)XX=XLOSS
      TEMP=PRECIP(I)
      PRECIP(I)=PRECIP(I)-XX
      XXSUM=XXSUM+XX
      IF(KKST.EQ.2 .AND. KRUN.EQ.2)WRITE(KWT,22)I,TEMP,XX,PRECIP(I)
 22   FORMAT(10X,I3,13X,F6.4,11X,F6.4,14X,F6.4)
 300  CONTINUE
      GO TO 350
C
C  HORTON EQUATION
C
 260  TIME=0.
      REM=ABSD
      IF(KRUN.EQ.1)GO TO 450
      WRITE(KWT,406)FC,FO,FC,COEH
 406  FORMAT(2X,'HORTON INFILTRATION EQUATION',/,3X,'F = ',
     C F6.2,' + ( ',F6.2,' - ',F6.2,' ) * EXP ( -1. * TIME(HOUR) * ',
     C F5.2,')')
      IF(KKST.EQ.2)WRITE(KWT,23)
      IF(KKST.EQ.2)WRITE(KWT,21)
 450  DO 320 I=1,480
      ABS=0.
      RAIN=PRECIP(I)
      SOIL=0.
      IF(RAIN.GT.0.)TIME=TIME+.05
      IF(TIME.EQ.0.)GO TO 321
      SOIL=FC+(FO-FC)*EXP(-COEH*TIME)
      SOIL=SOIL/20.
 321  IF(REM.EQ.0.)GO TO 322
      IF(ABSD.EQ.0.)GO TO 322
      IF(ABSP.NE.0.)RAIN=RAIN*ABSP
```

```
      IF(REM.LE.RAIN)GO TO 325
      REM=REM-RAIN
      ABS=RAIN
      IF(KOPT.EQ.3)RAIN=RAIN/ABSP
      GO TO 322
 325  ABS=REM
      REM=0.
      IF(KOPT.EQ.3)RAIN=RAIN/ABSP
 322  XLOSS=AMAX1(SOIL,ABS)
      IF(XLOSS.GT.RAIN)XX=RAIN
      IF(XLOSS.LE.RAIN)XX=XLOSS
      TEMP=PRECIP(I)
      PRECIP(I)=PRECIP(I)-XX
      XXSUM=XXSUM+XX
      IF(KKST.EQ.2 .AND. KRUN.EQ.2)WRITE(KWT,22)I,TEMP,XX,PRECIP(I)
 320  CONTINUE
      GO TO 350
C
C  LOSS = C1 * (RAINFALL INTENSITY) ** C2
C
 270  REM=ABSD
      IF(KRUN.EQ.1)GO TO 460
      WRITE(KWT,407)C1,C2
 407  FORMAT(2X,'GENERAL EXPONENTIAL EQUATION',/,3X,'F = ',
     C F6.2,' * (RAINFALL INTENSITY) ** ',F6.2)
      IF(KKST.EQ.2)WRITE(KWT,23)
      IF(KKST.EQ.2)WRITE(KWT,21)
 460  DO 330 I=1,480
      ABS=0.
      RAIN=PRECIP(I)
      SOIL=C1*(RAIN*20.)**C2
      SOIL=SOIL/20.
      IF(REM.EQ.0.)GO TO 331
      IF(ABSD.EQ.0.)GO TO 331
      IF(ABSP.NE.0.)RAIN=RAIN*ABSP
      IF(REM.LE.RAIN)GO TO 335
      REM=REM-RAIN
      ABS=RAIN
      IF(KOPT.EQ.3)RAIN=RAIN/ABSP
      GO TO 331
 335  ABS=REM
      REM=0.
      IF(KOPT.EQ.3)RAIN=RAIN/ABSP
 331  XLOSS=AMAX1(SOIL,ABS)
      IF(XLOSS.GT.RAIN)XX=RAIN
      IF(XLOSS.LE.RAIN)XX=XLOSS
      TEMP=PRECIP(I)
      PRECIP(I)=PRECIP(I)-XX
      XXSUM=XXSUM+XX
      IF(KKST.EQ.2 .AND. KRUN.EQ.2)WRITE(KWT,22)I,TEMP,XX,PRECIP(I)
 330  CONTINUE
      GO TO 350
C
C  S.C.S. AND MODIFIED S.C.S. MODELS
C
```

```
110  IF(KRUN.EQ.1)WRITE(KWT,65)
65   FORMAT(24(/),'  *** S.C.C. MODEL OPTIONS: ***',/,
    1 2X,'1 = S.C.S. MODEL',/,2X,'2 = MODIFIED S.C.S. MODEL')
     IF(KRUN.EQ.1)WRITE(KWT,99)
     READ(KRD,*)KOPTT
     IF(KOPTT.LT.1 .OR. KOPTT.GT.2)GO TO 110
     IF(KRUN.EQ.1)WRITE(10,*)KOPTT
     FMAX=1000.
     FMIN=0.
361  IF(KRUN.EQ.1)WRITE(KWT,66)
66   FORMAT(/,2X,'ENTER THE CURVE NUMBER BETWEEN [0] AND [100]')
     IF(KRUN.EQ.1)WRITE(KWT,99)
     READ(KRD,*)CN
     IF(CN.LT.0. .OR. CN.GT.100.)GO TO 361
     IF(KRUN.EQ.1)WRITE(10,*)CN
362  IF(KRUN.EQ.1)WRITE(KWT,67)
67   FORMAT(/,2X,'ENTER THE COEFFICIENT k ( Ia = k*S)',
    1 ' BETWEEN [0] AND [.9]')
     IF(KRUN.EQ.1)WRITE(KWT,99)
     READ(KRD,*)XFACT
     IF(XFACT.LT.0. .OR. XFACT.GT.0.9)GO TO 362
     IF(KRUN.EQ.1)WRITE(10,*)XFACT
     IF(KOPTT.EQ.1)GO TO 360
363  IF(KRUN.EQ.1)WRITE(KWT,68)
68   FORMAT(/,2X,'ENTER THE MAXIMUM LOSS RATE (INCHES/HOUR)')
     IF(KRUN.EQ.1)WRITE(KWT,99)
     READ(KRD,*)FMAX
     IF(FMAX.LT.0.)GO TO 363
     IF(KRUN.EQ.1)WRITE(10,*)FMAX
     FMAX1=FMAX-.01
365  IF(KRUN.EQ.1)WRITE(KWT,69)FMAX1
69   FORMAT(/,2X,'ENTER THE MINIMUM LOSS RATE (INCHES/HOUR)',
    1 /,2X,'(ALLOWABLE VALUE BETWEEN [',F6.3,'] AND [0])')
     IF(KRUN.EQ.1)WRITE(KWT,99)
     READ(KRD,*)FMIN
     IF(FMIN.LT.0. .OR. FMIN.GT.FMAX1)GO TO 365
     IF(KRUN.EQ.1)WRITE(10,*)FMIN
360  S=1000./CN-10.
     SUMP=0.
     ABSD=S*XFACT
     R2=0.
     IF(KRUN.EQ.1)GO TO 410
     IF(KOPTT.EQ.1)WRITE(KWT,408)
408  FORMAT(2X,'S.C.S. MODEL')
     IF(KOPTT.EQ.2)WRITE(KWT,409)
409  FORMAT(2X,'MODIFIED S.C.S. MODEL')
     WRITE(KWT,411)CN
411  FORMAT(3X,'CURVE NUMBER = ',F4.0)
     WRITE(KWT,412)XFACT
412  FORMAT(3X,'COEFFICIENT OF INITIAL ABSTRACTION = ',F6.3)
     IF(KOPTT.EQ.1)GO TO 415
     WRITE(KWT,413)FMAX
413  FORMAT(3X,'MAXIMUM LOSS RATE (INCHES/HOUR) = ',F6.3)
     WRITE(KWT,414)FMIN
414  FORMAT(3X,'MINIMUM LOSS RATE (INCHES/HOUR) = ',F6.3)
```

```
 415  IF(KKST.EQ.2)WRITE(KWT,23)
      IF(KKST.EQ.2)WRITE(KWT,21)
C
 410  DO 370 I=1,480
      RAIN=PRECIP(I)
      SUMP=SUMP+RAIN
      R1=0.
      IF(SUMP.LE.ABSD)GO TO 371
      XX=SUMP-ABSD
      R1=XX*XX/(XX+S)
      IF(KOPT.EQ.1)GO TO 371
      F=(RAIN-R1+R2)*20.
      IF(F.LT.FMIN)F=FMIN
      IF(F.GT.FMAX)F=FMAX
      R1=RAIN+R2-.05*F
 371  IF(I.NE.1)XX=R1-R2
      IF(I.EQ.1)XX=R1
      PRECIP(I)=XX
      R2=R1
      XX=RAIN-XX
      IF(PRECIP(I).LT.0.)PRECIP(I)=0.
      XXSUM=XXSUM+XX
      IF(KKST.EQ.2 .AND. KRUN.EQ.2)WRITE(KWT,22)I,RAIN,XX,PRECIP(I)
 370  CONTINUE
 350  CONTINUE
      IF(KRUN.EQ.1)RETURN
      IF(KKST.EQ.2)WRITE(KWT,21)
      XXSUM=XXSUM/12.*AREA
      WRITE(KWT,425)XXSUM
 425  FORMAT('   TOTAL SOIL-LOSS VOLUME (ACRE-FEET) = ',F12.4)
C
      RETURN
      END
```

peak discharge estimation

4

4.1. Introduction

Some design problems must be concerned with the time dependency in watershed, channel, and reservoir storages. Such design problems require hydrograph analyses, which is the subject of Chapter 5. Other types of design problems only require estimates of maximum flow rates (or peak discharges) and the time characteristics of the runoff. These design problems can be handled using the methods such as those outlined in this chapter.

The time from the beginning of rainfall excess to the occurrence of the peak runoff rate is important both for determining the duration of the rainfall and in balancing discharge rates from subareas of a watershed. The most important time parameter is the time of concentration, which is discussed in the first part of this chapter. The time of concentration is a necessary input to most hydrologic models, including the rational method, which is the subject of the remaining parts of this chapter. In addition to basic rational method computations, methods for designing on subdivided watersheds are included.

4.2. Time of Concentration Estimation

The time of concentration (Tc) is usually defined as the duration required for runoff at the point of concentration to become a maximum under a uniform and constant rainfall intensity. This occurs when all parts of the drainage area are contributing to the flow. Generally, the time of concentration is the interval of time from the beginning of rainfall for water from the hydraulically most remote point of the drainage area to reach the point of concentration (e.g., a drainage structure). The Tc is a function of several variables including the length of the flow path from the most remote point of the watershed to the concentration point, the slope of the flowpath, characteristics of natural and improved channels within the drainage area, the infiltration properties of the soil, and the extent and type of development.

Recognizing the importance of the time of concentration as input to hydrologic models, a wide array of methods have been proposed for estimating Tc. Most methods are a function of two or more of the following factors: (1) flow length, (2) flowpath slope, (3) land use or a representative surface roughness, and (4) the intensity of the rainfall. Since times of concentration are required for all types of flow regimes, including overland flow, street flow, pipe flow, and channel flow, the inputs for estimating Tc should reflect the primary flow regime in the flowpath being analyzed. Given the wide array of methods available, it is important to select for design a Tc method that uses inputs that correspond directly to the character of the flowpath being analyzed. The following paragraphs describe the methods most commonly used for design.

Over two dozen overland flow formulas have been proposed in the literature to estimate the initial subarea Tc. A variation of the Kirpich (1940) formula that is widely used has the form

$$Tc = k(L^3/H)^E \qquad (4.1)$$

where

L = length of initial subarea flowpath (feet)
H = drop in elevation along flowpath (feet)
k = coefficient depending on development type
E = constant exponent.

A nomograph for the solution of Eq. (4.1) with E = 0.20 is given in Fig. 4.1. Due to the inherent inaccuracy in the determination of a generalized overland flow Tc, Eq. (4.1) should only be used on initial subareas of less than about 10 acres.

An alternative version of the Kirpich formula is:

$$Tc = 0.00013 L_f^{0.77} S_f^{-0.385} \qquad (4.2)$$

in which L_f is the hydraulic length in feet, S_f is the slope in ft/ft, and Tc is the time of concentration in hours. Equation (4.2) is based on watersheds in Tennessee that have areas from 1 to 112 acres and slopes from 3% to 10%. The computed times of concentration should be multiplied by 0.4 and 0.2 for watersheds where the overland flow path is either concrete or asphalt and the channel is lined, respectively.

The simplest method is to estimate a Tc at the point of concentration by using a generalized overland flow formula or corresponding nomograph. For example, Fig. 4.2 provides a nomograph for the California Culvert Practice (1942) formula of Eq. (4.1) with k = 0.0078.

In comparison, the Federal Aviation Agency (1970) proposed

$$Tc = 1.8(1.1 - C)\, L^{0.50}\, S^{-0.333} \qquad (4.3)$$

where

C = runoff coefficient for the rational method
S_f = average surface slope (percent)
L = characteristic flow length.

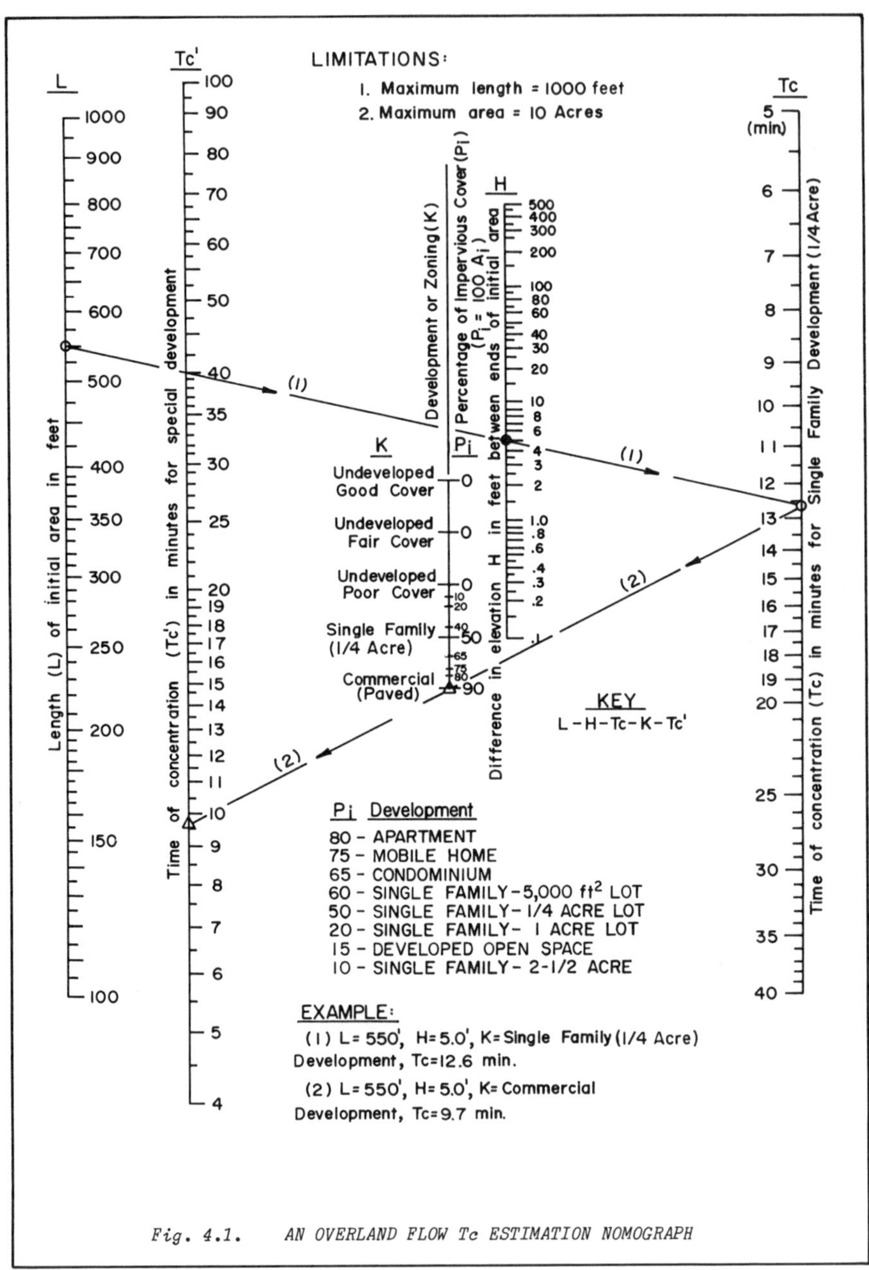

Fig. 4.1. AN OVERLAND FLOW Tc ESTIMATION NOMOGRAPH

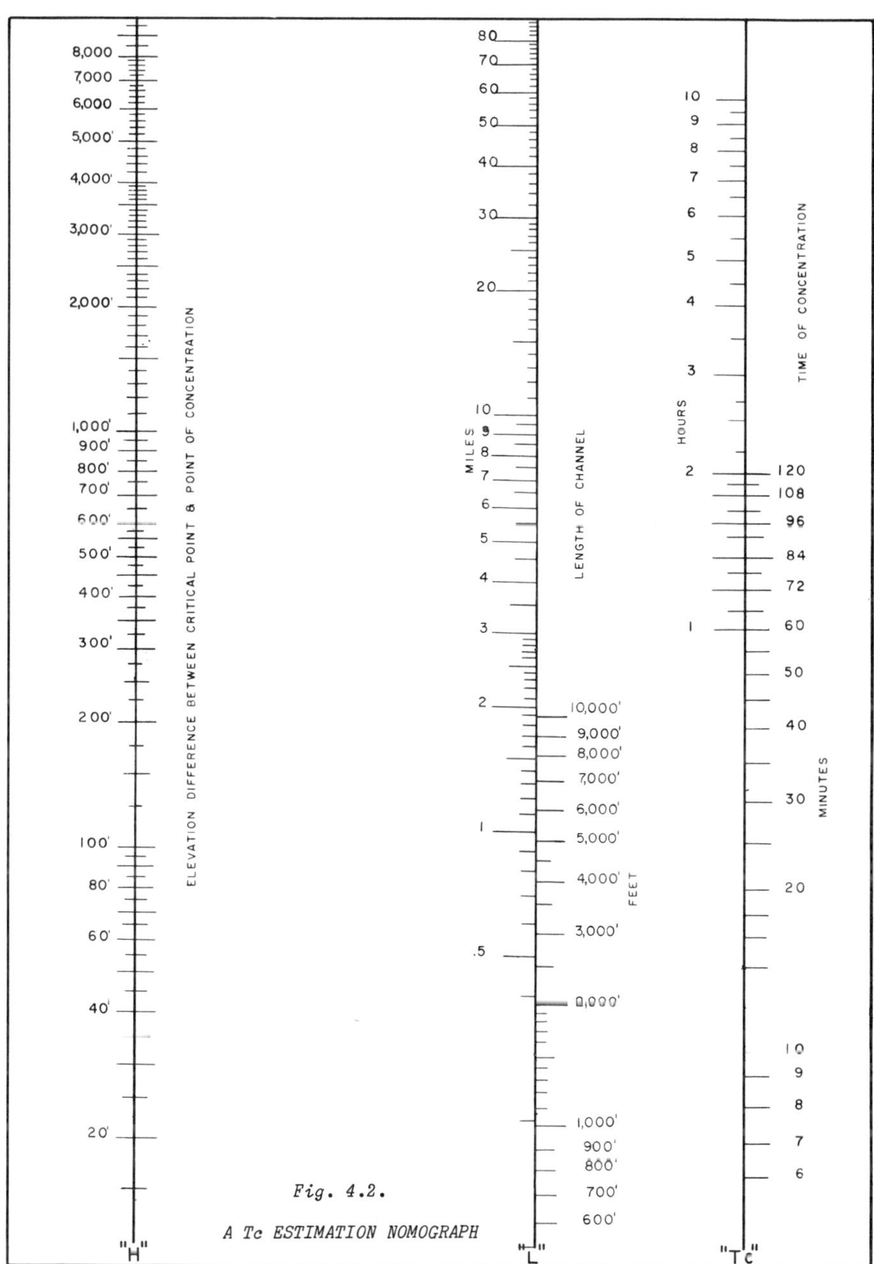

Fig. 4.2.

A Tc ESTIMATION NOMOGRAPH

The kinematic wave equation is widely used for very small flow lengths:

$$Tc = 0.94\ L^{0.6}\ n^{0.6}\ i^{-0.4}\ S_f^{-0.3} \tag{4.4}$$

in which L is the flow length in feet, n is Manning's roughness coefficient, i is the excess rainfall rate (in/hour), and S_f is the slope in ft/ft. Equation (4.4) is valid only for very shallow sheet flow over lengths of 300 feet or less. It is especially useful for Tc estimation of gutter flow in urban areas.

For watersheds having flow paths of mixed land covers, the time of concentration (Tc) can be estimated by summing the runoff travel times (Tt) through the several flow regimes as the flood peak travels downstream to the watershed outlet. These flow regimes include overland flow, streetflow, pipeflow, and open channel flow in natural or improved channels, and must include the effects of the flood peak rate (Q) increasing in magnitude as the tributary area to the main collection stream increases. The mixed velocity method is applied to watersheds that have flow paths of mixed land covers. For upland flow paths an equation such as Eq. (4.1) or (4.4) can be used. As the runoff enters an area of more concentrated flow the flow velocity can be estimated with Manning's equation or Fig. 4.3. The travel time for the flow path with velocity V(feet/sec) can be computed with:

$$Tt = L/V \tag{4.5a}$$

where Tt is the travel time in seconds and L is the length in feet. The total travel time for the watershed is the sum of the individual travel times:

$$Tt = \sum_{i=1}^{n} L_i/V_i \tag{4.5b}$$

in which n is the number of flow paths. Any travel times computed with Eq. (4.1) or Eq. (4.4) would have to be added to the summation of Eq. 4.5b.

For larger watersheds where conditions are not homogeneous, the watershed should be divided into subareas and the times of concentration computed for each subarea. The main flowpath is identified such that the watershed can be subdivided into m subareas with each subarea tributary to the collection stream (Fig. 4.9, of example 4.2). The main flowpath is segmented into reaches that are relatively homogeneous in runoff characteristics. From the figure, it is noted that the initial (the most upstream) subarea is relatively small (about 10 acres) and has an associated overland flowpath length of less than about 1000 feet. The subareas gradually increase in size in the downstream direction

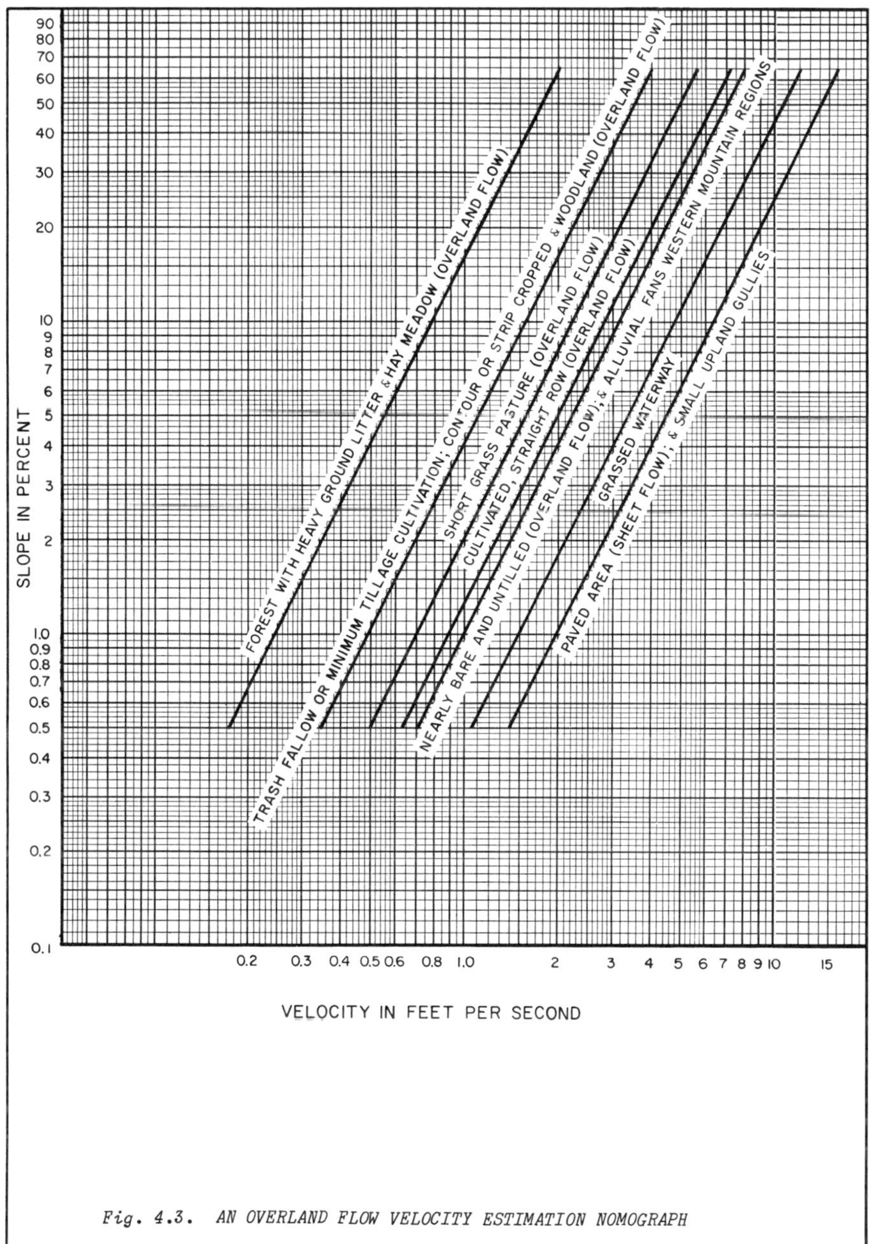

Fig. 4.3. AN OVERLAND FLOW VELOCITY ESTIMATION NOMOGRAPH

along the collection stream. Additionally, nodal points (i = 1,2,...,m) are defined along the main stream so that each subarea has an associated upstream and downstream node number. The initial subarea time of concentration for the overland flow between node numbers 1 and 2 is estimated by some overland flow formula or an assumed average flow velocity for the runoff traveling along the main flowpath within the initial subarea. Subsequent Tc values are determined by

$$Tc(i+1) = Tc(i) + Tt(i,i+1) \tag{4.6}$$

where

Tc(i+1) = Tc at node number i+1
Tc(i) = Tc at node number i
Tt(i,i+1) = traveltime of Q between nodes i and i+1

The travel time for each segment of the flowpath is then computed. To estimate travel time values $T_t(i,i+1)$ the Manning's formula is used to calculate a normal depth for the runoff flowing in the channel linking nodes i and i+1 and the corresponding flow velocity is used to estimate the time for the peak Q to move from node i to node i+1. Then the time of concentration computed by the velocity method is:

$$Tt(i,i+1) = L(i,i+1)/V(i,i+1) \tag{4.7}$$

where

L(i,i+1) = length of channel linking nodes i and i+1
V(i,i+1) = normal depth flow velocity for Q(i)

4.3. Rational Method

The most widely used hydrologic model for estimating watershed peak runoff rates is the rational method. Currently, this approach is typically used to estimate runoff rates from small urban areas of variable size. Some older versions of this method have been directly applied to watersheds with sizes in excess of several square miles. Modern versions of this approach generally limit the watershed size to about 1 square mile.

The rational method equation relates rainfall intensity, a runoff coefficient, and drainage area size to the direct peak runoff rate. This relationship is expressed by the equation

$$Q = CIA \tag{4.8}$$

where

- Q = the peak runoff rate in cubic feet per second (cfs) at the point of concentration
- C = a runoff coefficient representing the area-averaged ratio of runoff to rainfall rates
- I = the time-averaged rainfall intensity in inches per hour corresponding to the time of concentration
- A = the drainage area (acres)

The values of the runoff coefficient and rainfall intensity are based on a study of drainage area characteristics such as the type and condition of the runoff surfaces and the time of concentration. These factors and the limitations of the rational method equation are discussed in the following sections.

Data required for the computation of peak discharge by the rational method include (1) rainfall intensity for a storm of specified duration and selected return frequency; (2) drainage area characteristics of size, shape, slope; and (3) a land use index that reflects the amount of rainfall that will appear as direct runoff. The drainage area may be determined by planimetering a suitable topographic map of the tributary watershed area. The duration of the storm rainfall required in the rational method equation is based on the time of concentration of the tributary drainage area. Rainfall intensity (I) is determined from the local precipitation intensity-duration curves of the desired return frequency (see Fig. 2.3). Since one acre-inch/hour is equal to 1.008 cfs, the rational method is generally assumed to estimate a peak flowrate in cfs.

Intensity duration curves for a particular region can be developed using the log-log paper of Fig. 2.3, plotting the area-averaged point rainfall value for the one hour duration, and drawing a straight line through the one hour value with a slope based on shorter duration rainfall intensity values.

4.4. Runoff Coefficient

The runoff coefficient (C) is the ratio of peak rate of runoff to the rate of rainfall at an average intensity when the total drainage area is contributing runoff to the point of concentration. The selection of the runoff coefficient depends on drainage area slope, type and amount of vegetative cover, distribution and magnitude of the soil infiltration capacity, and various other factors.

For calculation purposes, the runoff coefficient is most often defined to be either (1) a constant value depending on soil cover type and quality, or (2) a function of rainfall intensity, soil cover type, and quality. Table 4.1 lists typical C values for use with the rational method.

The second class of runoff coefficient representations relate the C value to the rainfall intensity. One approach used for urban design purposes is to assume that the watershed loss rate is equal to the infiltration loss rate which corresponds to the limiting value of the infiltration capacity curve (see Chapter 3). For design storm conditions, it can be argued that the impervious area runoff rate is independent of the rainfall intensity and that the pervious area infiltration loss rate is a constant which is determined from equations such as Eq. (3.9). For urban design studies, the runoff coefficient is sometimes assumed to be a function of the impervious and pervious area fractions, a characteristic infiltration rate (Fp) for the pervious area fraction, and the effects of watershed detention in the estimation of travel time of the peak runoff rate through the watershed channel system. Then estimates for runoff coefficients are developed using a relationship of the form

$$Cm = 0.85\ (Ai + (I - Fp)Ap/I) \tag{4.9}$$

where

Cm = modified runoff coefficient
I = rainfall intensity (inch/hour)
Fp = infiltration rate for pervious area fraction
Ai = impervious area fraction
Ap = pervious area fraction
0.85 = calibrated (or assumed) coefficient to correlate rainfall and runoff frequencies.

The infiltration rate for the pervious area (Fp) can be estimated for various combinations of soil type, cover, and antecedent moisture conditions using the methods of Chapter 3. For the most common types of urban development and soil covers, typical runoff coefficient curves based on Eq.(4.9) are shown in Figs. 4.4a through 4.4d for SCS soil groups A through D, respectively.

When the drainage area is composed of several types of runoff surfaces, an area-averaged runoff coefficient can be developed as demonstrated by the following example problem.

Example 4.1. Area-Averaged Runoff Coefficient
The watershed is composed of 3.5 acres of parking lot pavement and the associated street system, 35.6 acres of a condominium development, and 12.5 acres of a neighboring apartment complex. The area-averaged runoff coefficient is estimated by tabulating each area fraction's contribution:

TABLE 4.1
RUNOFF COEFFICIENTS FOR THE RATIONAL FORMULA FOR A HYDROLOGIC SOIL GROUP AND SLOPE RANGE

Land use	A			B			C			D		
	0-2%	0-6%	6%+	0-2%	2-6%	6%+	0-2%	0-6%	6%+	0-2%	2-6%	6%+
Cultivated land----	0.08[1]	0.13	0.16	0.11	0.15	0.21	0.14	0.19	0.26	0.18	0.23	0.31
	0.14[2]	0.18	0.22	0.16	0.21	0.28	0.20	0.25	0.34	0.24	0.29	0.41
Pasture------------	0.12	0.20	0.30	0.18	0.28	0.37	0.24	0.34	0.44	0.30	0.40	0.50
	0.15	0.25	0.37	0.23	0.34	0.45	0.30	0.42	0.52	0.37	0.50	0.62
Meadow-------------	0.10	0.16	0.25	0.14	0.22	0.30	0.20	0.28	0.36	0.24	0.30	0.40
	0.14	0.22	0.30	0.20	0.28	0.37	0.26	0.35	0.44	0.30	0.40	0.50
Forest-------------	0.05	0.08	0.11	0.08	0.11	0.14	0.10	0.13	0.16	0.12	0.16	0.20
	0.08	0.11	0.14	0.10	0.14	0.18	0.12	0.16	0.20	0.15	0.20	0.25
Residential lot	0.25	0.28	0.31	0.27	0.30	0.35	0.30	0.33	0.38	0.33	0.36	0.42
size 1/8 acre-----	0.33	0.37	0.40	0.35	0.39	0.44	0.38	0.42	0.49	0.41	0.45	0.54
Residential lot	0.22	0.26	0.29	0.24	0.29	0.33	0.27	0.31	0.36	0.30	0.34	0.40
size 1/4 acre-----	0.30	0.34	0.37	0.33	0.37	0.42	0.36	0.40	0.47	0.38	0.42	0.52
Residential lot	0.19	0.23	0.26	0.22	0.26	0.30	0.25	0.29	0.34	0.28	0.32	0.39
size 1/3 acre-----	0.28	0.32	0.35	0.30	0.35	0.39	0.33	0.38	0.45	0.36	0.40	0.50
Residential lot	0.16	0.20	0.24	0.19	0.23	0.28	0.22	0.27	0.32	0.26	0.30	0.37
size 1/2 acre-----	0.25	0.29	0.32	0.28	0.32	0.36	0.31	0.35	0.42	0.34	0.38	0.48
Residential lot	0.14	0.19	0.22	0.17	0.21	0.26	0.20	0.25	0.31	0.24	0.29	0.35
size 1 acre-------	0.22	0.26	0.29	0.24	0.28	0.34	0.28	0.32	0.40	0.31	0.35	0.46
Industrial---------	0.67	0.68	0.68	0.68	0.68	0.69	0.68	0.69	0.69	0.69	0.69	0.70
	0.85	0.85	0.86	0.85	0.86	0.86	0.86	0.86	0.87	0.86	0.86	0.88
Commercial---------	0.71	0.71	0.72	0.71	0.72	0.72	0.72	0.72	0.72	0.72	0.72	0.72
	0.88	0.88	0.89	0.89	0.89	0.89	0.89	0.89	0.90	0.89	0.89	0.90
Streets------------	0.70	0.71	0.72	0.71	0.72	0.74	0.72	0.73	0.76	0.73	0.75	0.78
	0.76	0.77	0.79	0.80	0.82	0.84	0.84	0.85	0.89	0.89	0.91	0.95
Open space---------	0.05	0.10	0.14	0.08	0.13	0.19	0.12	0.17	0.24	0.16	0.21	0.28
	0.11	0.16	0.20	0.14	0.19	0.26	0.18	0.23	0.32	0.22	0.27	0.39
Parking------------	0.85	0.86	0.87	0.85	0.86	0.87	0.85	0.86	0.87	0.85	0.86	0.87
	0.95	0.96	0.97	0.95	0.96	0.97	0.95	0.96	0.97	0.95	0.96	0.97

1 Runoff coefficients for storm recurrence intervals less than 25 years.

2 Runoff coefficients for storm recurrence intervals of 25 years or more.

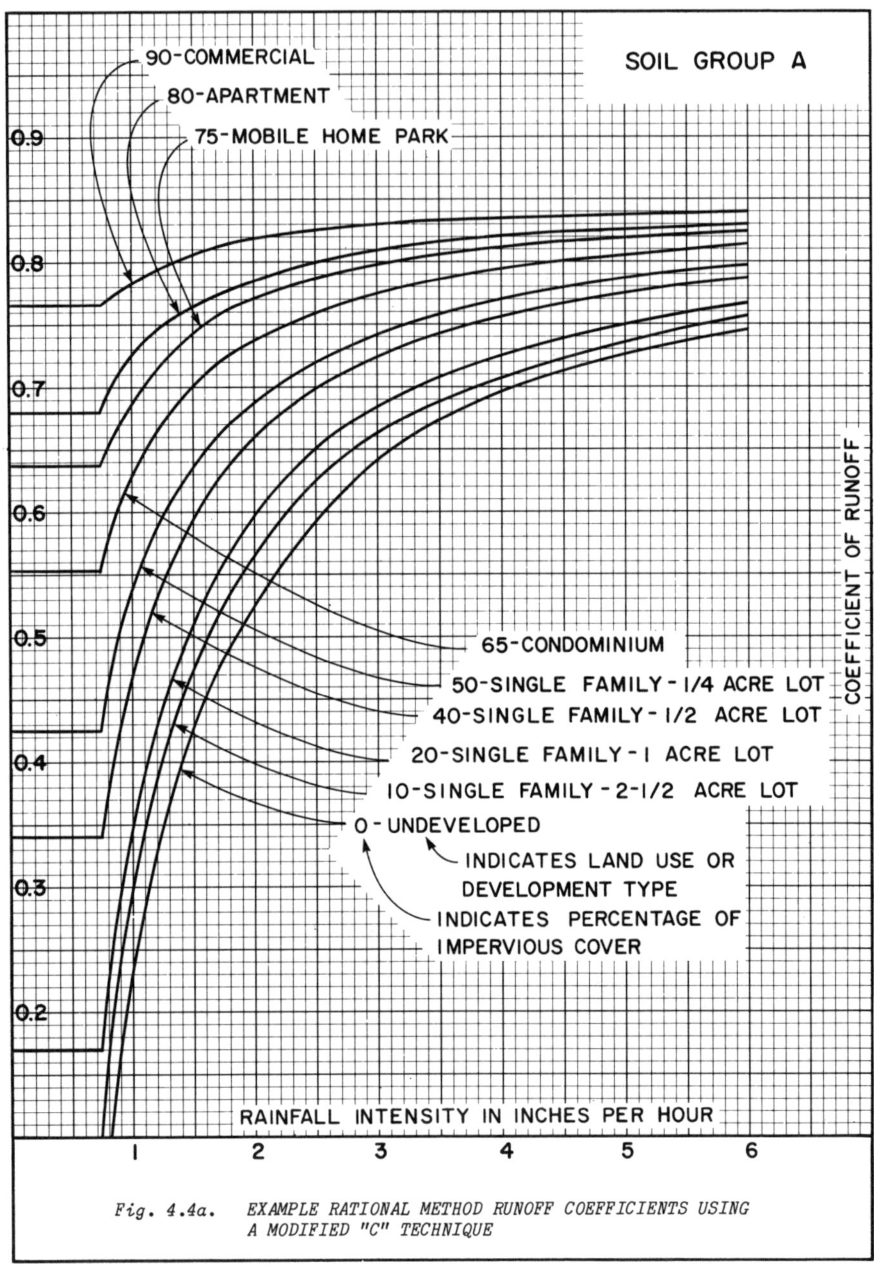

Fig. 4.4a. *EXAMPLE RATIONAL METHOD RUNOFF COEFFICIENTS USING A MODIFIED "C" TECHNIQUE*

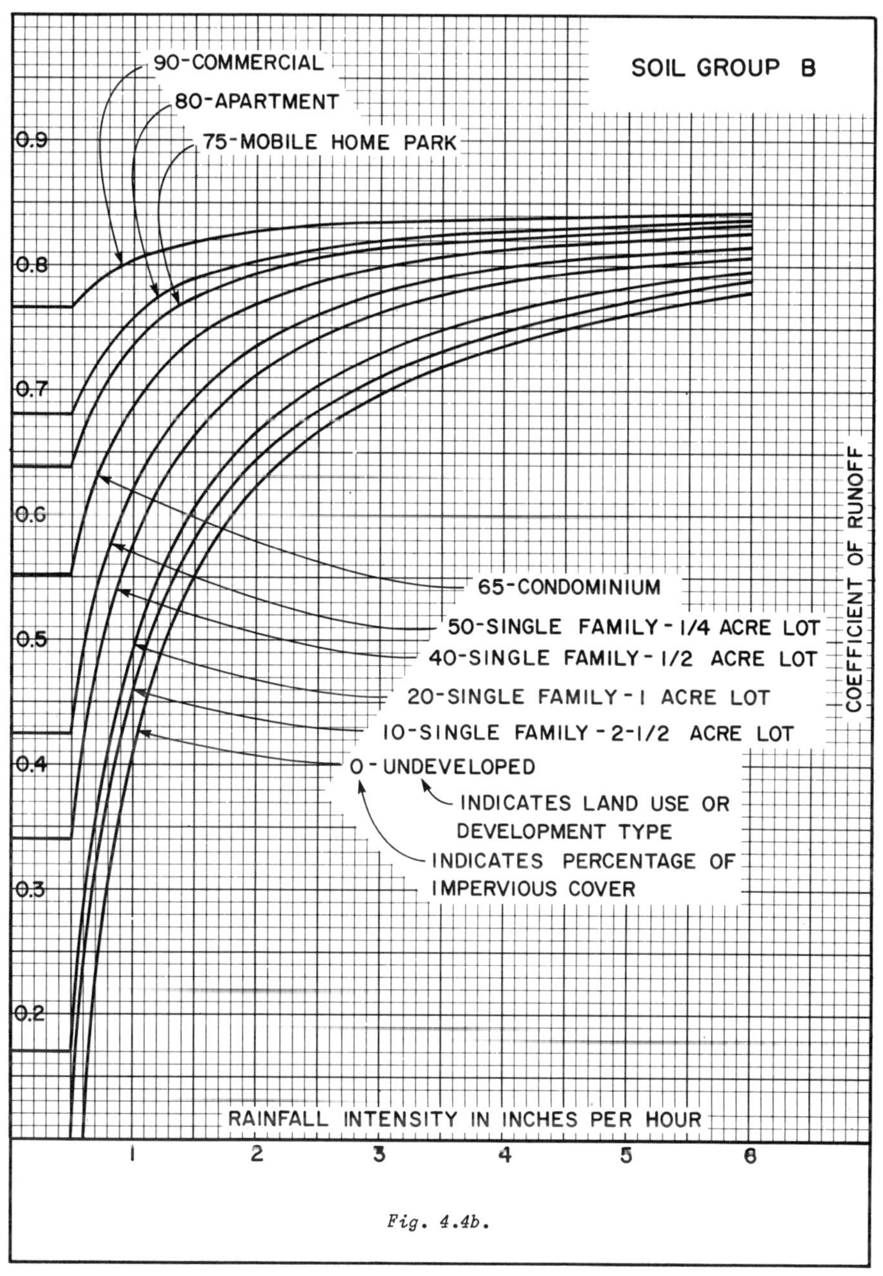
SOIL GROUP B
90-COMMERCIAL
80-APARTMENT
75-MOBILE HOME PARK
65-CONDOMINIUM
50-SINGLE FAMILY-1/4 ACRE LOT
40-SINGLE FAMILY-1/2 ACRE LOT
20-SINGLE FAMILY-1 ACRE LOT
10-SINGLE FAMILY-2-1/2 ACRE LOT
0-UNDEVELOPED
INDICATES LAND USE OR DEVELOPMENT TYPE
INDICATES PERCENTAGE OF IMPERVIOUS COVER
COEFFICIENT OF RUNOFF
0.9
0.8
0.7
0.6
0.5
0.4
0.3
0.2
RAINFALL INTENSITY IN INCHES PER HOUR
1
2
3
4
5
6

Fig. 4.4b.

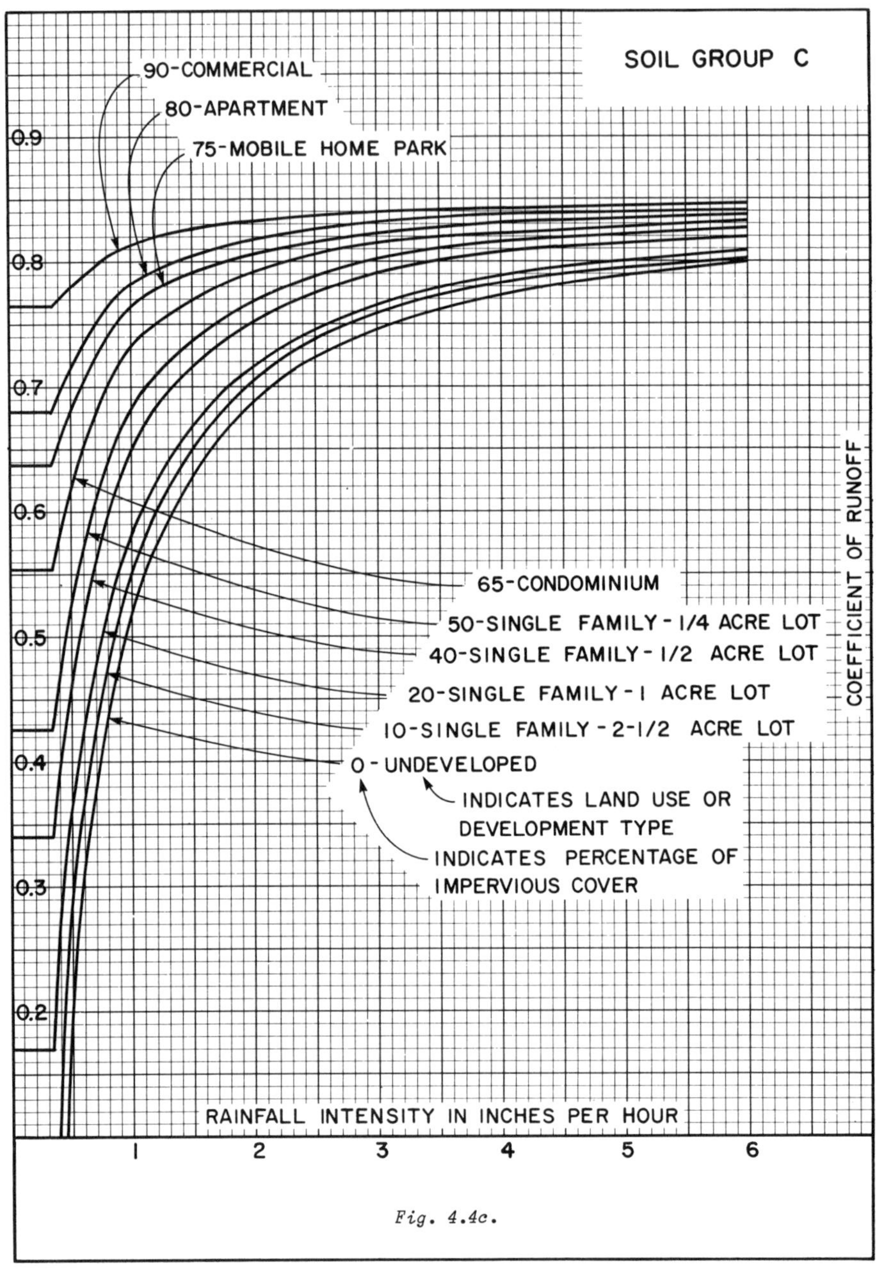
SOIL GROUP C
90-COMMERCIAL
80-APARTMENT
75-MOBILE HOME PARK
0.9
0.8
0.7
0.6
0.5
0.4
0.3
0.2
65-CONDOMINIUM
50-SINGLE FAMILY-1/4 ACRE LOT
40-SINGLE FAMILY-1/2 ACRE LOT
20-SINGLE FAMILY-1 ACRE LOT
10-SINGLE FAMILY-2-1/2 ACRE LOT
0-UNDEVELOPED
INDICATES LAND USE OR DEVELOPMENT TYPE
INDICATES PERCENTAGE OF IMPERVIOUS COVER
COEFFICIENT OF RUNOFF
RAINFALL INTENSITY IN INCHES PER HOUR
1
2
3
4
5
6

Fig. 4.4c.

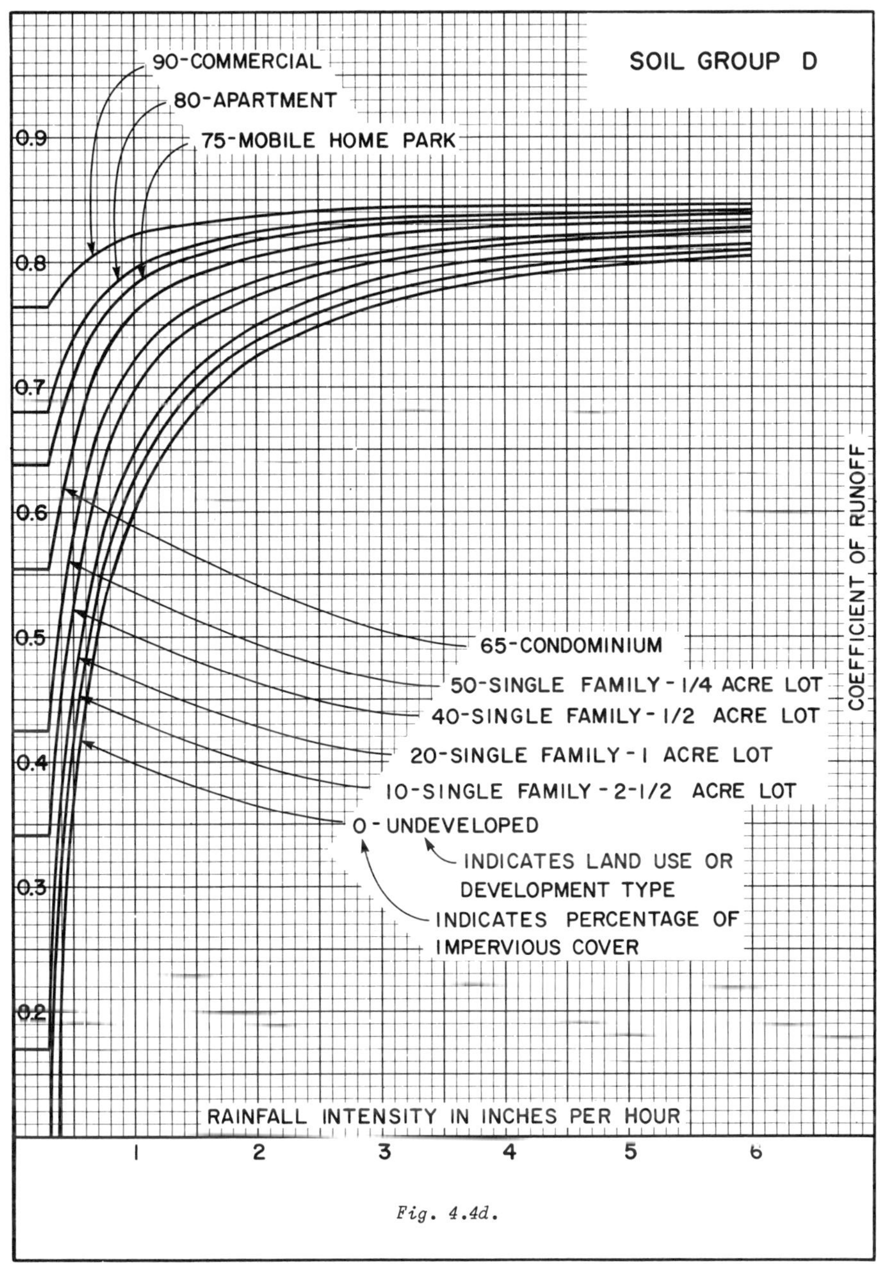

SOIL GROUP D
90-COMMERCIAL
80-APARTMENT
75-MOBILE HOME PARK
65-CONDOMINIUM
50-SINGLE FAMILY - 1/4 ACRE LOT
40-SINGLE FAMILY - 1/2 ACRE LOT
20-SINGLE FAMILY - 1 ACRE LOT
10-SINGLE FAMILY - 2-1/2 ACRE LOT
0 - UNDEVELOPED
INDICATES LAND USE OR DEVELOPMENT TYPE
INDICATES PERCENTAGE OF IMPERVIOUS COVER
COEFFICIENT OF RUNOFF
RAINFALL INTENSITY IN INCHES PER HOUR
0.9
0.8
0.7
0.6
0.5
0.4
0.3
0.2
1
2
3
4
5
6

Fig. 4.4d.

Area (acres)	Type of Surface	C	CA
3.5	concrete pavement	1.0	3.50
35.6	condominiums	0.67	23.85
12.5	apartments	0.77	9.63
51.6 (sum)			36.98

area-averaged C = 36.98/51.6 = 0.72

4.5. Limitations of the Rational Method

The relationship expressed by the rational method equation holds true only if certain assumptions are reasonably correct and limitations are observed. Four basic assumptions are that (1) the frequency of the storm runoff is the same as the return frequency of rainfall producing the runoff (that is, a 25-year recurrence interval rainfall will result in a 25-year recurrence interval storm runoff); (2) the peak runoff rate occurs when all parts of the drainage area are contributing to the runoff; (3) the design rainfall is uniform over the watershed area tributary to the point of concentration; and (4) the rainfall intensity is essentially constant during the storm duration, which is equal to the time of concentration.

The rational method is only applicable where the rainfall intensity can be assumed uniformly distributed over the drainage area at a uniform rate throughout the storm duration. This assumption applies fairly well to small drainage areas of less than about one square mile. Beyond this limit, the rainfall distribution may vary considerably from the point values given in rainfall isohyetal maps.

The selection of the runoff coefficient is another major limitation of the method. For small urban areas, the runoff coefficient can be reasonably estimated from field investigations and studies of aerial photographs. For larger areas, the determination of the runoff coefficient is to be based on vegetation type, cover density, the infiltration capacity of the soil, and the slope of the drainage area. For larger areas, an estimate of the runoff coefficient may be subject to a much greater error due to the variability of the drainage area characteristics, watershed storage, and the greater importance of hydraulic flow characteristics. Rainfall losses due to evaporation, transpiration, and both depression and channel storage cannot be properly evaluated and may appreciably affect the estimate of the watershed peak rate of runoff.

4.6. Modeling with the Rational Method

The preceding discussion of the rational method is oriented toward design problems for a single drainage area. Where it is necessary to subdivide a watershed because of nonhomogeneities in hydrologic processes or to evaluate the effects of development of a part of the watershed, then the method of modeling is a bit more involved than just applying Eq. 4.8. The following paragraphs describe alternative ways of modeling with the rational method.

4.7. The Link-Node Method

This approach estimates the peak runoff at a watershed point of concentration by the following steps:

(1). Subdivide the watershed into m subareas such as shown in Fig. 4.9. The subareas are chosen such that the initial subarea is relatively small and subsequent subareas gradually increase in size in the downstream direction. Each subarea has an associated runoff coefficient C_i and a tributary drainage area A_i.

(2). Estimate a time of concentration Tc' at the point of concentration of the watershed (that is, the most downstream nodal point, node m).

(3). Using Tc', determine a corresponding rainfall intensity (I') from the local precipitation intensity-duration curves.

(4). If the C_i are assumed to be functions of rainfall intensity, determine appropriate C_i' values for the intensity I'.

(5). Calculate a total watershed peak runoff (Q') by

$$Q' = (C_1'A_1 + C_2'A_2 + \ldots + C_m'A_m)I'$$

(6). Distribute Q' throughout the watershed according to the area proportion of runoff $Q_i' = C_i'A_iI'$, where Q_i' is the assumed runoff estimate at nodal point i in the estimation of the peak Q for node m.

(7). Estimate the time of concentration at node 1 for the initial subarea, Tc(1).

(8). In the next downstream subarea, calculate the travel time Tt(1,2) for the runoff Q' to flow to the next nodal point and determine Tc(2) = Tc(1) + Tt(1,2).

(9). In each subsequent downstream subarea, use Q_i' to estimate the travel time Tt(i,i+1) between nodes i and i+1, and estimate Tc(i+1) = Tc(i) + Tt(i,i+1).

(10). Using step 9, determine the final node Tc(m).

(11). Compare Tc(m) to the estimated Tc'.

(12). Calculate a new Tc'' by Tc'' = (Tc(m) + Tc')/2.

(13). Return to step 3 where Tc'' is substituted for Tc'.

This link-node model is based on the rational method attempts to estimate a Tc(m) by accounting for the several flow regimes through which the runoff peak flow rate must travel through and is thought to provide an improvement over the nomograph method when studying larger watersheds.

4.8. The Nodal Point Method

This approach attempts to reduce the calculation effort required by the Link-Node Method. The procedure for the Nodal Point method is as follows:

(1). Subdivide the watershed into m subareas such as shown in Fig. 4.9. Similar to the Link-Node Method, the subareas are selected such that the initial subarea is less than 10 acres and the subsequent downstream subareas gradually increase in size in order to reduce the computational effort in dealing with small subareas.

(2). Estimate an initial subarea Tc(1) for the overland flow concentrating at node 1.

(3). Using Tc(1), estimate the corresponding rainfall intensity $I_1 = I(Tc(1))$ and the runoff coefficient C_1. Then $Q(1) = C_1 I_1 A_1$.

(4). Using Q(1), estimate the travel time $T_t(1,2)$ between nodes 1 and 2 of the next downstream subarea.

(5). Calculate Tc(2) = Tc(1) + Tt(1,2). Determine the rainfall intensity $I_2 = I(Tc(2))$. Using I_2, determine an area-averaged runoff factor for the entire watershed tributary to node 2 by

$$(CA)_2 = (C_1 A_1 + C_2 A_2)$$

Then $Q(2) = (CA)_2 I_2$.

(6). Repeat steps 4 and 5 for each subsequent downstream subarea as the study proceeds in the downstream direction. At each node, the area-averaged runoff factor $(CA)_i$ is calculated based on the new Tc and I values.

A computational advantage for the Nodal-point method over the Link-Node Method is that the entire watershed is analyzed for peak flow rate estimates at each watershed nodal point with only one pass of the method. Consequently, computational effort is considerably reduced. A disadvantage of the model is that it is possible to estimate a downstream peak runoff rate (at a node i+1) which is less than the preceeding nodal point (at node i). This is due to the variation of the runoff factor $(CA)_i$ with rainfall intensity.

4.9. The Subarea Summation Method

This rational method modeling approach is widely used due to the simplicity in application, and the capability for estimating peak runoff rates throughout the interior of a study watershed. In this respect, it is analogous to the Nodal-Point method. The procedure for the Subarea Summation Model is as follows:

(1). Subdivide the watershed into m subareas with the initial subarea being approximately 10 acres in size, and the subsequent subareas gradually increasing in size. Assign upstream and downstream nodal point numbers to each subarea in order to correlate calculations to the watershed map (see Fig. 4.9).

(2). Estimate a Tc(1) by using a nomograph or overland flow velocity estimation.

(3). Using Tc(1), determine the corresponding values of I_1 and C_1. Then $Q(1) = C_1 I_1 A_1$.

(4). Using Q(1), estimate the travel time between nodes 1 and 2 by Manning's equation as applied to the particular channel or conduit linking nodes 1 and 2.

(5). Then Tc(2) = Tc(1) + Tt(1,2). Using Tc(2), estimate the rainfall intensity $I_2 = I(Tc(2))$ and the runoff coefficient corresponding to both I_2 and the properties of the subarea between nodes 1 and 2. Then

$$Q(2) = Q(1) + C_2 I_2 A_2$$

(6). Repeat steps 4 and 5 as the analysis proceeds in the downstream direction along the principal collection stream.

Of the models, the Subarea Summation Model is generally the easiest to use and formulate into a digital computer program. Computer applications of this model are easily used for the master plannings of large cities in urbanized regions. Because the calculations proceed in the downstream direction exclusively, the entire watershed tributary to each nodal point is characterized by only three variables: Q(i), Tc(i), and total area. The Nodal-point method is also easily programmable for use in master planning and design purposes.

4.10. Confluence of Streams (Junction Analysis)

Each of the above rational method modeling approaches determines peak runoff rates for major collection streams within a watershed. At the confluence of two or more collection streams, a procedure for adjusting the total summation of peak flow rates is required in order to account for each stream's time of concentration at the junction. The following procedure provides an estimate of the confluence peak flow rate assuming that each stream's runoff hydrograph is triangular in shape (Riverside County Hydrology Manual, 1978).

Let Qa, Ta, Ia be the peak runoff flowrate, time of concentration, and rainfall intensity that corresponds to the collection stream with the longer time of concentration. Let Qb, Tb, Ib correspond to the collection stream with the shorter time of concentration. Let Qp and Tp correspond to the confluence peak discharge and time of concentration, respectively. Then the following situations are possible:

(1). If the collection streams have the same time of concentration, then the values are directly summed,

$$Qp = Qa + Qb; \; Tp = Ta = Tb$$

(2). If the collections streams have different times of concentration, the tributary discharge values may be adjusted as follows:

(i) The most frequent case is where the collection stream with the longer time of concentration has the larger discharge. Then the smaller discharge value is adjusted by the ratio of rainfall intensities

$$Qp = Qa + Qb(Ia/Ib); \; Tp = Ta$$

(ii) In some cases, the collection stream with the shorter time of concentration has the larger discharge. Then the smaller discharge is adjusted by a ratio of the Tc values

$$Qp = Qb + Qa(Tb/Ta); \; Tp = Tb$$

4.11. Presentation of Product

Of interest to many civil engineers in both the private and public sectors is a standardized means of presenting rational method calculations. Many local governmental flood control agencies at the city and county levels use a standard tabulation form such as shown in Fig. 4.5. Consequently, a computer program prepared to perform master planning and design studies within a flood control district should be designed to produce a product which exactly satisfies the local agency requirements for study submittals. A typical study format is shown later in this chapter (Fig. 4.8) which includes the necessary hydrologic and geographic data requirements as well as the overall presentation of the product.

STUDY NAME: ______

_______YEAR STORM 1-HOUR RAINFALL (inch)=_____; INTENSITY SLOPE = _____

CALCULATED BY:
CHECKED BY:
PAGE NUMBER OF

CONCENTRATION POINT NUMBER	SOIL TYPE	DEV. TYPE	AREA (ACRES)	I in/h	C	Q (SUB)	Q TOTAL	SLOPE ft/ft	SECTION	V fps.	PATH ft.	T min.	TC min.	HYDRAULICS AND NOTES

Fig. 4.5. RATIONAL METHOD COMPUTATION FORM

4.12. A Regional Master Plan of Drainage (MPD)

With land development, the issues of how to control the change in runoff quantities have resulted in the development of policy statements that stipulate the design criteria for flood control. Due to fractional development, however, where only portions of a watershed are developed at any one time, the global flood control issues must be handled at a governmental level such as a city, county, state, or Federal agency. Capital must be obtained to develop the entire mainline flood control system when only a small portion of the entire watershed is currently developed.

In order to implement a regional flood control system, an assessment based on property holdings (or other policy statement) is used to initiate a long term bond to finance the construction of the project. This assessment often includes the cost of major drainage facilities and also the cost of storm drain collector systems that drain small portions of the catchment, yet are still too costly for a small development project to finance. This layout of the collector systems and the ultimate regional flood control facilities is called the master plan of drainage or MPD.

The MPD study develops a schematic of the entire flood control system and includes all major capital expenses including land costs, construction, engineering, maintenance over the life span of a facility, and the cost of storm drains down to a nominal size such as a 36-inch diameter pipe. Based on the MPD, a total regional system cost can be estimated (and adjusted for inflation) in order to develop an acreage fee to charge for development building permits. That is, all future development would be required to pay an equitable drainage fee based on land holding size (or other policy). The drainage fee oftentimes can be offset by the actual construction of necessary off-site flood control facilities for the specific development project. Should the off-site costs exceed the acreage fee charges, the agency reimburses the developer when other acreage fees can be safely directed accordingly. The use of an MPD acreage fee enables large regions, watersheds, or cities to modernize flood control systems, remove system deficiences, or construct new facilities.

Fundamental to the MPD, of course, is the planning and development of the MPD system schematic. A new and convenient approach is to develop a computerized link-node model of the entire MPD system by means of a rational method technique based on the subarea summation model or the nodal point method. Such an MPD schematic can be stored on a few microcomputer data discs, which enables the MPD to be updated by a simple editing of the data file, resulting in an updated MPD acreage fee.

The batch version of FORTRAN codes provided in this chapter enables such a MPD to be developed. Although the MPD is based on the rational method, which has theoretical difficulties for large areas, the main purpose of the MPD is to provide an equitable drainage fee which, due to mass-produced pipeline sizes, the errors in the rational method typically do not significantly impact a change in design pipe sizes.

4.13. Application: A City Master Plan of Drainage

The City of Rancho Cucamonga, California, developed a MPD for the entire city using a popular microcomputer and a proprietary Knowledge-Based System (KBS) version of the batch programs provided in this chapter. The MPD preparation process included (1) data collection and synthesis; (2) collection of existing condition and ultimate planned condition development patterns; (3) development of major facility watershed boundaries; (4) development of the MPD link-node schematics; (5) development of the MPD link-node models based on the subarea addition method; (6) calibration of the rational method peak flow rate equation (Eq. 4.9) to local rainfall frequency and runoff frequency data; and (7) preparation of MPD system sizing, maps, atlasses, and the MPD acreage fee.

4.13.1. MPD Data Collection

Information collected for the MPD includes local rain gage information for the development of intensity-duration curves for several return frequencies (e.g., 10-, 25-, 50-, and 100-year return frequencies); regionalized rainfall isohyetal information (see Fig. 4.6) to compare regionalized intensity-duration relationships to the local intensity-duration relationships; SCS soil group information providing locations of SCS type A,B,C, and D soils; ultimate condition development and planning information; topographic information such as recent aerial surveys and U.S.G.S. topographic maps; right-of-way information showing easements or other public land holdings; stream gage data and pertinent watershed characteristics information for use in rational method model calibration; local policy statements regarding levels of flood protection; utility information that may obstruct MPD facility construction; and other information that may impact the MPD development (e.g., financing availability, costs, etc.) This information is usually summarized on maps, one map for each parameter or consideration.

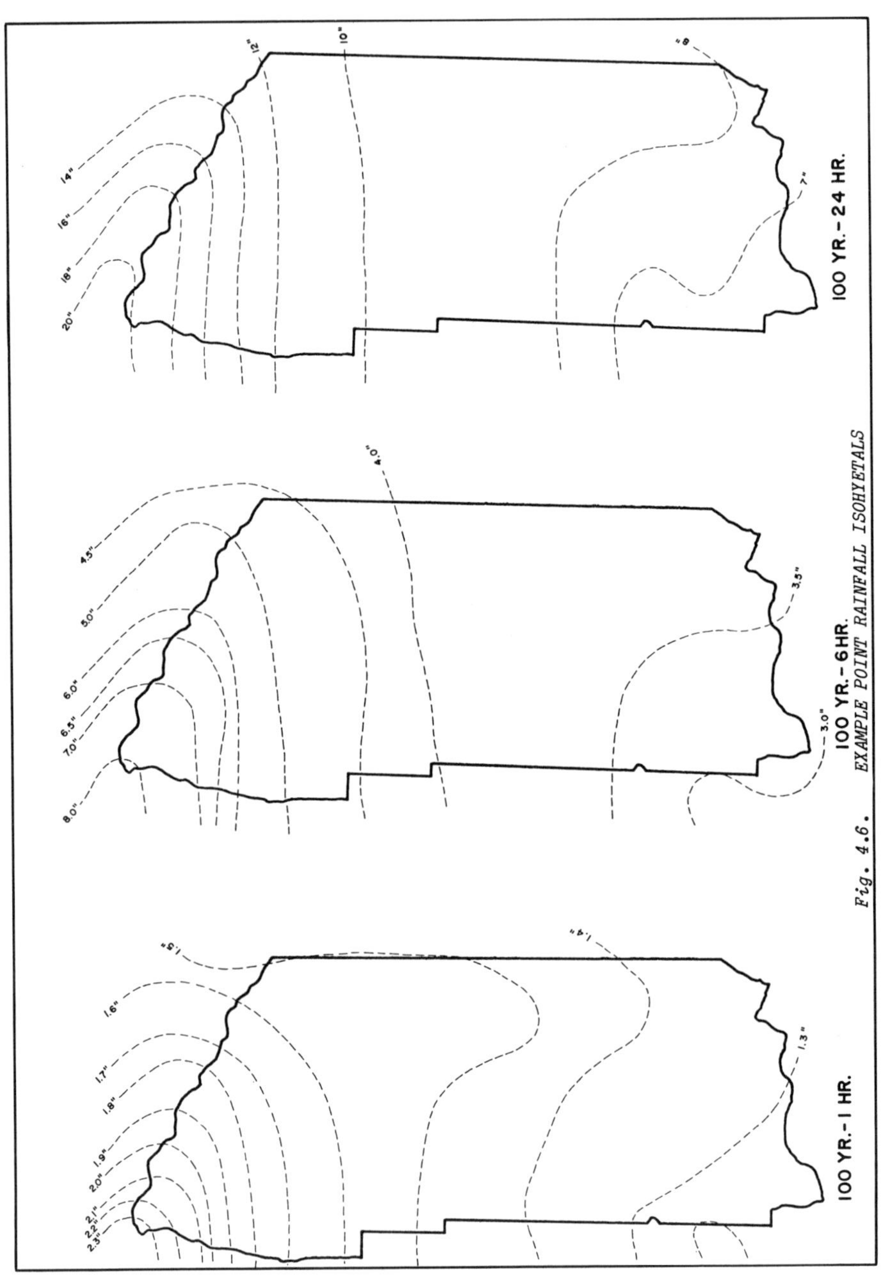

Fig. 4.6. EXAMPLE POINT RAINFALL ISOHYETALS

4.13.2. Existing Conditions and Ultimate Development Conditions

In order to calibrate the rational method peak flow rate equation to represent local flood frequency tendencies, hydrologic studies have to be prepared for historic and current catchment conditions in order to correlate modeling data to the available runoff data. These modeling studies will aid in the selection of suitable values for the rational method runoff coefficient based on land use, soil type and cover, and other pertinent variables.

After calibrating the model to existing or historic conditions of the catchments (where runoff data is available), the parameters are modified (usually an increase in the runoff coefficient due to an increase in imperviousness, and a decrease in Tc due to channel improvements) to represent the ultimate development planned for the total watershed being studied.

4.13.3. Development of Major Facility Watershed Boundaries

The city MPD can generally be subdivided into several major flood control systems that oftentimes serve several square miles of area. These watersheds are analyzed as separate problems and lead to the convenient representation of a separate MPD link-node schematic for each major system. Using topographic information supplemented by field inspection, and after analyzing the existing storm drain system layouts, the watershed boundaries for each major facility can be developed and a separate catchment map prepared.

4.13.4. Development of MPD Link-Node Schematics

Each of the major system catchments can now be independently studied in detail. After the hydrologic and ultimate development land use patterns are plotted on the catchment map, the link-node schematic is developed by subdividing the watershed area into small subareas tributary to their respective flowpaths (Fig. 4.7). Runoff flowpaths are identified and "links" defined through successive subareas such as to result in a rational method link-node schematic (see example problem 4.2). Node numbers are usually numbered sequentially, increasing in value along the flowpath in the downstream direction. It is convenient to number separate streams by hundreds (e.g., 101, 102, 103, etc. for stream #1; 201, 202, 203, etc. for stream #2, and so forth), and separate major facilities in the thousands (or ten-thousands).

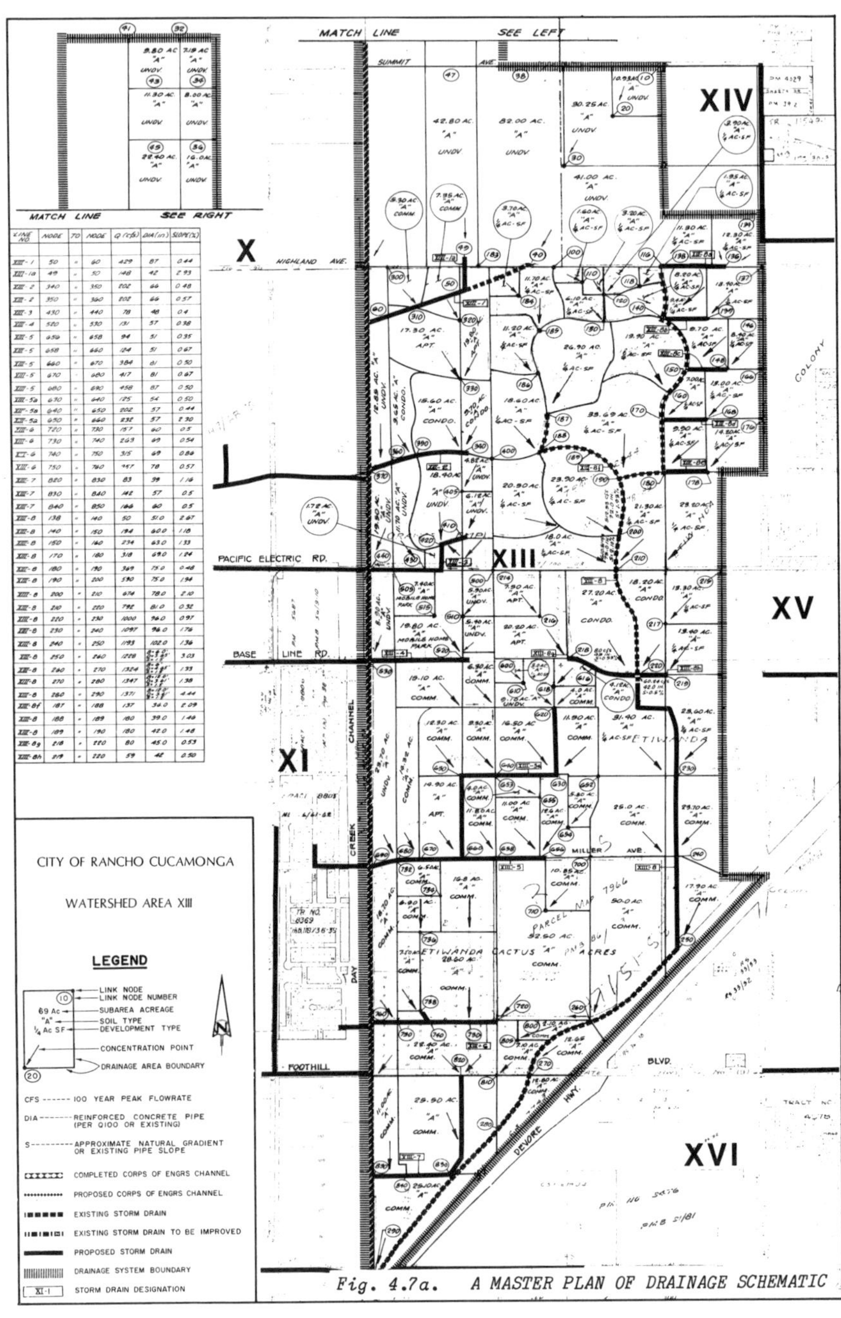

LINE NO.	NODE	TO	NODE	Q (cfs)	DIA (in)	SLOPE (%)
XIII-1	50	"	60	429	87	0.44
XIII-1a	49	"	50	148	42	2.93
XIII-2	340	"	350	202	66	0.48
XIII-2	350	"	360	202	66	0.57
XIII-3	430	"	440	78	48	0.4
XIII-4	520	"	530	131	57	0.38
XIII-5	656	"	658	94	51	0.35
XIII-5	658	"	660	124	51	0.67
XIII-5	660	"	670	384	81	0.50
XIII-5	670	"	680	417	81	0.67
XIII-5	680	"	690	458	87	0.50
XIII-5a	630	"	640	125	54	0.50
XIII-5a	640	"	650	202	57	0.44
XIII-5a	650	"	660	232	57	2.30
XIII-6	720	"	730	157	60	0.5
XIII-6	730	"	740	263	69	0.54
XIII-6	740	"	750	315	69	0.86
XIII-6	750	"	760	357	78	0.57
XIII-7	820	"	830	83	39	1.16
XIII-7	830	"	840	142	57	0.5
XIII-7	840	"	850	166	60	0.5
XIII-8	138	"	140	50	51.0	2.67
XIII-8	140	"	150	194	60.0	1.18
XIII-8	150	"	160	234	63.0	1.33
XIII-8	170	"	180	318	69.0	1.24
XIII-8	180	"	190	369	75.0	0.48
XIII-8	190	"	200	590	75.0	1.94
XIII-8	200	"	210	674	78.0	2.10
XIII-8	210	"	220	792	81.0	0.32
XIII-8	220	"	230	1000	96.0	0.97
XIII-8	230	"	240	1097	96.0	1.76
XIII-8	240	"	250	1193	102.0	1.36
XIII-8	250	"	260	1228	[illegible]	3.03
XIII-8	260	"	270	1324	[illegible]	1.33
XIII-8	270	"	280	1347	[illegible]	1.38
XIII-8	280	"	290	1371	[illegible]	4.44
XIII-8f	187	"	188	137	36.0	2.09
XIII-8	188	"	189	180	39.0	1.46
XIII-8	189	"	190	180	42.0	1.48
XIII-8g	218	"	220	80	45.0	0.53
XIII-8h	219	"	220	59	42	0.50

Fig. 4.7a. A MASTER PLAN OF DRAINAGE SCHEMATIC

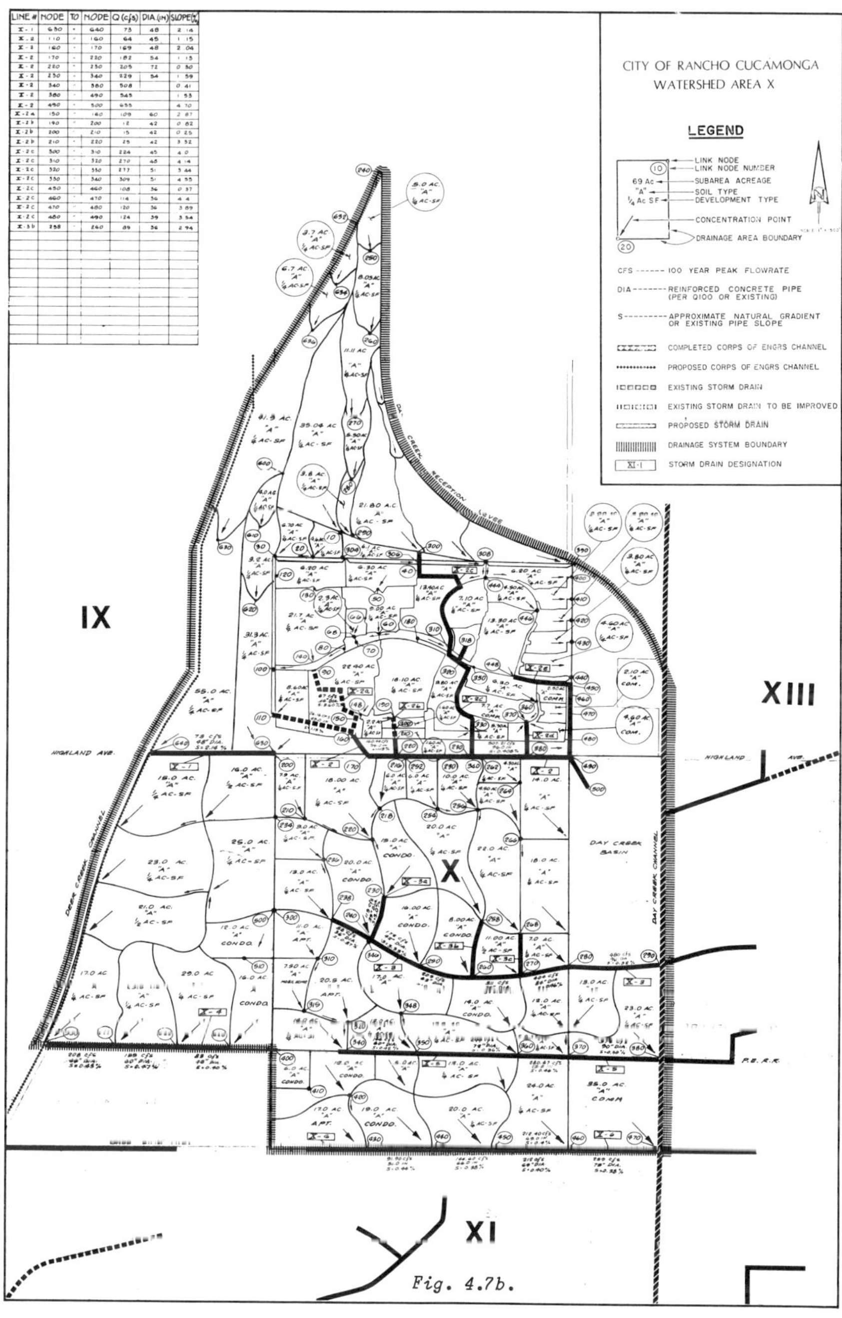

LINE #	NODE	TO	NODE	Q (cfs)	DIA (IN)	SLOPE (%)
X-1	630	"	640	73	48	2.14
X-2	110	"	160	64	45	1.15
X-2	160	"	170	169	48	2.04
X-2	170	"	220	182	54	1.13
X-2	220	"	230	205	72	0.30
X-2	230	"	340	229	54	1.59
X-2	340	"	380	508		0.41
X-2	380	"	490	545		1.53
X-2	490	"	500	655		4.70
X-2a	150	"	160	109	60	2.87
X-2b	190	"	200	12	42	0.82
X-2b	200	"	210	15	42	0.25
X-2b	210	"	220	25	42	3.32
X-2c	300	"	310	224	45	4.0
X-2c	310	"	320	270	48	4.14
X-2c	320	"	330	277	51	3.44
X-2c	330	"	340	309	51	4.55
X-2c	450	"	460	108	36	0.37
X-2c	460	"	470	114	36	4.4
X-2c	470	"	480	120	36	3.89
X-2c	480	"	490	124	39	3.54
X-3b	258	"	260	89	36	2.94

Fig. 4.7b.

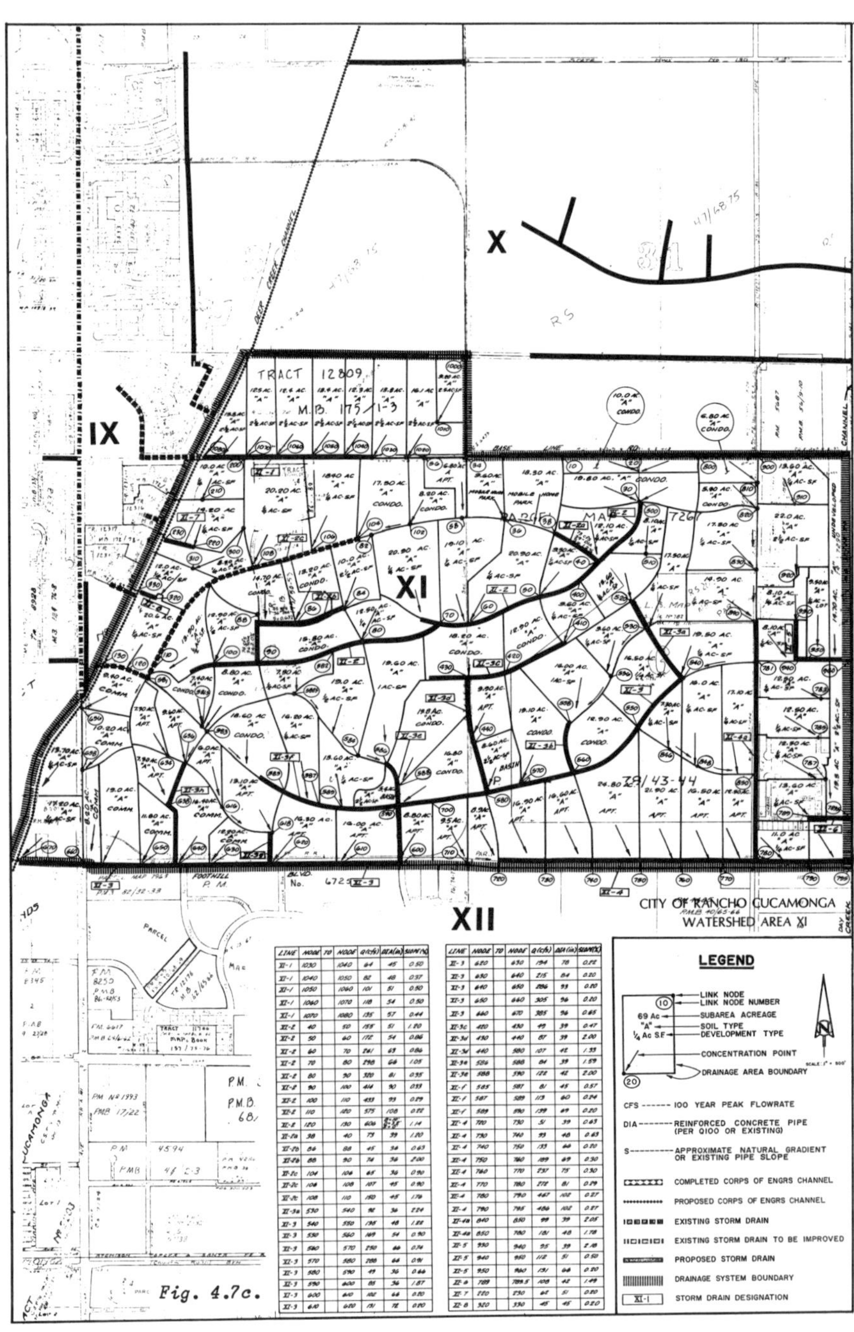

LINE	NODE	TO	NODE	Q(cfs)	DIA(in)	SLOPE(%)
XI-1	1030		1040	64	45	0.50
XI-1	1040		1050	82	48	0.37
XI-1	1050		1060	101	51	0.50
XI-1	1060		1070	118	54	0.50
XI-1	1070		1080	135	57	0.44
XI-2	40		50	155	51	1.20
XI-2	50		60	172	54	0.86
XI-2	60		70	261	63	0.86
XI-2	70		80	298	66	1.05
XI-2	80		90	320	81	0.35
XI-2	90		100	414	90	0.33
XI-2	100		110	433	93	0.29
XI-2	110		120	575	108	0.22
XI-2	120		130	606	[illegible]	1.14
XI-2a	38		40	73	39	1.20
XI-2b	84		88	45	36	0.63
XI-2b	88		90	74	36	2.00
XI-2c	104		106	65	36	0.90
XI-2c	106		108	107	45	0.90
XI-2c	108		110	150	45	1.76
XI-3a	530		540	92	36	2.24
XI-3	540		550	195	48	1.22
XI-3	550		560	169	54	0.90
XI-3	560		570	250	66	0.74
XI-3	570		580	288	66	0.91
XI-3	580		590	49	36	0.66
XI-3	590		600	85	36	1.87
XI-3	600		610	102	66	0.20
XI-3	610		620	131	72	0.20

LINE	NODE	TO	NODE	Q(cfs)	DIA(in)	SLOPE(%)
XI-3	620		630	194	78	0.22
XI-3	630		640	215	84	0.20
XI-3	640		650	286	93	0.20
XI-3	650		660	305	96	0.20
XI-3	660		670	385	96	0.65
XI-3c	420		430	49	39	0.47
XI-3d	430		440	87	39	2.00
XI-3d	440		580	107	42	1.33
XI-3e	586		588	84	39	1.59
XI-3e	588		590	122	42	2.00
XI-f	585		587	81	45	0.57
XI-f	587		589	113	60	0.24
XI-f	589		590	139	69	0.20
XI-4	720		730	51	39	0.63
XI-4	730		740	93	48	0.63
XI-4	740		750	133	66	0.20
XI-4	750		760	189	69	0.30
XI-4	760		770	237	75	0.30
XI-4	770		780	272	81	0.29
XI-4	780		790	467	102	0.27
XI-4	790		795	486	102	0.27
XI-4a	840		850	99	39	2.05
XI-4a	850		780	181	48	1.78
XI-5	930		940	95	39	2.18
XI-5	940		950	112	51	0.50
XI-5	950		960	131	66	0.20
XI-6	789		789.5	108	42	1.49
XI-7	220		230	62	51	0.20
XI-8	320		330	45	45	0.20

Fig. 4.7c.

4.13.5. Development of the Link-Node Models

Each subarea should have the appropriate hydrologic data written on the link-node schematic map to facilitate an easy data entry to the computer program. Data includes subarea size, development type, SCS soil group, upstream and downstream node number and associated topographic elevations, flowpath length, and necessary hydraulic information such as Manning's friction factor, channel basewidth, and so forth.

The hydrologic information is entered into the data base starting from the upstream subarea and proceeding in a downstream direction to a confluence point where two or more streams merge. After all streams tributary to the confluence have been entered into the model, the confluence is solved to estimate a representative peak flow rate (Q) for the entire catchment tributary to the confluence point. Once the confluence analysis is completed, the study resumes in the downstream direction from the now solved confluence point. In this fashion, the link-node schematic is built up in a manner analogous to an analysis prepared "by hand" in that successive reaches are analyzed and the MPD facility (e.g., pipe or channel, streetflow, etc.) is sized and approved before proceeding to the next link in the schematic.

It is noted that the computer program should normally be operated such as to have only one confluence point in the computer memory at any one time. After all streams which are tributary to the confluence have been entered into the confluence submodel and the analysis completed, the computer memory is freed of the confluence stream data enabling identification of another confluence point. With this technique, the link-node schematic is constructed in a proper sequence with link sizing always completed before proceeding downstream.

4.13.6. Calibration of the Rational Method

After link-node schematics have been prepared for the ultimate watershed planned condition, similar link-node schematics are prepared for catchments where stream gage data is sufficient to prepare reliable flood frequency curves. Because the flood frequency data represents historic and/or existing conditions, the link node model must represent those conditions. Additionally, care must be taken to identify the effective catchment area that principally contributes to the catchment peak flow as detention effects are not included in the peak flow equation. The rational method C value is then estimated by correlating model peak flow estimates based on T-year rainfall to T-year runoff flood frequency peak flow rates. The resulting C-values developed from several catchments are then correlated to the catchment characteristics such as development type, soil group, and so forth. In the general case, however, there is seldom sufficient data to conclude C values for the range of development conditions anticipated for the MPD, and estimates must be used such as Table 4.1 for the use in Eq. (4.8), or use of modified C values such as estimated using Eq. (4.9).

4.13.7. Preparation of MPD System Information

After incorporating the calibrated or estimated hydrologic parameters in the several link-node schematics, the MPD system can be sized. Usually, the smaller pipeline systems that drain to the major open channel facilities are a major MPD cost. Additionally, changes in MPD alignment or development patterns typically are reflected in changes in the smaller pipelines with minor changes in the open channels. Consequently, the link-node model computer program includes an option for the computer program to "size" pipelines (see Program 4.5). Preparing the MPD with computer estimated pipesizes greatly increases analysis efficiency due to the ability to quickly "update" the MPD from minor changes in the link-node schematic.

Once the MPD is developed and the data file stored for future editing, all the MPD pipeline, channel, engineering, right-of-way, and other costs can be estimated on a facility-by-facility basis. The resulting cost, adjustable for inflation and contingency formulas, can now be applied on an acreage basis or other policy statement in order to secure moneys for major MPD facility improvements.

4.14. A Rational Method Planning/Design Computer Program

Each of the rational method subarea modeling approaches utilize identical submodels for estimating (1) the initial time of concentration, (2) channel or pipeflow travel time, (3) runoff coefficients, (4) rainfall intensity values, (5) and confluence values at the junction of two or more collection streams. Therefore, once computer program subroutines are developed for each of these submodels, a main driver program can be developed that manipulates the individual hydrologic processes to formulate a link-node model of the watershed based on the rational method strategy desired. In Table 4.2, descriptions are listed for the computer programs used to approximate the hydrologic processes which occur in a rational method study of an urban watershed. Combining these subroutines using a simple main menu that branches to the selected submodel will result in a totally design-interactive computer program. Programs 4.2, 4.3, and 4.4 follow

TABLE 4.2
RATIONAL METHOD PROGRAM SUBROUTINES

Program Number	Description
4.1	main driver program
4.2	utilizes the Kirpich formula for estimating the initial subarea time of concentration Tc (Fig.4.1)
4.3 (no input)	calculates rainfall intensities by log-log interpolation (Fig. 4.10)
4.4 (no input)	estimates a runoff coefficient from Figs. 4.4a,b,c,d
4.5	estimates pipeflow travel time for a user-specified or computer estimated pipe size
4.6	estimates travel time in a trape-zoidal channel
4.7	estimates travel time in a street section of arbitrary size
4.8	estimates travel time in a pavement V-gutter
4.9	estimates confluence values
4.10	allows entry of specified data at a node
4.11	permits addition of subarea runoff to the mainline collection stream

4.5 estimates pipeflow travel time by computing the peak flow normal depth and determining the time of travel based upon the normal depth flow velocity. Flows which result in a normal depth greater than 0.82 of the pipe diameter are assumed to cause the pipe to flow full. If the pipe size is not specified, this program estimates a pipesize in 3- and 6-inch increments by utilizing a pipeflow with a normal depth less than or equal to 0.82 of the pipe diameter. Pipe slope is based on the gradient computed from the ground surface elevations entered concurrently with the subarea's upstream and downstream node numbers. However, a factor is included (set by the user) such that the natural gradient of the land is reduced (usually by about 10 percent) in order to account for minor losses within the pipe system. The pipe sizes are estimated by assuming this adjusted gradient of the topography between two nodal points to equal the slope of the pipe for normal depth flows. Program 4.8 estimates trapezoidal channel flow travel time based upon the normal depth flow velocity. Program 4.7 examines street flow travel time for two conditions: (1) all flow on one side of the street section, including the splitflow effects when the flow depth exceeds the street crown, and (2) equal flow on both sides of the street centerline. All flows outside of the street curbs are assumed negligible (that is, the water is in a ponded condition). Program 4.9 models a confluence with up to 5 independent collection streams. It is based upon the linear confluence formula presented in the text.

The study approach is to subdivide the watershed into subareas such as shown in Fig. 4.9. Nodal points are defined along the main flowpath at the upstream and downstream points of each subarea. Computer results are correlated to the watershed schematic by means of these nodal points. The programs are intended to be combined into a menu-driven program system in which the user interacts with the main menu. Starting at the most upstream nodal point of a collection stream, the program user selects which submodel is to be first employed. Usually, the first model used is the initial subarea program and the user enters the appropriate hydrologic data such as the subarea development type, soil group, area size, upstream and downstream node elevations, and length of the main flowpath. The subroutine computes the initial subarea Tc, the corresponding runoff coefficient and rainfall intensity, and the initial subarea runoff. The program should display this information on the CRT for the user to review and accept or reject. If the information is acceptable, the entered hydrologic data is permanently stored in a data file; if the computer results are unacceptable, the user rejects the submodel results and the computer program returns to the previous nodal point and the main menu for process selection.

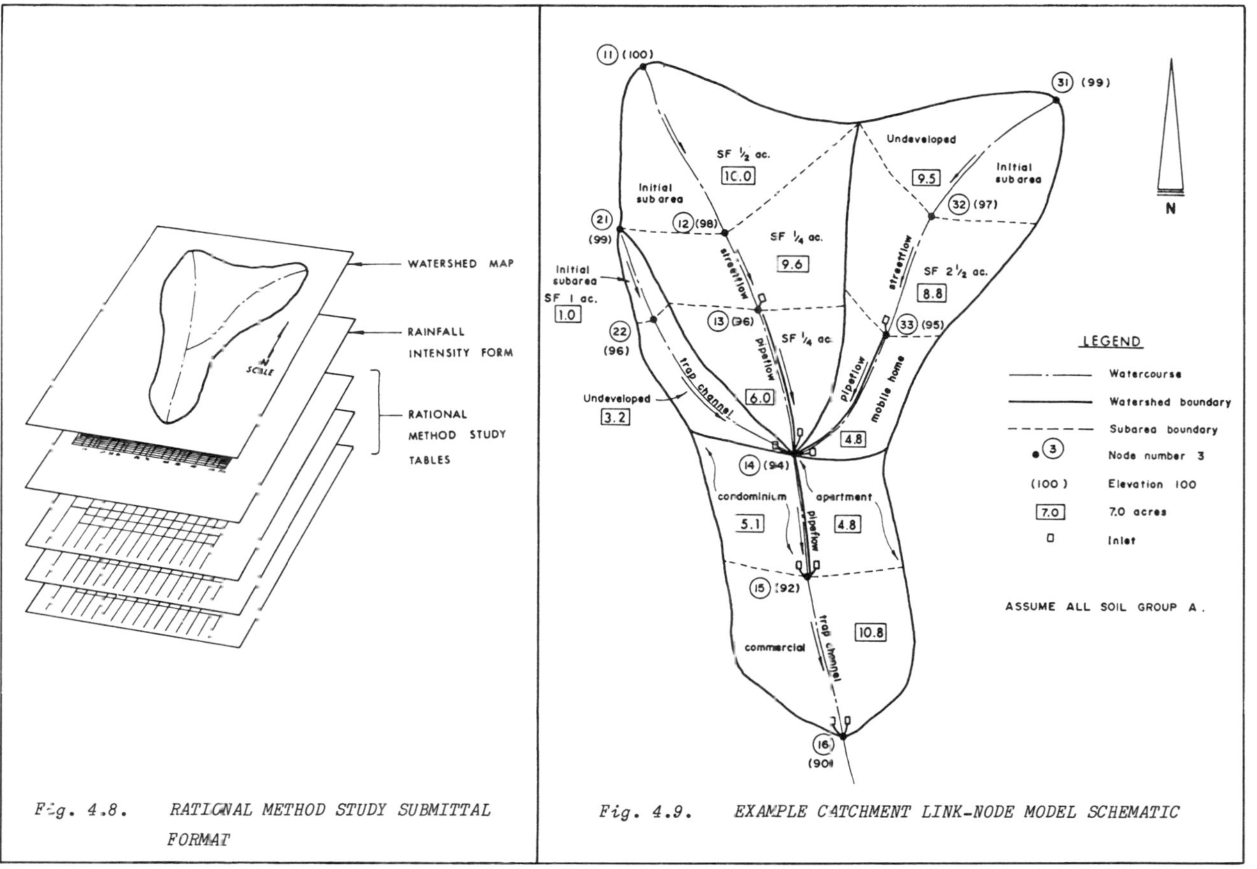

Fig. 4.8. RATIONAL METHOD STUDY SUBMITTAL FORMAT

Fig. 4.9. EXAMPLE CATCHMENT LINK-NODE MODEL SCHEMATIC

If the user had accepted the most recently computed information, the main program returns to the menu display for the user to select the next hydrologic submodel. The main program should store the recently computed Q, Tc, and the total area. In this manner, should the user now elect to employ the channel travel time program, the normal depth computed will be based on the stored peak Q value, and the travel time will be directly added to the stored Tc value, providing the time of concentration at the downstream point of the channel. Thus, the computer program follows the rational method modeling approach interactively rather than the user creating a data file to be operated upon by the program in a batch mode. Using such a menu-driven approach allows the watershed to be studied on the first pass, and in addition, the entered hydrologic data are stored for subsequent editing and master plan updating.

In the following pages, computer listings are provided for each of the discussed submodels. The language used is FORTRAN, and the codes can be keyed into most of the currently available microcomputers. The data entry requirements are presented in the form of screen text pages that contain suggested user-friendly data entry prompts, as well as other user-friendly program commands and features. Details of these screen text pages are discussed in the following section.

4.15. Computer-Aided Design Interaction

The computer programs were developed to aid the engineer in a computer-aided interactive mode rather than the batch mode that is associated to water resources software. In this fashion, the software is formulated on a system level where the individual submodels are employed as selected by the engineer from the main-menu, and the computer results reviewed by the engineer prior to proceeding to the next hydrologic process. This type of programming approach can be directly applied to other link-node models where the links direct the logic process in one direction only. For example, the rational method planning/design program system proceeds in the "downstream" direction with the entire watershed tributary to a node completely described by three characteristic variables: peak runoff rate, time of concentration, and total area. Thus the hydrologic process used to link to the next downstream node acts only upon the most recently computed values of the three characteristic variables.

Because the main purpose of studying the watershed is to determine an appropriate flood control system to safely contain the design peak flow rates, each link of the link-node model can be properly sized and evaluated as to "success or failure" prior to proceeding to the next link or hydrologic process.

In comparison, the various submodels can be combined into a batch mode of operation where the engineer builds a data file containing all the necessary data for each hydrologic process or link used to develop a link-node model of the watershed. The program system then operates upon the data file to generate the model solutions. The user then reviews the computed results for unacceptable design conditions (e.g., such as streetflow above the top of curb, or excessively high flow velocities in a user-specified pipe size linking two nodal points, etc.) and identifies the necessary alterations in the link-node model data file to remedy the unacceptable condition. The program is re-executed and another review of the computed results is performed. This procedure is repeated until the entire link-node model provides an acceptable flood control system design.

The batch mode of operation requires considerably more computational effort, time expenditure, and frustration to the engineer than the interactive mode. Therefore, the engineer should develop the main driver branching program using the basic user-friendly environment as discussed in Hromadka et al. (1983b). The main program data entry sequences for each submodel should be developed such that the communication/presentation (C/P) provides an easy-to-use and self-teaching environment. Some of the major requirements for such a user-friendly environment are as follows:

(1). The C/P should present all data entry prompts and computed results in such a way that any engineer could readily evaluate the information. Avoid use of abbreviations and other 'short-cuts'.

(2). All engineering units should be displayed.

(3). Any program system flow logic should be clearly described in the program in order to reduce the first-time user learning curve.

(4). All program system commands should be consistently displayed between submodels (and between separate computer programs) so that the user can operate special data entry or editing features without confusion.

(5). All data file management operations (such as opening, closing, and saving data files) should be programmed interior to the system program in order to provide ease of use.

The submodel data entry prompts for the provided program listings are presented in a typical C/P for use on currently available microcomputers. The viewing displays are constructed as pages that contain sets of data entry prompts grouped together according to the submodel process selected. Each of the pages contain the following set of operational commands located at the bottom of the CRT screen:

(1). TOP. Clears the screen, redisplays the page information, and returns the program to the first data entry prompt of the page.

(2). BACK. Returns the program to the previous page (if one exists), and positions the program to the first data entry prompt.

(3). MAIN. This command performs several important tasks. First, the program system data file is properly saved and closed so that all data entries are protected, and the data file is available for later use. Second, the command terminates the submodel process. Third, the command returns the program system to the main driver program menu.

(4). EXIT. This command is identical to the MAIN command, except the program system is terminated.

It should be noted that these four commands can be entered at any time and at any data entry prompt within the program system. Thus, if the user should wish to exit the program while entering the data needed to solve for pipeflow travel time between two nodal points (Program 4.5), then the user simply enters the word EXIT at any data entry prompt. It should also be noted that the C/P pages contain a description of each data entry as well as the allowable value range for data entry. Each data entry is checked for range limits prior to proceeding to the next data entry prompt. If the entered data is outside of the allowable value range, an error message is displayed to the user and the program returns to the invalid data entry point for another data entry attempt. In this way, the data file development is error free with the first pass of the data entry sequence.

Unfortunately, there are still wide differences between computers and peripheral devices, such as the CRT terminals of different manufacturers. Additionally, the internal operating systems of the computers differ. Developing a user-friendly environment for the computer program system is therefore dependent on the computer system hardware and associated software. The necessary steps in opening files, closing files, subroutine employment, overlay structuring, cursor addressing, clearing the screen, program execution, and other operations must be obtained from the "user-guides" of the appropriate operation manuals.

4.16. Example 4.2. Rational Method Program Application

The following example problem illustrates the use of the Subarea Summation Model for rational method hydrologic studies of urban watersheds. The example problem presentation contains the following information:

Figure Number	Description
4.9	example problem drainage system
4.10	intensity-duration point rainfall plot
4.11	example problem computer program results and example tabulation form output

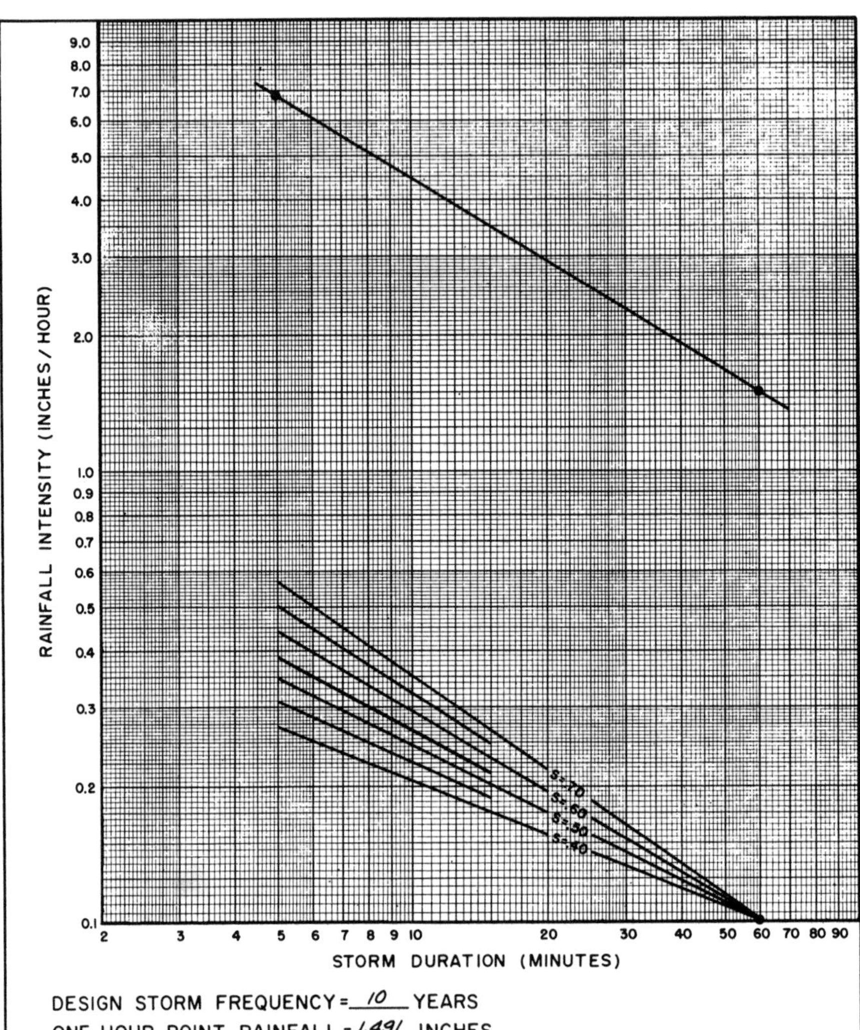

Fig. 4.10. EXAMPLE PROBLEM RAINFALL INTENSITY-DURATION PLOT

```
============================================================================
             RATIONAL METHOD HYDROLOGY COMPUTER PROGRAM

============================================================================

<<<<<<<<<<<<<<<<<<<<<<<<<<<<<<<<<<<<<<>>>>>>>>>>>>>>>>>>>>>>>>>>>>>>>>>>>>>>
           (C) Copyright 1983 Advanced Engineering Software [AES]





<<<<<<<<<<<<<<<<<<<<<<<<<<<<<<<<<<<<<<>>>>>>>>>>>>>>>>>>>>>>>>>>>>>>>>>>>>>>

*********DESCRIPTION OF RESULTS*********************************************
*                                                                          *
*                            EXAMPLE PROBLEM                               *
*                                                                          *
****************************************************************************

----------------------------------------------------------------------------
   USER SPECIFIED HYDROLOGY AND HYDRAULIC MODEL INFORMATION:
----------------------------------------------------------------------------
   USER SPECIFIED STORM EVENT(YEAR) =   25.00
   SPECIFIED MINIMUM PIPE SIZE(INCH) =  24.00
   SPECIFIED PERCENT OF GRADIENTS(DECIMAL) TO USE FOR FRICTION SLOPE =  .90
   10-YEAR STORM 60-MINUTE INTENSITY(INCH/HOUR) =  1.270
   100-YEAR STORM 60-MINUTE INTENSITY(INCH/HOUR) =  1.900
   COMPUTED RAINFALL INTENSITY DATA:
   STORM EVENT =   25.00   1-HOUR INTENSITY(INCH/HOUR) = 1.4910
   SLOPE OF INTENSITY DURATION CURVE =  .6000

   SBC HYDROLOGY MANUAL "C"-VALUES USED

<<<<<<<<<<<<<<<<<<<<<<<<<<<<<<<<<<<<<<>>>>>>>>>>>>>>>>>>>>>>>>>>>>>>>>>>>>>>

                    Advanced Engineering Software [AES]
                              SERIAL No. L0008
                    REV. 3.0      RELEASE DATE: 5/17/83

<<<<<<<<<<<<<<<<<<<<<<<<<<<<<<<<<<<<<<>>>>>>>>>>>>>>>>>>>>>>>>>>>>>>>>>>>>>>


****************************************************************************
   FLOW PROCESS FROM NODE    11.00 TO NODE    12.00 IS CODE =   2
----------------------------------------------------------------------------
   >>>>>RATIONAL METHOD INITIAL SUBAREA ANALYSIS<<<<<
============================================================================
          ASSUMED INITIAL SUBAREA UNIFORM
          DEVELOPMENT IS SINGLE FAMILY(1/2 ACRE)
   TC = K*[(LENGTH**3)/(ELEVATION CHANGE)]**.2
   INITIAL SUBAREA FLOW-LENGTH =  800.00
   UPSTREAM ELEVATION =     100.00
   DOWNSTREAM ELEVATION =      98.00
   ELEVATION DIFFERENCE =       2.00
   TC =  .422*[(  800.00**3)/(     2.00)]**.2 =   20.380
     25.00 YEAR RAINFALL INTENSITY(INCH/HOUR) =  2.858
```

Fig. 4.11. RATIONAL METHOD HYDROLOGY COMPUTER PROGRAM

```
   SOIL CLASSIFICATION IS "A"
   SINGLE-FAMILY(1/2 ACRE LOT) RUNOFF COEFFICIENT = .7180
   SUBAREA RUNOFF(CFS) =      20.52
   TOTAL AREA(ACRES) =      10.00   TOTAL RUNOFF(CFS) =      20.52

****************************************************************************
   FLOW PROCESS FROM NODE    12.00 TO NODE    13.00 IS CODE =   6
----------------------------------------------------------------------------
   >>>>>COMPUTE STREETFLOW TRAVELTIME THRU SUBAREA<<<<<
============================================================================
   UPSTREAM ELEVATION =      98.00   DOWNSTREAM ELEVATION =      96.00
   STREET LENGTH(FEET) =    350.00   CURB HEIGTH(INCHES) = 8.
   STREET HALFWIDTH(FEET) = 22.00    STREET CROSSFALL(DECIMAL) = .0200
   SPECIFIED NUMBER OF HALFSTREETS CARRYING RUNOFF =  2
          **TRAVELTIME COMPUTED USING MEAN FLOW(CFS) =   30.03
              ***STREET FLOWING FULL***
          STREET FLOWDEPTH(FEET) =    .61
          HALFSTREET FLOODWIDTH(FEET) =    22.00
          AVERAGE FLOW VELOCITY(FEET/SEC.) =     2.86
          PRODUCT OF DEPTH&VELOCITY =     1.74
   STREETFLOW TRAVELTIME(MIN) =   2.04   TC(MIN) =  22.32

     25.00 YEAR RAINFALL INTENSITY(INCH/HOUR) =  2.699
   SOIL CLASSIFICATION IS "A"
   SINGLE-FAMILY(1/4 ACRE LOT) RUNOFF COEFFICIENT = .7335
   SUBAREA AREA(ACRES) =     9.60   SUBAREA RUNOFF(CFS) =    19.00
   SUMMED AREA(ACRES) =    19.60   TOTAL RUNOFF(CFS) =      39.53
   END OF SUBAREA STREETFLOW HYDRAULICS:
   DEPTH(FEET) =  .65   HALFSTREET FLOODWIDTH(FEET) = 22.00
   FLOW VELOCITY(FEET/SEC.) = 3.24   DEPTH*VELOCITY =   2.10

****************************************************************************
   FLOW PROCESS FROM NODE    13.00 TO NODE    14.00 IS CODE =   3
----------------------------------------------------------------------------
   >>>>>COMPUTE PIPEFLOW TRAVELTIME THRU SUBAREA<<<<<
   >>>>>USING COMPUTER-ESTIMATED PIPESIZE (NON-PRESSURE FLOW)<<<<<
============================================================================
   DEPTH OF FLOW IN  39.0 INCH PIPE IS  29.2 INCHES
   PIPEFLOW VELOCITY(FEET/SEC.) =    5.9
   UPSTREAM NODE ELEVATION =     96.00
   DOWNSTREAM NODE ELEVATION =     94.00
   FLOWLENGTH(FEET) =    650.00   MANNINGS N =   .013
   ESTIMATED PIPE DIAMETER(INCH) =   39.00    NUMBER OF PIPES =   1
   PIPEFLOW THRU SUBAREA(CFS) =      39.53
   TRAVEL TIME(MIN.) =   1.83    TC(MIN.) =  24.14

****************************************************************************
   FLOW PROCESS FROM NODE    14.00 TO NODE    14.00 IS CODE =   8
----------------------------------------------------------------------------
   >>>>>ADDITION OF SUBAREA TO MAINLINE PEAK FLOW<<<<<
============================================================================
     25.00 YEAR RAINFALL INTENSITY(INCH/HOUR) =  2.575
   SOIL CLASSIFICATION IS "A"
   SINGLE-FAMILY(1/4 ACRE LOT) RUNOFF COEFFICIENT = .7278
   SUBAREA AREA(ACRES) =     6.00   SUBAREA RUNOFF(CFS) =    11.24
   TOTAL AREA(ACRES) =    25.60   TOTAL RUNOFF(CFS) =    50.77
   TC(MIN) =  24.14

****************************************************************************
```

Fig. 4.11 cont.

```
   FLOW PROCESS FROM NODE    14.00 TO NODE    14.00 IS CODE =   1
----------------------------------------------------------------------------
   >>>>>DESIGNATE INDEPENDENT STREAM FOR CONFLUENCE<<<<<
============================================================================
   CONFLUENCE VALUES USED FOR INDEPENDENT STREAM  1 ARE:
   TIME OF CONCENTRATION(MINUTES) =  24.14
   RAINFALL INTENSITY (INCH./HOUR) =   2.57
   TOTAL STREAM AREA (ACRES) =   25.60
   TOTAL STREAM RUNOFF(CFS) AT CONFLUENCE =     50.77

****************************************************************************
   FLOW PROCESS FROM NODE    21.00 TO NODE    22.00 IS CODE =   2
----------------------------------------------------------------------------
   >>>>>RATIONAL METHOD INITIAL SUBAREA ANALYSIS<<<<<
============================================================================
          ASSUMED INITIAL SUBAREA UNIFORM
          DEVELOPMENT IS SINGLE FAMILY(1-ACRE)
   TC = K*[(LENGTH**3)/(ELEVATION CHANGE)]**.2
   INITIAL SUBAREA FLOW-LENGTH =   400.00
   UPSTREAM ELEVATION =      99.00
   DOWNSTREAM ELEVATION =      96.00
   ELEVATION DIFFERENCE =       3.00
   TC =  .469*[(  400.00**3)/(      3.00)]**.2 =   13.717
     25.00 YEAR RAINFALL INTENSITY(INCH/HOUR) =  3.614
   SOIL CLASSIFICATION IS "A"
   SINGLE-FAMILY(1-ACRE LOT) RUNOFF COEFFICIENT = .7108
   SUBAREA RUNOFF(CFS) =      2.57
   TOTAL AREA(ACRES) =      1.00   TOTAL RUNOFF(CFS) =       2.57

****************************************************************************
   FLOW PROCESS FROM NODE    22.00 TO NODE    14.00 IS CODE =   5
----------------------------------------------------------------------------
   >>>>>COMPUTE TRAPEZOIDAL-CHANNEL FLOW<<<<<
   >>>>>TRAVELTIME THRU SUBAREA<<<<<
============================================================================
   UPSTREAM NODE ELEVATION =      96.00
   DOWNSTREAM NODE ELEVATION =      94.00
   CHANNEL LENGTH THRU SUBAREA(FEET) =    850.00
   CHANNEL BASE(FEET) =      .50    "Z" FACTOR =   1.000
   MANNINGS FACTOR = .030   MAXIMUM DEPTH(FEET) =   2.00
   CHANNEL FLOW THRU SUBAREA(CFS) =       2.57
   FLOW VELOCITY(FEET/SEC) =   1.51   FLOW DEPTH(FEET) =   1.08
   TRAVEL TIME(MIN.) =   9.38   TC(MIN.) =  23.10

****************************************************************************
   FLOW PROCESS FROM NODE    14.00 TO NODE    14.00 IS CODE =   8
----------------------------------------------------------------------------
   >>>>>ADDITION OF SUBAREA TO MAINLINE PEAK FLOW<<<<<
============================================================================
     25.00 YEAR RAINFALL INTENSITY(INCH/HOUR) =  2.644
   SOIL CLASSIFICATION IS "A"
   UNDEVELOPED WATERSHED RUNOFF COEFFICIENT = .6121
   SUBAREA AREA(ACRES) =    3.20   SUBAREA RUNOFF(CFS) =     5.18
   TOTAL AREA(ACRES) =    4.20   TOTAL RUNOFF(CFS) =     7.75
   TC(MIN) =  23.10

****************************************************************************
   FLOW PROCESS FROM NODE    14.00 TO NODE    14.00 IS CODE =   1
----------------------------------------------------------------------------
```

```
   >>>>>DESIGNATE INDEPENDENT STREAM FOR CONFLUENCE<<<<<
==============================================================================
   CONFLUENCE VALUES USED FOR INDEPENDENT STREAM  2 ARE:
   TIME OF CONCENTRATION(MINUTES) =  23.10
   RAINFALL INTENSITY (INCH./HOUR) =   2.64
   TOTAL STREAM AREA (ACRES) =    4.20
   TOTAL STREAM RUNOFF(CFS) AT CONFLUENCE =      7.75

******************************************************************************
   FLOW PROCESS FROM NODE    31.00 TO NODE    32.00 IS CODE =   2
------------------------------------------------------------------------------
   >>>>>RATIONAL METHOD INITIAL SUBAREA ANALYSIS<<<<<
==============================================================================
          ASSUMED INITIAL SUBAREA UNIFORM
          DEVELOPMENT IS: UNDEVELOPED WITH GOOD COVER
   TC = K*[(LENGTH**3)/(ELEVATION CHANGE)]**.2
   INITIAL SUBAREA FLOW-LENGTH =   750.00
   UPSTREAM ELEVATION =      99.00
   DOWNSTREAM ELEVATION =      97.00
   ELEVATION DIFFERENCE =       2.00
   TC =  .937*[(  750.00**3)/(      2.00)]**.2 =   43.326
     25.00 YEAR RAINFALL INTENSITY(INCH/HOUR) =  1.813
   SOIL CLASSIFICATION IS "A"
   UNDEVELOPED WATERSHED RUNOFF COEFFICIENT = .5030
   SUBAREA RUNOFF(CFS) =       8.66
   TOTAL AREA(ACRES) =       9.50   TOTAL RUNOFF(CFS) =       8.66

******************************************************************************
   FLOW PROCESS FROM NODE    32.00 TO NODE    33.00 IS CODE =   6
------------------------------------------------------------------------------
   >>>>>COMPUTE STREETFLOW TRAVELTIME THRU SUBAREA<<<<<
==============================================================================
   UPSTREAM ELEVATION =      97.00   DOWNSTREAM ELEVATION =      95.00
   STREET LENGTH(FEET) =   550.00   CURB HEIGTH(INCHES) = 8.
   STREET HALFWIDTH(FEET) = 22.00   STREET CROSSFALL(DECIMAL) = .0200
   SPECIFIED NUMBER OF HALFSTREETS CARRYING RUNOFF =  2
          **TRAVELTIME COMPUTED USING MEAN FLOW(CFS) =    12.53
          STREET FLOWDEPTH(FEET) =    .50
          HALFSTREET FLOODWIDTH(FEET) =    17.31
          AVERAGE FLOW VELOCITY(FEET/SEC.) =     1.97
          PRODUCT OF DEPTH&VELOCITY =      .99
   STREETFLOW TRAVELTIME(MIN) =   4.66   TC(MIN) =  47.99

     25.00 YEAR RAINFALL INTENSITY(INCH/HOUR) =  1.705
   SOIL CLASSIFICATION IS "A"
   SINGLE-FAMILY(2.5-ACRE LOT) RUNOFF COEFFICIENT = .5179
   SUBAREA AREA(ACRES) =     8.80   SUBAREA RUNOFF(CFS) =     7.77
   SUMMED AREA(ACRES) =    18.30   TOTAL RUNOFF(CFS) =     16.43
   END OF SUBAREA STREETFLOW HYDRAULICS:
   DEPTH(FEET) =  .55   HALFSTREET FLOODWIDTH(FEET) = 19.81
   FLOW VELOCITY(FEET/SEC.) =  2.00   DEPTH*VELOCITY =   1.11

******************************************************************************
   FLOW PROCESS FROM NODE    33.00 TO NODE    14.00 IS CODE =   3
------------------------------------------------------------------------------
   >>>>>COMPUTE PIPEFLOW TRAVELTIME THRU SUBAREA<<<<<
   >>>>>USING COMPUTER-ESTIMATED PIPESIZE (NON-PRESSURE FLOW)<<<<<
==============================================================================
   DEPTH OF FLOW IN  33.0 INCH PIPE IS  23.7 INCHES
   PIPEFLOW VELOCITY(FEET/SEC.) =   3.6
```

```
   UPSTREAM NODE ELEVATION =     95.00
   DOWNSTREAM NODE ELEVATION =     94.00
   FLOWLENGTH(FEET) =    700.00   MANNINGS N =  .013
   ESTIMATED PIPE DIAMETER(INCH) =  33.00    NUMBER OF PIPES =   1
   PIPEFLOW THRU SUBAREA(CFS) =      16.43
   TRAVEL TIME(MIN.) =   3.25    TC(MIN.) =  51.24

****************************************************************************
   FLOW PROCESS FROM NODE    14.00 TO NODE    14.00 IS CODE =   8
----------------------------------------------------------------------------
   >>>>>ADDITION OF SUBAREA TO MAINLINE PEAK FLOW<<<<<
============================================================================
     25.00 YEAR RAINFALL INTENSITY(INCH/HOUR) =  1.639
   SOIL CLASSIFICATION IS "A"
   MOBILE HOME PARK DEVELOPMENT RUNOFF COEFFICIENT = .7541
   SUBAREA AREA(ACRES) =     4.80   SUBAREA RUNOFF(CFS) =      5.93
   TOTAL AREA(ACRES) =    23.10   TOTAL RUNOFF(CFS) =    22.36
   TC(MIN) =  51.24

****************************************************************************
   FLOW PROCESS FROM NODE    14.00 TO NODE    14.00 IS CODE =   1
----------------------------------------------------------------------------
   >>>>>DESIGNATE INDEPENDENT STREAM FOR CONFLUENCE<<<<<
   >>>>>AND COMPUTE VARIOUS CONFLUENCED STREAM VALUES<<<<<
============================================================================
   CONFLUENCE VALUES USED FOR INDEPENDENT STREAM  3 ARE:
   TIME OF CONCENTRATION(MINUTES) =  51.24
   RAINFALL INTENSITY (INCH./HOUR) =    1.64
   TOTAL STREAM AREA (ACRES) =   23.10
   TOTAL STREAM RUNOFF(CFS) AT CONFLUENCE =      22.36

   CONFLUENCE INFORMATION:
   STREAM   RUNOFF       TIME   INTENSITY
   NUMBER    (CFS)      (MIN.) (INCH/HOUR)
----------------------------------------------------------------------------
     1       50.77     24.14       2.575
     2        7.75     23.10       2.644
     3       22.36     51.24       1.639

   RAINFALL INTENSITY AND TIME OF CONCENTRATION RATIO
   FORMULA(SBC) USED FOR  3 STREAMS.
   VARIOUS CONFLUENCED RUNOFF VALUES ARE AS FOLLOWS:
         68.85    66.41    59.49
   COMPUTED CONFLUENCE ESTIMATES ARE AS FOLLOWS:
   RUNOFF(CFS) =      68.85   TIME(MINUTES) =   24.142
   TOTAL AREA(ACRES) =      52.90

****************************************************************************
   FLOW PROCESS FROM NODE    14.00 TO NODE    15.00 IS CODE =   3
----------------------------------------------------------------------------
   >>>>>COMPUTE PIPEFLOW TRAVELTIME THRU SUBAREA<<<<<
   >>>>>USING COMPUTER-ESTIMATED PIPESIZE (NON-PRESSURE FLOW)<<<<<
============================================================================
   DEPTH OF FLOW IN  45.0 INCH PIPE IS  36.7 INCHES
   PIPEFLOW VELOCITY(FEET/SEC.) =   7.1
   UPSTREAM NODE ELEVATION =     94.00
   DOWNSTREAM NODE ELEVATION =     92.00
   FLOWLENGTH(FEET) =    550.00   MANNINGS N =  .013
   ESTIMATED PIPE DIAMETER(INCH) =  45.00    NUMBER OF PIPES =   1
   PIPEFLOW THRU SUBAREA(CFS) =      68.85
```

```
  TRAVEL TIME(MIN.) =    1.28    TC(MIN.) =  25.43

****************************************************************************
  FLOW PROCESS FROM NODE    15.00 TO NODE    15.00 IS CODE =   8
----------------------------------------------------------------------------
  >>>>>ADDITION OF SUBAREA TO MAINLINE PEAK FLOW<<<<<
============================================================================
    25.00 YEAR RAINFALL INTENSITY(INCH/HOUR) =  2.496
  SOIL CLASSIFICATION IS "A"
  CONDOMINIUM DEVELOPMENT RUNOFF COEFFICIENT = .7618
  SUBAREA AREA(ACRES) =     5.10   SUBAREA RUNOFF(CFS) =     9.70
  TOTAL AREA(ACRES) =    58.00   TOTAL RUNOFF(CFS) =    78.55
  TC(MIN) =  25.43

****************************************************************************
  FLOW PROCESS FROM NODE    15.00 TO NODE    15.00 IS CODE =   8
----------------------------------------------------------------------------
  >>>>>ADDITION OF SUBAREA TO MAINLINE PEAK FLOW<<<<<
============================================================================
    25.00 YEAR RAINFALL INTENSITY(INCH/HOUR) =  2.496
  SOIL CLASSIFICATION IS "A"
  APARTMENT DEVELOPMENT RUNOFF COEFFICIENT = .7996
  SUBAREA AREA(ACRES) =     4.80   SUBAREA RUNOFF(CFS) =     9.58
  TOTAL AREA(ACRES) =    62.80   TOTAL RUNOFF(CFS) =    88.13
  TC(MIN) =  25.43

****************************************************************************
  FLOW PROCESS FROM NODE    15.00 TO NODE    16.00 IS CODE =   5
----------------------------------------------------------------------------
  >>>>>COMPUTE TRAPEZOIDAL-CHANNEL FLOW<<<<<
  >>>>>TRAVELTIME THRU SUBAREA<<<<<
============================================================================
  UPSTREAM NODE ELEVATION =        92.00
  DOWNSTREAM NODE ELEVATION =        90.00
  CHANNEL LENGTH THRU SUBAREA(FEET) =    700.00
  CHANNEL BASE(FEET) =     2.00    "Z" FACTOR =    2.000
  MANNINGS FACTOR =  .015   MAXIMUM DEPTH(FEET) =    3.00
  CHANNEL FLOW THRU SUBAREA(CFS) =       88.13
  FLOW VELOCITY(FEET/SEC) =    5.97    FLOW DEPTH(FEET) =    2.26
  TRAVEL TIME(MIN.) =    1.95    TC(MIN.) =  27.38

****************************************************************************
  FLOW PROCESS FROM NODE    16.00 TO NODE    16.00 IS CODE =   8
----------------------------------------------------------------------------
  >>>>>ADDITION OF SUBAREA TO MAINLINE PEAK FLOW<<<<<
============================================================================
    25.00 YEAR RAINFALL INTENSITY(INCH/HOUR) =  2.387
  SOIL CLASSIFICATION IS "A"
  COMMERCIAL DEVELOPMENT RUNOFF COEFFICIENT = .8237
  SUBAREA AREA(ACRES) =    10.80   SUBAREA RUNOFF(CFS) =    21.24
  TOTAL AREA(ACRES) =    73.60   TOTAL RUNOFF(CFS) =  109.36
  TC(MIN) =  27.38
============================================================================
  END OF RATIONAL METHOD ANALYSIS
```

STUDY NAME:

25.0 YEAR STORM 1-HOUR RAINFALL(inch)= 1.49;INTENSITY SLOPE = .600

CALCULATED BY:
CHECKED BY:
PAGE NUMBER OF

[A D V A N C E D E N G I N E E R I N G S O F T W A R E]

CONCENTRATION POINT NUMBER	SOIL TYPE	DEV. TYPE	AREA (ACRES)	I in/h	C	Q (SUB)	Q TOTAL	SLOPE ft/ft	SECTION	V fps.	PATH ft.	T min.	Tc min.	HYDRAULICS AND NOTES
								.0025			800	20.3		INITIAL SUBAREA
12.00	1	6	10.0	2.86	.718	20.5							20.3	
							20.5	.0057	street	3.2	350	2.0		*Qavg= 30.0cfs Dn= .61 W= 22.0 X-fall= .02000
44.ft STREET flow to PT.# 13.00	1	5	9.6	2.70	.733	19.0							22.3	
							39.5	.0031	D= 39."	5.9	650	1.8		n=.0130 Dn= 2.4
14.00	1	5	6.0	2.57	.728	11.2							24.1	
14.00			25.6	2.57			50.8						24.1	Stream Summary
								.0075			400	13.7		INITIAL SUBAREA
22.00	1	7	1.0	3.61	.711	2.6							13.7	
							2.6	.0024	B= .5	1.5	850	9.4		n=.0300 Dn= 1.1 Z= 1.0 Fr= .33
14.00	1	9	3.2	2.64	.612	5.2							23.1	
14.00			4.2	2.64			7.7						23.1	Stream Summary
								.0027			750	43.3		INITIAL SUBAREA
32.00	1	11	9.5	1.81	.503	8.7							43.3	
							8.7	.0036	street	2.0	550	4.7		*Qavg= 12.5cfs Dn= .50 W= 17.3 X-fall= .02000
44.ft STREET flow to PT.# 33.00	1	8	8.8	1.70	.518	7.8							48.0	
							16.4	.0014	D= 33."	3.6	700	3.2		n=.0130 Dn= 2.0
14.00	1	3	4.8	1.64	.754	5.9							51.2	
CONFLUENCE ANALYSIS FOR POINT# 14.00	TC#1= 24.1 Q#1= 53.8 I#1= 2.57 Q1 = 63.9		TC#2= 23.1 Q#2= 7.7 I#2= 2.64 Q2 = 66.4			TC#3= 51.2 Q#3= 22.4 I#3= 1.64 Q3 = 59.5		TC#4= .0 Q#4= .0 I#4= .00 Q4 = .0	TC#5= .0 Q#5= .0 I#5= .00 Q5 = .0					LARGEST CONFLUENCE Q= 68.9
							68.9	.0036	D= 45."	7.1	550	1.3		n=.0130 Dn= 3.1

DEVELOPMENT TYPES:1=COM,2=APT,3=MH,4=CONDO,5=SF(1/4-AC),6=SF(1/2-AC) SOIL TYPES:1=A,2=B,3=C,4=D,
7=SF(1-AC),8=SF(2.5-AC),9=UNDEV(POOR COV),10=UNDEV(FAIR COV),11=UNDEV(GOOD COV) 0,5=SPECIFIED COEFFICIENT

STUDY NAME:

25.0 YEAR STORM 1-HOUR RAINFALL(inch)= 1.49;INTENSITY SLOPE = .600

CALCULATED BY:
CHECKED BY:
PAGE NUMBER OF

[A D V A N C E D   E N G I N E E R I N G   S O F T W A R E]

CONCENTRATION POINT NUMBER	SOIL TYPE	DEV. TYPE	AREA (ACRES)	I in/h	C	Q (SUB)	Q TOTAL	SLOPE ft/ft	SECTION	V fps.	PATH ft.	T min.	Tc min.	HYDRAULICS AND NOTES
15.00	1	4	5.1	2.50	.762	9.7	------	-----	-------	----	----	----	25.4	-------------
15.00	1	2	4.8	2.50	.800	9.6	------	-----	-------	----	----	----	25.4	-------------
------------	----	----	-------	----	----	-----	88.1	.0029	B= 2.0	6.0	700	2.0	-----	n=.0150 Dn= 2.3 Z= 2.0 Fr= .91
16.00	1	1	10.8	2.39	.824	21.2	------	-----	-------	----	----	----	27.4	-------------
16.00			73.6	2.39			109.4						27.4	Stream Summary

DEVELOPMENT TYPES:1=COM,2=APT,3=MH,4=CONDO,5=SF(1/4-AC),6=SF(1/2-AC) SOIL TYPES:1=A,2=B,3=C,4=D,
7=SF(1-AC),8=SF(2.5-AC),9=UNDEV(POOR COV),10=UNDEV(FAIR COV),11=UNDEV(GOOD COV) 0,5=SPECIFIED COEFFICIENT

4.17. **PROGRAM 4.1: DATA ENTRY**

```
Enter Rational Method storm event year.............. ===>    "YR"
:ALLOWABLE VALUES ARE [1] TO [1000  ]


Enter Logarithm slope of intensity-duration curve... ===>    "RSLOPE"
:ALLOWABLE VALUES ARE [.01    ] TO [1.0    ]








-------------------------------------------------------------------------------
    TYPE: EXIT to leave program ; TOP  to go to top of page
```

```
Enter 10-year storm 60-minute rainfall(INCHES)...... ===>    "R60T"
:ALLOWABLE VALUES ARE [.1     ] TO [9.95   ]



Enter 100-year storm 60-minute rainfall(INCHES)..... ===>    "R60H"
:ALLOWABLE VALUES ARE [.1     ] TO [15.0   ]



-------------------------------------------------------------------------------
    TYPE: EXIT to leave program ; TOP  to go to top of page
                                ; BACK to go back one page
```

```
The computer program estimates design pipe sizes.
Enter MINIMUM pipe size acceptable(INCHES).......... ===>    "PIPMIN"
:ALLOWABLE VALUES ARE [3] TO [240   ]

The computer program estimates a design pipe size through
a subarea using the Mannings equation, with the friction
slope set equal to the gradient of the land.  The program
determines a CONSTRUCTABLE pipe size such that non-pressure
flow occurs.  The user can specify a percentage of the
land-gradient to be used for the pipeflow friction slope.
(SUGGESTION:
            Use [.95] for pipesystems with FEW minor losses
            Use [.85] for pipesystems with CONSIDERABLE minor
                      losses

Enter percentage of subarea land-gradient to be used for
the pipeflow friction slope(DECIMAL)................ ===>    "PERCNT"
:ALLOWABLE VALUES ARE [.001  ] TO [1.00  ]

--------------------------------------------------------------------------------
    TYPE: EXIT to leave program ; TOP  to go to top of page
                                ; BACK to go back one page
```

```
SUBAREA RUNOFF-COEFFICIENT OPTIONS:
    1: Use SCS soil group (A) thru (D) runoff
       coefficients
    2: Use a constant (default) runoff-coefficient
       for the entire study
Select option desired............................... ===>    "CV"

Enter default runoff-coefficient(DECIMAL)........... ===>    "CVAL"
:ALLOWABLE VALUES ARE [.05   ] TO [.99   ]

--------------------------------------------------------------------------------
    TYPE: EXIT to leave program ; TOP  to go to top of page
                                ; BACK to go back one page
```

```
Enter upstream node number.......................... ===>    "ZN1"
:ALLOWABLE VALUES ARE [0.00  ] TO [9999.99   ]

Enter downstream node number........................ ===>    "ZN2"
:ALLOWABLE VALUES ARE [0.00  ] TO [9999.99   ]

SUBAREA HYDROLOGIC PROCESSES:
    1: Confluence analysis at node
    2: Initial subarea analysis
    3: Pipeflow traveltime (COMPUTER ESTIMATED pipesize)
    4: Pipeflow traveltime ....(USER SPECIFIED pipesize)
    5: Trapezoidal channel travel time
    6: Street-Flow analysis thru subarea
    7: User-specified information at node
    8: Addition of subarea runoff to mainline
    9: V-gutter flow thru subarea
Select subarea hydrologic process .................. ===>    "KODE"

--------------------------------------------------------------------------------
    TYPE: EXIT to leave program ; TOP  to go to top of page
          MAIN to go to main menu
```

PROGRAM 4.1

```
C ------------------------------------------------------------------
C       PROGRAM RATION
C ------------------------------------------------------------------
C
C  EXECUTIVE DRIVER FOR RATIONAL BATCH SYSTEM
C
       COMMON/CALC/YR,XI,OPTN,NR,C,CVAL,V,Y,IGUESS,DS,XN
       COMMON/CALC1/E1,E2,XL,K,IS,AREA,XKI,XP,KI,RIYR,RSLOPE
       COMMON/CALC2/TCON(5),TOTALA(5),XINT(5),QJN(5),QBAR(5),ZCON(5)
C  INITIALIZE
       NUT=6
       DO 3 I=1,5
       TCON(I)=0.
       TOTALA(I)=0.
       XINT(I)=0.
       QJN(I)=0.
       QBAR(I)=0.
       ZCON(I)=0.
3      CONTINUE
C  OPEN FILES
       OPEN 5,"RAT.DAT"
       OPEN 6,"RAT.ANS"
C
C  PROCESS DATA FILE
C
C  PROBLEM CONTROLS FIRST
       READ FREE(5) YR,RSLOPE,R60T,R60H,PIPMIN,PERCNT,CVAL
C
C  SUBAREAS
100    READ FREE(5,END=1000)ZN1,ZN2,KODE
       WRITE(NUT,601)
       WRITE(NUT,600)ZN1,ZN2,KODE
       WRITE(NUT,601)
C  PROCESSES
       IERR=0
       IF(KODE.EQ.1)CALL CONFB(Q,TC,SUMA,IERR)
       IF(KODE.EQ.2)CALL INITRU(Q,TC,SUMA,IERR)
       IF(KODE.EQ.3)CALL PIPEFR(Q,TC,SUMA,IERR,PIPMIN,PERCNT)
       IF(KODE.EQ.4)CALL PIPEFR(Q,TC,SUMA,IERR,PIPMIN,PERCNT)
       IF(KODE.EQ.5)CALL TRAPZR(Q,TC,SUMA,IERR)
       IF(KODE.EQ.6)CALL STREET(Q,TC,SUMA,IERR)
       IF(KODE.EQ.7)CALL USEROR(Q,TC,SUMA,IERR)
       IF(KODE.EQ.8)CALL SUBTOB(Q,TC,SUMA,IERR)
       IF(KODE.EQ.9)CALL VGUTTR(Q,TC,SUMA,IERR)
       IF(KODE.EQ.999)GO TO 1000
       IF(IERR.NE.0)GO TO 50
       GO TO 100
C
C  ERROR PROCESSING
50     WRITE(NUT,602)
1000   CONTINUE
C  FORMATS
600    FORMAT(3X,'FLOW PROCESS FROM NODE ',F8.2,' TO NODE ',F8.2,
     . ' IS CODE = ',I3)
601    FORMAT(1X,76('*'))
602    FORMAT(1X'***FATAL ERROR - CHECK DATA INPUT***')
       RETURN
       END
```

```
C ------------------------------------------------------------------
      SUBROUTINE ADJUST(A,B,YR)
C ------------------------------------------------------------------
      IF(A.LE.0.)A=.01
      B=A
      IF(YR.LE.2.OR.YR.GE.50)GOTO 200
C
      IF(YR.GE.2.OR.YR.LE.5.)B=1.+(YR-2.)*.008467
      IF(YR.GT.5..AND.YR.LE.10.)B=1.0254-(YR-5.)*.00308
      IF(YR.GT.10..AND.YR.LE.25)B=1.01-(YR-10.)*.00197
      IF(YR.GT.25..AND.YR.LT.50.)B=.98052+(YR-25.)*.00078
      B=B*A
200   CONTINUE
C
      RETURN
      END
```

4.18. **PROGRAM 4.2: DATA ENTRY**

```
---DATA ENTRY FOR INITIAL SUBAREA ANALYSIS---PAGE 1


Enter upstream node     1.00 elevation(FEET)........ ===>    "E1"
:ALLOWABLE VALUES ARE [.01    ] TO [99999.99  ]

Enter downstream node     2.00 elevation(FEET)...... ===>    "E2"
:ALLOWABLE VALUES ARE [0      ] TO [99999.99  ]

Enter runoff travel-length through subarea(FEET).... ===>    "XL"
:ALLOWABLE VALUES ARE [.01    ] TO [10000 ]
(NOTE: SUGGESTED RANGE(FEET) IS [0] TO [1000])







--------------------------------------------------------------------------------
TYPE: EXIT to leave program ; TOP  to go to top of page
```

```
---DATA ENTRY FOR INITIAL SUBAREA ANALYSIS---PAGE 2

SUBAREA LAND USE OR DEVELOPMENT TYPE:

    1= Commercial
    2= Apartment
    3= Mobile home park
    4= Condominium
    5= Single family (1/4 ACRE lot)
    6= Single family (1/2 ACRE lot)
    7= Single family (  1-ACRE lot)
    8= Single family (2.5-ACRE lot)
    9= Undeveloped with POOR cover
   10= Undeveloped with FAIR cover
   11= Undeveloped with GOOD cover

Select subarea development classification number.... ===>    "K"



--------------------------------------------------------------------------------
TYPE: EXIT to leave program ; TOP  to go to top of page
                            ; BACK to go back one page
```

```
---DATA ENTRY FOR INITIAL SUBAREA ANALYSIS---PAGE 3

SUBAREA RUNOFF COEFFICIENT OPTIONS:

    1= Assume soil group A
    2= Assume soil group B
    3= Assume soil group C
    4= Assume soil group D
    5= User to specify runoff coefficient

Select runoff coefficient option number............ ===>    "FSOIL"

Enter SUBAREA runoff coefficient.................... ===>    "C"
:ALLOWABLE VALUES ARE [.01   ] TO [.99   ]

Enter subarea area(ACRES)........................... ===>    "AREA"
:ALLOWABLE VALUES ARE [0] TO [1000  ]
(NOTE: SUGGESTED AREA IS BETWEEN [0] AND [10]

--------------------------------------------------------------------------------
TYPE: EXIT to leave program ; TOP  to go to top of page
                            ; BACK to go back one page
```

PROGRAM 4.2

```
C  ------------------------------------------------------------------
         SUBROUTINE INITRU(Q,TC,SUMA,IERR)
C  ------------------------------------------------------------------
C
C    INITIAL SUBAREA ANALYSIS
C
         COMMON /NUT/NUT
         COMMON/CALC/YR,XI,OPTN,NR,C,CVAL,V,Y,IGUESS,DS,XN
         COMMON/CALC1/E1,E2,XL,K,IS,AREA,XKI,XP,KI,RIYR,RSLOPE
C
C
C..READ DATA INPUT
         READ FREE(5)E1,E2,XL,K,FSOIL,AREA
         IS=FSOIL
         C=FSOIL-5.0
C
         WRITE(NUT,1007)
1007     FORMAT(3X,'>>>>>RATIONAL METHOD INITIAL SUBAREA ANALYSIS<<<<<')
         WRITE(NUT,403)
C
         CALL INITAR(E1,E2,XL,K,TC,NUT)
C
         IGUESS=0
C
         CALL INTENR(YR,TC,XI,IGUESS,ERROR,NUT,RIYR,RSLOPE)
         IF(ERROR.NE.1.)GO TO 9990
         IERR=9
         GO TO 10000
C
9990     CALL COEFAR(XI,IS,K,C,IGUESS,CVAL,NUT)
C
         DELQ=C*XI*AREA
         SUMA=AREA
         Q=DELQ
         WRITE(NUT,1010)DELQ,SUMA,Q
1010     FORMAT(3X,'SUBAREA RUNOFF(CFS) = ',F9.2,/,
      C  3X,'TOTAL AREA(ACRES) = ',F9.2,3X,'TOTAL RUNOFF(CFS) = ',F9.2)
403      FORMAT(1X,76('='))
C
10000    CONTINUE
C
         RETURN
         END
C
C
C  ------------------------------------------------------------------
         SUBROUTINE INITAR(E1,E2,XL,K,TC,NUT)
C  ------------------------------------------------------------------
C
         DIMENSION XXK(7,11),KK(8),RK(11)
         DATA XXK/'COMM','ERCI','AL  ','    ','    ','    ','    ',
     .         'APAR','TMEN','T   ','    ','    ','    ','    ',
     .         'MOBI','LE H','OME ','PARK','    ','    ','    ',
     .         'COND','OMIN','IUM ','    ','    ','    ','    ',
     .         'SING','LE F','AMIL','Y (1','/4 A','CRE)','    ',
     .         'SING','LE F','AMIL','Y (1','/2 A','CRE)','    ',
     .         'SING','LE F','AMIL','Y (1','-ACR','E)  ','    ',
     .         'SING','LE F','AMIL','Y (2','.5-A','CRE)','    ',
     .         'UNDE','VELO','PED ','WITH',' POO','R CO','VER ',
     .         'UNDE','VELO','PED ','WITH',' FAI','R CO','VER ',
     .         'UNDE','VELO','PED ','WITH',' GOO','D CO','VER '/
         DATA RK/.3031,.3227,.3359,.3592,.3926,.4221,.4693,.5,.5326,
     .   .7094,.9374/
         DATA KK/'DE','VE','LO','PM','EN','T ','IS',': '/
C
C
         WRITE(NUT,6)
6        FORMAT(10X,'ASSUMED INITIAL SUBAREA UNIFORM')
```

```
      H=E1-E2
      XK=RK(K)
      WRITE(NUT,100)KK,(XXK(J,K),J=1,7)
100   FORMAT(10X,8A2,7A4)
C
      TC=XK*((XL**3.)/H)**0.2
C
      WRITE(NUT,250)
250   FORMAT(3X,'TC = K*[(LENGTH**3)/(ELEVATION CHANGE)]**.2')
      WRITE(NUT,201)XL,E1,E2,H
201   FORMAT(3X,'INITIAL SUBAREA FLOW-LENGTH = ',F8.2,/,
     13X,'UPSTREAM ELEVATION = ',F9.2,/,
     13X,'DOWNSTREAM ELEVATION = ',F9.2,/,
     13X,'ELEVATION DIFFERENCE = ',F9.2)
      WRITE(NUT,255)XK,XL,H,TC
255   FORMAT(3X,'TC = ',F5.3,'*[(',F8.2,'**3)/(',F9.2,')]**.2 = ',
     C F8.3)
C
      IF(TC.GT.5.)GOTO 300
      WRITE(NUT,310)
310   FORMAT(3X,'COMPUTED TIME OF CONCENTRATION INCREASED TO 5 MIN.')
      TC=5.
C
300   CONTINUE
C
      RETURN
      END
C
```

C 4.19. PROGRAM 4.3

```
C ---------------------------------------------------------------------
      SUBROUTINE INTENR(YR,TC,XI,IGUESS,ERROR,NUT,RIYR,RSLOPE)
C ---------------------------------------------------------------------
C
C
      ERROR=0.
      XI=-RSLOPE*ALOG(TC)+ALOG(RIYR)+RSLOPE*4.094345
      XI=EXP(XI)
      IF(XI.LT. .01)XI=.01
      IF(XI.EQ. .01 .AND. IGUESS.NE.1)WRITE(NUT,101)
101   FORMAT(3X,'** MINIMUM DEFAULT RAINFALL INTENSITY USED **')
      IF(IGUESS.NE.1)WRITE(NUT,100)YR,XI
100   FORMAT(3X,F7.2,' YEAR RAINFALL INTENSITY(INCH/HOUR) = ',F6.3)
C
      RETURN
      END
```

4.20. PROGRAM 4.4

```
C ---------------------------------------------------------------------
      SUBROUTINE COEFAR(XXI,ISOIL,KK,C,IGUESS,CVAL,NUT)
C ---------------------------------------------------------------------
C
      DIMENSION XXK(12,9),RF(4),RK(9),MK(4)
      DATA XXK/'COMM','ERCI','AL D','EVEL','OPME','NT R','UNOF',
     .         'F CO','EFFI','CIEN','T = ','    ',
     .         'APAR','TMEN','T DE','VELO','PMEN','T RU','NOFF',
     .         ' COE','FFIC','IENT',' =  ','    ',
     .         'MOBI','LE H','OME ','PARK',' DEV','ELOP','MENT',
     .         ' RUN','OFF ','COEF','FICI','ENT=',
     .         'COND','OMIN','IUM ','DEVE','LOPM','ENT ','RUNO',
     .         'FF C','OEFF','ICIE','NT =','    ',
     .         'SING','LE-F','AMIL','Y(1/','4 AC','RE L','OT) ',
     .         'RUNO','FF C','OEFF','ICIE','NT =',
     .         'SING','LE-F','AMIL','Y(1/','2 AC','RE L','OT) ',
     .         'RUNO','FF C','OEFF','ICIE','NT =',
     .         'SING','LE-F','AMIL','Y(1-','ACRE',' LOT',') RU',
```

```
     .          'NOFF',' COE','FFIC','IENT',' =  ',
     .          'SING','LE-F','AMIL','Y(2.','5-AC','RE L','OT) ',
     .          'RUNO','FF C','OEFF','ICIE','NT =',
     .          'UNDE','VELO','PED ','WATE','RSHE','D RU','NOFF',
     .          ' COE','FFIC','IENT',' =  ',','    '/
          DATA RF/32.,56.,69.,75./
          DATA RK/.1,.2,.25,.35,.5,.6,.8,.9,1./
          DATA MK/'A ','B ','C ','D '/
          XI=XXI
          IF(XI.LT. .5)XI=.5
          K=KK
          IF(K.GT.9)K=9
          IF(CVAL.EQ.0.)GOTO 20
          C=CVAL
          IF(IGUESS.NE.0)GO TO 1000
          WRITE(NUT,21)
21        FORMAT(3X,'*USER SPECIFIED(GLOBAL):')
          GOTO 900
20        CONTINUE
          IF(ISOIL.NE.5)GO TO 23
          IF(IGUESS.EQ.0)WRITE(NUT,24)
24        FORMAT(3X,'*USER SPECIFIED(SUBAREA):')
          GO TO 900
23        A=.35
          IF(K.GT.0)A=RK(K)
30        FORMAT(3X,'**NOTE: SOIL GROUP "D" USED AS DEFAULT VALUE**')
          F=0.
          IF(ISOIL.GT.0)F=RF(ISOIL)
          IF(F.EQ.0 .AND. IGUESS.EQ.0)WRITE(NUT,30)
          IF(F.EQ.0)F=75.
          SP=-8.+800./F
          F=(XI*SP)/(SP+6.*XI)
          B=1.-A
          C=0.9*((B+(XI-F)/XI*A))
          IF(IGUESS.EQ.1)GO TO 1000
          IF(ISOIL.GT.0)WRITE(NUT,100)MK(ISOIL)
100       FORMAT(3X,'SOIL CLASSIFICATION IS "',A1,'"')
900       IF(K.GT.0)WRITE(NUT,110)(XXK(J,K),J=1,12),C
110       FORMAT(3X,12A4,1X,F5.4)
1000      CONTINUE
          RETURN
          END
```

4.21.

PROGRAM 4.5: DATA ENTRY

```
---DATA ENTRY FOR PIPE-FLOW TRAVEL TIME THROUGH SUBAREA---PAGE 1

Enter upstream node     1.00 elevation(FEET)........ ===>    "E1"
:ALLOWABLE VALUES ARE [.01    ] TO [99999.99  ]

Enter downstream node     2.00 elevation(FEET)...... ===>    "E2"
:ALLOWABLE VALUES ARE [0      ] TO [99999.99  ]

Enter pipe length through subarea(FEET)............. ===>    "XL"
:ALLOWABLE VALUES ARE [.01    ] TO [10000 ]

Enter Mannings friction factor for pipe............. ===>    "XN"
(NOTE: FOR RCP USE n = .013
       FOR CSP(or CMP) USE n = .024)
:ALLOWABLE VALUES ARE [.005   ] TO [.9999 ]

--------------------------------------------------------------------------------
   TYPE: EXIT to leave program ; TOP  to go to top of page
```

```
---DATA ENTRY FOR PIPE-FLOW TRAVEL TIME THROUGH SUBAREA---PAGE 2

Enter diameter of given pipe size(INCHES)........... ===>    "DIAM"
:ALLOWABLE VALUES ARE [      3.00] TO [240       ]

Enter number of pipes in subarea reach.............. ===>    "NUMPI"
:ALLOWABLE VALUES ARE GREATER THAN [0     ]

--------------------------------------------------------------------------------
   TYPE. EXIT to leave program ; TOP  to go to top of page
                               ; BACK to go back one page
```

PROGRAM 4.5

```
C  ----------------------------------------------------------------
       SUBROUTINE PIPEFR(Q,TC,SUMA,IERR,PIPMIN,PERCNT)
C  ----------------------------------------------------------------
C
C  PIPE-FLOW ANALYSIS
C
       COMMON /NUT/NUT
       COMMON/CALC/YR,XI,OPTN,NR,C,CVAL,V,Y,IGUESS,DS,XN
       COMMON/CALC1/E1,E2,XL,K,IS,AREA,XKI,XP,KI,RIYR,RSLOPE
C
C..READ DATA INPUT
       READ FREE(5) KODE,E1,E2,XL,XN
C  READ MORE IF USER SPECIFIED PIPESIZE
       IF(KODE.EQ.4)READ FREE(5) DIAM,NUMPI
C
C      PROCESS FLOW
C
       IF(KODE.EQ.4)GO TO 2052
2050   WRITE(NUT,2105)
2105  FORMAT(3X,'>>>>>COMPUTE PIPEFLOW TRAVELTIME THRU SUBAREA<<<<<',/,
     C  3X,'>>>>>USING COMPUTER-ESTIMATED PIPESIZE (NON-PRESSURE FLOW)',
     .  '<<<<<')
       GO TO 2053
C
2052   WRITE(NUT,2106)
2053   CONTINUE
C
2106  FORMAT(3X,'>>>>>COMPUTE PIPEFLOW TRAVELTIME THRU SUBAREA<<<<<',/,
     C  3X,'>>>>>USING USER-SPECIFIED PIPESIZE<<<<<')
       WRITE(NUT,403)
       DIAM=DIAM/12.
       S1=(E1-E2)/XL*PERCNT
C
       IF(KODE.EQ.4)GO TO 2200
C
       CALL LETFLA(S1,Q,SIZE1,NUMPI,XN)
       DIAM=SIZE1*12.
       IF(DIAM.LT.PIPMIN)WRITE(NUT,2110)PIPMIN
2110   FORMAT(3X,'ESTIMATED PIPE DIAMETER(INCH) INCREASED TO ',F6.3)
       IF(DIAM.LT.PIPMIN)DIAM=PIPMIN
       DIAM=DIAM/12.
C
2200   QQ=NUMPI
       QQ=Q/QQ
C
       CALL PNORMA(DIAM,QQ,S1,V,XN,NUT)
C
       DIAM=DIAM*12.
       WRITE(NUT,2300)E1,E2,XL,XN
2300   FORMAT(3X,'UPSTREAM NODE ELEVATION = ',F8.2,/,
     C  3X,'DOWNSTREAM NODE ELEVATION = ',F8.2,/,
     C  3X,'FLOWLENGTH(FEET) = ',F8.2,3X,'MANNINGS N = ',F5.3)
       IF(KODE.EQ.4)GO TO 2057
C
       WRITE(NUT,2315)DIAM,NUMPI
       GO TO 2058
2315   FORMAT(3X,'ESTIMATED PIPE DIAMETER(INCH) = ',F6.2,4X,
     C  'NUMBER OF PIPES = ',I3)
2057   WRITE(NUT,2310)DIAM,NUMPI
2310   FORMAT(3X,'GIVEN PIPE DIAMETER(INCH) = ',F6.2,4X,
```

```
     C  'NUMBER OF PIPES = ',I3)
2058    TRAVEL=XL/60./V
        TC=TC+TRAVEL
        WRITE(NUT,2317)Q
2317    FORMAT(3X,'PIPEFLOW THRU SUBAREA(CFS) = ',F9.2)
        WRITE(NUT,2320)TRAVEL,TC
2320    FORMAT(3X,'TRAVEL TIME(MIN.) = ',F6.2,4X,'TC(MIN.) = ',F6.2)
C
C
3020    CONTINUE
C
C  FORMATS
C
403     FORMAT(1X,76('='))
C
        RETURN
        END
C
C
C  ---------------------------------------------------------------
        SUBROUTINE PNORMA(D,Q,S0,V,XXN,NUT)
C  ---------------------------------------------------------------
C
C       TEST FOR FULL PIPE-FLOW
C
        DN=D*12.
        XK=.4631646/XXN*S0**.5
        QFULL=XK*D**(8./3.)
        IF(Q.GE.QFULL)GO TO 200
C
        XK=1.486/XXN*(S0**0.5)
        KTEST=0
        THETAU=2.2643743
        THETAL=0.
100     THETA=(THETAU+THETAL)/2.
        KTEST=KTEST+1
        IF(KTEST.GE.101)WRITE(NUT,110)
110     FORMAT(3X,'ASSUME FULL-FLOWING PIPELINE')
        IF(KTEST.GE.101) GO TO 200
        XF=THETA*2.
        XS=SIN(XF)
        AREA=D*D/4.*(THETA-0.5*XS)
        RH=D/4.*(1.-(XS/XF))
        QTEST=XK*AREA*RH**(.66667)
        FACTOR=Q-QTEST
        IF(FACTOR.GE.0.)THETAL=THETA
        IF(FACTOR.LE.0.)THETAU=THETA
        IF(ABS(FACTOR).LT..005)GO TO 150
        GO TO 100
150     IF(THETA.GE.1.570796)DN=D/2.+(SIN(THETA-1.570796))*D/2.
        IF(THETA.LT.1.570796)DN=D/2.-(COS(THETA))*D/2.
        DN=DN*12.
        D1=D*12.
        WRITE(NUT,120)D1,DN
120     FORMAT(3X,'DEPTH OF FLOW IN ',F5.1,' INCH PIPE IS ',F5.1,
     C  ' INCHES')
200     CONTINUE
        AREAF=.7854*D*D
        IF(Q.GE.QFULL.OR.KTEST.GE.101)V=Q/AREAF
        IF(Q.LT.QFULL)V=Q/AREA
```

```
      WRITE(NUT,300)V
300   FORMAT(3X,'PIPEFLOW VELOCITY(FEET/SEC.) = ',F5.1)
C
      RETURN
      END
C
C
C ------------------------------------------------------------------
      SUBROUTINE LETFLA(S0,QQ,SIZE1,NUMPI,XN)
C ------------------------------------------------------------------
C
C
      NTEST=0
      QQQ=QQ
      NUMPI=1
700   Q=QQQ
      XK=Q/(S0**0.5)
      XD=(XK/.4631644*XN)**0.375
      XINCH=XD*12.
      TEST=XINCH/3.
      ITEST=TEST
      XTEST=ITEST
      DEL=1.
      XDEL=TEST-XTEST
      IF(XDEL.EQ.0.)DEL=0.
      PIPE=XTEST+DEL
      SIZE=PIPE*3.
      SIZE1=SIZE/12.
      IF(SIZE1.LE.8.)GO TO 500
      TEST=XINCH/6.
      ITEST=TEST
      XTEST=ITEST
      DEL=1.
      IF(XDEL.EQ.0.)DEL=0.
      PIPE=XTEST+DEL
      SIZE=PIPE*6.
      SIZE1=SIZE/12.
500   CONTINUE
      IF(SIZE1.LE.12.)GO TO 1000
      NTEST=1
      NUMPI=NUMPI+1
      Z=NUMPI
      QQQ=QQ/Z
      GO TO 700
1000  CONTINUE
C
      RETURN
      END
```

4.22. **PROGRAM 4.6: DATA ENTRY**

```
---DATA ENTRY FOR TRAPEZOIDAL CHANNEL TRAVEL TIME---PAGE 1

Enter upstream node      1.00 elevation(FEET)........ ===>     "E1"
:ALLOWABLE VALUES ARE [.01    ] TO [99999.99  ]

Enter downstream node     2.00 elevation(FEET)...... ===>     "E2"
:ALLOWABLE VALUES ARE [0      ] TO [99999.99  ]

Enter channel length through subarea(FEET).......... ===>     "XL"
:ALLOWABLE VALUES ARE [.01    ] TO [10000 ]

Enter Mannings friction factor for channel.......... ===>     "XN"
(NOTE: FOR CONCRETE SECTIONS,   USE n = .015
       FOR GOOD EARTH CHANNELS, USE n = .03
 SEE COUNTY MANUAL FOR MORE INFORMATION)
:ALLOWABLE VALUES ARE [.005   ] TO [.9999 ]

-------------------------------------------------------------------------------
    TYPE: EXIT to leave program ; TOP  to go to top of page
```

```
---DATA ENTRY FOR TRAPEZOIDAL CHANNEL TRAVEL TIME---PAGE 2

Enter horizontal base width of channel(FEET)........ ===>     "B"
:ALLOWABLE VALUES ARE [0] TO [1000. ]

Enter uniform symmetrical channel "Z" factor........ ===>     "Z"
(NOTE: THE CHANNEL "Z" FACTOR IS THE SIDE SLOPE
       RATIO OF [HORIZONTAL/VERTICAL]
 EXAMPLE: FOR A 2:1 SIDE SLOPE, "Z"=2)
:ALLOWABLE VALUES ARE [0] TO [99.99 ]

Enter maximum allowable depth of flow
in channel(FEET)....................                 ===>     "DMAX"
:ALLOWABLE VALUES ARE [.001   ] TO [500    ]

-------------------------------------------------------------------------------
    TYPE: EXIT to leave program ; TOP  to go to top of page
                                ; BACK to go back one page
```

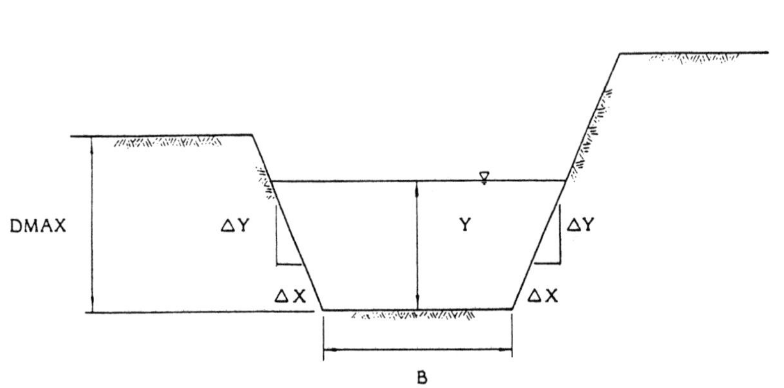

$$Z = \Delta X / \Delta Y$$

Flow Approximation:

$$Q = \frac{\frac{(1.486)}{n}((B + Z(Y))Y)^{5/3} S^{1/2}}{(B + 2Y(Z^2 + 1)^{1/2})^{2/3}}$$

Fig. 4.12. TRAPEZOIDAL CHANNEL MODEL GEOMETRY VARIABLE DEFINITIONS

PROGRAM 4.6

```
C ---------------------------------------------------------------------
      SUBROUTINE TRAPZR(Q,TC,SUMA,IERR)
C ---------------------------------------------------------------------
C
C  TRAPEZOIDAL CHANNEL HYDRAULICS (TRAVEL TIME)
C
      COMMON /NUT/NUT
      COMMON/CALC/YR,XI,OPTN,NR,C,CVAL,V,Y,IGUESS,DS,XN
      COMMON/CALC1/E1,E2,XL,K,IS,AREA,XKI,XP,KI,RIYR,RSLOPE
C
C
C..READ DATA INPUT
      READ FREE(5)E1,E2,XL,XN,B,Z,DMAX
C
C  PROCESS TRAPEZOIDAL CHANNEL TRAVEL TIME
C
      WRITE(NUT,3190)
3190  FORMAT(3X,'>>>>>COMPUTE TRAPEZOIDAL-CHANNEL FLOW<<<<<',
     C /,3X,'>>>>>TRAVELTIME THRU SUBAREA<<<<<')
      WRITE(NUT,403)
      QQ=Q
      S1=(E1-E2)/XL
      WRITE(NUT,3200)E1,E2,XL,B,Z,XN,DMAX,QQ
3200  FORMAT(3X,'UPSTREAM NODE ELEVATION = ',F9.2,/,3X,
     C 'DOWNSTREAM NODE ELEVATION = ',F9.2,/,3X,
     C 'CHANNEL LENGTH THRU SUBAREA(FEET) = ',F8.2,/,3X,
     C 'CHANNEL BASE(FEET) = ',F7.2,3X,'"Z" FACTOR = ',F7.3,/,
     C 3X,'MANNINGS FACTOR = ',F5.3,3X,'MAXIMUM DEPTH(FEET) = ',
     C F6.2,/,3X,'CHANNEL FLOW THRU SUBAREA(CFS) = ',F9.2)
C
      CALL TNORMA(S1,QQ,Z,B,XN,V,NUT,DN,DMAX)
C
      TRAVEL=XL/60./V
      TC=TC+TRAVEL
      WRITE(NUT,3300)V,DN,TRAVEL,TC
3300  FORMAT(3X,'FLOW VELOCITY(FEET/SEC) = ',F6.2,3X,
     C 'FLOW DEPTH(FEET) = ',F6.2,/,3X,'TRAVEL TIME(MIN.) = ',
     C F6.2,3X,'TC(MIN.) = ',F6.2)
      IF(DN.GE.DMAX)WRITE(NUT,3350)
3350  FORMAT(/,3X,'==>FLOWDEPTH EXCEEDS MAXIMUM ALLOWABLE DEPTH',/)
C
3020  CONTINUE
C
C FORMATS
C
403   FORMAT(1X,76('='))
C
C
      RETURN
      END
C
C
C ---------------------------------------------------------------------
      SUBROUTINE TNORMA(S0,Q,Z,B,XN,V,NUT,DN,DMAX)
C ---------------------------------------------------------------------
C
C
      DN=DMAX/2.
      YMIN=0.
      YMAX=DMAX
```

```
      FUN=1.-Q*XN*(B+2.*DMAX*SQRT(Z*Z+1.))**.6667/(1.486*((B+Z*DMAX)
     C *DMAX)**1.6667*SQRT(S0))
      IF(FUN)450,500,500
450   WRITE(NUT,460)
460   FORMAT(/,10X,'==>>ERROR: FLOW IN CHANNEL EXCEEDS CHANNEL',/,
     C 14X,'CAPACITY( NORMAL DEPTH EQUAL TO SPECIFIED MAXIMUM ',/,
     C 14X,'ALLOWABLE DEPTH).',
     C /,14X,'AS AN APPROXIMATION, FLOWDEPTH IS SET AT MAXIMUM',/,
     C14X,'ALLOWABLE DEPTH AND IS USED FOR TRAVELTIME CALCULATIONS.',/)
      DN=DMAX
      GOTO 570
500   I=0
510   I=I+1
      IF(I.GT.30)GO TO 560
      FUN=1.-Q*XN*(B+2.*DN*SQRT(Z*Z+1))**.6667/(1.486*((B+Z*DN)
     C*DN)**1.6667*SQRT(S0))
      IF(FUN)520,540,530
520   YMIN=DN
      GO TO 540
530   YMAX=DN
540   DN1=(YMIN+YMAX)/2.
      IF(ABS(DN-DN1).LT..01)GO TO 570
      DN=DN1
550   GO TO 510
560   WRITE(NUT,2010)
2010  FORMAT(/,10X,'==>>ERROR: NO CONVERGENCE IN PROGRAM TO NORMAL ',
     . 'DEPTH.',
     C /,14X,'AS AN APPROXIMATION, FLOWDEPTH IS SET AT MAXIMUM ',
     . 'ALLOWABLE',/,
     C 14X,'DEPTH AND IS USED FOR TRAVELTIME CALCULATIONS.',/)
      DN=DMAX
570   CONTINUE
      TA=(DN*Z+B)*DN
      V=Q/TA
3000  CONTINUE
C
      RETURN
      END
```

4.23. **PROGRAM 4.7: DATA ENTRY**

```
---DATA ENTRY FOR STREET FLOW THROUGH SUBAREA---PAGE 1


Enter upstream node      1.00 elevation(FEET)........ ===>     "E1"
:ALLOWABLE VALUES ARE [.01    ] TO [99999.99  ]

Enter downstream node      2.00 elevation(FEET)...... ===>     "E2"
:ALLOWABLE VALUES ARE [0      ] TO [99999.99  ]

Enter runoff travel-length through subarea(FEET).... ===>     "XL"
:ALLOWABLE VALUES ARE [.01    ] TO [10000 ]

--------------------------------------------------------------------------------
TYPE: EXIT to leave program ; TOP  to go to top of page
```

```
---DATA ENTRY FOR STREET FLOW THROUGH SUBAREA---PAGE 2

SUBAREA LAND USE OR DEVELOPMENT TYPE:

    1= Commercial
    2= Apartment
    3= Mobile home park
    4= Condominium
    5= Single family (1/4 ACRE lot)
    6= Single family (1/2 ACRE lot)
    7= Single family (  1-ACRE lot)
    8= Single family (2.5-ACRE lot)
    9= Undeveloped


Select subarea development classification number.... ===>     "K"


--------------------------------------------------------------------------------
TYPE: EXIT to leave program ; TOP  to go to top of page
                            ; BACK to go back one page
```

```
---DATA ENTRY FOR STREET FLOW THROUGH SUBAREA---PAGE 3

SUBAREA RUNOFF COEFFICIENT OPTIONS:

    1= Assume soil group A
    2= Assume soil group B
    3= Assume soil group C
    4= Assume soil group D
    5= User to specify runoff coefficient

Select runoff coefficient option number............. ===>    "FSOIL"

Enter SUBAREA runoff coefficient.................... ===>    "C"
:ALLOWABLE VALUES ARE [.01   ] TO [.99   ]

Enter subarea area(ACRES)........................... ===>    "AREA"
:ALLOWABLE VALUES ARE [0] TO [1000  ]



-------------------------------------------------------------------------------
TYPE: EXIT to leave program ; TOP  to go to top of page
                            ; BACK to go back one page
```

```
---DATA ENTRY FOR STREET FLOW THROUGH SUBAREA---PAGE 4


Enter standard curb height(INCHES).................. ===>    "CURB"
:ALLOWABLE VALUES ARE [6] OR [8]

Enter symmetrical street crossfall[DECIMAL NOTATION] ===>    "XFALL"
:ALLOWABLE VALUES ARE [.001   ] TO [.9999 ]

Enter the symmetrical street halfwidth(FEET)........ ===>    "HWIDTH"
:ALLOWABLE VALUES ARE [5     ] TO [99.99 ]

STREETFLOW OPTIONS:
    1: Runoff flows on one side of the street
    2: Runoff flows evenly on both sides of the street
Select streetflow option desired.................... ===>    "IWIDTH"



-------------------------------------------------------------------------------
TYPE: EXIT to leave program ; TOP  to go to top of page
                            ; BACK to go back one page
```

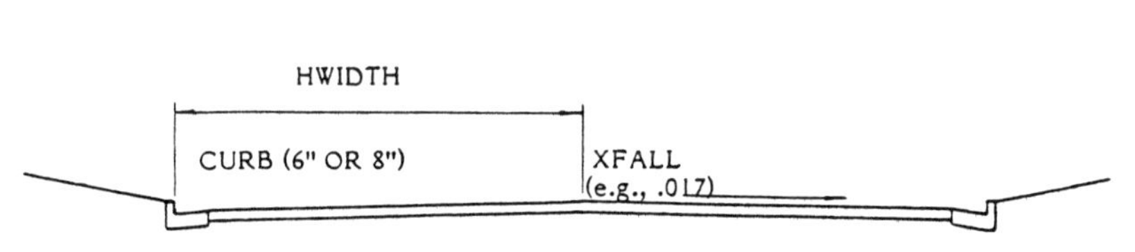

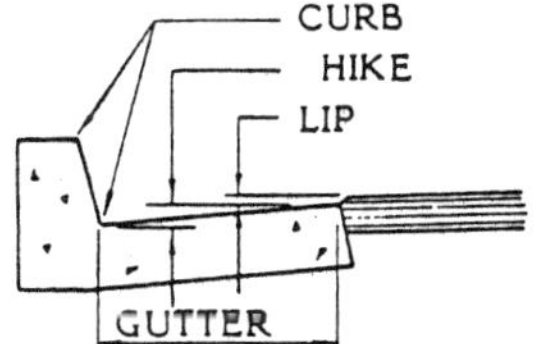

ASSUMED VALUES:

	6" CURB	8" CURB
Curb	6.0"	8.0"
Gutter	1.5'	2.0'
Hike	0.125'	0.167'
Lip	.03125'	.03125'
Manning's n	.015	.015

Fig. 4.13. STREET FLOW MODEL GEOMETRY VARIABLE DEFINITIONS

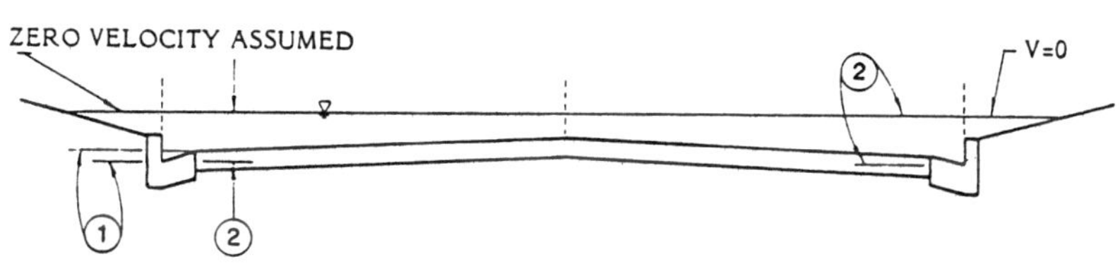

$$Q_j = \left(\frac{1.486}{.015}\right)A_j(A_j/WP_j)^{2/3}\ S^{1/2}$$

Where

A_1 = GUTTER (LIP+HIKE+LIP) /2

WP_1 = 2(LIP) + HIKE + GUTTER

A_2 = GUTTER (Y-HIKE/2)+1/2(Y-HIKE-LIP)^2/XFALL

WP_2 = Y^1+GUTTER+LIP+(Y-HIKE-LIP)/XFALL

Where

Y^1 = { Y, for Y less than CURB,
CURB, otherwise

Fig. 4.14. STREET FLOW MODEL FLOW APPROXIMATIONS

PROGRAM 4.7

```
C  -----------------------------------------------------------------
         SUBROUTINE STREER(Q,TC,SUMA,IERR)
C  -----------------------------------------------------------------
C
C  STREET FLOW THRU SUBAREA
C
         COMMON /NUT/NUT
         COMMON/CALC/YR,XI,OPTN,NR,C,CVAL,V,Y,IGUESS,DS,XN
         COMMON/CALC1/E1,E2,XL,K,IS,AREA,XKI,XP,KI,RIYR,RSLOPE
C
C
C..READ DATA INPUT
         READ FREE(5)E1,E2,XL,K,FSOIL,AREA,CURB,XFALL,HWIDTH,
         IWIDTH
         IS=FSOIL
         C=FSOIL-5.0
C
C
         WRITE(NUT,4093)
4093  FORMAT(3X,'>>>>>COMPUTE STREETFLOW TRAVELTIME THRU SUBAREA<<<<<')
         WRITE(NUT,403)
         WRITE(NUT,4095)E1,E2,XL,CURB,HWIDTH,XFALL,IWIDTH
4095  FORMAT(3X,'UPSTREAM ELEVATION = ',F9.2,
     C  3X,'DOWNSTREAM ELEVATION = ',F9.2,/,
     C  3X,'STREET LENGTH(FEET) = ',F8.2,
     C  3X,'CURB HEIGTH(INCHES) = ',F2.0,/,
     C  3X,'STREET HALFWIDTH(FEET) =',F6.2,
     C  3X,'STREET CROSSFALL(DECIMAL) =',F6.4,/,
     C  3X,'SPECIFIED NUMBER OF HALFSTREETS CARRYING RUNOFF = ',I2)
4103     QAVG=Q/SUMA
         QQ=Q+QAVG/2.*AREA
         S1=(E1-E2)/XL
         ITER=0
         IGUESS=1
C
4100     IF(CURB.EQ.6.)CALL SNORMA(S1,QQ,.5,XFALL,HWIDTH,.015,.03125,
     C  1.5,.125,IWIDTH,FWIDTH,V,DN,ERR,NUT,ITER,DATA,SD1,SW1)
         IF(CURB.EQ.8.)CALL SNORMA(S1,QQ,.6667,XFALL,HWIDTH,.015,
     C.03125,2.,.167,IWIDTH,FWIDTH,V,DN,ERR,NUT,ITER,DATA,SD1,SW1)
C
         IF(ITER.EQ.2)GO TO 4150
         TRAVEL=XL/60./V
         TCA=TC+TRAVEL
C
         IF(ITER.EQ.1)WRITE(NUT,4137)TRAVEL,TCA
C
         CALL INTENR(YR,TCA,XI,IGUESS,ERR,NUT,RIYR,RSLOPE)
         IF(ERR.NE.1.)GO TO 4105
         IERR=9
         GO TO 3022
C
4105     CALL COEFAR(XI,IS,K,C,IGUESS,CVAL,NUT)
C
         IF(ITER.EQ.1.OR.ITER.EQ.2)GOTO 4130
         IGUESS=0
         ITER=1
         QQ=C*XI*AREA/2.+Q
         GOTO 4100
4130     IF(ITER.EQ.2)GOTO 4150
C
```

```
      QAVG=QQ
      SD1=DN
      SD2=FWIDTH
C
      TC=TCA
C
4137  FORMAT(3X,'STREETFLOW TRAVELTIME(MIN) = ',F6.2,3X,
     C 'TC(MIN) = ',F6.2,/)
      DELQ=XI*C*AREA
      Q=Q+DELQ
      SUMA=SUMA+AREA
      WRITE(NUT,4140)AREA,DELQ,SUMA,Q
4140  FORMAT(3X,'SUBAREA AREA(ACRES) = ',F7.2,3X,'SUBAREA RUNOFF',
     .'(CFS) = ',
     CF7.2,/,3X,'SUMMED AREA(ACRES) = ',F7.2,3X,'TOTAL RUNOFF(CFS) = ',
     C F9.2)
      ITER=2
      IGUESS=1
      QQ=Q
      GOTO 4100
C
4150  DV=DN*V
      WRITE(NUT,4160)DN,FWIDTH,V,DV
4160  FORMAT(3X,'END OF SUBAREA STREETFLOW HYDRAULICS:',/,3X,
     C'DEPTH(FEET) = ',F4.2,3X,'HALFSTREET FLOODWIDTH(FEET) = ',F5.2,/,
     C3X,'FLOW VELOCITY(FEET/SEC.) = ',F5.2,3X,'DEPTH*VELOCITY = ',F6.2)
C
3019  CONTINUE
      GO TO 3020
3021  IERR=3
      GO TO 3020
3022  IERR=9
      GO TO 3020
C
C
3020  CONTINUE
C
C
403   FORMAT(1X,76('='))
4110  FORMAT('**ERROR-NO SOLUTION FROM STREETFLOW MODEL: RECHECK',
     . ' YOUR DATA')
C
C
      RETURN
      END
C
C
C ---------------------------------------------------------------------
      SUBROUTINE SNORMA(S0,QQ,CURB,XFALL,HWIDTH,XN,XLIP,GUTTER,
     . GHIKE,IWIDTH,FWIDTH,V,DN,ERROR,NUT,ITER)
C ---------------------------------------------------------------------
C
C
C
      IF(ITER.EQ.1)WRITE(NUT,3120)QQ
3120  FORMAT(10X,'**TRAVELTIME COMPUTED USING MEAN FLOW(CFS) = ',
     . F7.2)
      ERROR=0.
      FACTOR=SQRT(1.+XFALL*XFALL)
      ISPLIT=0
```

```
          A1=GUTTER*(XLIP+GHIKE+XLIP)/2.
          WP1=XLIP+GHIKE+SQRT(GUTTER*GUTTER+GHIKE*GHIKE)+XLIP
          HR=A1/WP1
          XK1=1.486/XN*A1*HR**.6667
          CROWN=(HWIDTH-GUTTER)*XFALL+XLIP+GHIKE
          A2=GUTTER*(CROWN-GHIKE/2.)+(HWIDTH-GUTTER)*.5*(CROWN-GHIKE-XLIP)
          WP2=WP1+CROWN-XLIP-GUTTER+(HWIDTH-GUTTER)*FACTOR
          HR=A2/WP2
          XK2=1.486/XN*A2*HR**.6667
          Q=QQ
          IF(IWIDTH.EQ.2)Q=QQ/2.
          TESTK=Q/S0**.5
          IF(TESTK.GT.XK1)GO TO 300
C-------GUTTER FLOW MODEL
          V=Q/A1
          FWIDTH=GUTTER
          DN=XLIP+GHIKE
          GO TO 3000
300       IF(TESTK.GT.XK2)GO TO 400
C-------FLOW IS LESS THAN CROWN
          YTEST1=0.
          YMAX=CROWN
          SET=XLIP+GHIKE
          YMIN=SET
          I=0
350       I=I+1
          YTEST=(YMAX+YMIN)*.5
          A=GUTTER*(YTEST-GHIKE/2.)+((YTEST-SET)**2.)/XFALL/2.
          WP=WP1+YTEST-SET+((YTEST-SET)/XFALL)*FACTOR
          HR=A/WP
          XK=1.486/XN*A*HR**.6667
          TEST=TESTK-XK
          IF(TEST)360,380,365
360       YMAX=YTEST
          GO TO 370
365       YMIN=YTEST
370       TEST=ABS(YTEST1-YTEST)
          IF(TEST.LT..01)GO TO 380
          IF(I.GT.50)GO TO 2900
          YTEST1=YTEST
          GO TO 350
380       V=Q/A
          DN=YTEST
          FWIDTH=GUTTER+(DN-SET)/XFALL
          IF(ISPLIT.EQ.1)GOTO 437
          GO TO 3000
400       CONTINUE
C-------FLOW EXCEEDS CROWN
          IF(IWIDTH.EQ.2)GO TO 440
          CFULLK=2.*XK2
          IF(TESTK.LT.CFULLK)GO TO 430
          Q=Q/2.
          TESTK=Q/S0**.5
          GO TO 440
430       DN=CROWN
C-------FLOW SPLITS AND IS LESS THAN FULL(CROWN) STREET
          FWIDTH=HWIDTH
          Q1=XK2*S0**.5
          V1=Q1/A2
          IF(ITER.EQ.1)WRITE(NUT,435)
```

```
435     FORMAT(14X,'***STREETFLOW SPLITS OVER STREET-CROWN***')
        ISPLIT=1
        Q=Q-Q1
        TESTK=Q/S0**.5
        GOTO 300
437     IF(ITER.EQ.1)WRITE(NUT,438)CROWN,HWIDTH,V1,DN,FWIDTH,V
438     FORMAT(10X,'FULL DEPTH(FEET) = ',F7.2,3X,'FLOODWIDTH(FEET) = ',
     .  F7.2,
     C  /,10X,'FULL HALF-STREET VELOCITY(FEET/SEC.) = ',F7.2,/,
     C  10X,'SPLIT DEPTH(FEET) = ',F7.2,3X,'SPLIT FLOODWIDTH(FEET) = ',
     .  F7.2,
     C  /,10X,'SPLIT VELOCITY(FEET/SEC.) = ',F7.2)
        V=V1
        DN=CROWN
        FWIDTH=HWIDTH
        GOTO 3000
440     IF(ITER.EQ.1)WRITE(NUT,441)
441     FORMAT(14X,'***STREET FLOWING FULL***')
        DMAX=CROWN+5.
        X=DMAX-CROWN
        A=HWIDTH*X+A2
        HR=A/WP2
        XK=1.486/XN*A*HR**.6667
        IF(TESTK.GE.XK)GOTO 3001
        DMIN=CROWN
        DO 450 I=1,12
        TEST=.5*(DMAX+DMIN)
        X=TEST-CROWN
        DN=CROWN+X
        A=HWIDTH*X+A2
        WP=WP2
        IF(DN.LE.CURB)WP=X+WP2
        HR=A/WP
        XK=1.486/XN*A*HR**.6667
        XZ=XK-TESTK
        XD=DN-DMAX
        IF(ABS(XD).LT..01)GOTO 3090
        IF(XZ)443,3090,445
443     DMIN=TEST
        GOTO 450
445     DMAX=TEST
450     CONTINUE
        GO TO 3000
480     FWIDTH=HWIDTH
        V=Q/A
        GO TO 3000
2900    IF(ITER.EQ.1)WRITE(NUT,2901)
2901    FORMAT(/,10X,'==>>ERROR:"BELOW-CROWN" STREETFLOW MODEL FAILS.')
        GOTO 3006
3000    CONTINUE
        GOTO 3100
3001    WRITE(NUT,3005)
3005    FORMAT(10X,'==>>ERROR: FLOW DEPTH EXCEEDS 5-FEET ABOVE STREET ',
     .  'CROWN')
3006    IF(ITER.EQ.1)WRITE(NUT,3007)
3007    FORMAT(14X,'AS AN APPROXIMATION, TRAVELTIME ESTIMATES ARE BASED'
     C,/,14X,'ON A DEFAULT VELOCITY OF [10] FEET PER SECOND.'/)
        V=10.
        DN=0.
        FWIDTH=0.
```

```
          GOTO 4000
3090      FWIDTH=HWIDTH
          V=Q/A
          DV=V*DN
3100      CONTINUE
          IF(DN.LE.CURB)GOTO 3109
          IF(ITER.EQ.1)WRITE(NUT,3107)
3107      FORMAT(10X,'NOTE: STREETFLOW EXCEEDS TOP OF CURB.',/,
     C    10X,'THE FOLLOWING STREETFLOW RESULTS ARE BASED ON THE ',
     .    'ASSUMPTION',
     C/,10X,'THAT NEGLIBLE FLOW OCCURS OUTSIDE OF THE STREET CHANNEL.'
     .    ,/,
     C    10X,'THAT IS, ALL FLOW ALONG THE PARKWAY, ETC., IS NEGLECTED.')
3109      CONTINUE
          DV=DN*V
          IF(ITER.EQ.1)WRITE(NUT,3110)DN,FWIDTH,V,DV
3110      FORMAT(
     C    10X,'STREET FLOWDEPTH(FEET) = ',F5.2,/,
     C    10X,'HALFSTREET FLOODWIDTH(FEET) = ',F7.2,/,
     C    10X,'AVERAGE FLOW VELOCITY(FEET/SEC.) = ',F7.2,/,
     C    10X,'PRODUCT OF DEPTH&VELOCITY = ',F7.2)
4000      CONTINUE
C
          RETURN
          END
```

4.24. **PROGRAM 4.8: DATA ENTRY**

```
---DATA ENTRY FOR PAVEMENT "V" DITCH FLOW THROUGH SUBAREA---PAGE 1


Enter upstream node     1.00 elevation(FEET)........ ===>     "E1"
:ALLOWABLE VALUES ARE [.01    ] TO [99999.99  ]

Enter downstream node     2.00 elevation(FEET)...... ===>     "E2"
:ALLOWABLE VALUES ARE [0      ] TO [99999.99  ]

Enter runoff travel-length through subarea(FEET).... ===>     "XL"
:ALLOWABLE VALUES ARE [.01    ] TO [10000 ]

--------------------------------------------------------------------------------
TYPE: EXIT to leave program ; TOP  to go to top of page
```

```
---DATA ENTRY FOR PAVEMENT "V" DITCH FLOW THROUGH SUBAREA---PAGE 2

SUBAREA LAND USE OR DEVELOPMENT TYPE:

    1= Commercial
    2= Apartment
    3= Mobile home park
    4= Condominium
    5= Single family (1/4 ACRE lot)
    6= Single family (1/2 ACRE lot)
    7= Single family (  1-ACRE lot)
    8= Single family (2.5-ACRE lot)
    9= Undeveloped


Select subarea development classification number.... ===>     "K"


--------------------------------------------------------------------------------
TYPE: EXIT to leave program ; TOP  to go to top of page
                            ; BACK to go back one page
```

```
---DATA ENTRY FOR PAVEMENT "V" DITCH FLOW THROUGH SUBAREA---PAGE 3

SUBAREA RUNOFF COEFFICIENT OPTIONS:

    1= Assume soil group A
    2= Assume soil group B
    3= Assume soil group C
    4= Assume soil group D
    5= User to specify runoff coefficient

Select runoff coefficient option number............ ===>    "FSOIL"

Enter SUBAREA runoff coefficient................... ===>    "C"
:ALLOWABLE VALUES ARE [.01    ] TO [.99    ]

Enter subarea area(ACRES).......................... ===>    "AREA"
:ALLOWABLE VALUES ARE [0] TO [1000   ]

-----------------------------------------------------------------------------
TYPE: EXIT to leave program ; TOP  to go to top of page
                            ; BACK to go back one page
```

```
---DATA ENTRY FOR PAVEMENT "V" DITCH FLOW THROUGH SUBAREA---PAGE 4

Enter "V" gutter width(FEET)....................... ===>    "W"
:ALLOWABLE VALUES ARE [.001   ] TO [100    ]

Enter "V" gutter-hike(FEET)........................ ===>    "HIKE"
:ALLOWABLE VALUES ARE [.001   ] TO [10     ]

Enter pavement lip(FEET)........................... ===>    "XLIP"
:ALLOWABLE VALUES ARE [.001   ] TO [10     ]

Enter assumed uniform Mannings friction factor...... ===>    "XN"
(NOTE: SUGGESTED VALUE FOR MANNINGS n IS [.015])
:ALLOWABLE VALUES ARE [.005   ] TO [.9999 ]

Enter symmetric pavement crossfall[DECIMAL NOTATION] ===>    "XFALL"
:ALLOWABLE VALUES ARE [.00001] TO [.9999 ]

Enter maximium allowable depth of flow(FEET)........ ===>    "DMAX"
:ALLOWABLE VALUES ARE [.001        ] TO [100         ]

-----------------------------------------------------------------------------
TYPE: EXIT to leave program ; TOP  to go to top of page
                            ; BACK to go back one page
```

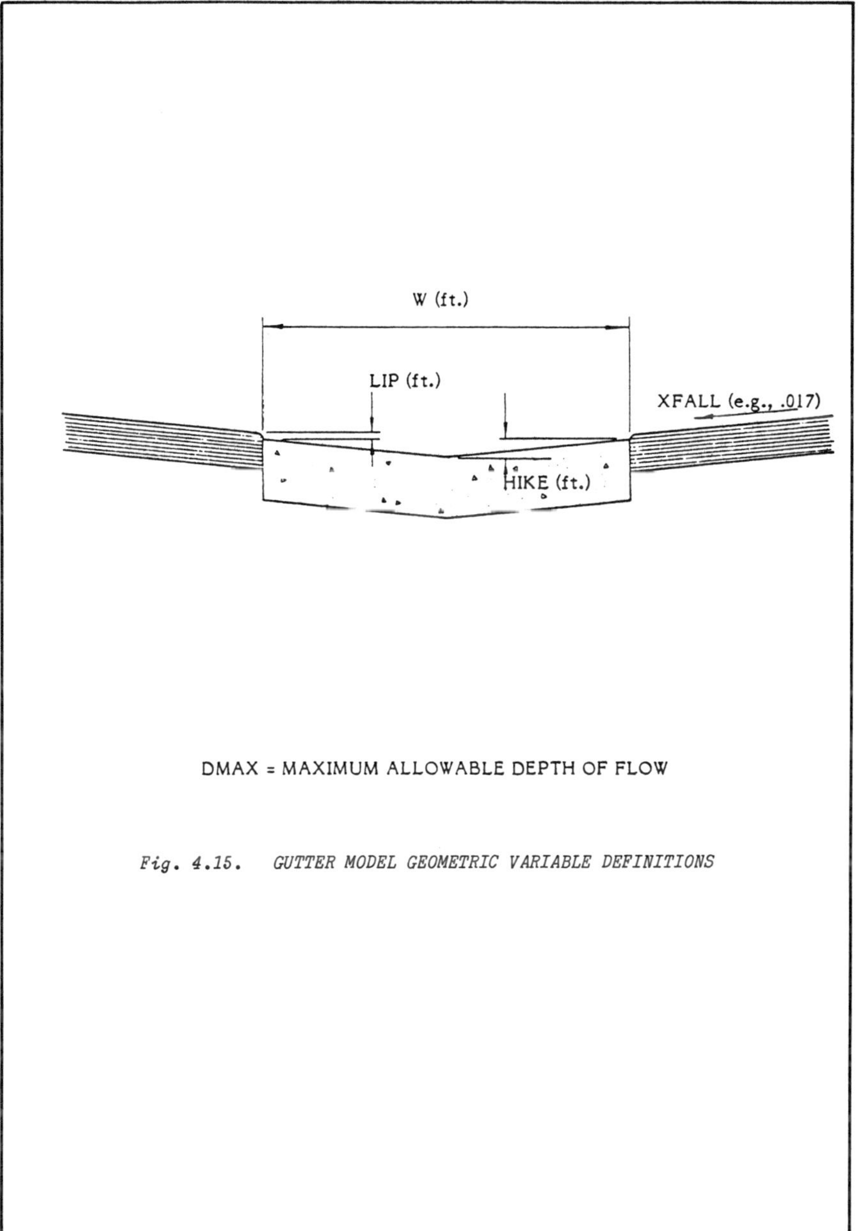

Fig. 4.15. GUTTER MODEL GEOMETRIC VARIABLE DEFINITIONS

PROGRAM 4.8

```
C ---------------------------------------------------------------
      SUBROUTINE VGUTTR(Q,TC,SUMA,IERR)
C ---------------------------------------------------------------
C
C  PAVEMENT "V" DITCH FLOW
C
      COMMON/NUT/NUT
      COMMON/CALC/YR,XI,OPTN,NR,C,CVAL,V,Y,IGUESS,DS,XN
      COMMON/CALC1/E1,E2,XL,K,IS,AREA,XKI,XP,KI,RIYR,RSLOPE
C
C
C..READ DATA INPUT
      READ FREE(5)E1,E2,XL,K,FSOIL,AREA,W,HIKE,XLIP,XN,XFALL,DMAX
      IS=FSOIL
      C=FSOIL-5.0
C
C  PROCESS FLOW THRU SUBAREA
C
      WRITE(NUT,7080)
7080  FORMAT(3X,'>>>>>COMPUTE "V" GUTTER FLOW TRAVELTIME THRU SUBAREA',
     . '<<<<<')
      WRITE(NUT,403)
      WRITE(NUT,7105)E1,E2,XL
7105  FORMAT(3X,'UPSTREAM NODE ELEVATION = ',F9.2,/,3X,
     . 'DOWNSTREAM NODE ELEVATION = ',F9.2,/,3X,
     . 'CHANNEL LENGTH THRU SUBAREA(FEET) = ',F8.2)
      WRITE(NUT,7100)W,HIKE,XLIP,XN,XFALL,DMAX
7100  FORMAT(3X,'"V" GUTTER WIDTH(FEET) = ',F6.2,3X,'GUTTER HIKE(FEET)',
     .' = ',
     CF6.3,/,3X,'PAVEMENT LIP(FEET) = ',F6.3,3X,'MANNINGS N = ',F5.4,/,
     C 3X,'PAVEMENT CROSSFALL(DECIMAL NOTATION) = ',F6.2,/,
     C 3X,'MAXIMUM DEPTH(FEET) = ',F6.2)
      QAVG=Q/SUMA
      ITER=0
      QQ=Q+QAVG/2.*AREA
      S1=(E1-E2)/XL
      IGUESS=1
7140  DS=0.
C
      CALL GNORMA(W,HIKE,XLIP,XN,XFALL,S1,QQ,DMAX,Y,DS,V,IGUESS,
     . NUT)
C
      IF(ITER.EQ.2)GO TO 7200
C
      TRAVEL=XL/60./V
      TCA=TC+TRAVEL
C
      CALL INTENR(YR,TCA,XI,IGUESS,ERROR,NUT,RIYR,RSLOPE)
      IF(ERROR.NE.1.)GO TO 7203
      IERR=9
      GO TO 3020
C
7203  CALL COEFAR(XI,IS,K,C,IGUESS,CVAL,NUT)
      IF(ITER.EQ.1.OR.ITER.EQ.2)GOTO 7200
      IGUESS=0
      ITER=1
      QQ=C*XI*AREA/2.+Q
      GOTO 7140
C
7200  IF(ITER.EQ.2)GOTO 7300
```

```
C
      QAVG=QQ
      SD1=Y
      SW1=W
C
      IF(DS.LE.W)DS=0.
      DS=W+2.*DS
      WRITE(NUT,7250)V,Y,DS
7250 FORMAT(3X,'TRAVELTIME THRU SUBAREA BASED ON VELOCITY(FEET/SEC) = '
    C  ,F6.2,/,3X,'AVERAGE FLOWDEPTH(FEET) = ',F6.2,3X,'FLOODWIDTH',
    C '(FEET) = ',F7.2)
      TC=TCA
      WRITE(NUT,7260)TRAVEL,TC
7260  FORMAT(3X,'"V" GUTTER FLOW TRAVEL TIME(MIN) = ',F6.2,3X,
    C  'TC(MIN) = ',F6.2)
      DELQ=C*XI*AREA
      Q=Q+DELQ
      SUMA=SUMA+AREA
      WRITE(NUT,4140)AREA,DELQ,SUMA,Q
4140  FORMAT(3X,'SUBAREA AREA(ACRES) = ',F7.2,3X,'SUBAREA RUNOFF',
    .'(CFS) = ',
    .F7.2,/,3X,'SUMMED AREA(ACRES) = ',F7.2,3X,'TOTAL RUNOFF(CFS) = ',
    .  F9.2)
      ITER=2
      IGUESS=1
      QQ=Q
      GOTO 7140
7300  DV=Y*V
      IF(DS.LE.W)DS=0.
      DS=W+2.*DS
      WRITE(NUT,7350)Y,DS,V,DV
7350  FORMAT(3X,'END OF SUBAREA "V" GUTTER HYDRAULICS:',/,3X,
    C  'DEPTH(FEET) = ',F5.2,3X,'FLOODWIDTH(FEET) = ',F5.2,/,
    C3X,'FLOW VELOCITY(FEET/SEC.) = ',F6.2,3X,'DEPTH*VELOCITY = ',F6.2)
C
3019  CONTINUE
      GO TO 3020
3021  IERR=3
C
3020  CONTINUE
C
C
C  FORMATS
C
403   FORMAT(1X,76('='))
7150  FORMAT('**ERROR-FLOW DEPTH EXCEEDS ALLOWABLE DEPTH: RECHECK ',
    .  ,'YOUR DATA')
7160  FORMAT('**ERROR-NO SOLUTION FROM GUTTER MODEL: RECHECK YOUR',
    .  ' DATA')
C
      RETURN
      END
C
C
C  ---------------------------------------------------------------
      SUBROUTINE GNORMA(W,GHIKE,XLIP,XN,SLOPE,S0,Q,DMAX,Y,DS,
    .  V,IGUESS,NUT)
C  ---------------------------------------------------------------
C
C
```

```
      A1=.5*GHIKE*W
      WP1=2.*SQRT(W*W/4.+GHIKE*GHIKE)
      HR=A1/WP1
      XK1=1.486/XN*A1*HR**.6667
      WP2=W + 2.*XLIP
      A2=A1+XLIP*W
      HR=A2/WP2
      XK2=1.486/XN*A2*HR**.6667
      TESTK=Q/S0**.5
      IF(TESTK.GT.XK1)GO TO 100
      QFULL=XK1*S0**.5
      DS=W
      Y=GHIKE
      V=QFULL/A1
      WRITE(NUT,90)
 90   FORMAT(/,10X,'NOTE:TRAVELTIME ESTIMATES BASED ON NORMAL DEPTH',/,
     C 10X,'IN A FLOWING-FULL GUTTER(NORMAL DEPTH = GUTTER HIKE)',/)
      GO TO 1000
 100  IF(TESTK.GT.XK2)GO TO 200
      QFULL=XK2*S0**.5
      V=QFULL/A2
      DS=W
      Y=GHIKE+XLIP
      WRITE(NUT,190)
 190  FORMAT(/,10X,'NOTE:TRAVELTIME ESTIMATES BASED ON NORMAL',/,
     C 10X,'DEPTH EQUAL TO [GUTTER-HIKE + PAVEMENT LIP]')
      GO TO 1000
 200  I=0
      YMAX=DMAX
      Y1=0.
      SET=GHIKE+XLIP
      YMIN=SET
      DS=(DMAX-SET)/SLOPE
      AS=.5*DS*(DMAX-SET)
      HR=AS/SQRT((DMAX-SET)*(DMAX-SET)+DS*DS)
      XKA=1.486/XN*AS*HR**.6667
      AM=(DMAX-SET)*W+A2
      HR=AM/WP2
      XKM=1.486/XN*AM*HR**.6667
      XKT=2.*XKA+XKM
      IF(TESTK.GT.XKT)GO TO 600
 201  I=I+1
      IF(I.GT.50)GO TO 500
      Y=.5*(YMAX+YMIN)
      DS=(Y-SET)/SLOPE
      AS=.5*DS*(Y-SET)
      HR=AS/SQRT((Y-SET)*(Y-SET)+DS*DS)
      XKA=1.486/XN*AS*HR**.6667
      AM=(Y-SET)*W+A2
      HR=AM/WP2
      XKM=1.486/XN*AM*HR**.6667
      XKT=2.*XKA+XKM
      IF(TESTK-XKT)300,900,400
 300  YMAX=Y
      GO TO 450
 400  YMIN=Y
 450  TEST=ABS(Y-Y1)
      IF(TEST.LT..01)GO TO 900
      Y1=Y
      GO TO 201
```

```
500     WRITE(NUT,501)
501     FORMAT(/,10X,'==>>ERROR:NO CONVERGENCE IN V-GUTTER PROGRAM.',/,
     C  14X,'AS AN APPROXIMATION, TRAVELTIME CALCULATIONS ARE BASED',/,
     C  14X,'ON FLOWDEPTH EQUAL TO THE SPECIFIED MAXIMUM ALLOWABLE ',
     .  'DEPTH.',/)
        GOTO 650
600     WRITE(NUT,610)
610    FORMAT(/,10X,'==>>ERROR:FLOW EXCEEDS CAPACITY OF CHANNEL WITH',/,
     C14X,'NORMAL DEPTH EQUAL TO SPECIFIED MAXIMUM ALLOWABLE DEPTH.',/,
     C  14X,'AS AN APPROXIMATION, TRAVELTIME CALCULATIONS ARE BASED',/,
     C  14X,'ON FLOWDEPTH EQUAL TO THE SPECIFIED MAXIMUM ALLOWABLE ',
     .  'DEPTH.',/)
650     Y=DMAX
        DS=(DMAX-SET)/SLOPE*2.+W
        V=Q/(AM+2.*AS)
        GOTO 1000
900     AT=2.*AS+AM
        V=Q/AT
1000    CONTINUE
C
        RETURN
        END
```

4.25.

PROGRAM 4.9: DATA ENTRY

```
DATA ENTRY FOR CONFLUENCE OF INDEPENDENT STREAMS AT A NODAL POINT


Enter the total number of independent streams
to confluence.................................... ===>    "NUMBER"
:ALLOWABLE VALUES ARE [1] TO [5     ]

Enter a number for THIS independent stream.......... ===>    "N"
(NOTE: STREAM NUMBERS MUST BE USED IN INCREASING
 ORDER FROM 1 TO 5)
:ALLOWABLE VALUES ARE [1] TO [5     ]

-----------------------------------------------------------------------------
TYPE: EXIT to leave program ; TOP  to go to top of page
```

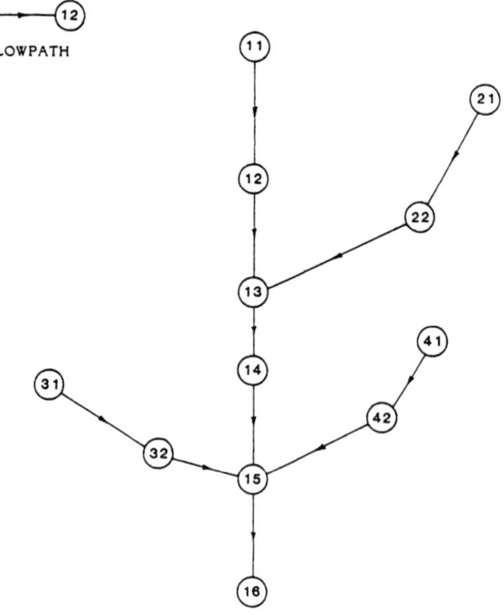

Fig. 4.16. MAIN-LINE CONFIGURATION REQUIRING ONE CONFLUENCE

PROGRAM 4.9

```
C ---------------------------------------------------------------
      SUBROUTINE CONFB(Q,TC,SUMA,IERR)
C ---------------------------------------------------------------
C
C     CONFLUENCE ANALYSIS
C
      COMMON /NUT/NUT
      COMMON/CALC/YR,XI,OPTN,NR,C,CVAL,V,Y,IGUESS,DS,XN
      COMMON/CALC1/E1,E2,XL,K,IS,AREA,XKI,XP,KI,RIYR,RSLOPE
      COMMON/CALC2/TCON(5),TOTALA(5),XINT(5),QJN(5),QBAR(5),ZCON(5)
C
C
C..READ DATA INPUT
      READ FREE(5)N,NUMBER
C
C-------PROCESS CONFLUENCE ANALYSIS
C
      IF(N.LT.NUMBER)WRITE(NUT,531)
531   FORMAT(3X,'>>>>>DESIGNATE INDEPENDENT STREAM FOR CONFLUENCE<<<<<')
      IF(N.EQ.NUMBER)WRITE(NUT,532)
532   FORMAT(3X,'>>>>>DESIGNATE INDEPENDENT STREAM FOR CONFLUENCE<<<<<',
     C /,3X,'>>>>>AND COMPUTE VARIOUS CONFLUENCED STREAM VALUES<<<<<')
      WRITE(NUT,403)
      TCON(N)=TC
      IGUESS = 1
C
      CALL INTENR(YR,TC,XI,IGUESS,ERROR,NUT,RIYR,RSLOPE)
      IF(ERROR.EQ.1.)IERR=9
      IF(ERROR.EQ.1.)GOTO 3019
C
      XINT(N)=XI
      TOTALA(N)=SUMA
      QJN(N)=Q
      WRITE(NUT,550)N,TC,XI,SUMA,Q
550   FORMAT(3X,'CONFLUENCE VALUES USED FOR INDEPENDENT STREAM ',I2,
     C ' ARE:',/,3X,'TIME OF CONCENTRATION(MINUTES) = ',F6.2,/,
     C 3X,'RAINFALL INTENSITY (INCH./HOUR) = ',F6.2,/,
     C 3X,'TOTAL STREAM AREA (ACRES) = ',F7.2,/,
     C 3X,'TOTAL STREAM RUNOFF(CFS) AT CONFLUENCE = ',F9.2)
C
      IF(N.LT.NUMBER)GOTO 2000
552   CONTINUE
      WRITE(NUT,560)
560   FORMAT(/,3X,'CONFLUENCE INFORMATION:',/,3X,'STREAM',3X,
     C 'RUNOFF',6X,'TIME',3X,'INTENSITY',/,3X,'NUMBER',4X,
     . '(CFS)',5X,'(MIN.)',1X,'(INCH/HOUR)')
      WRITE(NUT,401)
      WRITE(NUT,570)(I,QJN(I),TCON(I),XINT(I),I=1,NUMBER)
570   FORMAT(3X,I3,3X,F9.2,4X,F6.2,6X,F6.3)
      SUMAJN=0.
C
      DO 575 I=1,NUMBER
      SUMAJN=SUMAJN+TOTALA(I)
575   QBAR(I)=0.
C
      DO 580 I=1,NUMBER
      DO 580 J=1,NUMBER
      IF(XINT(J).LE.0.)GOTO 597
      FACTTC=TCON(I)/TCON(J)
      FACTOR=XINT(I)/XINT(J)
```

```
         IF(FACTOR.LE.0..OR. FACTTC.LE.0.)GO TO 597
         IF(FACTOR.GT.1.)FACTOR=1.
         IF(FACTTC.GT.1.)FACTTC=1.
         QBAR(I)=QBAR(I)+FACTOR*FACTTC*QJN(J)
580      CONTINUE
C
         WRITE(NUT,585)NUMBER
585     FORMAT(/,3X,'RAINFALL INTENSITY AND TIME OF CONCENTRATION RATIO',
     .  /,3X,'FORMULA USED FOR ',I2,' STREAMS.')
         WRITE(NUT,587)(QBAR(I),I=1,NUMBER)
587      FORMAT(3X,'VARIOUS CONFLUENCED RUNOFF VALUES ARE AS FOLLOWS:',
     C  /,5X,5F9.2,/)
         MAX=1
         QMAX=QBAR(1)
C
         DO 590 I=1,NUMBER
         QMAX=QBAR(MAX)
         IF(QBAR(I).GT.QMAX)MAX=I
590      CONTINUE
C
         TC=TCON(MAX)
         Q=QBAR(MAX)
         SUMA=SUMAJN
         WRITE(NUT,595)Q,TC,SUMA
595      FORMAT(3X,'COMPUTED CONFLUENCE ESTIMATES ARE AS FOLLOWS:',/,
     C   3X,'RUNOFF(CFS) = ',F9.2,3X,'TIME(MINUTES) = ',F8.3,/,3X,
     C   'TOTAL AREA(ACRES) = ',F9.2)
301      II=NUMBER+1
         IF(II.GT.5)GO TO 2000
302      CONTINUE
303      CONTINUE
C
597      WRITE(NUT,598)
598      FORMAT(3X,'[FATAL ERROR: INVALID CONFLUENCE VALUES: SEE USERS ',
     .   'MANUAL]')
C
2000     CONTINUE
3019     CONTINUE
3020     CONTINUE
C
C  FORMATS
C
403      FORMAT(1X,76('='))
401      FORMAT(1X,76('-'))
C
         RETURN
         END
```

4.26. **PROGRAM 4.10: DATA ENTRY**

```
---DATA ENTRY FOR SPECIFICATION OF HYDROLOGY DATA AT A NODE---

Enter user-specified time of concentration(MIN.).... ===>    "TC"
:ALLOWABLE VALUES ARE [5] TO [1000  ]

Enter user-specified total area(ACRES) tributary
to node............................................. ===>    "SUMA"
:ALLOWABLE VALUES ARE [.0001 ] TO [10000 ]

Enter user-specified total runoff(CFS).............. ===>    "Q"
:ALLOWABLE VALUES ARE [.0001 ] TO [100000]

--------------------------------------------------------------------------------
TYPE: EXIT to leave program ; TOP  to go to top of page
```

PROGRAM 4.10

```
C  ---------------------------------------------------------------
         SUBROUTINE USEROR(Q,TC,SUMA,IERR)
C  ---------------------------------------------------------------
C
C  USER SPECIFICATION OF HYDROLOGY
C
         COMMON /NUT/NUT
         COMMON/CALC/YR,XI,OPTN,NR,C,CVAL,V,Y,IGUESS,DS,XN
         COMMON/CALC1/E1,E2,XL,K,IS,AREA,XKI,XP,KI,RIYR,RSLOPE
C
C
C...READ DATA INPUT
         READ FREE(5) TC,SUMA,Q
C
         WRITE(NUT,5143)
5143     FORMAT(3X,'>>>>>USER SPECIFIED HYDROLOGY INFORMATION AT NODE',
     .   '<<<<<')
         WRITE(NUT,403)
C
         CALL INTENR(YR,TC,XI,1,ERROR,NUT,RIYR,RSLOPE)
         IF(ERROR.NE.1.)GO TO 5150
         IERR=9
         GO TO 3019
C
5150     WRITE(NUT,5160)TC,XI,SUMA,Q
5160     FORMAT(3X,'USER-SPECIFIED VALUES ARE AS FOLLOWS:',/,3X,
     C   'TC(MIN) = ',F6.2,3X,'RAIN INTENSITY(INCH/HOUR) = ',F5.2,/,
     C   3X,'TOTAL AREA(ACRES) = ',F8.2,3X,'TOTAL RUNOFF(CFS) = ',F9.2)
C
3019     CONTINUE
C
3020     CONTINUE
C
C  FORMATS
C
403      FORMAT(1X,76('='))
C
         RETURN
         END
```

4.27.

PROGRAM 4.11: DATA ENTRY

```
---DATA ENTRY FOR ADDITION OF SUBAREA TO MAINLINE RUNOFF
   AT MAINLINE TIME OF CONCENTRATION---PAGE 1

SUBAREA LAND USE OR DEVELOPMENT TYPE:

    1: Commercial
    2: Apartment
    3: Mobile home park
    4: Condominium
    5: Single family(1/4 ACRE lot)
    6: Single family(1/2 ACRE lot)
    7: Single family(  1-ACRE lot)
    8: Single family(2.5-ACRE lot)
    9: Undeveloped



Specify assumed uniform subarea land use/development ===>     "K"



--------------------------------------------------------------------------------
TYPE: EXIT to leave program ; TOP  to go to top of page
```

```
---DATA ENTRY FOR ADDITION OF SUBAREA TO MAINLINE RUNOFF
   AT MAINLINE TIME OF CONCENTRATION---PAGE 2
SUBAREA RUNOFF COEFFICIENT OPTIONS:

    1= Assume soil group A
    2= Assume soil group B
    3= Assume soil group C
    4= Assume soil group D
    5= User to specify runoff coefficient

Select runoff coefficient option number............. ===>     "FSOIL"

Enter SUBAREA runoff coefficient.................... ===>     "C"
:ALLOWABLE VALUES ARE [.01    ] TO [.99    ]

Enter subarea area(ACRES)........................... ===>     "AREA"
:ALLOWABLE VALUES ARE [0] TO [1000   ]



--------------------------------------------------------------------------------
TYPE: EXIT to leave program ; TOP  to go to top of page
                            ; BACK to go back one page
```

PROGRAM 4.11

```
C ----------------------------------------------------------------------
      SUBROUTINE SUBTOB(Q,TC,SUMA,IERR)
C ----------------------------------------------------------------------
C
C  ADDITION OF SUBAREA TO MAINLINE RUNOFF
C
      COMMON /NUT/NUT
      COMMON/CALC/YR,XI,OPTN,NR,C,CVAL,V,Y,IGUESS,DS,XN
      COMMON /CALC1/E1,E2,XL,K,IS,AREA,XKI,XP,KI,RIYR,RSLOPE
C
C
C..READ DATA INPUT
      READ FREE(5)K,FSOIL,AREA
      IS=FSOIL
      C=FSOIL-5.0
C
C  PROCESS SUBAREA ADDITION TO MAINLINE FLOW
C
      WRITE(NUT,6160)
6160  FORMAT(3X,'>>>>>ADDITION OF SUBAREA TO MAINLINE PEAK FLOW<<<<<')
      WRITE(NUT,403)
C
      CALL INTENR(YR,TC,XI,0,ERROR,NUT,RIYR,RSLOPE)
      IF(ERROR.NE.1.)GO TO 6162
      IERR=9
      GO TO 3019
C
6162  CALL COEFAR(XI,IS,K,C,0,CVAL,NUT)
C
      DELQ=C*XI*AREA
      SUMA=SUMA+AREA
      Q=Q+DELQ
      WRITE(NUT,6200)AREA,DELQ,SUMA,Q,TC
6200  FORMAT(3X,'SUBAREA AREA(ACRES) = ',F7.2,3X,'SUBAREA RUNOFF',
     . '(CFS) = ',
     C F7.2,/,3X,'TOTAL AREA(ACRES) = ',F7.2,3X,'TOTAL RUNOFF(CFS) = ',
     C F7.2,/,3X,'TC(MIN) = ',F6.2)
C
3020  CONTINUE
C
C  FORMATS
C
403   FORMAT(1X,76('='))
C
      RETURN
      END
```

synthetic unit hydrograph methods

5

5.1. Introduction

The unit hydrograph method is a synthetic hydrograph approach initially advanced by L. K. Sherman (1932). The keystone of the method is the assumption that watershed discharge is related to the total volume of runoff, and that the time factors which affect the unit hydrograph shape are invariant. The basic unit hydrograph theory was extended by F. F. Snyder (1938) to transpose storm rainfall-runoff relationships from gaged watersheds to hydrologically and geographically similar watersheds that lack rainfall-runoff stream gage data. The basic assumptions used in this later work are that the watershed rainfall-runoff relationships are functions of watershed area, slope, and certain shape factors. The method is used to estimate a time distribution of runoff accumulating at the watershed downstream point of concentration when stream gage data is either unavailable or inadequate to provide a sound statistical analysis.

To determine the rainfall-runoff relationships to be transposed to ungaged watersheds, stream gage records are studied for various types and sizes of gaged watersheds. For example, the U.S. Army Corps of Engineers (Los Angeles Office) has determined several runoff time-distribution patterns for watersheds in the State of California. Such relationships provide a basis for transposing to ungaged watersheds a characteristic time distribution of runoff that is the average distribution for several similar watersheds. This approach is considered appicable when watersheds are physiographically and hydrologically similar. In Southern California, for example, the counties of Orange, Riverside, and San Bernardino (which together represent a vast spectrum of flood control conditions) have successfully utilized this approach in the development of county-wide flood control facilities. The same is true for most other areas of the United States.

Although there are several theoretical shortcomings associated with the unit hydrograph approach (such as the assumption of a linear system where runoff hydrographs resulting from a unit period of effective rainfall can be directly summed), the general approach continues to be widely used throughout the United States as a runoff synthesis method for ungaged watersheds (see Chapter 1). This chapter will examine the general unit hydrograph approach and several currently used variations of the method. A FORTRAN computer program for a severe design storm condition (Hromadka et al., 1983a), which is readily adaptable to currently available microcomputers, is provided.

The unit hydrograph design storm approach involves several assumptions that are imprecise approximations of the corresponding hydrologic processes. These basic assumptions are that (1) the critical storm rainfall pattern is uniformly distributed through the watershed; (2) there exists a direct proportionality between watershed runoff and the effective rainfall volume; (3) for any volume of effective rainfall occurring within a specified duration, the resulting runoff hydrograph is of a constant duration; and (4) the basin unit hydrograph is invariant throughout the critical design storm. The requirement that the watershed runoff is proportional to the effective rainfall has a direct analogy to a linear systems approach. Consequently, the unit hydrograph method can be considered a black-box modeling approach where the major characteristics of the model are determined by correlating the model output (runoff hydrograph) to the input data (rainfall records). Although the lumped-system model produces only approximations of the complex hydrologic characteristics of the watershed, its use continues to be widespread due to the ease of application and the wide range of applications available for master planning purposes.

5.2. Unit Hydrograph Terminology

In order to discuss the development of unit hydrographs for both gaged and ungaged watersheds, the following definitions are presented. Figure 5.1 illustrates each of the unit hydrograph concepts and definitions.

Effective Rainfall: The total rainfall less losses, which includes infiltration losses, evaporation, transpiration, absorption, and detention. This part of the rainfall runs off the watershed surface in a relatively brief time period. This is also referred to as rainfall excess, with the volume of rainfall excess equaling the volume of direct runoff.

Unit Hydrograph: Unit hydrograph for a point of concentration on a watershed is a hydrograph showing the time distribution of runoff that results from 1 inch of effective rainfall over the entire tributary watershed. The effective rainfall is assumed to occur as a constant rainfall with respect to both space and time throughout the unit duration.

Distribution Graph: The distribution graph is a unit hydrograph whose ordinates are expressed in terms of percent of the ultimate discharge. A distribution graph is generally

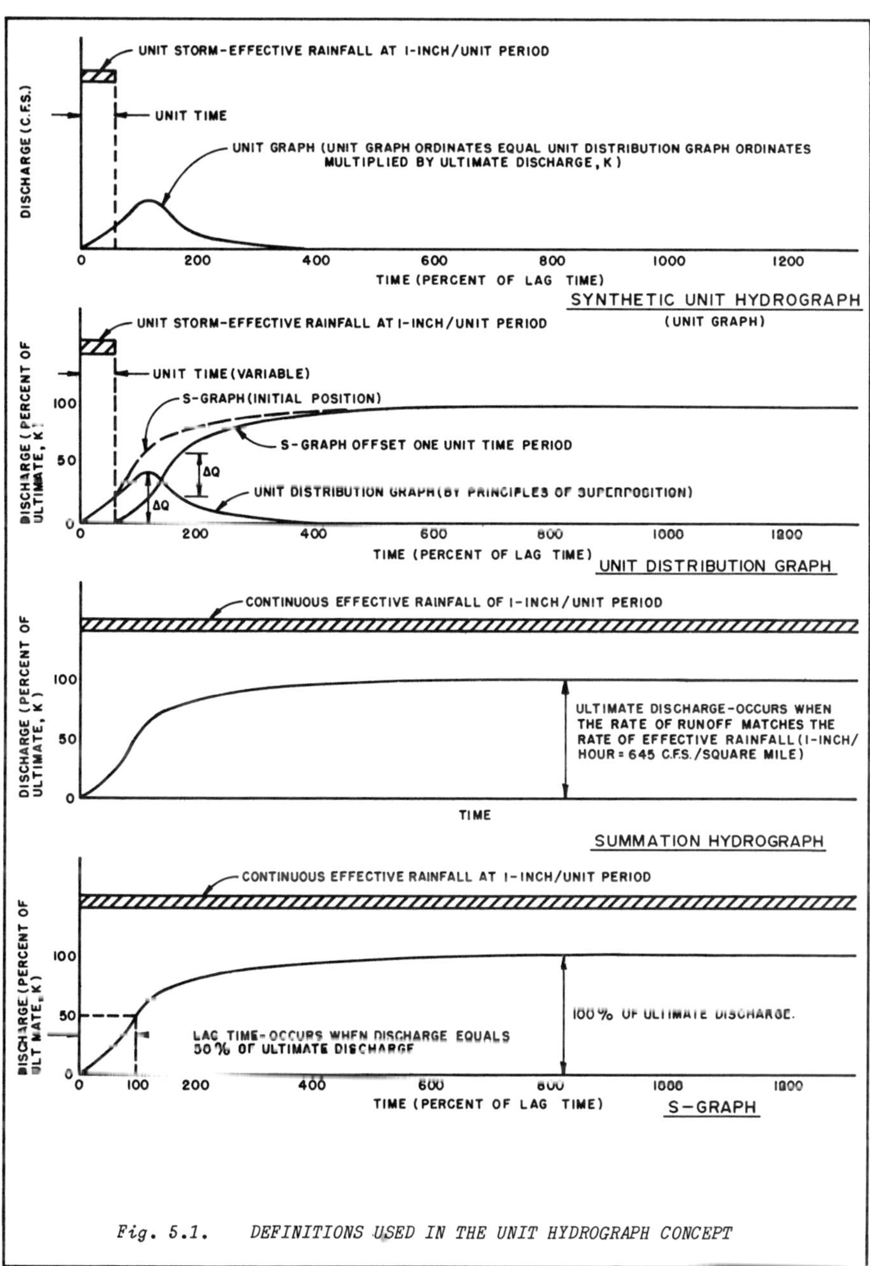

Fig. 5.1. DEFINITIONS USED IN THE UNIT HYDROGRAPH CONCEPT

developed as a block graph with each block representing its associated percent of unit runoff that occurs during the specified unit time period. The unit time period used in the distribution graph is identical to the unit time period specified for the unit hydrograph.

Summation Hydrograph: The summation hydrograph for a point of concentration is a hydrograph showing the time distribution of the rates of runoff that would result from a continuous series of unit effective rainfalls over the tributary watershed. The ordinates are expressed as rate of runoff in percent of the ultimate rate of runoff.

Lag: The watershed lag is the time (hours) from the beginning of a continuous series of unit effective rainfalls over the watershed to the instant when the rate of resulting runoff at the point of concentration reaches 50 percent of the ultimate rate of runoff. Another definition for lag is the time from the center of mass of the effective rainfall to the peak of the corresponding runoff hydrograph. Some hydrologists define lag as the time from the center of mass of effective rainfall to the center of mass of the runoff hydrograph. In this book, however, the first definition of lag will be used for both the theoretical and computer program development.

Ultimate Discharge: The ultimate discharge (or the ultimate rate of runoff) is the maximum rate of watershed runoff that can result from a specified effective rainfall intensity. Ultimate discharge occurs when the rate of runoff on the summation hydrograph is equivalent to the rate of effective rainfall. For a unit effective rainfall intensity of one inch occurring in a unit interval of one hour, the ultimate discharge is 645 cfs for every square mile of watershed.

S-Graph: The S-graph is a summation hydrograph developed by plotting watershed discharge in percent of the ultimate discharge as a function of time expressed in percent of the watershed lag.

5.3. Conceptual Evaluation of Unit Hydrographs

A simple example can be used to illustrate the concepts defined in the previous section. A 2-hour rainfall of 1.4 inches fell on a 1600 acre watershed, with the triangular direct runoff hydrograph shown in Fig. 5.2 resulting. Baseflow is assumed to be zero throughout the storm. Converting the area under the direct runoff hydrograph to units of inches (or area-inches) yields a direct runoff depth of 0.75 inches. According to the definition provided in the previous section, a unit hydrograph has a runoff depth of 1 inch; therefore, the unit hydrograph is found by dividing each ordinate by 0.75, which is also shown in Fig. 5.2. In order to find the duration of the unit hydrograph, it is

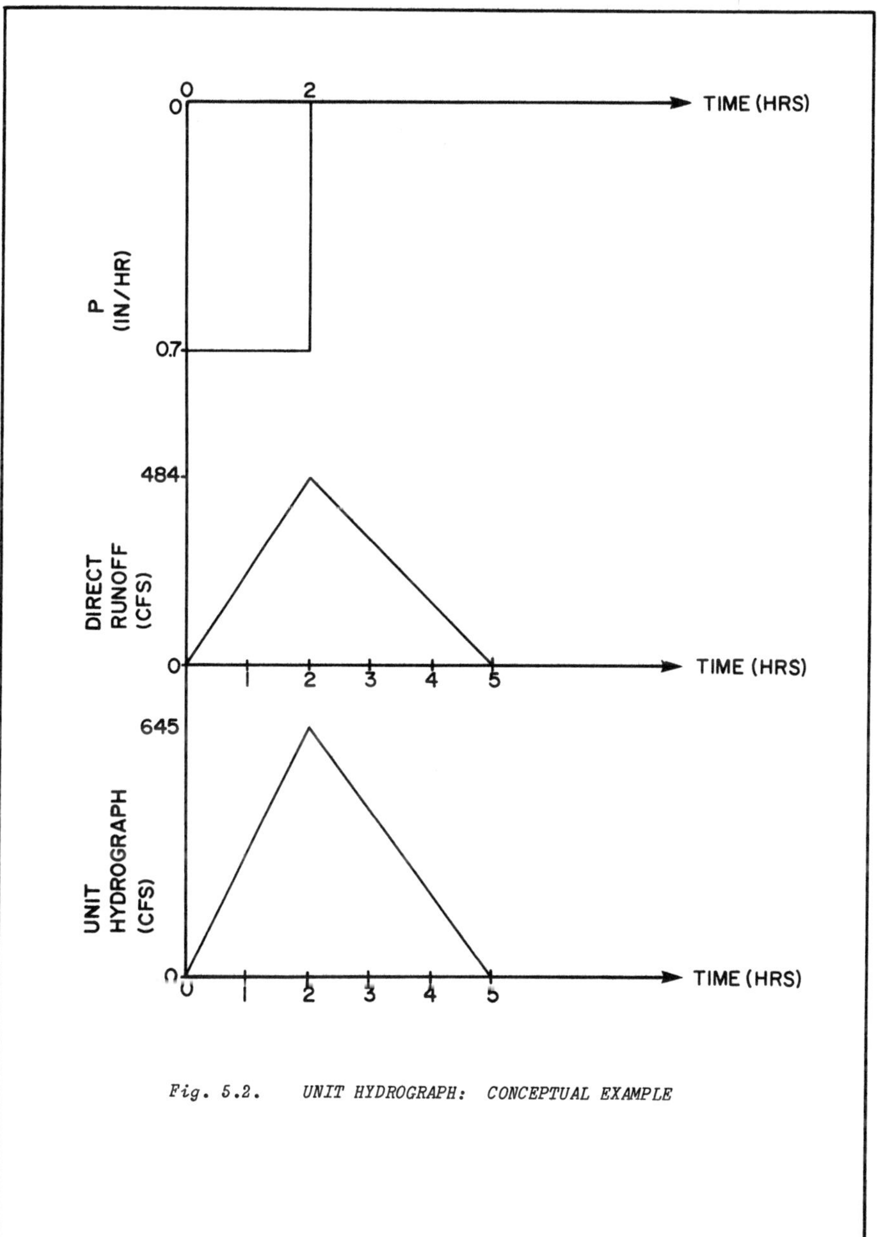

Fig. 5.2. UNIT HYDROGRAPH: CONCEPTUAL EXAMPLE

necessary to compute the time distribution of rainfall excess (i.e., effective rainfall). The phi-index method can be used to separate the rainfall into rainfall excess and losses. While the total rainfall is 1.4 inches and the intensity is 0.7 in/hr., the depth of rainfall excess must equal the depth of direct runoff, or 0.75 inches. Thus, the depth of losses is (1.4-0.75) = 0.65 inches over the 2-hour storm duration. Thus, the phi-index loss rate is 0.325 in/hr. Since the rainfall excess is constant over the 2-hour period, the unit hydrograph has a unit duration of 2 hours; it is referred to as a 2-hour unit hydrograph because the duration of the rainfall excess equals 2 hours.

5.4. Convolution

A unit hydrograph serves as the link between rainfall excess and direct runoff. Whereas watershed storage and the physical processes convert rainfall to runoff, the unit hydrograph is the modeling element that transforms a part of the total rainfall (i.e., the rainfall excess) into a part of the total runoff (i.e., the direct runoff). Thus, the unit hydrograph reflects the watershed storage and physical processes of the watershed. However, the definition of the unit hydrograph suggests that it represents the watershed response under a restrictive set of conditions, specifically a constant rate of rainfall excess occurring uniformly over the watershed during a specific time duration. Rarely, if ever, will actual storms meet the restrictive conditions, and certainly the characteristics of synthetic storms, such as those discussed in Chapter 2, will not correspond to the unit storm. Thus, a mechanism for converting a rainfall excess distribution that is different from the unit storm for which the unit hydrograph was developed is needed.

Convolution is the term that is used to represent the process of converting a rainfall excess distribution into a direct runoff hydrograph. Convolution involves the steps of multiplication, time translation, and addition. Each ordinate of the hydrograph is multiplied by the first ordinate of the rainfall excess distribution, which results in a direct runoff hydrograph having a duration equal to the duration of the unit hydrograph. The unit hydrograph is then translated by one time unit and again each ordinate is multiplied by the second ordinate of the rainfall excess distribution, which produces a second direct runoff hydrograph. The unit hydrograph is then translated another time unit and followed by the same multiplication operation. The translation/multiplication process is continued over all ordinates of the rainfall excess distribution. For each time increment, a separate direct runoff hydrograph is generated. Then for each time period, all of the ordinates of the individual direct runoff hydrographs are added to get a total direct runoff hydrograph. The process is illustrated in the example of Fig.5.3. The unit hydrograph is applied to each individual burst of the rainfall excess by multiplying each ordinate of the unit hydrograph. After each multiplication operation the unit hydrograph is translated one time unit so that the time at which the unit hydrograph begins coincides with the start of the burst of rainfall excess. After the multiplication-translation operation has been performed on each burst of rainfall, then the individual direct runoff hydrographs are summed to get the total direct runoff hydrograph.

In practice, the convolution process is used in modeling both actual measured storm events and synthetic design storms, such as those discussed in Chapter 2. The rainfall distribution is converted to a distribution of rainfall excess using one of the loss functions of Chapter 3. The rainfall excess distribution is then convoluted with a unit hydrograph to get the total direct runoff hydrograph. Baseflow can then be added, if appropriate.

5.5. Unit Hydrograph Analysis

The conceptual development of the previous section suggests that, if a unit hydrograph for a duration of T-hours is needed, it is necessary to find one or more storms having a uniform rainfall intensity for durations close to the desired duration of T hours. The above procedure can then be applied to each storm to derive unit hydrographs for each storm and an average watershed unit hydrograph computed. This is the procedure that was used prior to the development of more elaborate, computer assisted methods.

Since very few storms have a reasonably constant intensity over the duration of the storm, it is difficult to find measured storm event data that can be used to accurately estimate unit hydrographs by the above procedure. Thus, more elaborate methods have been developed so that complex storms with nonuniform intensities can be evaluated. While unit hydrographs could be developed by trial and error analyses of complex storms, this would be exceptionally difficult for all except very simple cases and the resulting unit hydrographs would probably not be very reproducible; that is, analyses made by different individuals would not lead to the same unit hydrograph. Other methods require the assumption of a particular model form, which is then fit by the method of moments. For example, Aron and White (1982) used a gamma function, which has the general shape of a unit hydrograph, and fit the parameters of the gamma function using the method of moments (which is the fitting of the two parameters of the gamma function using the mean and standard deviation of the hydrologic data). Other techniques such as least squares regression analysis

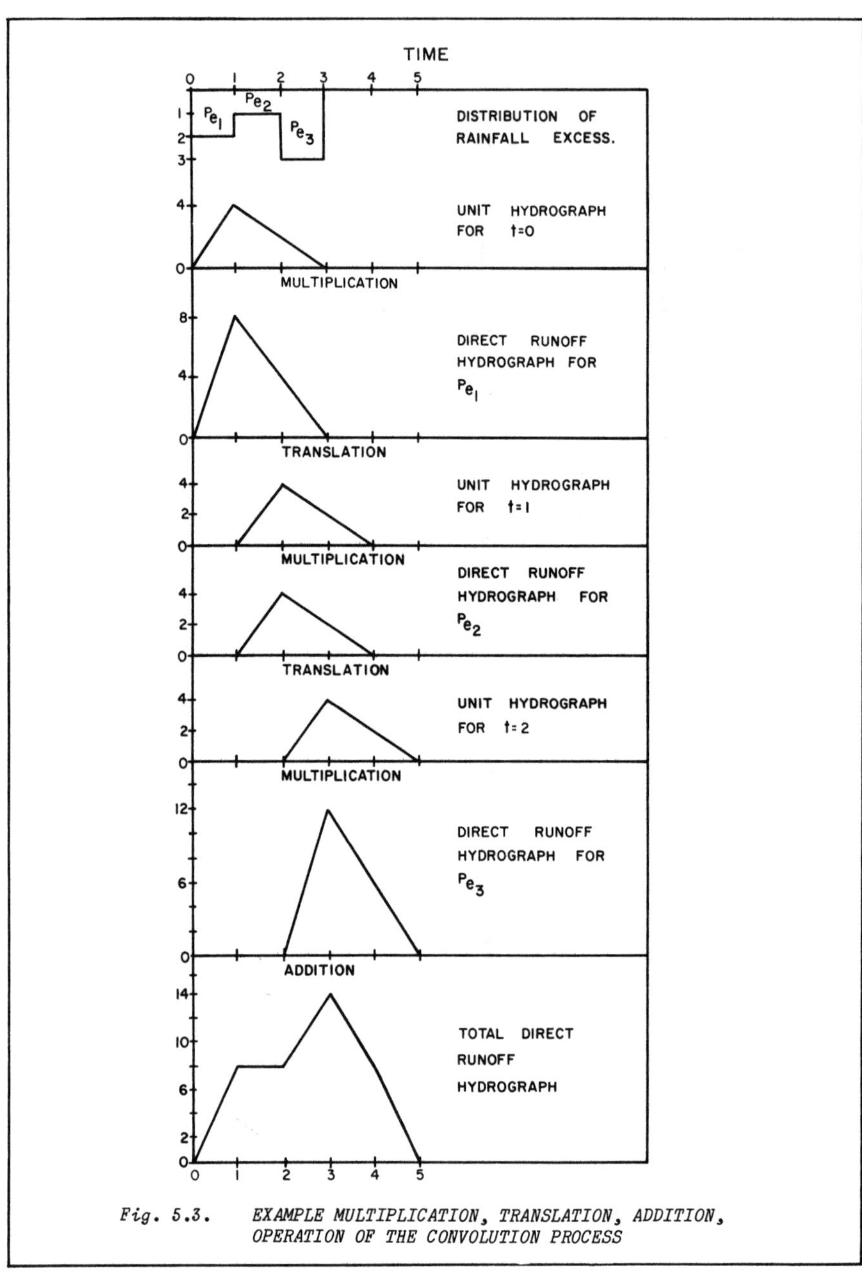

Fig. 5.3. EXAMPLE MULTIPLICATION, TRANSLATION, ADDITION, OPERATION OF THE CONVOLUTION PROCESS

have also been proposed. The calibration of unit hydrograph models are discussed in more detail, including examples, in Chapter 7.

5.6. Forming Synthetic Unit Hydrographs

The data required to analyze unit hydrographs is rarely available for developing a site specific unit hydrograph. This is especially true in small urban watersheds because data collection networks have been for the most part located on large drainage basins. Furthermore, the process of developing a unit hydrograph would be confounded because urban watersheds are constantly being changed and so the watershed characteristics over the duration of the data record would not be constant. The nonhomogeneity of the data records would effect the characteristics of the resulting unit hydrograph, including the magnitude and timing of the peak.

To overcome this problem of not being able to develop a site specific unit hydrograph, a number of synthetic unit hydrographs have been formulated. The Snyder and Clark synthetic hydrographs and the SCS dimensionless unit hydrographs are widely used for ungaged locations. These methods are based on many analyses, as briefly described in the previous section and in Chapter 7, of watersheds where data were available, and such methods are generally usable in most regions of the country.

In transposing a unit hydrograph (or S-graph) between watersheds for design work or when regionalizing unit hydrographs for a policy manual, it is assumed that the drainage areas within a given region are physiographically and hydrologically similar. Because no two drainage areas have identical hydrologic characteristics, the corresponding rainfall-runoff patterns are dissimilar and the distribution graphs may differ. However, most distribution graphs exhibit certain characteristics that appear to be related to watershed characteristics such as the drainage area and the timing factor of watershed lag. Based on such relationships, the generalized synthetic unit hydrographs can be dimensionalized for use at ungaged locations.

To illustrate the development and use of synthetic unit hydrographs, a method developed for southern California will be presented. It is important to recognize the two distinct phases involved: analysis and design. In the analysis phase, the form of the generalized unit hydrograph and the relationships necessary to dimensionalize it are developed through the analysis of measured rainfall and runoff data. The second phase involves dimensionalizing the generalized form using watershed information for the watershed where a design is required. The example is based on a watershed lag input. Watershed lag can be defined as the time from the beginning of the unit effective rainfall to the instant that the summation hydrograph reaches 50 percent of ultimate discharge. When the lags determined from summation hydrographs for several gaged watersheds are regressed on the hydrologic characteristics of the watersheds, a significant empirical relationship may result. Analysis of watershed lags estimated for several types of watersheds indicate that lag may be expressed by a calibrated formula such as

$$\text{lag} = C_t L^m L_c^m S^{-m/2} \tag{5.1}$$

where

- lag = watershed lag (hours)
- C_t = a calibration constant
- L = length of longest watercourse (miles)
- L_c = length along longest watercourse, measured from the point of concentration to a point opposite the watershed area centroid (miles)
- S = mean watershed slope along watercourse (feet/mile)
- m = a calibration constant.

For Southern California watersheds, studies have shown that

$$C_t = 24n^* \tag{5.2}$$

where

n^* = a basin factor (see Fig. 5.4a), and m = 0.38.

The basin factors of Fig. 5.4a were determined by the U.S. Army Corps of Engineers by calibration with major storm hydrographs from several Southern California watersheds. The watershed lag relationship of Eq. 5.1 is illustrated in the plot of Fig. 5.4b.

The second element of the synthetic unit hydrograph method is the dimensionless unit hydrograph, or S-graph. In the dimensionless form, the ordinate is expressed as a percent of the ultimate discharge and the abscissa as a percentage of the lag. Four S-graphs assumed applicable in Southern California watersheds are shown in Figs. 5.5 a,b,c,d. Each of the S-graphs were developed from calibration studies of measured significant rainfall-runoff events. These S-graphs are then assumed transferable for use in watersheds classified as valley, foothill, mountain, or desert. According to the definition, each S-graph reaches 50 percent of ultimate discharge at 100 percent of watershed lag, where the lag is determined using Eq. 5.1.

n*= 0.015

1. Drainage area has fairly uniform, gentle slopes
2. Most watercourses are either improved or along paved streets
3. Groundcover consists of some grasses - large % of area impervious
4. Main watercourse is improved channel or conduit

n*= 0.020

1. Drainage area has some graded and non-uniform, gentle slopes
2. Over half of the area watercourses are improved or paved streets
3. Groundcover consists of equal amount of grasses and impervious area
4. Main watercourse is partly improved channel or conduit and partly greenbelt (see n*= 0.025)

n*= 0.025

1. Drainage area is generally rolling with gentle side slopes
2. Some drainage improvements in the area - streets and canals
3. Groundcover consists mostly of scattered brush and grass and small % impervious
4. Main watercourse is straight channels which are turfed or with stony beds and weeds on earth bank (greenbelt type)

n*= 0.030 (Foothill Area)

1. Drainage area is generally rolling with rounded ridges and moderate side slopes
2. No drainage improvements exist in the area
3. Groundcover includes scattered brush and grasses
4. Watercourses meander in fairly straight, unimproved channels with some boulders and lodged debris

n*= 0.040 (Foothill Area)

1. Drainage area is composed of steep upper canyons with moderate slopes in lower canyons
2. No drainage improvements exist in the area
3. Groundcover is mixed brush and trees with grasses in lower canyons
4. Watercourses have moderate bends and are moderately impeded by boulders and debris with meandering courses

n*= 0.050 (Mountain Areas)

1. Drainage area is quite rugged with sharp ridges and steep canyons
2. No drainage improvements exist in the area
3. Groundcover, excluding small areas of rock outcrops, includes many trees and considerable underbrush
4. Watercourse meander around sharp bends, over large boulders and considerable debris obstruction

n*= 0.200

1. Drainage area has comparatively uniform slopes
2. No drainage improvements exist in the area
3. Groundcover consists of cultivated crops or substantial growths of grass and fairly dense small shrubs, cacti, or similar vegetation
4. Surface characteristics are such that channelization does not occur

Fig. 5.4a. BASIN FACTOR DEFINITION

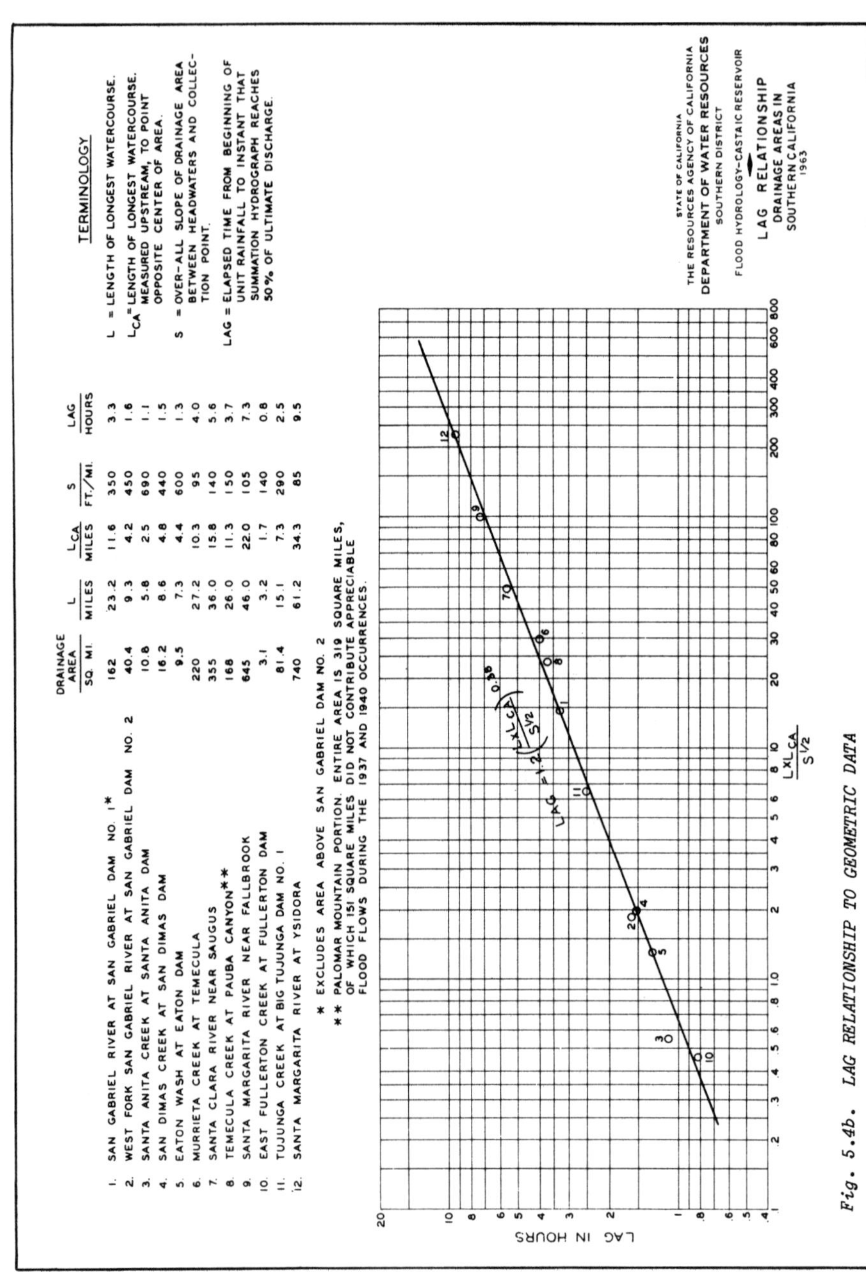

	DRAINAGE AREA SQ. MI.	L MILES	L_{CA} MILES	S FT./MI.	LAG HOURS
1. SAN GABRIEL RIVER AT SAN GABRIEL DAM NO. 1*	162	23.2	11.6	350	3.3
2. WEST FORK SAN GABRIEL RIVER AT SAN GABRIEL DAM NO. 2	40.4	9.3	4.2	450	1.6
3. SANTA ANITA CREEK AT SANTA ANITA DAM	10.8	5.8	2.5	690	1.1
4. SAN DIMAS CREEK AT SAN DIMAS DAM	16.2	8.6	4.8	440	1.5
5. EATON WASH AT EATON DAM	9.5	7.3	4.4	600	1.3
6. MURRIETA CREEK AT TEMECULA	220	27.2	10.3	95	4.0
7. SANTA CLARA RIVER NEAR SAUGUS	355	36.0	15.8	140	5.6
8. TEMECULA CREEK AT PAUBA CANYON**	168	26.0	11.3	150	3.7
9. SANTA MARGARITA RIVER NEAR FALLBROOK	645	46.0	22.0	105	7.3
10. EAST FULLERTON CREEK AT FULLERTON DAM	3.1	3.2	1.7	140	0.8
11. TUJUNGA CREEK AT BIG TUJUNGA DAM NO. 1	81.4	15.1	7.3	290	2.5
12. SANTA MARGARITA RIVER AT YSIDORA	740	61.2	34.3	85	9.5

* EXCLUDES AREA ABOVE SAN GABRIEL DAM NO. 2

** PALOMAR MOUNTAIN PORTION. ENTIRE AREA IS 319 SQUARE MILES, OF WHICH 151 SQUARE MILES DID NOT CONTRIBUTE APPRECIABLE FLOOD FLOWS DURING THE 1937 AND 1940 OCCURRENCES.

TERMINOLOGY

L = LENGTH OF LONGEST WATERCOURSE.

L_{CA} = LENGTH OF LONGEST WATERCOURSE, MEASURED UPSTREAM, TO POINT OPPOSITE CENTER OF AREA.

S = OVER-ALL SLOPE OF DRAINAGE AREA BETWEEN HEADWATERS AND COLLECTION POINT.

LAG = ELAPSED TIME FROM BEGINNING OF UNIT RAINFALL TO INSTANT THAT SUMMATION HYDROGRAPH REACHES 50% OF ULTIMATE DISCHARGE.

Fig. 5.4b. LAG RELATIONSHIP TO GEOMETRIC DATA

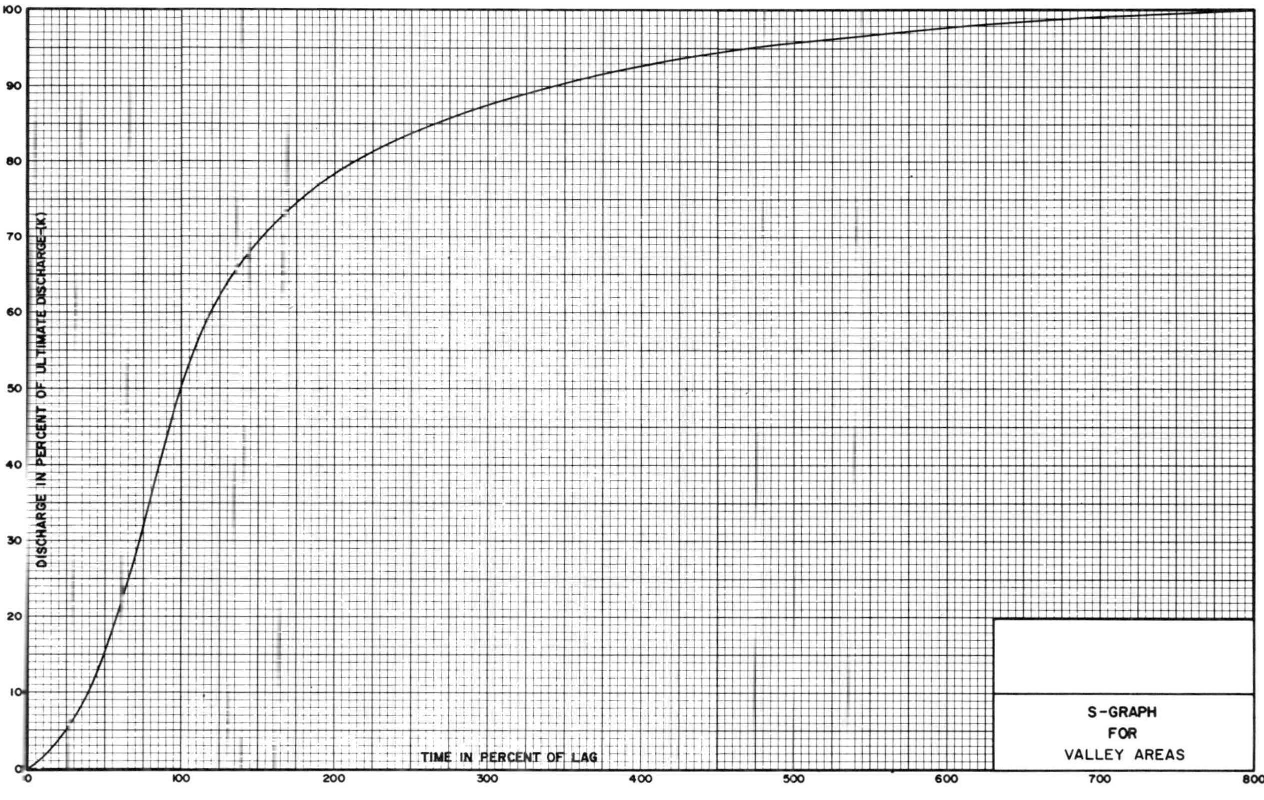

Fig. 5.5a. *S-GRAPH FOR VALLEY AREA*

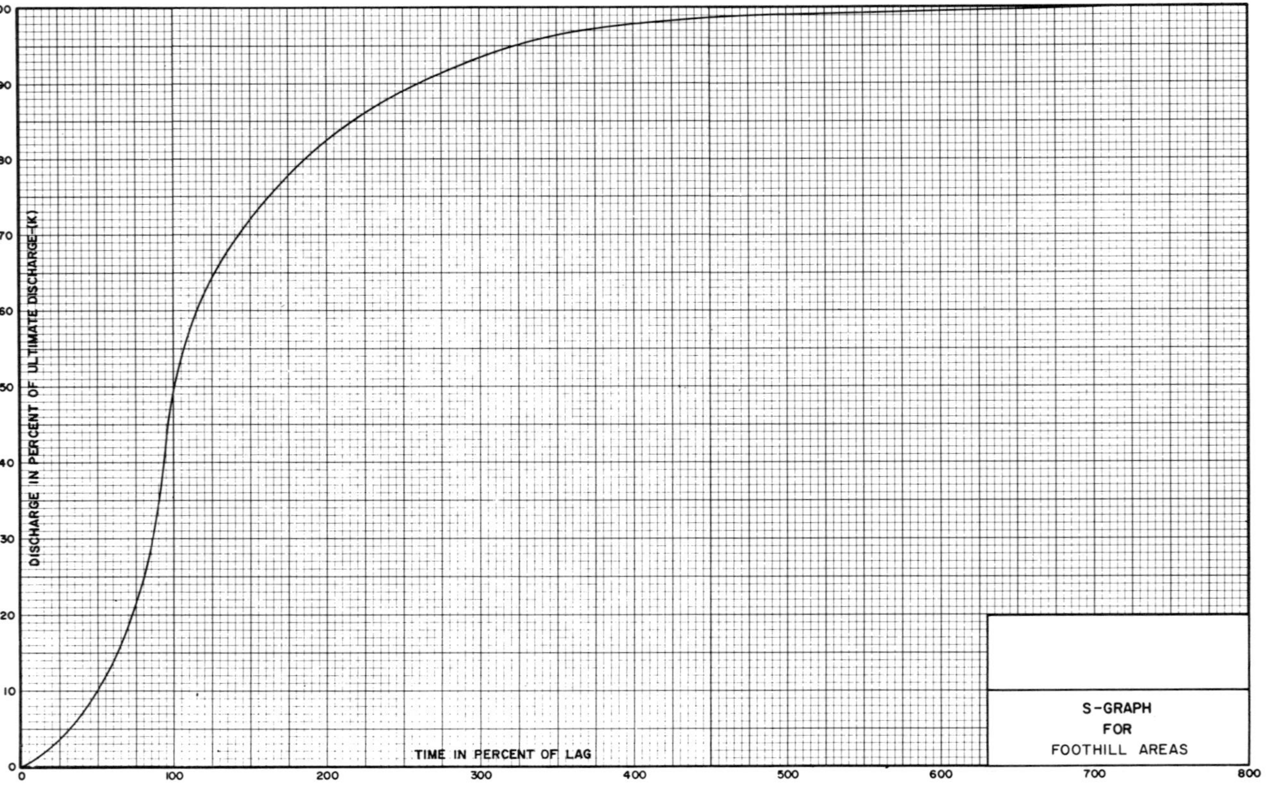

Fig. 5.5b. S-GRAPH FOR FOOTHILL AREA

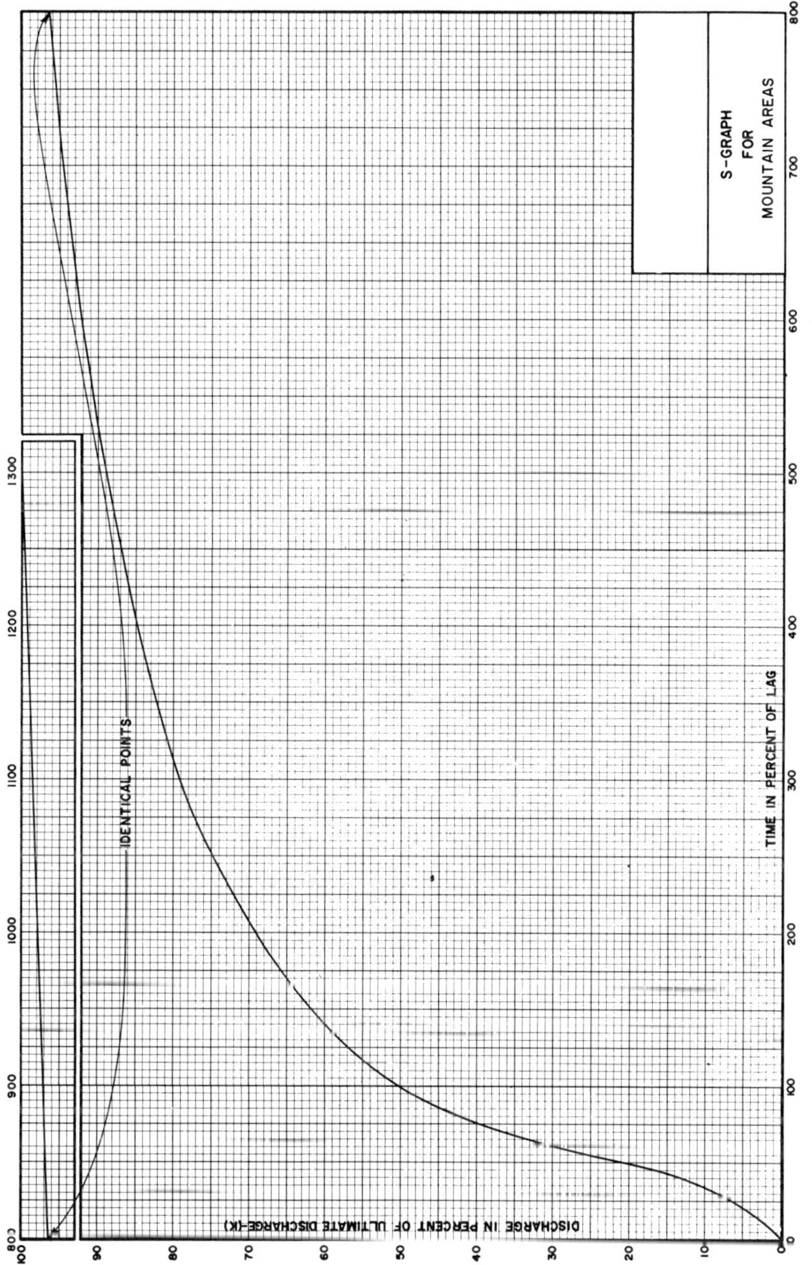

Fig. 5.5c. S-GRAPH FOR MOUNTAIN AREA

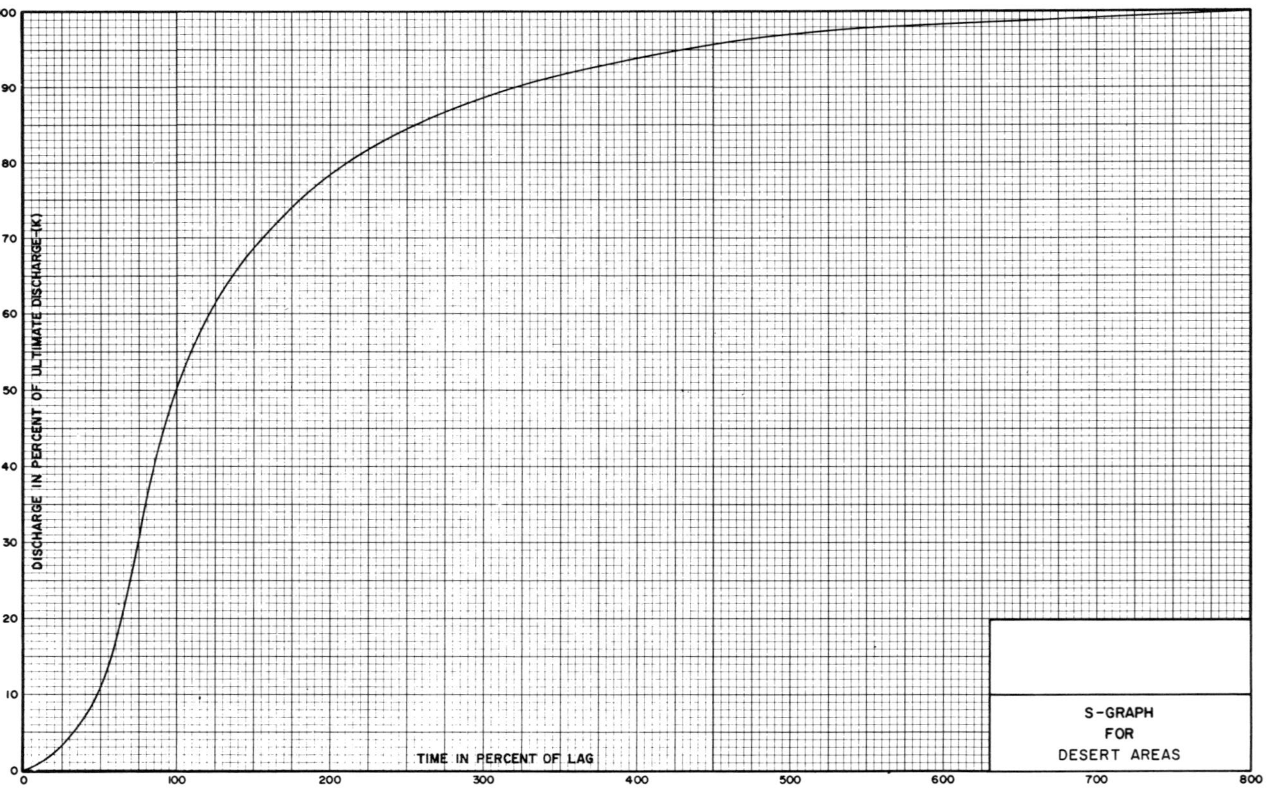

Fig. 5.5d. S-GRAPH FOR DESERT AREA

The final step in developing the synthetic unit hydrograph is to multiply the ordinates of the distribution block graph by the ultimate discharge defined by

$$K = 645A/T \tag{5.3}$$

where

K = the watershed ultimate discharge (cfs)
A = drainage area (square miles)
T = unit time period (hours)

To use the S-graphs, a watershed lag is estimated from Eq. 5.1 and Fig. 5.4. A duration is selected (usually at 15 to 25 percent of the watershed lag time) and amassed unit periods are expressed as accumulated percentages of the watershed lag. These percentages of lag are used for superimposing a block graph on the selected S-graph and the resulting block graph pattern is used in determining the accumulated mean percentage of ultimate discharge for each accumulated unit period. Finally, the incremental mean percentage of ultimate discharge for each unit period is estimated by a series of successive subtractions.

5.7. The SCS Unit Hydrograph

The above process for unit hydrograph analysis has been made a number of times for different data bases. Watershed and regionally averaged synthetic unit hydrographs have been developed for application at ungaged locations. The more data used to calibrate a synthetic unit hydrograph, the greater the accuracy that can be expected when the synthetic unit hydrograph is used at ungaged locations.

The Soil Conservation Service (SCS) calibrated a unit hydrograph using data from a large number of measured storm events. The analysis indicated that the unit hydrograph, which was reported in dimensionless form, requires the drainage area A (square miles), the runoff volume Q (inches), and the time to peak T_p (hours) in order to dimensionalize it. The dimensionless unit hydrograph relates the ratios of q/q_p and t/Tp, where q is the discharge at any time t and q_p is the peak discharge which occurs at time Tp. The peak discharge of the unit hydrograph is computed by the following:

$$q_p = 484\ AQ/Tp \tag{5.4}$$

The time to peak can be estimated from the time of concentration, Tc. SCS uses the following:

$$Tc = 5\ T_p/3 \tag{5.5a}$$

or

$$Tp = 3\ Tc/5 \tag{5.5b}$$

Given the drainage area, runoff volume, and time of concentration, the peak discharge and the time to peak can be computed from Eqs. 5.4 and 5.5b, respectively; these values are then used with the dimensionless unit hydrograph of Table 5.1 and Fig. 5.6 to compute the watershed unit hydrograph. For a unit hydrograph the volume of runoff equals 1 inch, which should be used with Eq. 5.4. A tabulation of the SCS dimensionless unit hydrograph ratios are given in Table 5.1. From the table, it is evident that whatever the condition or classification of the watershed, the ratio of the time to peak to total unit hydrograph duration is a constant. Additionally, it is evident that the volume of runoff under the rising limb of the unit hydrograph is a constant 37.5 percent of the total runoff volume and that the peak rate factor associated to the unit rainfall is a constant value of 484. The variability of the SCS unit hydrograph with respect to other peak rate factors was investigated by McCuen and Bondelid (1983). In their study, methods to estimate a peak rate factor for ungaged watersheds is proposed using a gamma function distribution for the unit hydrograph and the estimation of approximate watershed storage effects in the determination of the volume of runoff defined under the rising limb. As with many other unit hydrograph schemes, the assumption of linearity is once again utilized to develop a runoff hydrograph. Watershed peak rate factors can be estimated by time-area analysis and attempts to duplicate runoff hydrographs.

5.8. Variations on the SCS Unit Hydrograph

The concept of time to peak (T_p) is used and can be shown from Fig. 5.6 to be geometrically related to the watershed lag time. Here lag time follows the definition used in this chapter rather than the SCS definition of lag, which is the time from the center of mass of the effective rainfall to the peak of the runoff hydrograph. It has been shown that the time to peak T_p and watershed lag can be related by the general formula:

$$Tp = C_o\ Lag \tag{5.6a}$$

where C_o is a constant. Studies have shown that C_o varies from about 0.6 to 1.25. For example, the County of San Diego,

California (1975) uses the empirical watershed lag formulas of Eqs. (5.1) and (5.2) plus the relationship:

$$Tp = (0.862)\ Lag \tag{5.6b}$$

TABLE 5.1
RATIOS FOR THE SCS UNIT HYDROGRAPH

Time Ratio	Discharge Ratio	Mass Curve Ratio
0.00	0.000	0.00
0.10	0.030	0.001
0.20	0.100	0.006
0.30	0.190	0.012
0.40	0.310	0.035
0.50	0.470	0.065
0.60	0.660	0.107
0.70	0.820	0.163
0.80	0.930	0.228
0.90	0.990	0.300
1.00	1.000	0.375
1.10	0.990	0.450
1.20	0.930	0.522
1.30	0.860	0.589
1.40	0.780	0.650
1.50	0.680	0.700
1.60	0.560	0.751
1.70	0.460	0.790
1.80	0.390	0.822
2.00	0.280	0.871
2.20	0.207	0.908
2.40	0.147	0.934
2.60	0.107	0.953
2.80	0.077	0.967
3.00	0.055	0.977
3.20	0.040	0.984
3.40	0.029	0.989
3.60	0.021	0.993
3.80	0.015	0.995
4.00	0.011	0.997
4.50	0.005	0.999
5.00	0.000	1.000

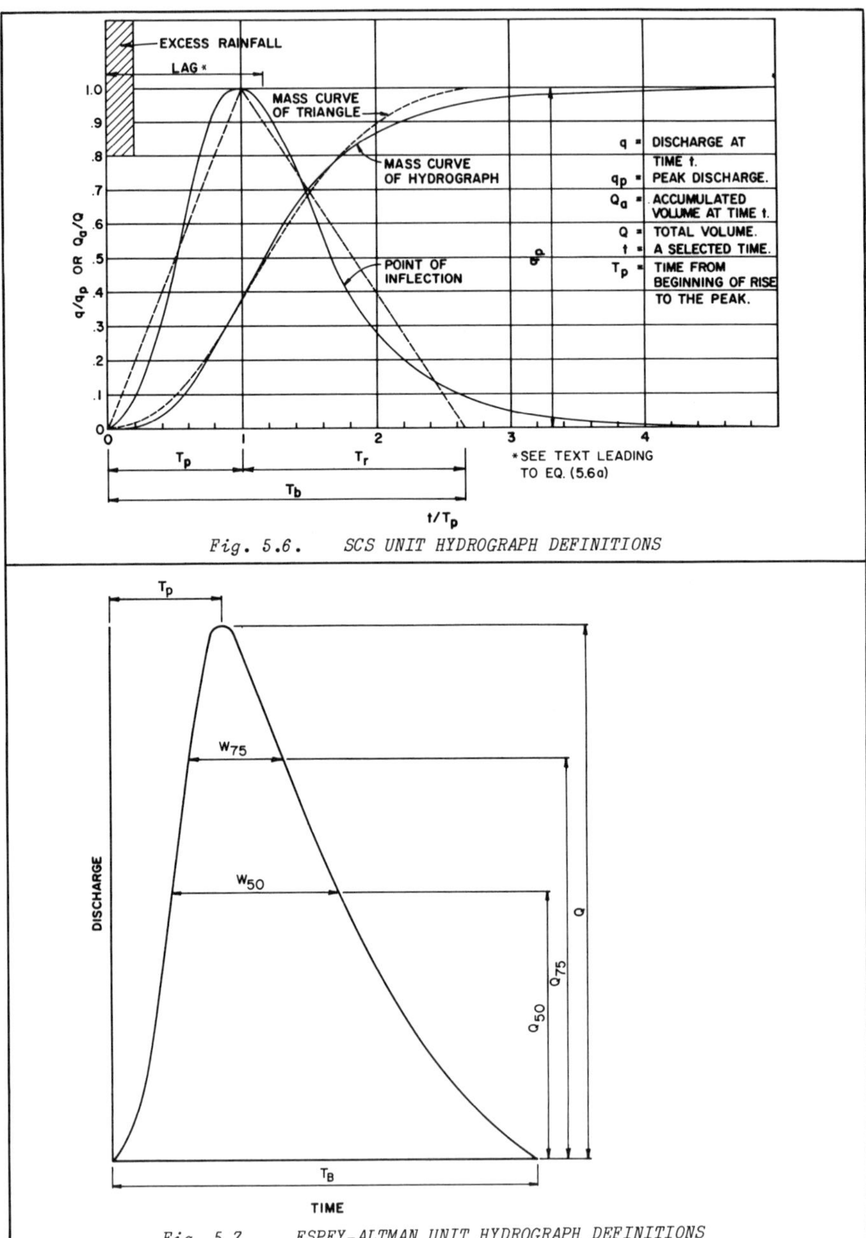

Fig. 5.6. SCS UNIT HYDROGRAPH DEFINITIONS

Fig. 5.7. ESPEY-ALTMAN UNIT HYDROGRAPH DEFINITIONS

Equation 5.6b is based on the geometric relationships for the unit hydrographs shown in Fig. 5.6 which includes the triangular version of the SCS curvilinear unit hydrograph. From Fig. 5.6, it is assumed that:

$$Tr = (1.67)Tp \tag{5.7}$$

$$Tb = (2.67)Tp, \text{ (triangular unit hydrograph)}$$

5.9. The Espey-Altman Unit Hydrograph

Possibly, more insight into the mechanics of the unit hydrograph is gained by working with the Espey and Altman (1978) approach. This version of the unit hydrograph involves a set of five parameters which are defined according to Fig. 5.7. The variable relationships are of the form:

$$T_p = (3.1)L^{0.23}S^{-0.25}I^{-0.18}K^{1.57}$$

$$T_B = (125890)AQ^{-0.95}$$

$$Q = (31620)A^{0.96}T_p^{-1.07}$$

$$W_{50} = (16220)A^{0.93}Q^{-0.92}$$

$$W_{75} = (3240)A^{0.79}Q^{-0.78} \tag{5.8}$$

where

L = main channel length (feet)
H = drop in elevation along channel
S = channel slope
I = impervious area fraction
A = watershed area (square miles)
T_p = time to peak (minutes)
T_B = base time (minutes)
K = a dimensionless channel conveyance factor
W_{50}, W_{75} = width at 50 percent and 75 percent Q

The Espey unit hydrograph is sensitive to the channel conveyance factor (K) in a fashion analogous to the sensitivity of the unit hydrograph lag parameter to the basin n^* factor. Each of the factors used in Eq. (5.8) require calibration in order to properly evaluate their respective correlation to local watershed characteristics. Additionally, the K factor requires calibration to local rainfall-runoff data in order to develop definitions similar to those used for the basin factor of Eqs. (5.1) and (5.2).

5.10. Design Storm Effective Rainfall

The design storm rainfalls (see Chapter 2) should be reduced by a watershed loss model which reflects the effects of the several losses considered important in the resulting runoff hydrograph. Chapter 3 examines several of the most often considered watershed losses and several techniques for computing the probable effects of these losses on the design storm rainfall. The Chapter 5 computer programs can be easily modified to use other watershed loss models by simply substituting the desired computer code into the program. Regardless of the loss model used, the total design storm model needs to be calibrated to local flood frequency curve information in order to ascertain the return frequency of runoff produced by the model for a specified return frequency of design storm. Detailed discussions of this calibration effort are contained in Chapter 7.

5.11. Synthetic Runoff Hydrograph Development

The several modeling elements of the design storm pattern, point rainfall determination, watershed lag estimation, and the unit hydrograph approximation are combined in the convolution of effective rainfall with the unit hydrograph. Figure 5.8 illustrates the convolution procedure for developing a synthetic runoff hydrograph. Each unit interval (or unit period) of rainfall from the design storm pattern is split into the effective unit rainfall and the corresponding watershed unit loss. The resulting effective rainfall storm pattern is then combined with the assumed watershed unit hydrograph by a process called convolution to produce the synthetic runoff hydrograph. Figure 5.8 shows a 12 unit period storm pattern and the corresponding unit period runoff hydrographs. Summing together the 12 unit period runoff hydrographs (i.e., assuming the system to respond linearly to the design storm input) results in the design storm runoff hydrograph. Example problem 5.1 contains a detailed application of this procedure. The convolution process is shown in tabulation form in Fig. 5.30.

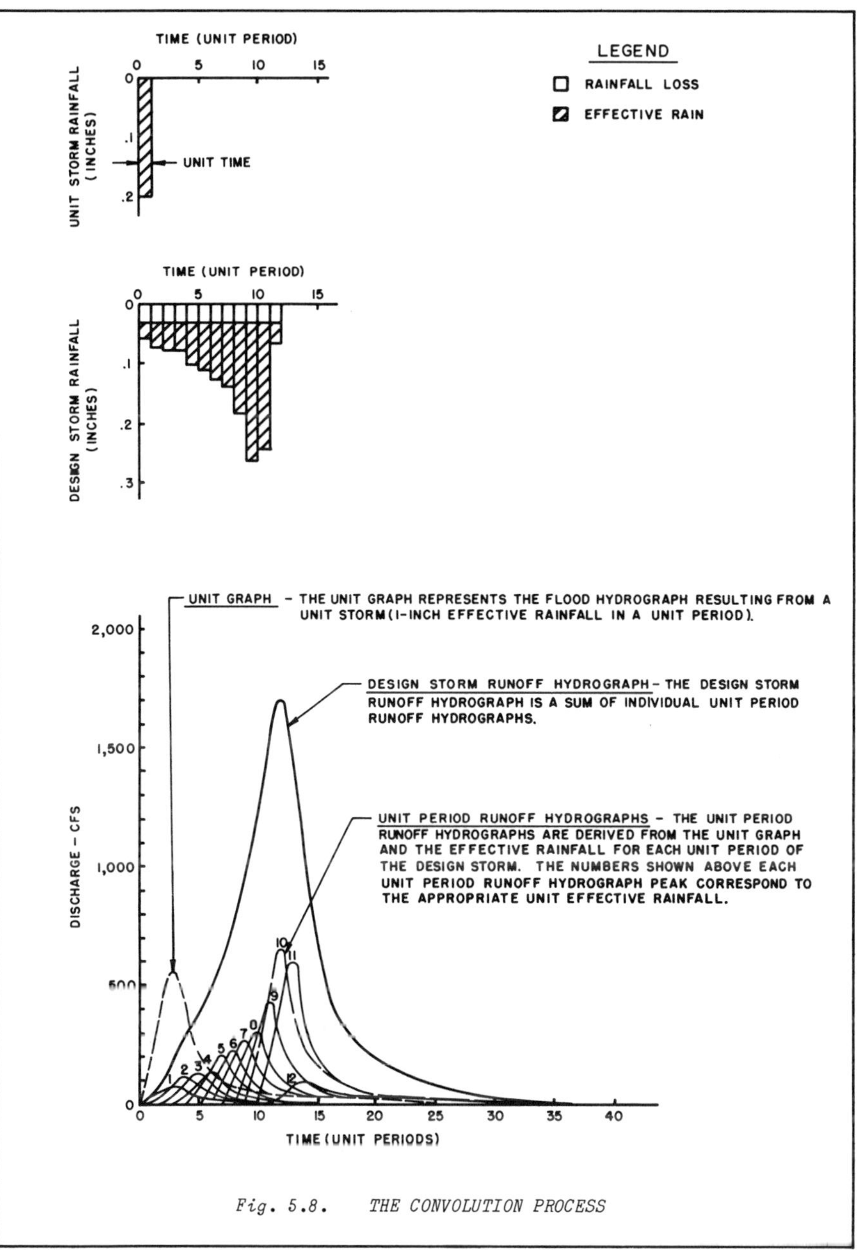

Fig. 5.8. THE CONVOLUTION PROCESS

It should be noted that in the convolution approach, each unit period runoff hydrograph has the same base period regardless of the magnitude of the associated unit effective rainfall. Also, each unit period runoff hydrograph has a constant time-to-peak (Tp). Each of the unit period runoff hydrographs are directly related to the volume of the associated effective rainfalls. Additionally, in the summation process it is assumed that there are negligible restrictions to flow for the summed runoff hydrographs. (These restrictions would occur due to channel capacity problems and watershed detention effects). Each of these concerns should be addressed after the watershed runoff hydrograph is generated in order to ascertain whether the computed results are reasonable for the watershed conditions being studied. In many study situations, a single runoff hydrograph generation would be inappropriate due to additional watershed storage conditions, and a more complex watershed model is required that submodels for simulating channel and basin routing storage effects (see Chapter 6).

Other contributions to the runoff hydrograph such as baseflow and snowmelt can be summed to the synthetic design storm runoff hydrograph. Generally, such considerations are neglected in the study of urban watersheds; however, detailed discussions of ancillary hydrologic processes are contained in standard texts.

5.12. The Rational Method as a Unit Hydrograph Method

The rational method can be interpreted as a design storm unit hydrograph method. The design storm pattern is developed by using a selected return frequency rainfall intensity-duration curve. At a point of concentration with time of concentration, Tc, the rational method design storm pattern is constructed from an intensity duration curve (see Chapter 2) by first determining the total amount of rainfall (i.e., unit rainfalls) that falls in several successive unit periods, each of duration Tc. The next step is to arrange these several unit rainfalls into the rational method design storm pattern of Fig. 5.9 by placing the largest unit rainfall as the first unit, followed by the second largest unit rainfall, and so forth until a sufficiently long design storm pattern is developed (usually about 1-hour in total length, but may be longer depending on the various stream confluence Tc values).

Using the area-averaged loss rate, Fm, the design storm unit effective rainfalls are calculated by subtracting the appropriate proportion of Fm from each unit rainfall. It is noted that the design storm unit rainfalls are given in units of inches of precipitation, whereas Fm is given as a rate (inch/hour).

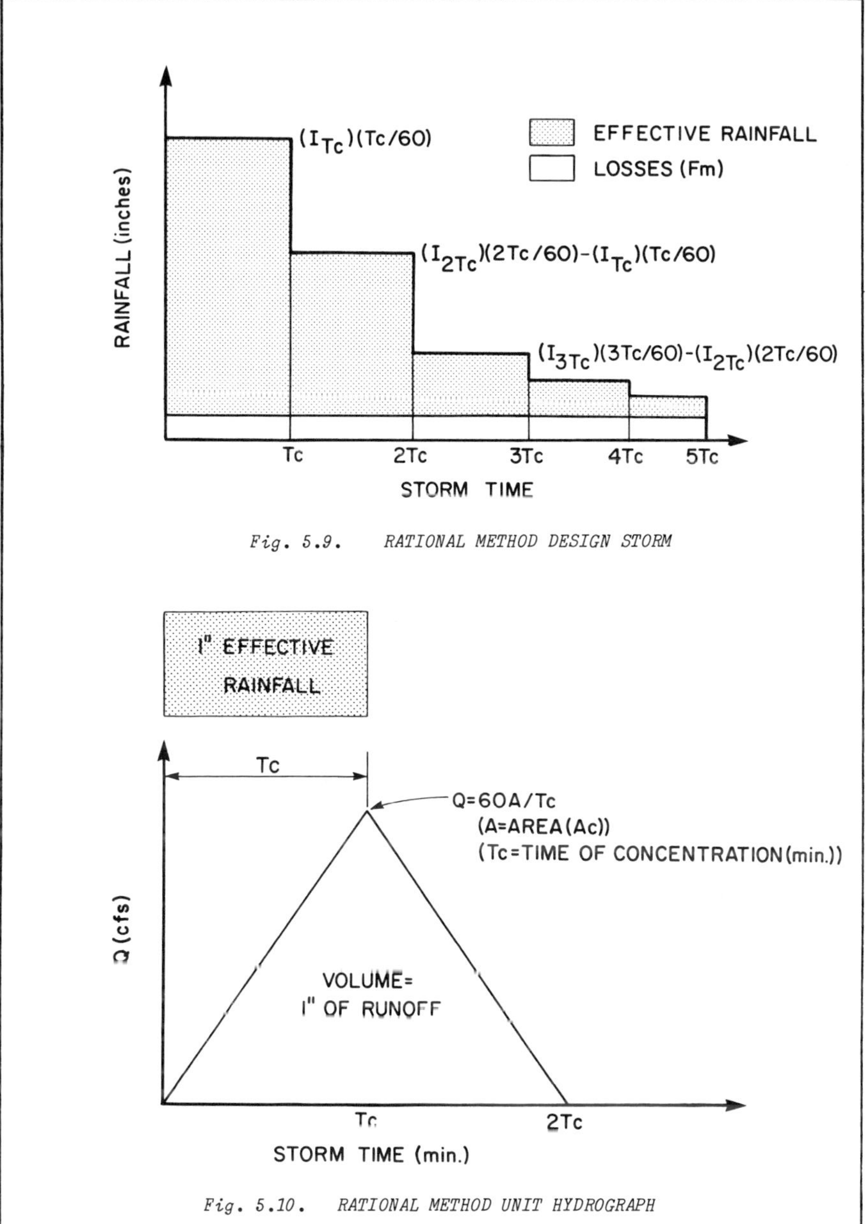

Fig. 5.9. RATIONAL METHOD DESIGN STORM

Fig. 5.10. RATIONAL METHOD UNIT HYDROGRAPH

The unit hydrograph corresponding to the rational method is a triangle with base 2Tc, and a peak occurring at time Tc (see Fig.5.10). For a unit period of duration equal to Tc and a unit effective rainfall of 1 inch, the associated unit period runoff hydrograph must have a peak flow rate of (60/Tc) cfs per acre where Tc is given in minutes. Similarly, a unit period effective rainfall of only 1/2-inch must have an associated unit period runoff hydrograph with a base of 2Tc and a peak flow rate of (1/2)(60/Tc) cfs per acre. The runoff hydrographs associated to each unit effective rainfall are determined similarly, and then arranged as shown in Fig. 5.11 so that the resulting unit period runoff hydrographs correspond in timing to the proper unit period effective rainfalls. The runoff hydrograph is developed by adding the flow contributions from the several unit period runoff hydrographs. Should the rational method design storm pattern be rearranged into another sequence of unit effective rainfalls, the resulting runoff hydrograph would reflect the change in timing (see Fig. 5.12). Such a procedure is useful for small area studies.

5.13. Presentation of Product

An important element in the preparation of hydrology studies is the presentation of the results. Figure 5.13 illustrates a typical format for preparing watershed studies and includes the general organization of the study results, assumed data, and estimation procedures and calculations.

5.14. A Design Storm Runoff Hydrograph Computer Program

In order to describe the salient features of the supplied FORTRAN computer program, a process flow chart (Table 5.2) is provided that describes each operation performed by the program. The computer program uses watershed S-graphs (defined by the user) to develop unit hydrographs based on the previously discussed lag factor definition. Additionally, the program uses the design storm development program of Chapter 2 and the effective rainfall development program provided in Chapter 3. Local watershed S-graph information, design storm depth-area relationships, and local loss rate data should be used in order to develop a runoff hydrograph generator program that represents local hydrologic characteristics. The basic elements of the program are as follows:

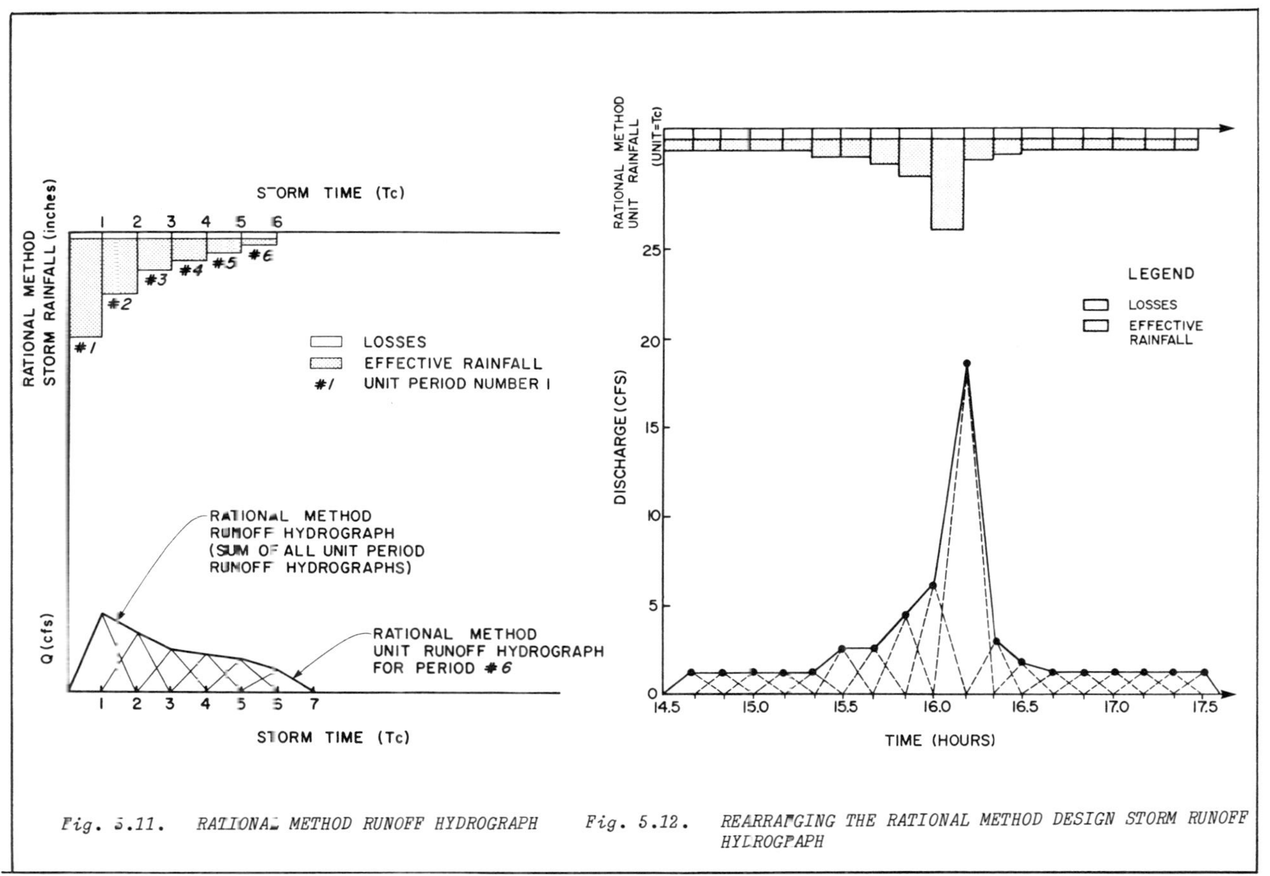

Fig. 5.11. RATIONAL METHOD RUNOFF HYDROGRAPH

Fig. 5.12. REARRANGING THE RATIONAL METHOD DESIGN STORM RUNOFF HYLROGPAPH

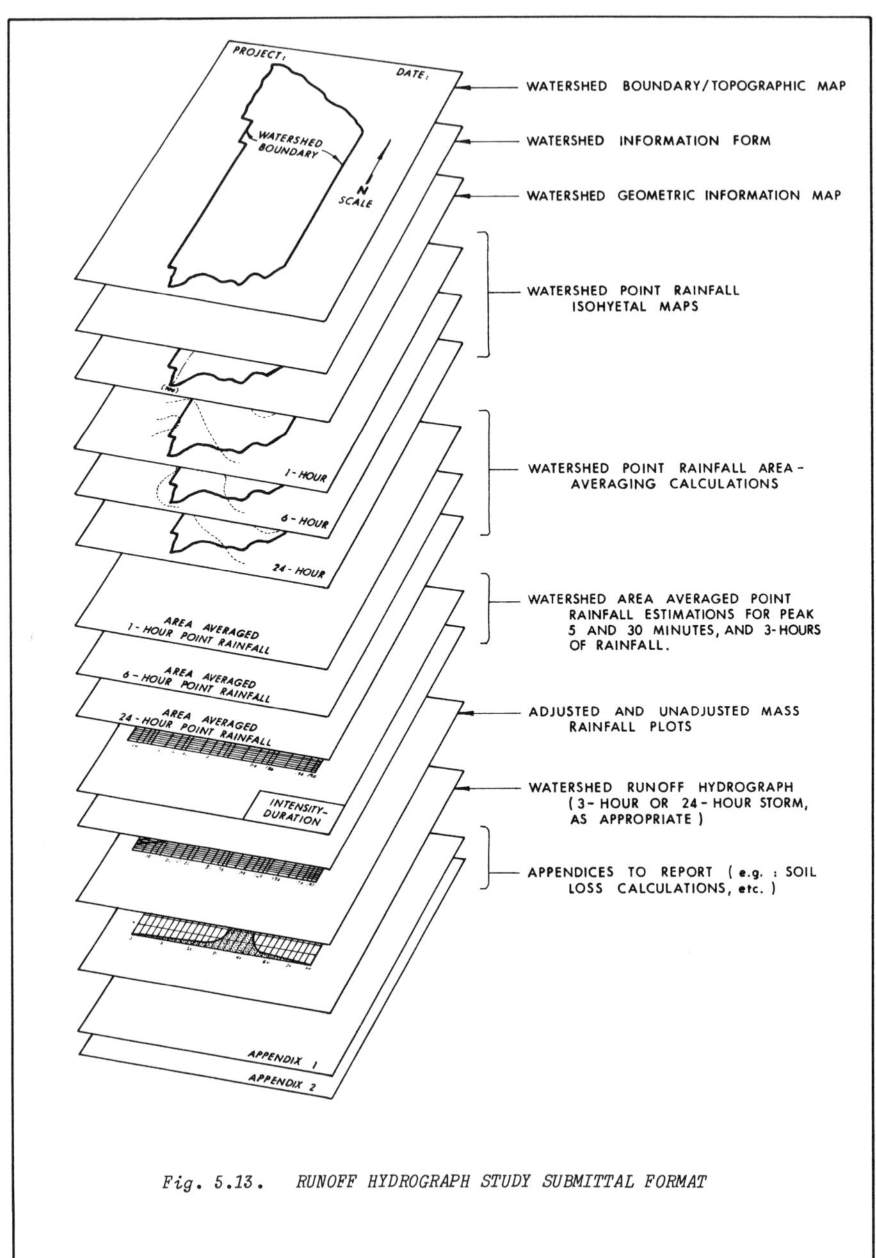

Fig. 5.13. RUNOFF HYDROGRAPH STUDY SUBMITTAL FORMAT

(1). A critical storm pattern is selected from the several options contained in the development of Chapter 2. The design storm is developed by the program using precipitation data and depth-area relationships supplied by the user.

(2). Unit hydrographs are developed from the user-defined S-graph information. The time base of the unit hydrograph is determined by the watershed lag relationship of Eqs. (5.1) and (5.2). The user may elect to define the watershed lag directly.

(3). Watershed losses are modeled by use of a selected loss rate model from the Chapter 3 computer program.

(4). The effective rainfall pattern is constructed using a unit period of 3 minutes. The unit hydrograph developed in step 2 is convoluted with the effective rainfall estimates to produce a runoff hydrograph.

TABLE 5.2
RUNOFF HYDROGRAPH PROGRAM SUBROUTINES

Program Number	Program Description
2.1, 3.1	Main program for developing the effective rainfall storm pattern distribution.
5.1	Synthetic unit hydrograph generator which develops a unit hydrograph from the user-defined S-curves.
5.2	Output plotting program which plots the runoff hydrograph flowrate and accumulated runoff volume as a function of storm time

Example 5.1. Design Storm Runoff Hydrograph Determination
The following example problem illustrates in detail the synthetic unit hydrograph procedure. The example problem presentation includes the following information:

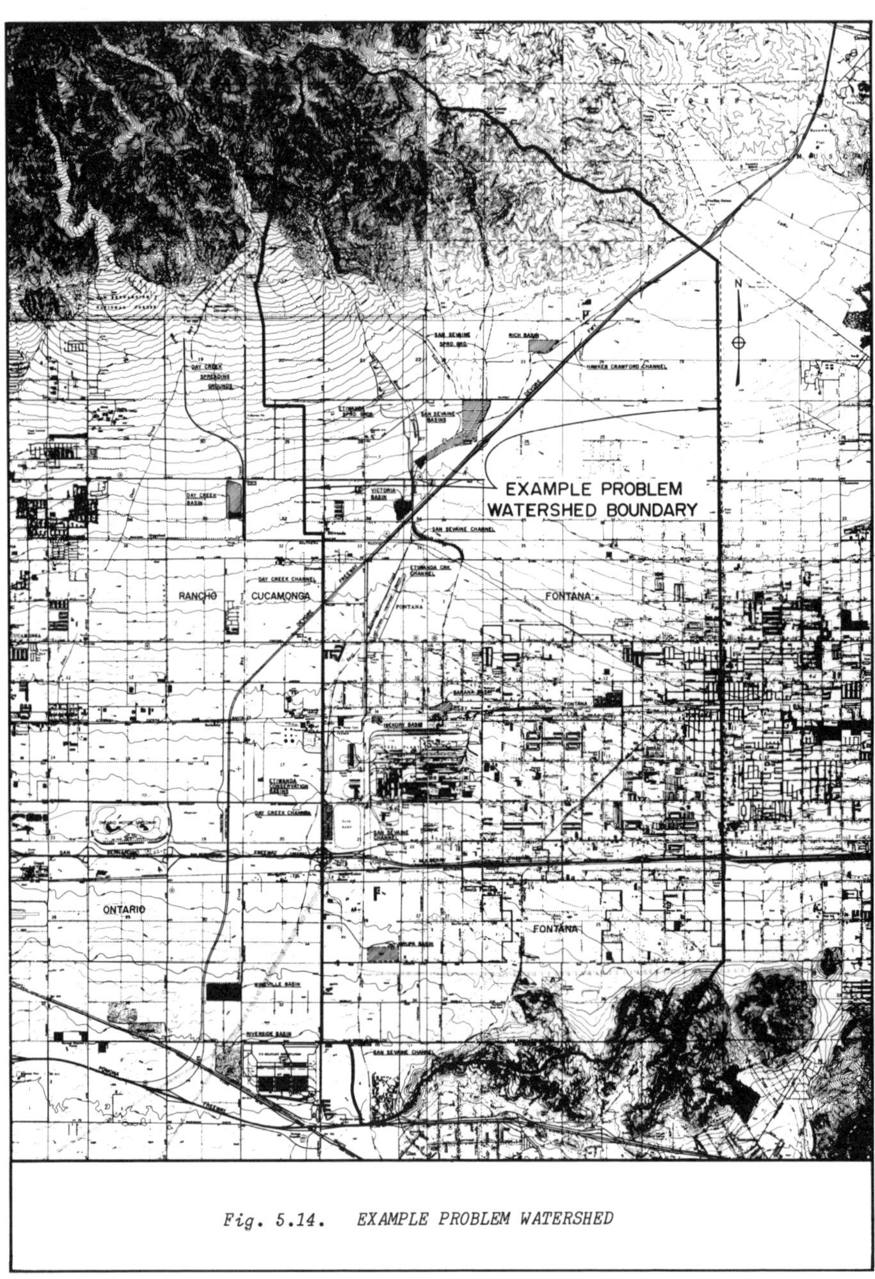

Fig. 5.14. EXAMPLE PROBLEM WATERSHED

PROJECT: EXAMPLE PROBLEM DATE: ______________

ENGINEER: ______________

1.	Enter the length of the longest watercourse (feet)	75,300
2.	Enter the length along the longest watershed watercourse measured from the point of concentration upstream to a point opposite the centroid of the watershed area (feet)	38,800
3.	Enter the difference in elevation between the most remote point in the watershed and the point of concentration (feet)	5,541
4.	Enter the basin factor (see Eq. (5.1), and Fig. 5.4a)	0.030
5.	Enter the watershed area (acres)	38,590
6.	Enter baseflow (cfs/square mile)	0.0
7.	Enter S-Graph proportions (decimal)	
	SCS[1]	0.0
	Valley	1.0
	Foothill	0.0
	Mountain	0.0
	Desert	0.0
8.	Enter adjusted loss rate (inch/hour)	0.30
9.	Enter low loss rate percentage (decimal)	0.90
10.	Enter watershed area-averaged 5-minute point rainfall (inches)*	0.55
	Enter watershed area-averaged 30-minute point rainfall (inches)*	1.13
	Enter watershed area-averaged 1-hour point rainfall (inches)*.	1.52
	Enter watershed area-averaged 3-hour point rainfall (inches)*	2.90
	Enter watershed area-averaged 6-hour point rainfall (inches)*	4.29
	Enter watershed area-averaged 24-hour point rainfall (inches)*	9.83
11.	Enter 24-hour storm interval (minutes)	30.0

[1]Note: use either SCS or Valley, Foothill, Mountain, Desert
*Note: enter values unadjusted by depth-area factors

Fig. 5.15. EXAMPLE PROBLEM PARAMETER DATA

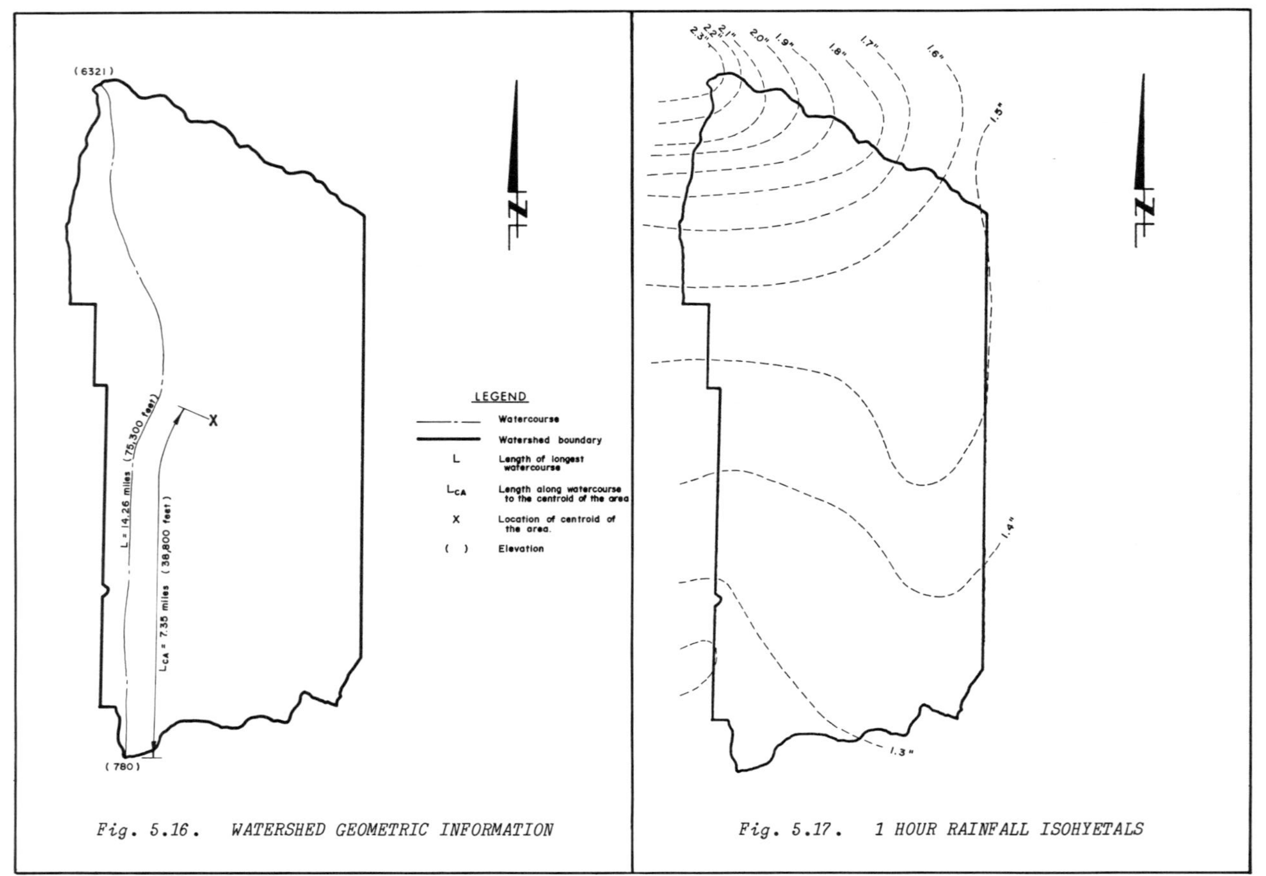

Fig. 5.16. WATERSHED GEOMETRIC INFORMATION

Fig. 5.17. 1 HOUR RAINFALL ISOHYETALS

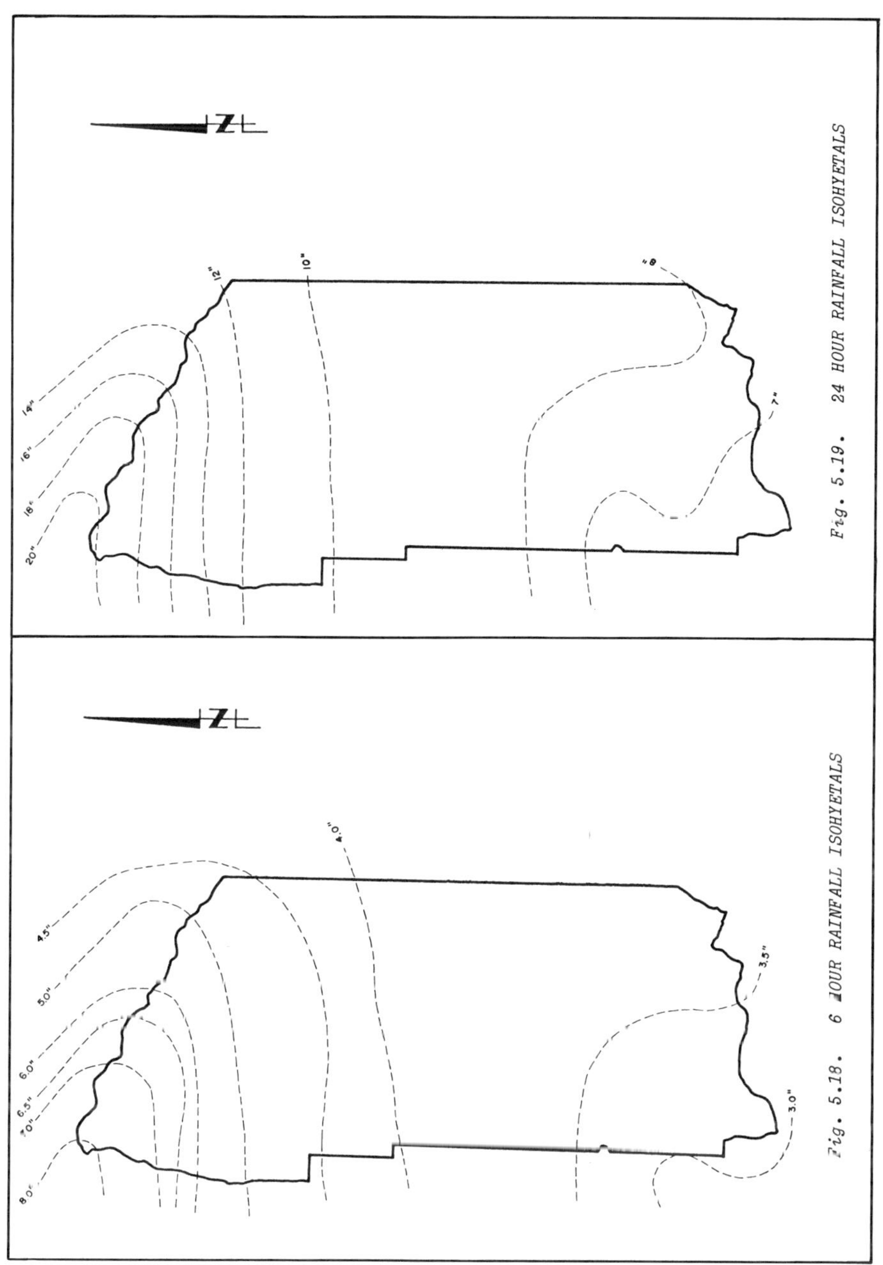

Fig. 5.18. 6 HOUR RAINFALL ISOHYETALS

Fig. 5.19. 24 HOUR RAINFALL ISOHYETALS

UNIT HYDROGRAPH STUDY:
EXAMPLE PROBLEM AREA-AVERAGED POINT PRECIPITATION DETERMINATION

1-Hour Isohyetal Area-Averaging

(1) Isohyetal (inches)	(2) Area (acres)	Weighting (1) · (2)
2.3	97	223.1
2.2	317	697.4
2.1	585	1,228.5
2.0	721	1,442.0
1.9	1,114	2,116.6
1.8	1,519	2,734.2
1.7	2,602	4,423.4
1.6	7,380	11,808.0
1.5	8,697	13,045.5
1.4	8,602	12,042.8
1.3	6,337	8,238.1
1.2	621	745.2
	Total Map Area = 38,592	Total Weighting = 58,744.8

Area-Averaged Value = 58,744.8 / 38,592 = 1.52

6-Hour Isohyetal Area-Averaging

(1) Isohyetal (inches)	(2) Area (acres)	Weighting (1) · (2)
8.0	441	3,528.0
7.0	1,384	9,688.0
6.5	1,092	7,098.0
6.0	1,727	10,362.0
5.0	4,593	22,965.0
4.5	5,302	23,859.0
4.0	9,780	39,120.0
3.5	12,399	43,396.5
3.0	1,872	5,616.0
	Total Map Area = 38,590	Total Weighting = 165,632.5

Area-Averaged Value = 165,632.5 / 38,590 = 4.29

24-Hour Isohyetal Area-Averaging

(1) Isohyetal (inches)	(2) Area (acres)	Weighting (1) · (2)
20.0	314	6,280.0
18.0	1,123	20,214.0
16.0	1,421	22,736.0
14.0	2,049	28,686.0
12.0	4,298	51,576.0
10.0	10,577	105,770.0
8.0	12,914	103,312.0
7.0	5,520	38,640.0
6.0	374	2,244.0
	Total Map Area = 38,590	Total Weighting = 379,458.0

Area-Averaged Value = 379,458.0 / 38,590 = 9.83

Fig. 5.20. AREA AVERAGING OF POINT RAINFALL DATA

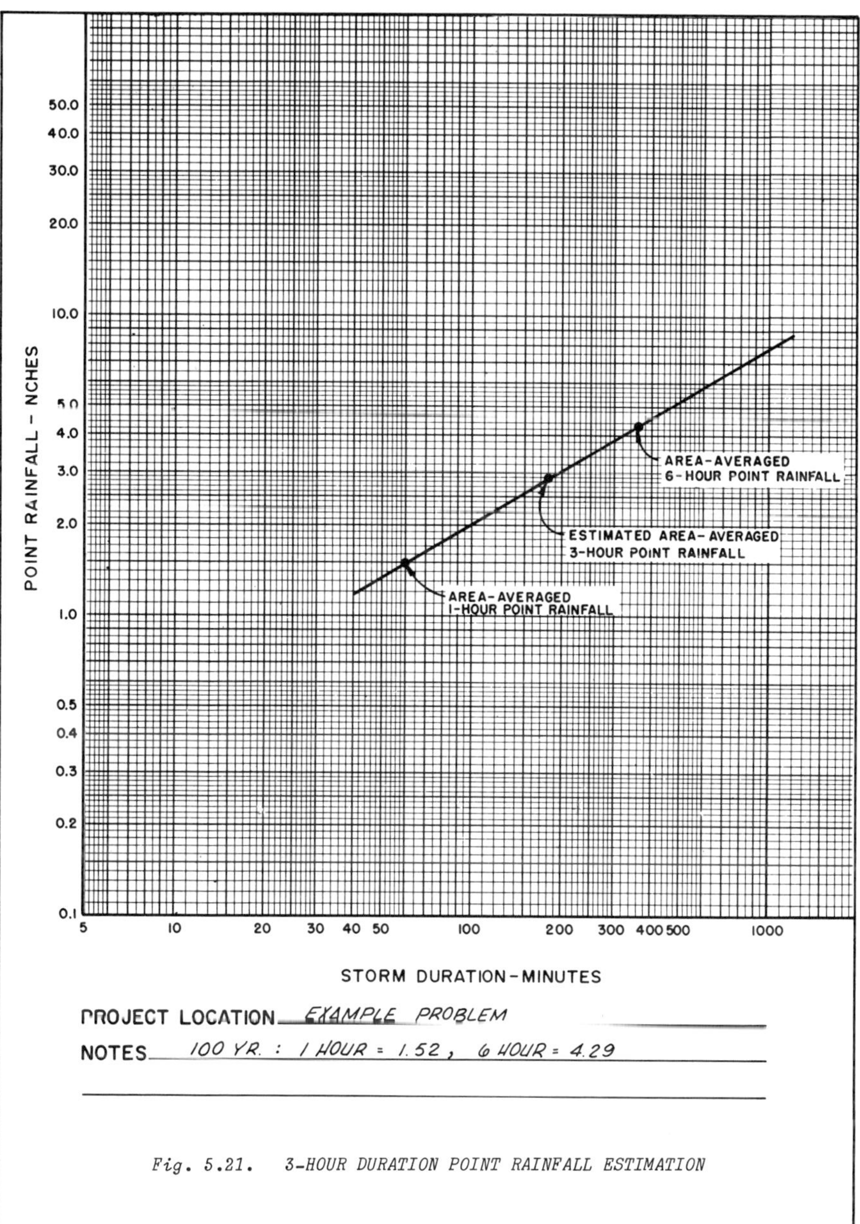

Fig. 5.21. 3-HOUR DURATION POINT RAINFALL ESTIMATION

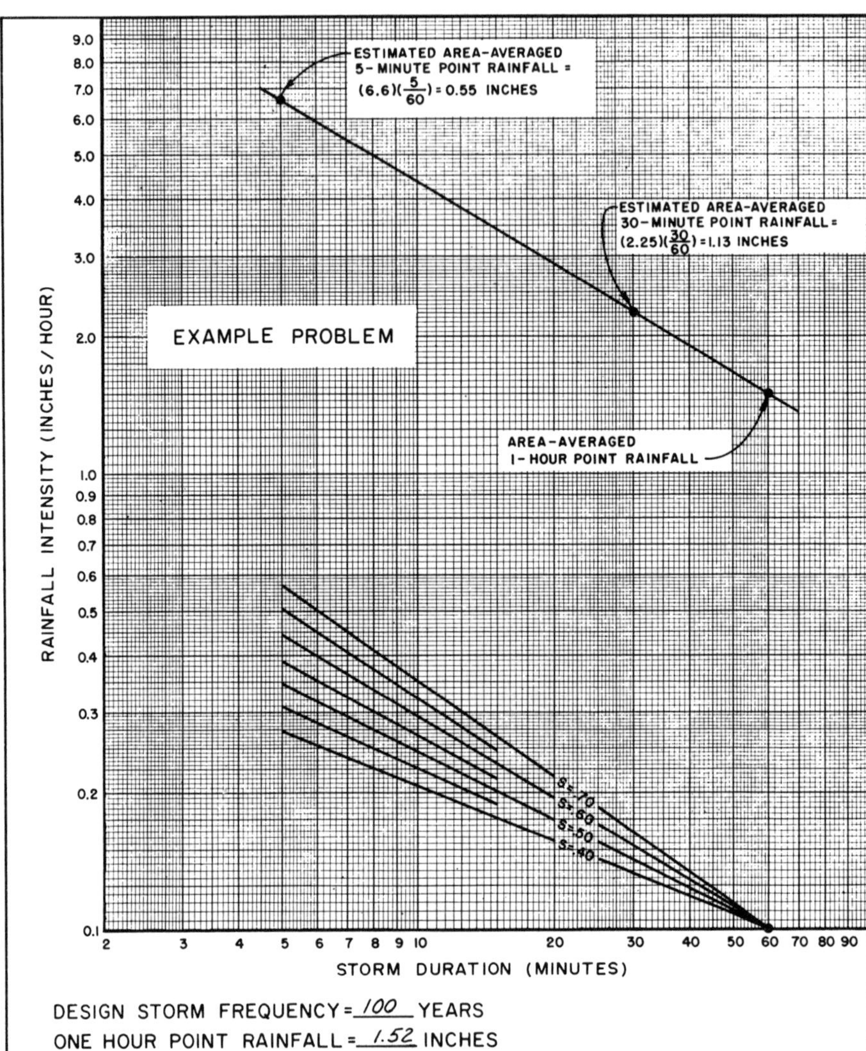

DESIGN STORM FREQUENCY = 100 YEARS

ONE HOUR POINT RAINFALL = 1.52 INCHES

LOG-LOG SLOPE = 0.60

PROJECT LOCATION = EXAMPLE PROBLEM

Fig. 5.22. SHORT DURATION POINT RAINFALL ESTIMATION

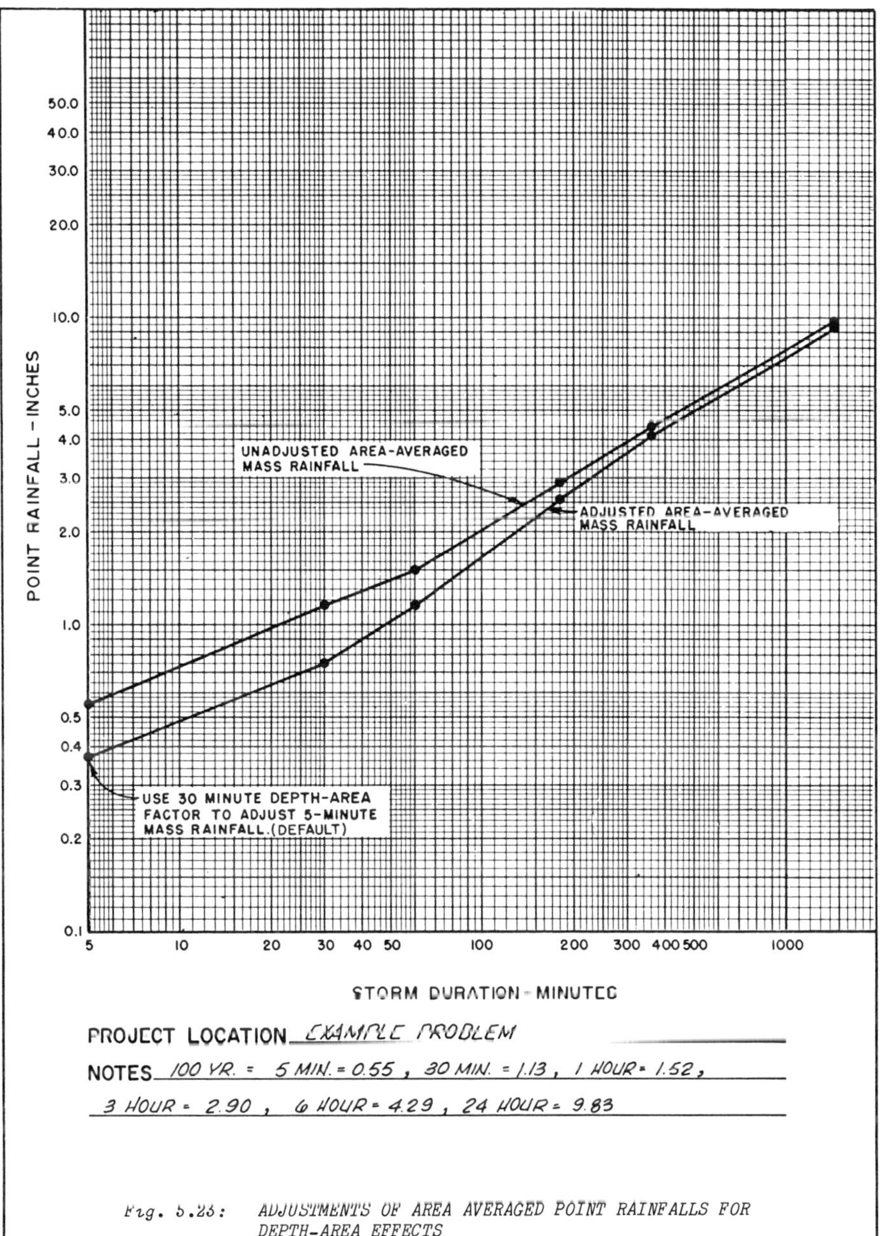

Fig. 5.23: ADJUSTMENTS OF AREA AVERAGED POINT RAINFALLS FOR DEPTH-AREA EFFECTS

UNIT HYDROGRAPH STUDY:
EXAMPLE PROBLEM UNIT RAINFALL DETERMINATION

(Example Unit Period = 30 minutes)

Peak Rainfall Unit Number	Adjusted Mass Rainfall (inches)	Unit Rainfall (inches)
1	.75	.75
2	1.19	.44
3	1.56	.36
4	1.92	.37
5	2.24	.32
6	2.56	.32
7	2.81	.25
8	3.06	.25
9	3.29	.23
10	3.52	.23
11	3.73	.21
12	3.94	.21
13	4.14	.20
14	4.34	.20
15	4.53	.19
16	4.72	.19
17	4.90	.18
18	5.08	.18
19	5.25	.17
20	5.42	.17
21	5.59	.17
22	5.76	.17
23	5.92	.16
24	6.08	.16
25	6.23	.15
26	6.38	.15
27	6.53	.15
28	6.68	.15
29	6.82	.14
30	6.96	.14
31	7.10	.14
32	7.24	.14
33	7.38	.14
34	7.52	.14
35	7.66	.14
36	7.80	.14
37	7.93	.13
38	8.06	.13
39	8.19	.13
40	8.32	.13
41	8.45	.13
42	8.58	.13
43	8.71	.13
44	8.84	.13
45	8.96	.12
46	9.08	.12
47	9.20	.12
48	9.32	.12
Time = 24 hours		Total = 9.32 inches

Fig. 5.24. DESIGN STORM UNIT RAINFALL ESTIMATION (30-MINUTE UNIT INTERVALS)

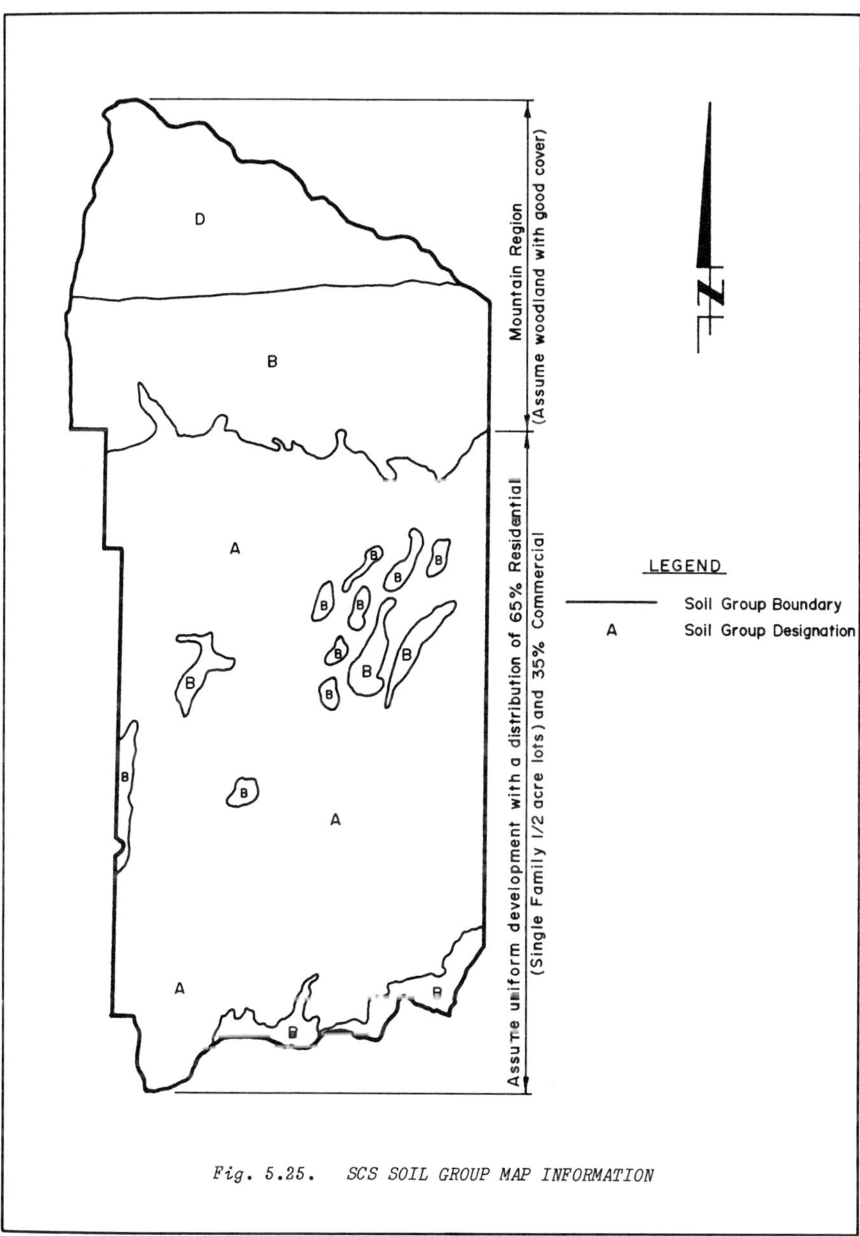

Fig. 5.25. SCS SOIL GROUP MAP INFORMATION

UNIT HYDROGRAPH STUDY:
EXAMPLE PROBLEM 24-HOUR STORM
EFFECTIVE RAINFALL DETERMINATION

Unit Period Number	Unit Rainfall (inches)	Unit Loss (inches)	Effective Rainfall (inches)
1	.12	.036	.084
2	.12	.036	.084
3	.12	.036	.084
4	.12	.036	.084
5	.13	.039	.091
6	.13	.039	.091
7	.13	.039	.091
8	.13	.039	.091
9	.13	.039	.091
10	.13	.039	.091
11	.13	.039	.091
12	.13	.039	.091
13	.14	.042	.098
14	.14	.042	.098
15	.14	.042	.098
16	.14	.042	.098
17	.14	.042	.098
18	.14	.042	.098
19	.15	.045	.105
20	.15	.045	.105
21	.16	.048	.112
22	.17	.051	.119
23	.17	.051	.119
24	.18	.054	.126
25	.19	.057	.133
26	.20	.060	.140
27	.21	.063	.147
28	.23	.069	.161
29	.25	.075	.175
30	.32	.096	.224
31	.36	.108	.252
32	.44	.132	.308
33	.75	.180 *	.570
34	.37	.111	.259
35	.32	.096	.224
36	.25	.075	.175
37	.23	.069	.161
38	.21	.063	.147
39	.20	.060	.140
40	.19	.057	.133
41	.18	.054	.126
42	.17	.054	.126
43	.17	.051	.119
44	.16	.048	.112
45	.15	.045	.105
46	.15	.045	.105
47	.14	.045	.105
48	.14	.042	.098
	9.32	2.75	6.57

*unit low loss exceeds unit adjusted loss

Fig. 5.26. EFFECTIVE RAINFALL DETERMINATION (UNIT LOSS VALUES BASED ON PHI-INDEX AND CONSTANT FRACTION LOSS FUNCTION)

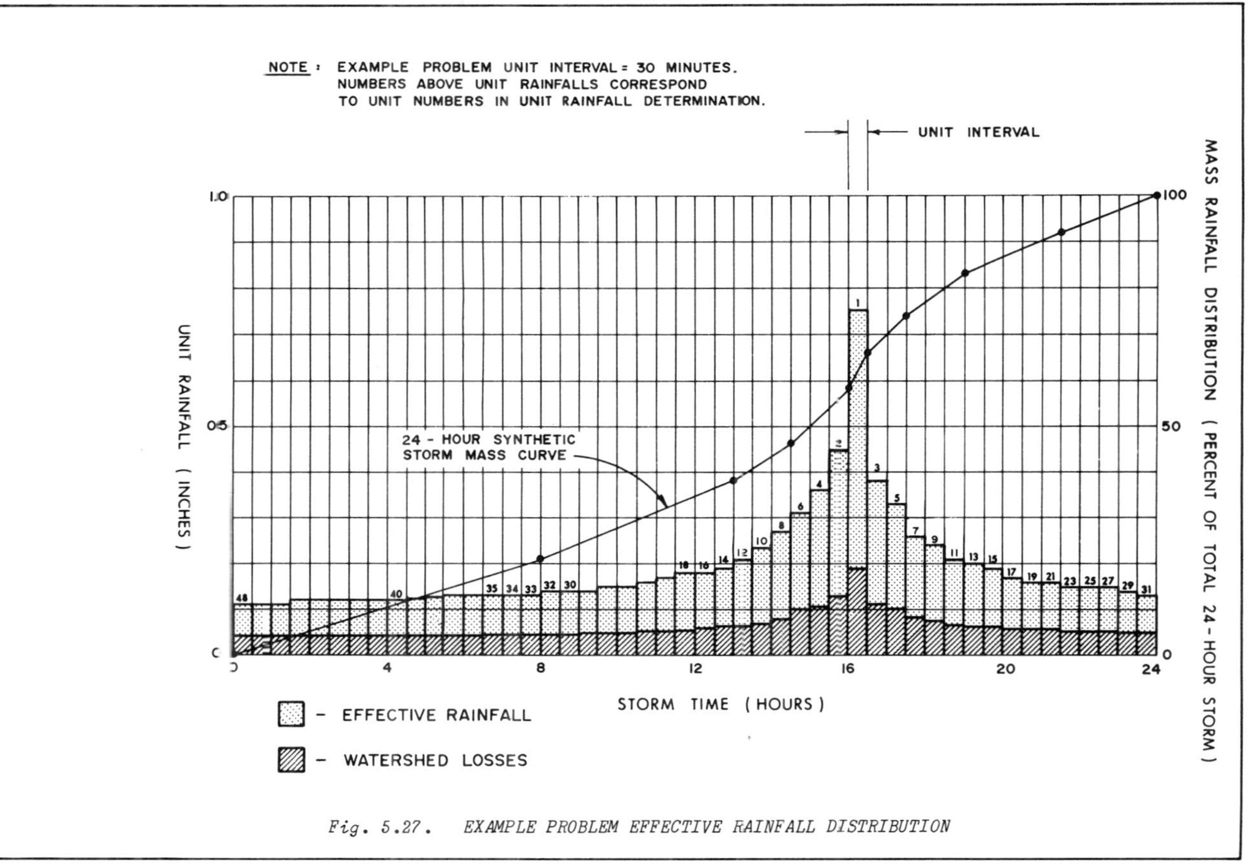

Fig. 5.27. EXAMPLE PROBLEM EFFECTIVE RAINFALL DISTRIBUTION

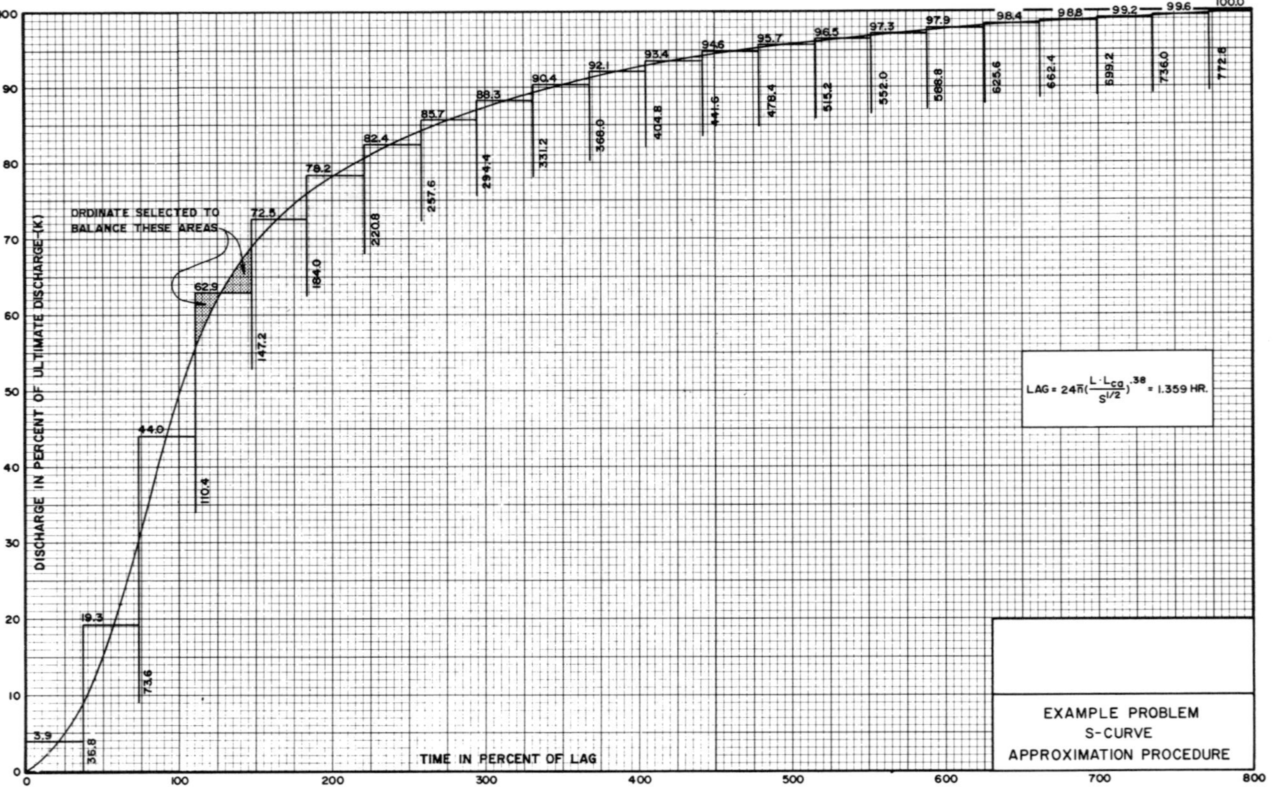

Fig. 5.28. BLOCK GRAPH REPRESENTATION OF S-GRAPH

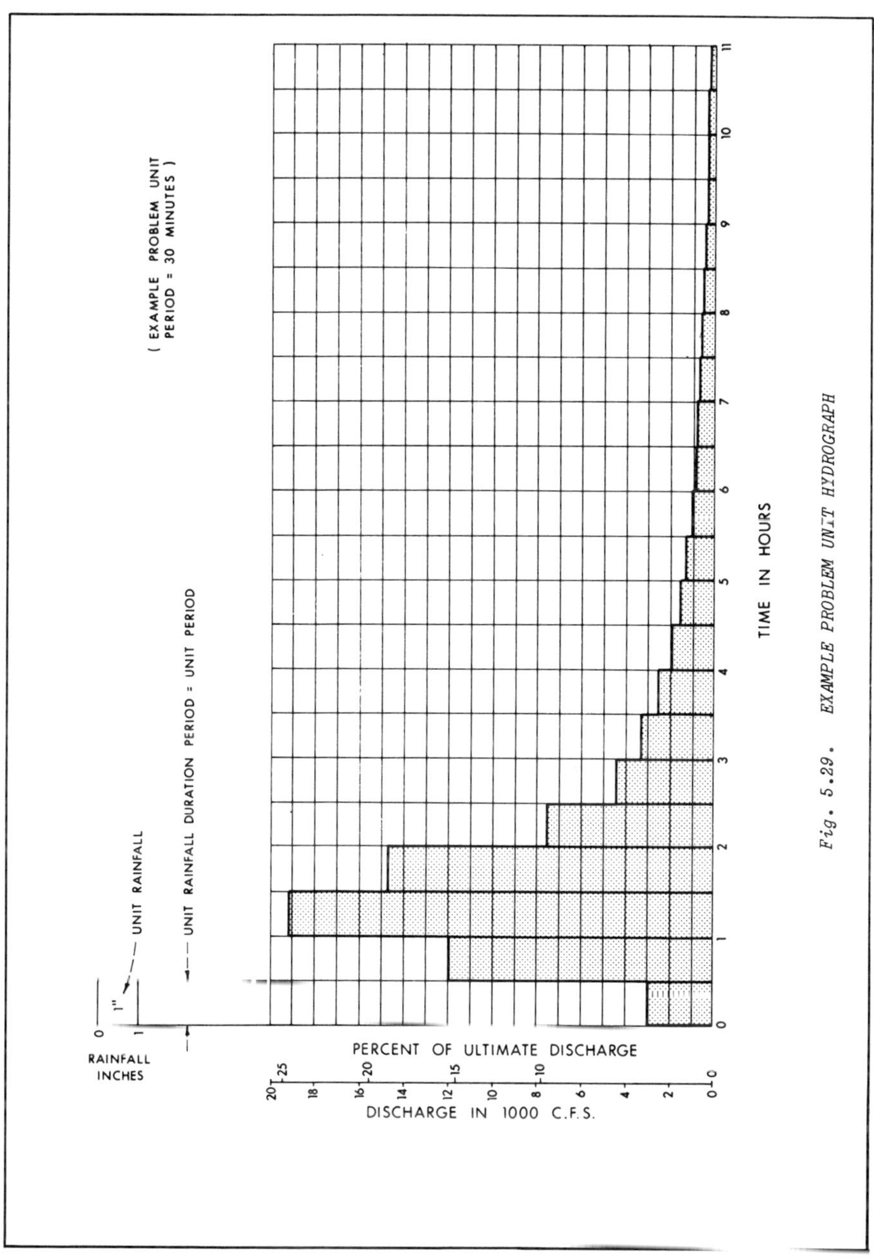

Fig. 5.29. EXAMPLE PROBLEM UNIT HYDROGRAPH

HYDROLOGY MANUAL		SYNTHETIC UNIT HYDROGRAPH METHOD — Flood Hydrograph Calculation Form														Project EXAMPLE PROBLEM — By A.E.S. Date — Checked Date										Sheet 1 / 2
	EFFEC RAIN	.084	.084	.084	.084	.091	.091	.091	.091	.091	.091	.091	.091	.098	.098	.098	.098	.098	.098	.105	.105	.112	.119	.119	.126	FLOOD HYDRO-GRAPH
UNIT GRAPH	UNIT PER.	1	2	3	4	5	6	7	8	9	10	11	12	13	14	15	16	17	18	19	20	21	22	23	24	
2996	1	252																								252
12,000	2	1008	252																							1260
19,200	3	1613	1008	252																						2873
14,720	4	1236	1613	1008	252																					4109
7,486	5	629	1236	1613	1008	273																				4759
4416	6	371	629	1236	1613	1092	273																			5214
3278	7	275	371	629	1236	1747	1092	273																		5623
2,567	8	216	275	371	629	1340	1747	1092	273																	5943
1,990	9	167	216	275	371	681	1340	1747	1092	273																6162
1,667	10	140	167	216	275	402	681	1340	1747	1092	273															6333
1,316	11	111	140	167	216	298	402	681	1340	1747	1092	273														6467
1,046	12	88	111	140	167	234	298	402	681	1340	1747	1092	273													6573
899	13	76	88	111	140	181	234	298	402	681	1340	1747	1092	294												6684
841	14	71	76	88	111	152	181	234	298	402	681	1340	1747	1176	294											6851
669	15	56	71	76	88	120	152	181	234	298	402	681	1340	1882	1176	294										7051
566	16	48	56	71	76	95	120	152	181	234	298	402	681	1443	1882	1176	294									7209
487	17	41	48	56	71	82	95	120	152	181	234	298	402	734	1443	1882	1176	294								7309
400	18	34	41	48	56	77	82	95	120	152	181	234	298	433	734	1443	1882	1176	294							7380
311	19	26	34	41	48	61	77	82	95	120	152	181	234	321	433	734	1443	1882	1176	315						7455
311	20	26	26	34	41	52	61	77	82	95	120	152	181	252	321	433	734	1443	1882	1260	315					7587
311	21	26	26	26	34	44	52	61	77	82	95	120	152	195	252	321	433	734	1443	2016	1260	336				7785
304	22	26	26	26	26	36	44	52	61	77	82	95	120	163	195	252	321	433	734	1546	2016	1344	357			8032
			26	26	26	28	36	44	52	61	77	82	95	129	163	195	252	321	433	786	1546	2150	1428	357		8313
				26	26	28	28	36	44	52	61	77	82	103	129	163	195	252	321	464	786	1649	2285	1428	377	8612
					26	28	28	28	36	44	52	61	77	88	103	129	163	195	252	344	464	838	1752	2285	1512	
						28	28	28	28	36	44	52	61	82	88	103	129	163	195	270	344	495	891	1752	2419	
							28	28	28	28	36	44	52	66	82	88	103	129	163	209	270	367	526	891	1855	
								28	28	28	28	36	44	55	66	82	88	103	129	175	209	288	390	526	943	
									28	28	28	28	36	48	55	66	82	88	103	138	175	223	305	390	556	
										28	28	28	28	39	48	55	66	82	88	110	138	187	237	305	413	
											28	28	28	30	39	48	55	66	82	94	110	147	198	237	323	
												28	28	30	30	39	48	55	66	88	94	117	157	198	251	
													28	30	30	30	39	48	55	70	88	101	124	157	210	
														30	30	30	30	39	48	59	70	94	107	124	166	
															30	30	30	30	39	51	59	75	100	107	132	
																30	30	30	30	42	51	63	80	100	113	
																	30	30	30	33	42	55	67	80	106	
																		30	30	33	33	45	58	67	84	
																			30	33	33	35	48	58	71	
																				32	33	35	37	48	61	
																					32	35	37	37	50	
																						34	37	37	39	
																							36	37	39	
																								36	39	
																									38	

Fig. 5.30a. CONVOLUTION OF EFFECTIVE RAINFALLS AND UNIT HYDROGRAPH

HYDROLOGY MANUAL	SYNTHETIC UNIT HYDROGRAPH METHOD Flood Hydrograph Calculation Form	Project EXAMPLE PROBLEM By A.E.S. Date Checked Date	Sheet 2 / 2

	EFFEC RAIN	.133	.140	.147	.161	.175	.224	.259	.308	.570	.259	.224	.175	.161	.147	.140	.133	.126	.126	.119	.112	.105	.105	.105	.098	FLOOD HYDRO-GRAPH
UNIT GRAPH	UNIT PER.	25	26	27	28	29	30	31	32	33	34	35	36	37	38	39	40	41	42	43	44	45	46	47	48	
2996	1	398																								8903
12,000	2	1596	419																							9251
19,200	3	2554	1680	440																						9667
14,720	4	1958	2688	1764	482																					10,138
7,486	5	996	2061	2822	1932	524																				10,712
4,416	6	587	1048	2164	3091	2100	671																			11,541
3,278	7	436	618	1100	2370	3360	2688	776																		12,861
2,567	8	341	459	649	1205	2576	4301	3108	923																	14,791
1,990	9	265	359	482	711	1310	3297	4973	3696	1708																17,811
1,667	10	222	279	377	528	773	1677	3812	5914	6840	776															22,025
1,316	11	175	233	293	413	574	989	1939	4534	10,944	3108	671														24,556
1,046	12	139	184	245	320	449	734	1144	2306	8390	4973	2688	524													22,665
899	13	120	146	193	268	348	575	849	1360	4267	3812	4301	2100	482												19,294
841	14	112	126	154	212	292	446	665	1010	2517	1939	3297	3360	1932	440											16,882
669	15	89	118	132	168	230	373	515	791	1868	1144	1677	2576	3091	1764	419										15,263
566	16	75	94	124	145	183	295	432	613	1463	849	989	1310	2370	2822	1680	398									14,088
487	17	65	79	98	135	157	234	341	513	1134	665	734	773	1205	2164	2688	1596	377								13,149
400	18	53	68	83	108	147	201	271	405	950	515	575	574	711	1100	2061	2554	1512	377							12,412
311	19	41	56	72	91	117	188	233	322	750	432	446	449	528	649	1048	1958	2419	1512	357						11,780
311	20	41	44	59	78	99	150	218	277	596	341	373	348	413	482	618	996	1855	2419	1428	336					11,246
311	21	41	44	46	64	85	127	173	259	512	271	295	292	320	377	459	587	943	1855	2285	1344	315				10,732
304	22	40	44	46	50	70	109	147	206	479	233	234	230	268	293	359	436	556	943	1752	2150	1260	315			10,220
			43	46	50	54	90	126	174	381	218	201	183	212	245	279	341	413	556	891	1649	2016	1260	315		9743
				45	50	54	70	104	150	323	173	188	157	168	193	233	265	323	413	526	838	1546	2016	1260	294	9389
					49	54	70	81	123	278	147	150	147	145	154	184	222	251	323	390	495	786	1546	2016	1176	8787
						53	70	81	96	228	126	127	117	135	132	146	175	210	251	305	367	464	786	1546	1882	7297
							68	81	96	177	104	109	99	108	124	126	139	166	210	237	288	344	464	786	1443	5169
								79	96	177	81	90	85	91	98	118	120	132	166	198	223	270	344	464	734	3566
									94	177	81	70	70	78	83	94	112	113	132	157	187	209	270	344	433	2704
										173	81	70	54	64	72	79	89	106	113	124	147	175	209	270	321	2147
											79	70	54	50	59	68	75	84	106	107	117	138	175	209	252	1643
												68	54	50	46	56	65	71	84	100	101	110	138	175	195	1313
													53	50	46	44	53	61	71	80	94	94	110	138	163	1057
														49	46	44	41	50	61	67	75	88	94	110	129	854
															45	44	41	39	50	58	63	70	88	94	103	695
																43	41	39	39	48	55	59	70	88	88	570
																	40	39	39	37	45	51	59	70	82	462
																		38	39	37	35	42	51	59	66	367
																			38	37	35	33	42	51	55	291
																				36	35	33	33	42	48	227
																					34	33	33	33	39	172
																						32	33	33	30	129
																							32	33	30	95
																								32	30	62
																									30	30

Fig. 5.30b

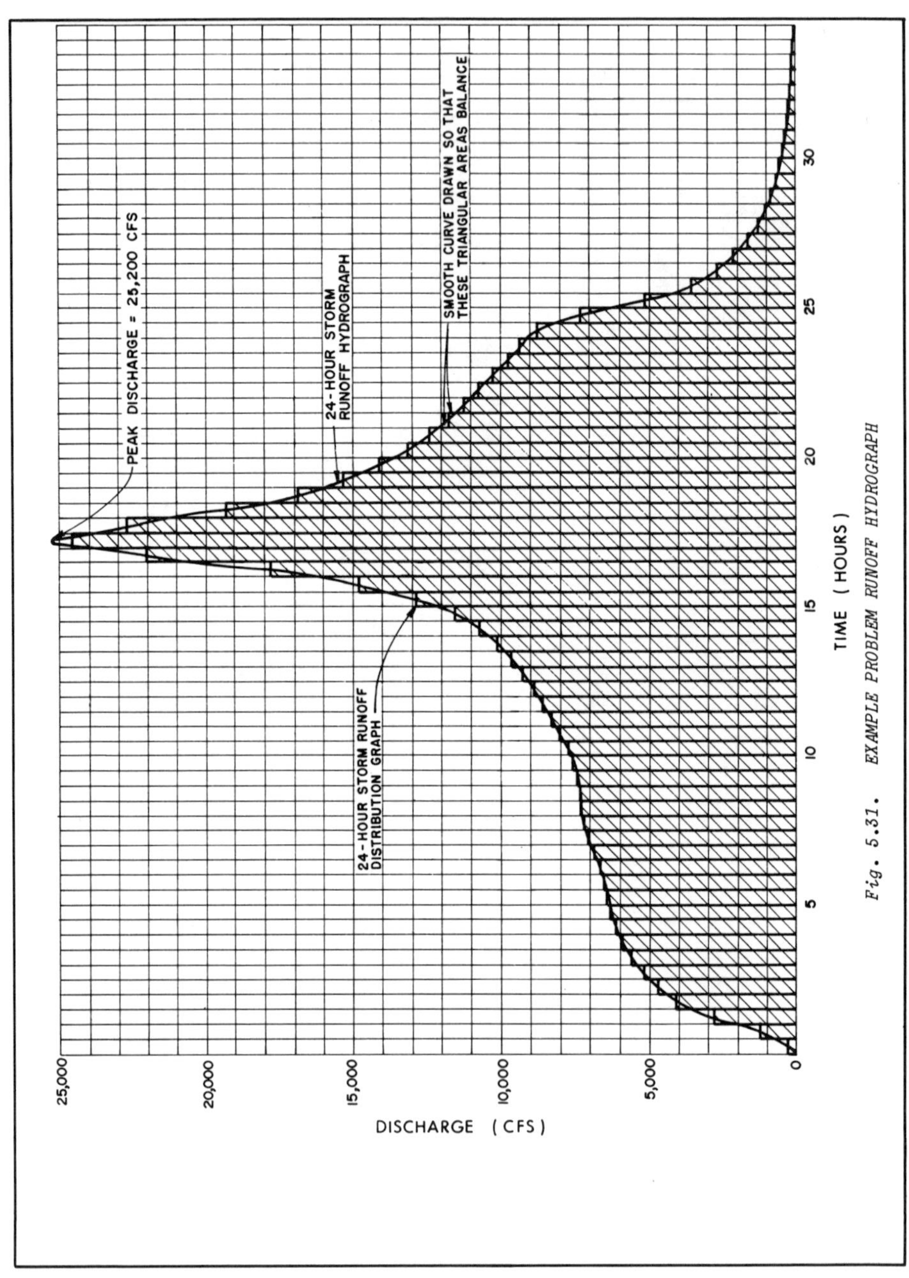

Fig. 5.31. EXAMPLE PROBLEM RUNOFF HYDROGRAPH

5.15. PROGRAM 5.1: Subroutine SGRAPH

The S-graph is a summation hydrograph developed by plotting watershed discharge in percent of ultimate discharge (or volume) as a function of time expressed in percent of watershed lag. The watershed lags can be estimated by empirical equations such as is previously shown in Eq. (5.1). For Southern California watersheds, it is often assumed that

n^* = a basin factor (see Fig. 5.4.)
$m = 0.38$
$C_t = 24\, n^*$

Another approach to estimating lag is using an empirical formula such as lag = C_oTc where Tc is the catchment time of concentration estimated by the sum of flow travel times through the catchment's drainage system (see Chapter 4 for estimating Tc using the rational method), and C_o is a calibration parameter. For example, a calibrated estimator for lag in Southern California catchments is lag = 0.80Tc.

In this program, a unit period is selected (see Table 5.3) such that a sufficient number of unit intervals are used to provide a reasonable representation of the unit hydrograph. By convolution of the unit hydrograph with respect to the effective rainfall (same unit interval as the unit hydrograph), the runoff hydrograph can be estimated.

TABLE 5.3
UNIT PERIOD SELECTION FOR UNIT HYDROGRAPH DETERMINATION

Unit Interval (minutes)	Watershed Lag Time (minutes)
3	Lag ≤ 30
6	30 < Lag ≤ 60
12	60 < Lag ≤ 120
15	120 < Lag ≤ 150
30	150 < Lag

A synthetic unit hydrograph is constructed for a 10 square-mile watershed by assuming an S-graph (Fig.5.5a) and using the watershed geometric information such as watercourse length = 25,000 ft, length from concentration point to centroid of the watershed = 12,500 ft and the averaged watershed basin factor = 0.035. Figure 5.32 depicts the resulting synthetic unit hydrograph. Figure 5.33 depicts the effective rainfall resulting from depth-area adjustment and a soil-loss function defined by 1 inch of initial abstraction and 10 percent of rainfall limited by a phi-index (0.1 inches/hour). Figure 5.34 illustrates the entire runoff hydrograph with the total runoff volume equaling 2210 acre-feet.

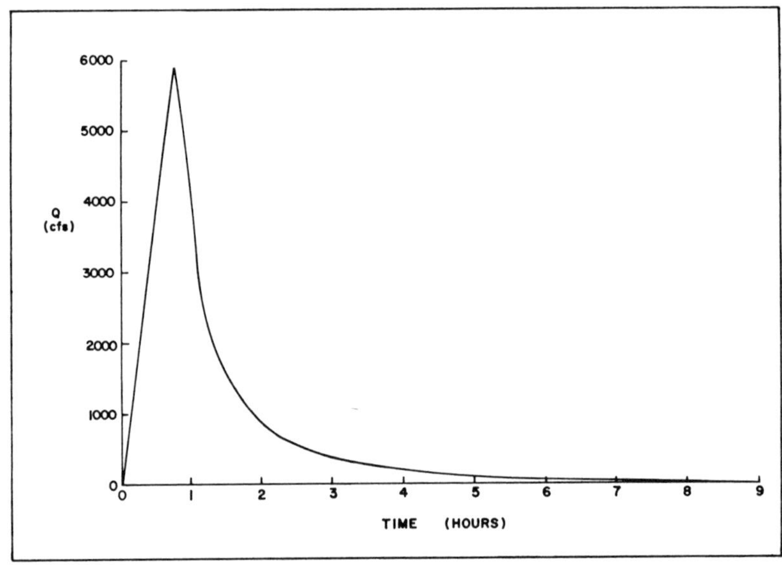

Fig. 5.32. SYNTHETIC UNIT HYDROGRAPH

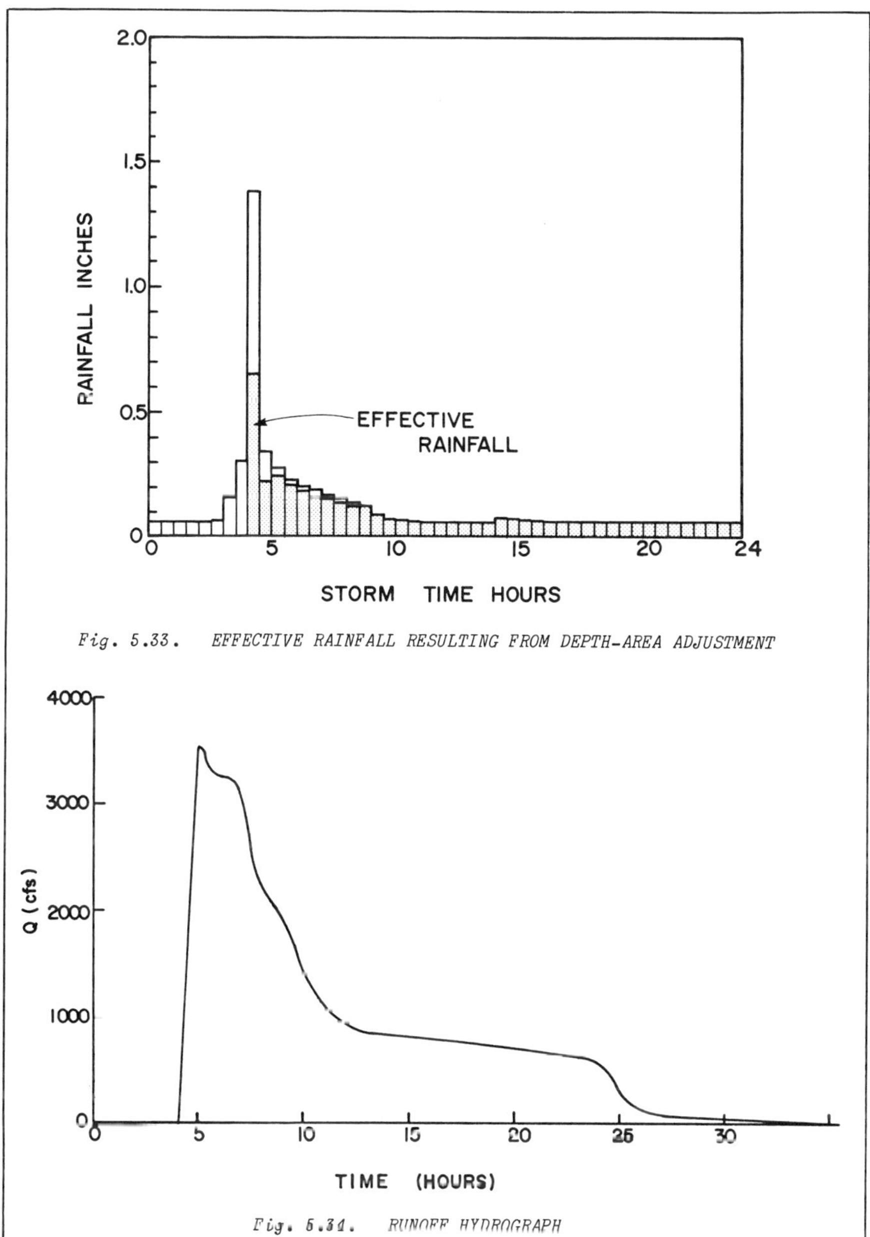

Fig. 5.33. EFFECTIVE RAINFALL RESULTING FROM DEPTH-AREA ADJUSTMENT

Fig. 5.34. RUNOFF HYDROGRAPH

5.16. Data Entry Sequences

5.16.1. Watercourse LAG Time Model

Two models are available for the watercourse LAG Time model.

```
*** WATERCOURSE "LAG" TIME MODEL OPTIONS: ***
   ENTER A [1] = CORPS OF ENGINEERS LAG FORMULA
      OR A [2] = USER SPECIFIED LAG VALUE
=====> 'ILAG'
ENTER WATERCOURSE LENGTH (FEET)
=====> 'XL'
ENTER LENGTH (FEET) ALONG THE LONGEST WATERSHED WATERCOURSE
 MEASURED FROM THE POINT OF CONCENTRATION UPSTREAM TO A POINT
 OPPOSITE THE CENTROID OF THE WATERSHED AREA
=====> 'XLAC'
ENTER ELEVATION (FEET) VARIATION ALONG WATERCOURSE
=====> 'HH'
```

```
BASIN    GENERAL DESCRIPTION OF DRAINAGE AREA
FACTOR   (Ref. U.S. Corps of Engineers)

n* = .015 >> Drainage area has fairly uniform gentle slopes.
             Most watercourses are either improved or paved streets.
n* = .020 >> Drainage area has some graded and non-uniform gentle
             slopes. Over half of the watercourses are improved or
             paved streets.
n* = .025 >> Drainage area is generally rolling with gentle side
             slopes. Some drainage improvements exist in the area -
             streets and canals.
n* = .030 >> Drainage area is generally rolling with rounded ridges
             and moderate side slopes. No drainage improvements
             exist in the area.
n* = .040 >> Drainage area is composed of steep upper canyons with
             moderate slopes in lower canyons. No drainage
             improvements exist in the area.
n* = .050 >> Drainage area is quite rugged with sharp ridges and
             steep canyons. No drainage improvements exist in the
             area.
n* = .200 >> Drainage area has comparatively uniform slopes. No
             drainage improvements in the area.

-----------------------------------------------------------------
ENTER THE BASIN FACTOR ALONG WATERCOURSE
=====> 'XN'
```

5.16.2. <u>S-Graph Development</u>

An S-graph can be selected from the data bank (see Chapter 8), or developed by use of proportions of defined S-graphs.

5.16.3. <u>Review of Computational Results</u>

After the runoff hydrograph is generated, the user can observe a portion, or the whole runoff hydrograph, on the CRT screen by the following entries:

```
ENTER [1] = WHOLE HYDROGRAPH
      [2] = PARTIAL HYDROGRAPH
=====> "2"
ENTER MODEL TIME (HOURS) FOR BEGINNING OF RESULTS
=====> "0"
ENTER MODEL TIME (HOURS) GREATER THAN [.050] FOR END OF RESULTS.
=====> "3"
```

5.16.4. <u>Accepting/Rejecting Computational Results and Output</u>

After reviewing the hydrograph, the user can either accept or reject the previous hydrograph results as follows:

```
ENTER A [1] TO ACCEPT THE HYDROGRAPH
   OR A [2] TO REJECT THE HYDROGRAPH
=====> "1"
ENTER UPSTREAM NODE NUMBER
=====> "101"
ENTER DOWNSTREAM NODE NUMBER
=====> "102"

SELECT PRINTING OPTIONS:

ENTER [1] = WHOLE HYDROGRAPH
      [2] = PARTIAL HYDROGRAPH
=====> "2"
ENTER MODEL TIME (HOURS) FOR BEGINNING OF RESULTS
=====> "0"
ENTER MODEL TIME (HOURS) GREATER THAN [.050] FOR END OF RESULTS
=====> "5"
ASSIGN A STREAM NUMBER THAT RECEIVES THE PREVIOUS HYDROGRAPH
ALLOWABLE VALUE IS BETWEEN [1] AND [5]
=====> "1"
```

After accepting or rejecting the hydrograph, the main menu will be displayed on the CRT screen for selection of the next process.

5.17. PROGRAM 5.1

```
C
C--------------------------------------------------------------
C     THIS SUBROUTINE DETERMINES THE PERCENTAGES OF DISCHARGE
C--------------------------------------------------------------
C
      SUBROUTINE SGRAPH(NS,INTERV,UNIT,KRUN,KRD,KWT)
C
      COMMON/BLK 1/ PRECIP(480)
      COMMON/BLK 8/ S(10,52),KNS(10)
      COMMON/BLK 9/ PERCENT(150)
      COMMON/BLK 10/ AREA
      COMMON/BLK 11/ H(1000)
C
      DIMENSION UH(150),SG(52,2)
C
      DATA SG/0.,2.,4.,6.,8.,10.,12.,14.,16.,18.,20.,22.,24.,26.,
     1 28.,30.,32.,34.,36.,38.,40.,42.,44.,46.,48.,50.,52.,54.,
     2 56.,58.,60.,62.,64.,66.,68.,70.,72.,74.,76.,78.,80.,82.,
     3 84.,86.,88.,90.,92.,94.,96.,98.,99.,100.,0.,0.,0.,0.,0.,0.,
     4 0.,0.,0.,0.,0.,0.,0.,0.,0.,0.,0.,0.,0.,0.,0.,0.,0.,0.,0.,
     5 100.,0.,0.,0.,0.,0.,0.,0.,0.,0.,0.,0.,0.,0.,0.,0.,0.,0.,0.,
     6 0.,0.,0.,0.,0.,0.,0.,0./
C
      DO 150 I=1,1000
 150  H(I)=0.
      DO 160 I=1,150
 160  PERCENT(I)=0.
C
C   ESTIMATION OF WATERCOURSE "LAG" TIME
C
 110  IF(KRUN.EQ.1)WRITE(KWT,51)
 51   FORMAT(24(/),2X,'*** WATERCOURSE "LAG" TIME MODEL OPTIONS: ',
     1 '***',/,5X,'ENTER A [1] = CORPS OF ENGINEERS LAG FORMULA',/,
     2 8X,'OR A [2] = USER SPECIFIED LAG VALUE')
      IF(KRUN.EQ.1)WRITE(KWT,99)
      READ(KRD,*)ILAG
      IF(KRUN.EQ.1)WRITE(10,*)ILAG
      IF(ILAG.EQ.1)GO TO 140
      IF(ILAG.EQ.2)GO TO 120
      GO TO 110
C
C   CORPS OF ENGINEERS FORMULA
C
 140  IF(KRUN.EQ.1)WRITE(KWT,13)
 13   FORMAT(/,2X,'ENTER WATERCOURSE LENGTH (FEET)')
      IF(KRUN.EQ.1)WRITE(KWT,99)
      READ(KRD,*)XL
      IF(XL.LT.100.)GO TO 140
```

```
      IF(KRUN.EQ.1)WRITE(10,*)XL
141   IF(KRUN.EQ.1)WRITE(KWT,14)
14    FORMAT(/,2X,'ENTER THE LENGTH (FEET) ALONG THE LONGEST',
     1 ' WATERSHED WATERCOURSE',/,'  MEASURED FROM THE POINT',
     2 ' OF CONCENTRATION UPSTREAM TO A POINT',/,'  OPPOSITE',
     3 ' THE CENTROID OF THE WATERSHED AREA')
      IF(KRUN.EQ.1)WRITE(KWT,99)
      READ(KRD,*)XLAC
      IF(XLAC.LT.50.)GO TO 141
      IF(KRUN.EQ.1)WRITE(10,*)XLAC
142   IF(KRUN.EQ.1)WRITE(KWT,15)
15    FORMAT(/,2X,'ENTER ELEVATION (FEET) VARIATION ALONG WATER',
     1 'COURSE')
      IF(KRUN.EQ.1)WRITE(KWT,99)
      READ(KRD,*)HH
      IF(HH.LT..1)GO TO 142
      IF(KRUN.EQ.1)WRITE(10,*)HH
143   IF(KRUN.EQ.1)WRITE(KWT,16)
16    FORMAT(24(/),'  BASIN    GENERAL DESCRIPTION OF DRAINAGE',
     C ' AREA',/,'  FACTOR    (Ref. U.S. Corps of Engineers)',//,
     C ' n*= .015 >> Drainage area has fairly uniform gentle',
     C ' slopes.',/,13X,'Most watercourses are either improved',
     C ' or paved streets.',/,' n*= .020 >> Drainage area has ',
     C 'some graded and non-uniform gentle slopes.',/,13X,'Over'
     C ,' half of the watercourses are improved or paved streets.'
     C ,/,' n*= .025 >> Drainage area is generally rolling ',
     C 'with gentle side slopes.',/,13X,'Some drainage improve',
     C 'ments exist in the area - streets and canals.',/,' n*=',
     C ' .030 >> Drainage area is generally rolling with round'
     C ,'ed ridges and moderate',/,13X,'side slopes.  No draina'
     C ,'ge improvements exist in the area.',/,' n*= .040 >> ',
     C 'Drainage area is composed of steep upper canyons with ',
     C 'moderate',/,13X,'slopes in lower canyons. No drainage',
     C ' improvements exist in the area.')
      IF(KRUN.EQ.1)WRITE(KWT,17)
17    FORMAT(' n*= .050 >> Drainage area is quite rugged with'
     C ,' sharp ridges and steep canyons.',/,13X,'No drainage ',
     C 'improvements exist in the area.',/,' n*= .200 >> Drain',
     C 'age area has comparatively uniform slopes.',/,13X,'No ',
     C 'drainage improvements exist in the area.',/,76(' '),
     C /,2X,'ENTER THE BASIN FACTOR ALONG WATERCOURSE')
      IF(KRUN.EQ.1)WRITE(KWT,99)
      READ(KRD,*)XN
      IF(XN.LT..001 .OR. XN.GT..99)GO TO 143
      IF(KRUN.EQ.1)WRITE(10,*)XN
      IF(KRUN.EQ.2)WRITE(KWT,22)XL,XLAC,HH,XN
22    FORMAT(2X,'WATERCOURSE LENGTH =',F11.3,' FEET',/,2X,
     C 'LENGTH FROM CONCENTRATION POINT TO CENTROID =',F11.3,
     C ' FEET',/,2X,'ELEVATION VARIATION ALONG WATERCOURSE =',
     C F11.3,' FEET',/,2X,'BASIN FACTOR ALONG WATERCOURSE =',F5.3,/)
      XL=XL/5280.
      XLAC=XLAC/5280.
      SL=HH/XL
      XLAGX=1.2*(XL*XLAC/SL**.5)**.38
      XLAG=20.*XN*XLAGX
```

```
      GO TO 130
C
C  USER SPECIFIED LAG
C
 120  IF(KRUN.EQ.1)WRITE(KWT,52)
 52   FORMAT(/'  ENTER WATERCOURSE LAG (HOUR) BETWEEN [.2] AND [5]')
      IF(KRUN.EQ.1)WRITE(KWT,99)
      READ(KRD,*)XLAG
      IF(XLAG.LT..2 .OR. XLAG.GT.5.)GO TO 120
      IF(KRUN.EQ.1)WRITE(10,*)XLAG
C
C  DESIGNATE UNIT INTERVALS
C
 130  XX=XLAG*60.
      IF(XX.LE.30.)UNIT=3.
      IF(XX.GT.30. .AND. XX.LE.60.)UNIT=6.
      IF(XX.GT.60. .AND. XX.LE.120.)UNIT=12.
      IF(XX.GT.120. .AND. XX.LE.150.)UNIT=15.
      IF(XX.GT.150. .AND. XX.LE.300.)UNIT=30.
      IF(XX.GT.300.)UNIT=60.
      TIMLAG=100.*UNIT/60./XLAG
      XL=XL*5280.
      XLAC=XLAC*5280.
      SQMI=AREA*43560./5280./5280.
      XK=645.*SQMI*60./UNIT
      XK=XK/100.
      IF(KRUN.EQ.2)WRITE(KWT,21)XLAG,UNIT,TIMLAG
 21   FORMAT(/,2X,'WATERCOURSE "LAG" TIME = ',F6.3,' HOURS',
     C /,2X,'UNIT HYDROGRAPH TIME UNIT = ',F6.2,' MINUTES',/,2X,
     C 'UNIT INTERVAL PERCENTAGE OF LAG-TIME = ',F6.2,/,76('-')/)
C
C  SELECTION OF S-GRAPH
C
      IF(NS.EQ.1)NNS=KNS(1)
      IF(NS.EQ.1)GO TO 550
 310  IF(KRUN.EQ.1)WRITE(KWT,56)
 56   FORMAT(24(/),'  *** S-GRAPH OPTIONS:',/,'  ENTER A [1] TO',
     1 ' SELECT A PREDEFINED S-GRAPH(FROM DATA BANK)'
     1 ,/,5X,'OR A [2] TO CONSTRUCT A COMBINED S-GRAPH')
      IF(KRUN.EQ.1)WRITE(KWT,99)
 99   FORMAT(2X,'=====> ',$)
      READ(KRD,*)KK
      IF(KK.LT.1 .OR. KK.GT.2)GO TO 310
      IF(KRUN.EQ.1)WRITE(10,*)KK
      IF(KK.EQ.2)GO TO 500
 320  IF(KRUN.EQ.1)WRITE(KWT,57)NS,(KNS(I),I=1,NS)
 57   FORMAT(/,2X,' [',I2,' ] S-GRAPHS EXIST',/,
     1 5X,'THE FOLLOWING S-GRAPHS ARE DEFINED :',/,
     2 7X,I2,:,I2,:,I2,:,I2,:,I2,:,I2,:,I2,:,I2,:,I2,:,I2)
      IF(KRUN.EQ.1)WRITE(KWT,59)
 59   FORMAT('  ENTER S-GRAPH NUMBER')
      IF(KRUN.EQ.1)WRITE(KWT,99)
      READ(KRD,*)NNS
      DO 505 I=1,NS
      IF(NNS.EQ.KNS(I))GO TO 507
```

```
 505   CONTINUE
       GO TO 320
 507   IF(KRUN.EQ.1)WRITE(10,*)NNS
 550   DO 510 I=1,52
       SG(I,2)=S(NNS,I)
 510   CONTINUE
       GO TO 600
 500   DO 520 I=1,NS
 560   IF(KRUN.EQ.1)WRITE(KWT,58)KNS(I)
 58    FORMAT(/2X,'ENTER FRACTION FOR S-GRAPH #',I2,
      1 /,'  (ALLOWABLE VALUE BETWEEN [0] AND [1])')
       IF(KRUN.EQ.1)WRITE(KWT,99)
       READ(KRD,*)PER
       IF(PER.LT.0. .OR. PER.GT.1.)GO TO 560
       IF(KRUN.EQ.1)WRITE(10,*)PER
       DO 530 J=1,52
       SG(J,2)=SG(J,2)+S(I,J)*PER
 530   CONTINUE
 520   CONTINUE
C
C  TIME=INCREMENTS OF LAG
C  NUMBER=INDEX OF PERCNT VECTOR
C
 600   ANEW=0.
       AOLD=0.
       K=1
       NUMBER=1
       TIME=TIMLAG
 10    K=K+1
       IF(NUMBER.GE.151)GO TO 1100
       N=K-1
C
C  INTEGRATE 'S' GRAPH IN ORDER TO DETERMINE UH(I)
C
       TEMP=0.5*(SG(K,1)+SG(N,1))*
      C (SG(K,2)-SG(N,2))
       ANEW=ANEW+TEMP
       IF(TIME.GT.SG(K,2))GO TO 10
       Y=SG(K,1)
       X=SG(N,1)
       DEL=(TIME-SG(N,2))/(SG(K,2)-SG(N,2))
       B=SG(K,2)-SG(N,2)
       DEL=DEL*(Y-X)
       XX=X+DEL
C
C  ADJUST INTEGRATION FOR INTERPOLATION
C
       DELA=0.5*(Y+XX)*(1.-DEL/(Y-X))*B
       ANEW=ANEW-DELA
       PERCENT(NUMBER)=(ANEW-AOLD)/TIMLAG
       NUMBER=NUMBER+1
       AOLD=ANEW
       ANEW=ANEW-.5*(X+XX)*DEL/(Y-X)*B
       TIME=TIME+TIMLAG
       K=K-1
```

```
      IF(PERCENT(NUMBER-1).GT.99.0)GO TO 1000
      GO TO 10
 1000 CONTINUE
C
C  FINISH-OFF HYDROGRAPH
C
      NUMBER=NUMBER-1
      IF(NUMBER.GE.150)GO TO 1100
      REM=100.-PERCENT(NUMBER)
      DEL=PERCENT(NUMBER)-PERCENT(NUMBER-1)
      DEL1=PERCENT(NUMBER-1)-PERCENT(NUMBER-2)
      IF(DEL.GE.REM*.6)GO TO 1110
      IF(DEL.GT.DEL1)DEL=DEL1
      KK=1
      NUMBER=NUMBER-1
 1130 PERCENT(NUMBER+KK)=PERCENT(NUMBER)+REAL(KK)*DEL
      IF(PERCENT(NUMBER+KK).GT.100.)PERCENT(NUMBER+KK)=100.
      IF(PERCENT(NUMBER+KK).EQ.100.)GO TO 1100
      IF(NUMBER+KK.GE.150)GO TO 1100
      KK=KK+1
      GO TO 1130
 1110 CONTINUE
      PERCENT(NUMBER+1)=PERCENT(NUMBER)+REM*.6
      PERCENT(NUMBER+2)=PERCENT(NUMBER+1)+REM*.3
      PERCENT(NUMBER+3)=100.
      KK=3
 1100 NUMBER=NUMBER+KK
      IF(NUMBER.GE.150 .AND. KRUN.EQ.2)WRITE(KWT,2001)NUMBER
 2001 FORMAT(7X,'UNIT-HYDROGRAPH TERMINATED AFTER',I4,' INTERVALS')
C
C  UNIT HYDROGRAPH DETERMINATION
C
      SUM1=0.
      DO 610 I=1,NUMBER
      UH(I)=(PERCENT(I)-SUM1)*XK
      SUM1=PERCENT(I)
      IF(UH(I).LT.0.)UH(I)=0.
 610  CONTINUE
C
      IF(KRUN.EQ.1)GO TO 640
      WRITE(KWT,2003)
 2003 FORMAT(/,27X,'UNIT HYDROGRAPH DETERMINATION')
      WRITE(KWT,2004)
 2004 FORMAT(76('-'))
      WRITE(KWT,2005)
 2005 FORMAT(5X,'INTERVAL',10X,'"S" GRAPH',10X,'UNIT HYDROGRAPH',/
     1 ,6X,'NUMBER',10X,'MEAN VALUES',10X,'ORDINATES(CFS)')
      WRITE(KWT,2004)
      WRITE(KWT,2006)(I,PERCENT(I),UH(I),I=1,NUMBER)
 2006 FORMAT(7X,I3,15X,F7.3,13X,F10.3)
C
C  ADJUST PRECIP ARRAY FOR LARGER UNIT INTERVALS
C  NI = NUMBER OF STORM HYDROGRAPH INTERVALS
C
 640  NI=480
```

```
      IF(UNIT.EQ.3.)GO TO 200
      IF(UNIT.EQ.6.)K=2
      IF(UNIT.EQ.12.)K=4
      IF(UNIT.EQ.15.)K=5
      IF(UNIT.EQ.30.)K=10
      NI=480/K
      DO 210 I=1,NI
      TEM=0.
      II=(I-1)*K
      DO 220 J=1,K
 220  TEM=TEM+PRECIP(II+J)
 210  PRECIP(I)=TEM
 200  CONTINUE
C
C  DETERMINE STORM RUNOFF HYDROGRAPH
C
      INTERV=NUMBER+NI-1
      DO 620 I=1,INTERV
      M-I
      N=1+M
      DO 630 J=1,M
      K=N-J
      IF(J.GT.NI)GO TO 630
      IF(K.GT.NUMBER)GO TO 630
      H(M)=H(M)+PRECIP(J)*UH(K)
 630  CONTINUE
 620  CONTINUE
C
      RETURN
      END
```

5.18. PROGRAM 5.2

```
C
C-------------------------------------------------------------
C     THIS SUBROUTINE DISPLAYS USER SPECIFIED HYDROGRAPH
C-------------------------------------------------------------
C
      SUBROUTINE HGRAPH(NA,NNN,KWT,UNIT,KODE,KRUN,KRD)
C
      COMMON/BLK 11/ H(1000)
      COMMON/BLK 16/ A(3000),B(3000)
C
      INTEGER LETM,BLANK,DOT,CROSS,DASH,LINE(41)
C
      DATA LETM,BLANK,DOT,CROSS,DASH,LINE/'V',' ','.','Q','I',41*' '/
C
      DO 140 I=1,1000
 140  A(I)=0.
C
      DO 150 I=1,NNN
 150  A(I)=H(I)
      INTERV=NNN
C
C  COMPUTE SUMMED HYDROGRAPH
C
      NUP=0
      NDOWN=0
      SUM=0.
      XMAX=0.
      DO 120 I=1,INTERV
      IF(A(I).LT.0.)A(I)=0.
      IF(A(I).GT.XMAX)XMAX=A(I)
      SUM=SUM+A(I)
 120  CONTINUE
C
      IF(SUM.GT.0. .OR. NA.EQ.0)GO TO 130
 110  IF(KRUN.EQ.1)WRITE(KWT,507)NA
 507  FORMAT(/,'  **** STREAM NUMBER',I2,' IS EMPTY ****'/,
     1 76('-'),/,'  ENTER A [1] TO CONTINUE')
      IF(KRUN.EQ.1)WRITE(KWT,99)
      IF(KRUN.EQ.1)READ(KRD,*)KX
      IF(KX.NE.1)GO TO 110
      RETURN
C
 101  FORMAT(4('******************'))
 102  FORMAT(4('------------------'))
 103  FORMAT(/14X,'HYDROGRAPH POINTS GIVEN IN THREE-MINUTE INTERVALS'/)
C
 130  IF(KODE.EQ.1)GO TO 505
      IF(KRUN.EQ.1)WRITE(KWT,501)
 501  FORMAT(/,2X,'ENTER UPSTREAM NODE NUMBER')
      IF(KRUN.EQ.1)WRITE(KWT,99)
 99   FORMAT('  =====> ',$)
      READ(KRD,*)NUP
      IF(KRUN.EQ.1)WRITE(7,*)NUP
```

```
      IF(KRUN.EQ.1)WRITE(KWT,502)
 502  FORMAT(/,'  ENTER DOWNSTREAM NODE NUMBER')
      IF(KRUN.EQ.1)WRITE(KWT,99)
      READ(KRD,*)NDOWN
      IF(KRUN.EQ.1)WRITE(7,*)NDOWN
 505  IF(KRUN.EQ.1 .AND. KODE.EQ.2)WRITE(KWT,517)
 517  FORMAT(/,'  SELECT PRINTING OPTIONS:')
      IF(KRUN.EQ.1)WRITE(KWT,503)
 503  FORMAT(/,2X,'ENTER A [1] = WHOLE HYDROGRAPH',/,5X,
     C 'OR A [2] = PARTIAL HYDROGRAPH')
      IF(KRUN.EQ.1)WRITE(KWT,99)
      IF(KRUN.EQ.1)READ(KRD,*)KK
      IF(KRUN.EQ.1 .AND. KODE.EQ.2)WRITE(7,*)KK
      IF(KRUN.EQ.2 .AND. KODE.EQ.2)READ(8,*)KK
      IF(KK.EQ.1)GO TO 510
      IF(KK.EQ.2)GO TO 520
      GO TO 505
 510  TIME1=0.
      TIME2=100.
      GO TO 530
 520  IF(KRUN.EQ.1)WRITE(KWT,504)
 504  FORMAT(/,'  ENTER MODEL TIME(HOURS) FOR BEGINNING OF',
     C ' RESULTS')
      IF(KRUN.EQ.1)WRITE(KWT,99)
      IF(KRUN.EQ.1)READ(KRD,*)TIME1
      IF(KRUN.EQ.2 .AND. KODE.EQ.2)READ(8,*)TIME1
      IF(TIME .LT. 0.)GO TO 520
      IF(KRUN.EQ.1 .AND. KODE.EQ.2)WRITE(7,*)TIME1
      TIME12=TIME1+.05
      IF(KRUN.EQ.1)WRITE(KWT,506)TIME12
 506  FORMAT(/,'  ENTER MODEL TIME(HOURS) GREATER THAN [ ',F8.3,
     C ' ] FOR END OF RESULTS')
      IF(KRUN.EQ.1)WRITE(KWT,99)
      IF(KRUN.EQ.1)READ(KRD,*)TIME2
      IF(KRUN.EQ.2 .AND. KODE.EQ.2)READ(8,*)TIME2
      IF(TIME2 .LE. TIME12)GO TO 520
      IF(KRUN.EQ.1 .AND. KODE.EQ.2)WRITE(7,*)TIME2
 530  CONTINUE
      IF(KRUN.EQ.1 .AND. KODE.EQ.2)RETURN
C
      SUM=SUM*UNIT*60./43560.
      STEP=3.*60./43560.
      XMAX=REAL(IFIX(XMAX/100.)+1)*100.
      F=40./XMAX
      FMASS=40./SUM
      T0=0.0
      T1=XMAX/4.
      T2=2.*T1
      T3=3.*T1
      WRITE(KWT,102)
      WRITE(KWT,101)
      IF(NA.NE.0)WRITE(KWT,106)NA
 106  FORMAT(/,2X,'*** STREAM NUMBER ',I2)
      IF(NUP.NE.0)WRITE(KWT,104)NUP,NDOWN
 104  FORMAT(/5X,'*** FLOW PROCESS FROM NODE ',I4,' TO NODE ',
```

```
     C I4,' WITH')
      WRITE(KWT,109)SUM
 109  FORMAT('    TOTAL STORM RUNOFF VOLUME(ACRE-FEET) = ',F12.4,/)
      WRITE(KWT,101)
      WRITE(KWT,103)
      WRITE(KWT,105)TO,T1,T2,T3,XMAX
 105  FORMAT('   TIME     VOLUME        Q     ',F2.0,1X,4F10.1,
     C /,' (HOURS)    (AF)       (CFS)')
      WRITE(KWT,102)
      XMASS=0.
C
C  OUTPUT HYDROGRAPH
C
      DO 200 I=1,INTERV
      TEST=A(I)*F
      XMASS=XMASS+A(I)*STEP
      LINE(1)=DOT
      LINE(11)=DOT
      LINE(21)=DOT
      LINE(31)=DOT
      LINE(41)=DOT
      J=TEST+1
      JMASS=XMASS*FMASS+1
      LINE(JMASS)=LETM
      LINE(J)=CROSS
      KI=I
      KNUM=IFIX(UNIT/3.)
      IF(KNUM.EQ.1)GO TO 400
C
      DO 350 K=1,KNUM
      KI=KNUM*(I-1)+K
      IF(K.EQ.1)GO TO 349
      XMASS=XMASS+A(I)*STEP
      JMASS=XMASS*FMASS+1
      LINE(JMASS)=LETM
      LINE(J)=CROSS
 349  TT=REAL(KI)*.05
      H(KI)=A(I)
      A1=1.-XMASS/SUM
      IF(A1 .LT. .0001)GO TO 350
      IF(TT.LT.TIME1 .OR. TT.GT.TIME2)GO TO 350
      WRITE(KWT,210)TT,XMASS,A(I),LINE
 350  LINE(JMASS)=BLANK
      GO TO 215
 400  CONTINUE
      TT=REAL(KI)*.05
      H(KI)=A(I)
      A1=1.-XMASS/SUM
      IF(A1 .LT. .0001)GO TO 215
      IF(TT.LT.TIME1 .OR. TT.GT.TIME2)GO TO 215
      WRITE(KWT,210)TT,XMASS,A(I),LINE
 210  FORMAT(1X,F6.2,F12.4,1X,F9.2,2X,41A1)
 215  LINE(J)=BLANK
      LINE(JMASS)=BLANK
 200  CONTINUE
```

```
 201  WRITE(KWT,101)
      NNN=KI
C
      IF(KODE.EQ.2 .OR. KRUN.NE.1 .OR. NA.EQ.0)RETURN
 108  WRITE(KWT,107)
 107  FORMAT(/,'  ENTER A [1] TO CONTINUE')
      WRITE(KWT,99)
      READ(KRD,*)KIJ
      IF(KIJ.NE.1)GO TO 108
      RETURN
      END
C
```

runoff hydrograph routing techniques

6

6.1. Introduction

The previous chapters have been concerned with watershed modeling techniques and rainfall-runoff correlation methods. Chapter 2 focused on the role played by rainfall in flood control planning and design studies, culminating in the design storm approach for analyzing the level of flood protection afforded by a particular catchment drainage system. Chapter 3 centered about the concept of losses which are a function of watershed storage characteristics. Watershed storage also controls discharge rates and runoff timing characteristics, both of which were discussed in Chapter 4. The shape, magnitude, and timing characteristics of unit hydrographs (Chapter 5) are also a function of watershed storage characteristics. As design parameters, peak discharges, travel times, and unit hydrographs are also greatly influenced by both channel storage and storage in lakes and ponds, whether or not they are natural or man-made. The effects of channel and reservoir storage on runoff characteristics are the topic of this chapter.

6.2. Benefits and Concerns of Detention Storage

A main purpose for inclusion of either a stormwater detention basin or a reservoir in a flood control system is to reduce the runoff peak flow rates generated from the tributary watershed. In the case of urban development the trade-off is to dedicate potentially developable land for detention basin purposes in exchange for a reduced flood control channel system downstream of the basin. Some of the advantages and disadvantages in using detention basins are listed in Table 6.1. The consideration of a detention basin system needs to address the various hydrologic, hydraulic, environmental, and flood control concerns listed in Table 6.1. Of special concern is the interplay between several detention basins utilized in a flood control system. Improperly placed basins can actually increase the total downstream runoff peak above what would be normally generated without the basins. Due to detention basins decreasing the peak flow rate, the basins outflow peak flow rate is distributed over a longer duration and can sum directly with downstream basin outflow peak flow rates. This direct summation of basin outflow peak flow rates can have a serious impact in a watershed drainage system where the downstream areas are controlled by detention basins rather than being drained prior to the arrival of the upstream peak flow rate. The computer program described at the end of this chapter enables the designer to evaluate watershed-wide effects to ensure that such peak compounding does not occur.

TABLE 6.1
DETENTION BASIN CONSIDERATIONS

Benefits	Potential Concerns
* reduces the peak rates of runoff to downstream reaches of the watershed	* detention basins do not reduce the total runoff volume unless basins are designed with dead storage (which is permanently stored within the basin)
* basins reduce the transport of sediments and debris carried by the floodwaters	* requires maintenance of storage capacity, inflow and outflow facilities
* inclusion of a detention basin in a flood control system may significantly reduce the downstream storm channel costs	* basins may discharge flows for long durations causing an increase in downstream channel erosion
* provides a location for groundwater recharge	* improperly sized and placed basins may aggravate rather than reduce downstream flooding potential
* provides a location to concentrate floodwaters for contaminant treatment	* sediment and debris occupies flood control storage volume, reducing the basin's benefit
* can provide aesthetically pleasing environments that increase the value of adjacent properties	* maintenance costs for debris removal
	* detention basins may be unsightly in an urban area

6.3. Detention Basin Routing Procedure (Modified Puls Method)

The modified Puls method is frequently used for flow-through detention basin modeling purposes. The reservoir routing equation is based upon the conservation formula:

$$I - O = dS/dT \tag{6.1}$$

where

I = detention basin inflow rate (cfs)
O = detention basin outflow rate (cfs)
dS = change in basin storage (cubic feet)
dT = time step (seconds)

Equation 6.1 can be approximated by replacing the variables I and O with time averaged values of

$$I = (I_1 + I_2)/2 \tag{6.2}$$

$$O = (O_1 + O_2)/2 \tag{6.3}$$

where the subscript 1 indicates the beginning of a time step, and the subscript 2 indicates the end of the timestep. Substituting the time averaged approximations for basin inflow and outflow into the conservation formula of Eq. (6.1) and rearranging gives:

$$(S_2 + O_2dT/2) = (S_1 - O_1dT/2) + (I_1 + I_2)dT/2 \tag{6.4}$$

The right-hand side of Eq. (6.4) is known from previously computed values of storage, S_1, outflow, O_1 and the average basin inflow $(I_1 + I_2)/2$ during the timestep, dT. The left-hand side of the equation contains the two unknowns S_2 and O_2. Therefore, Eq. (6.4) represents a single equation with two unknowns. The storage indication curve (discussed in the following text) represents the second equation.

Solution of the routing equation of Eq. (6.4.) requires the following information:

(1) Initial conditions of basin storage and outflow
(2) A routing timestep, dT
(3) An inflow hydrograph recorded for durations of dT
(4) A functional relationship between basin storage, outflow, and basin depth (i.e., the stage-storage-discharge relationship).

To aid in the calculation procedure, a storage indication curve is developed from the storage-discharge relationship by selecting various values of outflow and computing the left-hand side of Eq.(6.4), i.e.,(S + OdT/2). The solution procedure then consists of the following steps:

(1) Determine the average inflow volume from the inflow hydrograph and the selected unit period of dT by calculating the quantity $(I_1 + I_2)dT/2$.
(2) Determine $(S_1 - O_1dT/2)$ from the initial condition of the basin, or from the last computed values of S_1 and O_1.
(3) Use Eq. (6.4) to compute $(S_2 + O_2dT/2)$.
(4) Using the quantity determined in step 3, compute O_2 from the storage indication curve. The value O_2 is used to estimate the outflow volume during the timestep dT.
(5) Use O_2 to estimate S_2 from the basin storage versus outflow relationship.

These five steps are repeated for successive time steps using O_2 and S_2 as the new values of O_1 and S_1, respectively. The process is repeated until all of the inflow values are used and the basin outflow becomes negligible. The above procedure is illustrated by the following example problem.

Example 6.1. Flow-Through Detention Basin Routing

The assumed detention basin outflow and storage versus depth relationships are shown in Figs. 6.1 and 6.2, respectively. Using a timestep of dT = 60 minutes, the appropriate storage indication curves is developed in Table 6.2, and the resulting curve is plotted in Fig. 6.3. The example problem inflow hydrograph is tabulated in Table 6.3. Assuming an initial condition of zero for both the basin outflow and storage, a runoff hydrograph is routed through the basin using Eq. (6.4) in Table 6.3. The salient features of the routed hydrograph are shown in Fig. 6.4.

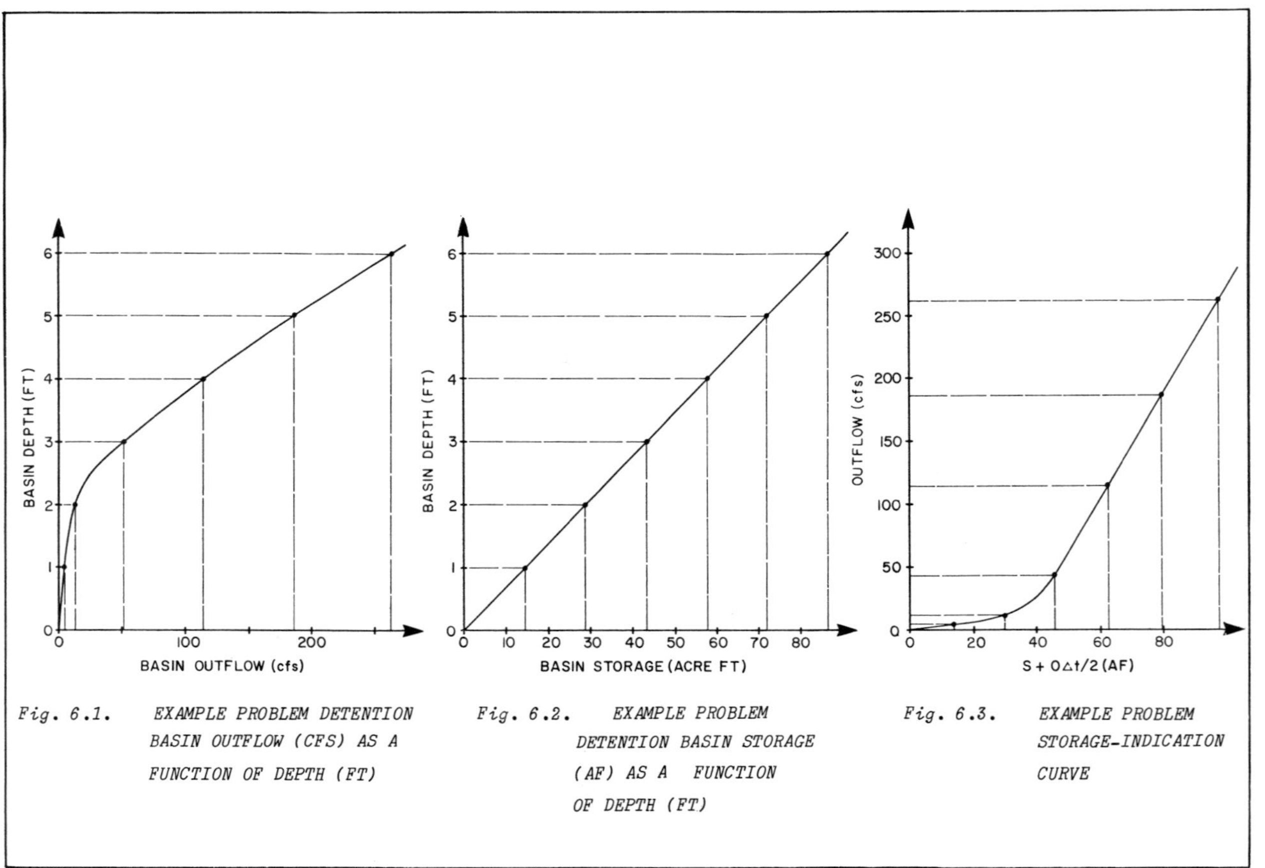

Fig. 6.1. EXAMPLE PROBLEM DETENTION BASIN OUTFLOW (CFS) AS A FUNCTION OF DEPTH (FT)

Fig. 6.2. EXAMPLE PROBLEM DETENTION BASIN STORAGE (AF) AS A FUNCTION OF DEPTH (FT)

Fig. 6.3. EXAMPLE PROBLEM STORAGE-INDICATION CURVE

TABLE 6.2
EXAMPLE PROBLEM STORAGE INDICATOR CURVE DEVELOPMENT

Depth (ft)	O (cfs)	S (AF)	S - OdT/2 (AF)	S + OdT/2 (AF)
0	0.0	0.0	0.0	0.0
1	4.2	14.4	14.22	14.57
2	12.0	28.8	28.30	29.30
3	51.7	43.2	41.06	45.34
4	114.7	57.6	52.86	62.34
5	186.8	72.0	64.28	79.72
6	263.2	86.4	75.52	97.28

TABLE 6.3
EXAMPLE PROBLEM BASIN ROUTING TABULATION

Time (min)	Inflow (cfs)	Mean Inflow	$(I_1+I_2)dT/2$ (AF)	$(S_1-O_1dT/2)$ (AF)	$(S_2+O_2dT/2)$ (AF)	Outflow (cfs)
0	0					0.0
		30	2.48	0.00	2.48	
60	60					0.7
		90	7.44	2.42	9.86	
120	120					2.8
		200	16.53	9.62	26.16	
180	280					10.3
		265	21.90	25.31	47.21	
240	250					58.6
		235	19.42	42.37	61.79	
300	220					112.7
		170	14.05	52.48	66.53	
360	120					132.1
		110	9.09	55.61	64.70	
420	100					124.5
		80	6.61	54.41	61.02	
480	60					109.8
		30	2.48	51.94	54.42	
540	0					85.4
		0	0.00	47.36	47.36	
600	0					59.2
		0	0.00	42.46	42.46	

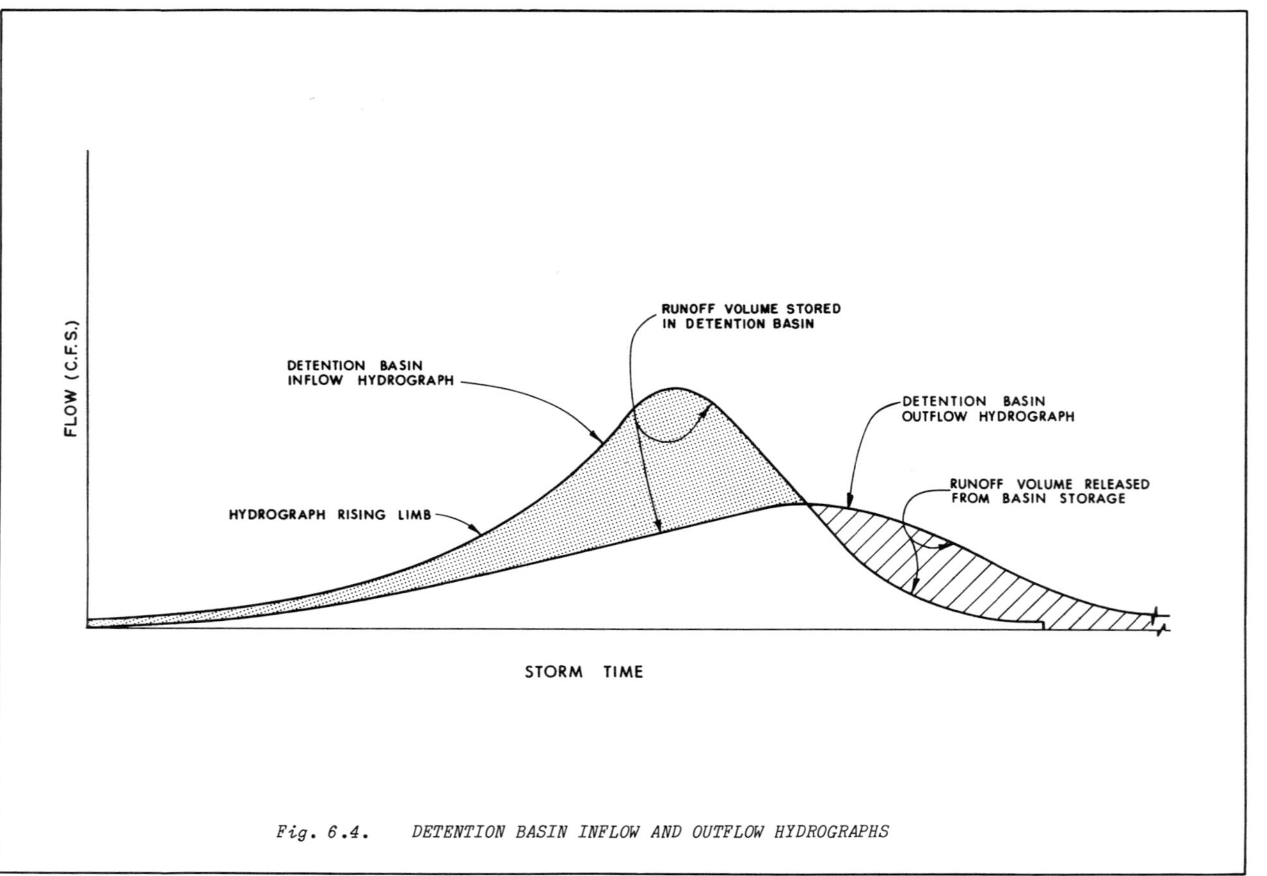

Fig. 6.4. DETENTION BASIN INFLOW AND OUTFLOW HYDROGRAPHS

6.4. Flow-by Channel Model (Runoff Hydrograph Separation)

The flow-by channel model is used to evaluate the effect of a channel flowing by a detention basin or structure which intercepts and diverts away from the channel all runoff flows in excess of some specified flowrate. Although simple in concept, use of a flow-by basin in a flood control system oftentimes provides a dramatic reduction in the runoff hydrograph peak flowrate with a minimum basin storage volume requirement. Such basins are usually designed to be considerably lower than the surrounding ground elevations and are often used for recreational purposes due to the infrequent use for flood control purposes. Figure 6.5 shows the main elements of a typical flow-by basin.

In most cases, the flow-by basin retains all overflow flood waters from the flow-by channel. Consequently, a multiple day storm may be required to guarantee that the basin has an adequate storage capacity remaining when a peak 24-hour storm event occurs.

In some cases, the excess flows are not stored but are diverted to another channel system. Therefore in order to accommodate both types of flow-by channel effects, the flow-by basin model needs to include a submodel for each of these two types of hydrologic conditions.

In flow-by basin design, an outlet pipe should be provided so that the basin may drain quickly. Otherwise, the basin storage capacity is essentially dead storage and may be exhausted by a prior sequence of storm events before the initiation of the severe 24-hour peak storm event. For a severe multiple day storm condition involving a flow-by basin system, the peak rainfall intensities may be incorporated within each 24 hour duration (up to the appropriate mass rainfall volume with the desired return frequency) such as is shown in Fig. 6.6.

6.5. The Modified Convex Channel Routing Method

There are several open channel flood routing models used for hydrologic study purposes, ranging from the complete solution of the St. Venant unsteady flow equations to a simple translation of the runoff hydrograph in time. A simple hydrologic flow routing model is the Convex Method (Mockus, 1972) which results (in a hydrologic sense) in both translation and peak flow attenuation of the runoff hydrograph due to channel storage effects. The provided computer program uses a modification of the convex method for channel routing approximation purposes. The governing relationship used in this approach is

$$O_{T+dT} = (1-C)O_T + CI_T \tag{6.5}$$

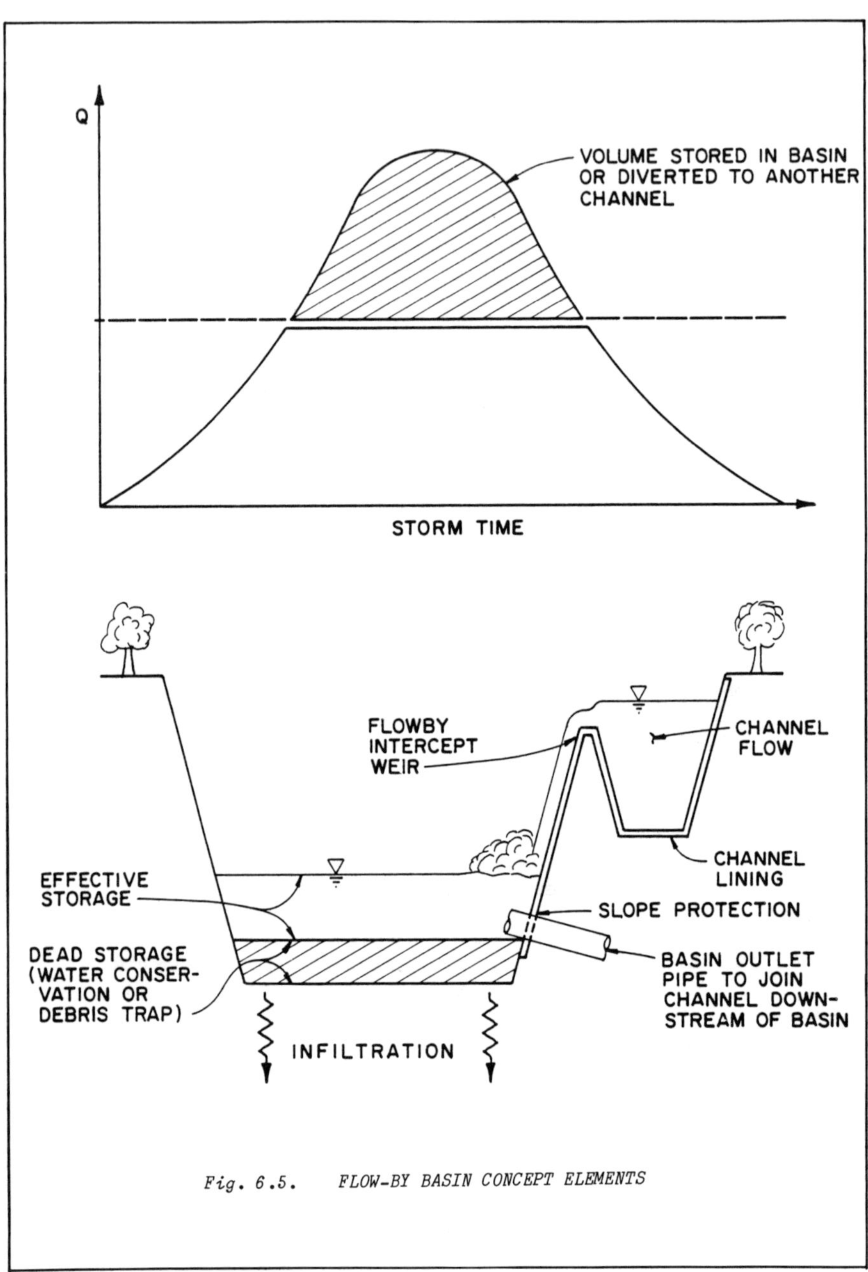

Fig. 6.5. FLOW-BY BASIN CONCEPT ELEMENTS

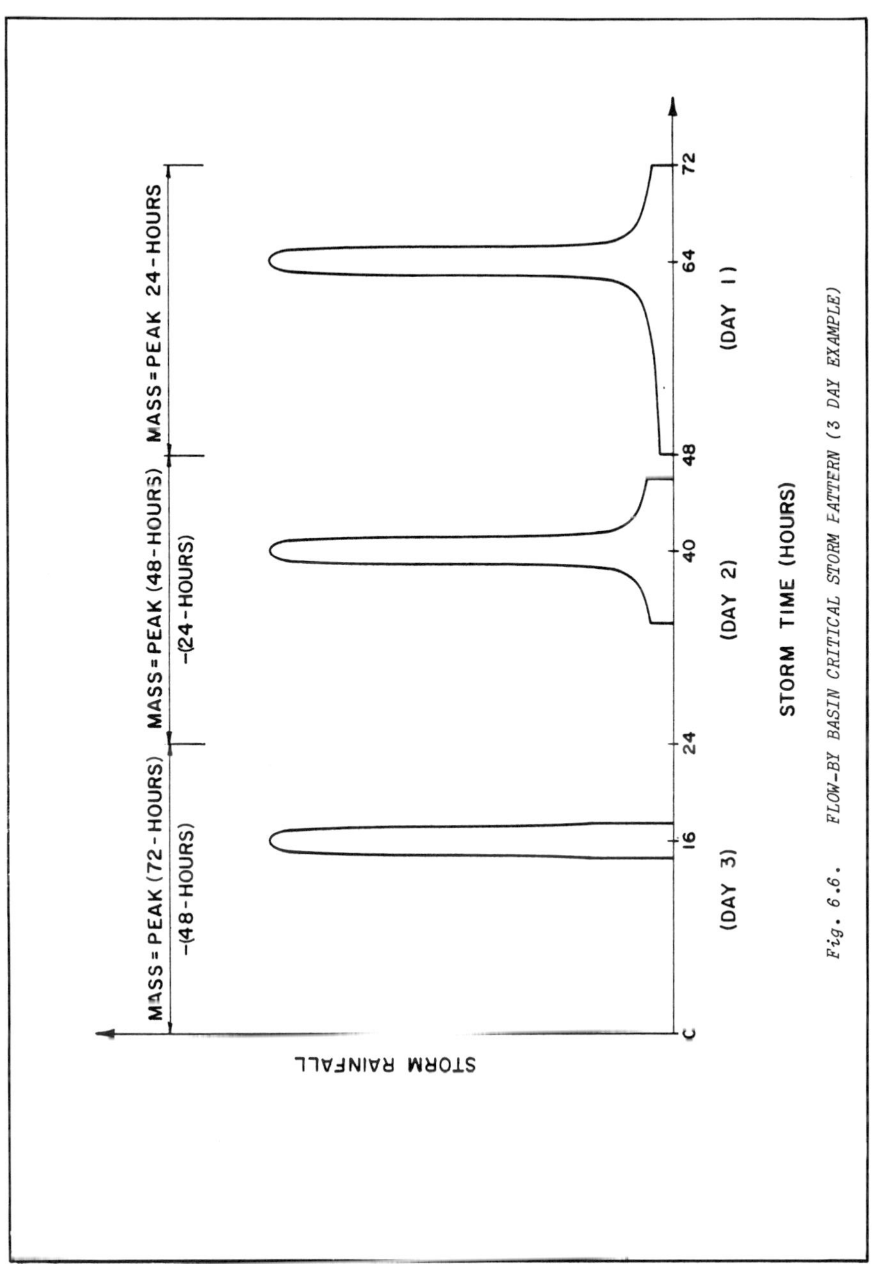

Fig. 6.6. FLOW-BY BASIN CRITICAL STORM PATTERN (3 DAY EXAMPLE)

where

I_T = hydrograph inflow at time T
O_T = channel outflow at time T
O_{T+dT} = channel outflow at time T+dT
C = a routing coefficient (where C is between 0 and 1)

Rearranging Eq. (6.5) gives the explicit statement:

$$O_{T+dT} = O_T + C\ (I_T - O_T) \tag{6.6}$$

and solving for C yields:

$$C = (O_{T+dT} - O_T)/(I_T - O_T) \tag{6.7}$$

Where measured streamflow data are not available to use with Eq. 6.7 for estimating C, the routing coefficient may be estimated by the following empirical relationship (Mockus, 1972):

$$C = V/(V + 1.7) \tag{6.8}$$

where V is a mean flow velocity computed from the inflow hydrograph. One method for computing V is to calculate the normal depth corresponding to the average flowrate of all unit flows greater than 50 percent of the inflow hydrograph peak flowrate. From this normal depth calculation, V is defined to be the corresponding flow velocity. Thus V represents an average velocity which is used to translate the inflow hydrograph along the total length of the channel. Obviously, other values for V can result depending on the choice of which average flowrate value to be used from the inflow hydrograph, and care must be taken to avoid offsetting the hydrograph peak flowrates at channel confluences by the selection of the channel V parameter.

Because the computer program unit period timestep is fixed at dT^* of 3 minutes, the unique dT value of Eq. (6.5) may differ from the model base dT^* unit period. Therefore, a method of modifying the routing coefficient is used (Mockus, 1972)

$$C^* = 1 - (1-C)^E \tag{6.9}$$

where

$$E = (dT^* + 0.5dT)/(1.5dT) \tag{6.10}$$

and $dT^* = 0.050$ hours.

To determine the dT which corresponds to Eq. (6.5), it is assumed that

$$(O_{T+dT} - O_T)/(I_T - O_T) = dT/K \tag{6.11}$$

where K is the channel reach travel time as estimated from the selection of the inflow hydrograph mean V value and the channel length, L. Figure 6.7 illustrates the geometric interpretation of the relationship given by Eq. (6.11). Thus,

$$C = dT/K \tag{6.12}$$

An examination of the convex routing method reveals that the entire routing approach is a function of the single routing parameter, C. Consequently, a watershed link-node model composed of m such channel links necessarily includes m channel routing parameters, each with an associated unknown uncertainty function. Additionally, the uncertainty involved in combining the m channel links is further aggravated by the fact that each channel routed hydrograph is also a function of the number and channel reach lengths used for each channel link. That is, the routed hydrograph through a channel with a length of 20,000 feet will differ from the results of routing a hydrograph through two successive reaches with a length of 10,000 feet, and so forth. In urban hydrology studies, channel routing processes usually involve relatively short reaches of improved channel where storage effects are minor or between confluences with other channels or pipes (where flows enter the main channel and a summation of hydrographs occurs). In these cases, the convex method is usually considered a reasonable model of the channel routing processes and produces approximation results that are usually within the level of accuracy that is provided by the runoff hydrograph approach (see Chapter 7).

Example 6.2. Channel Routing with the Modified Convex Routing Method

The example problem channel is a rectangular concrete section with a base width of 10 feet, a Manning friction factor of 0.015, a length of 3000 feet, and a mean slope of 0.005 feet/feet. The problem inflow hydrograph is tabulated in Table 6.4. From the

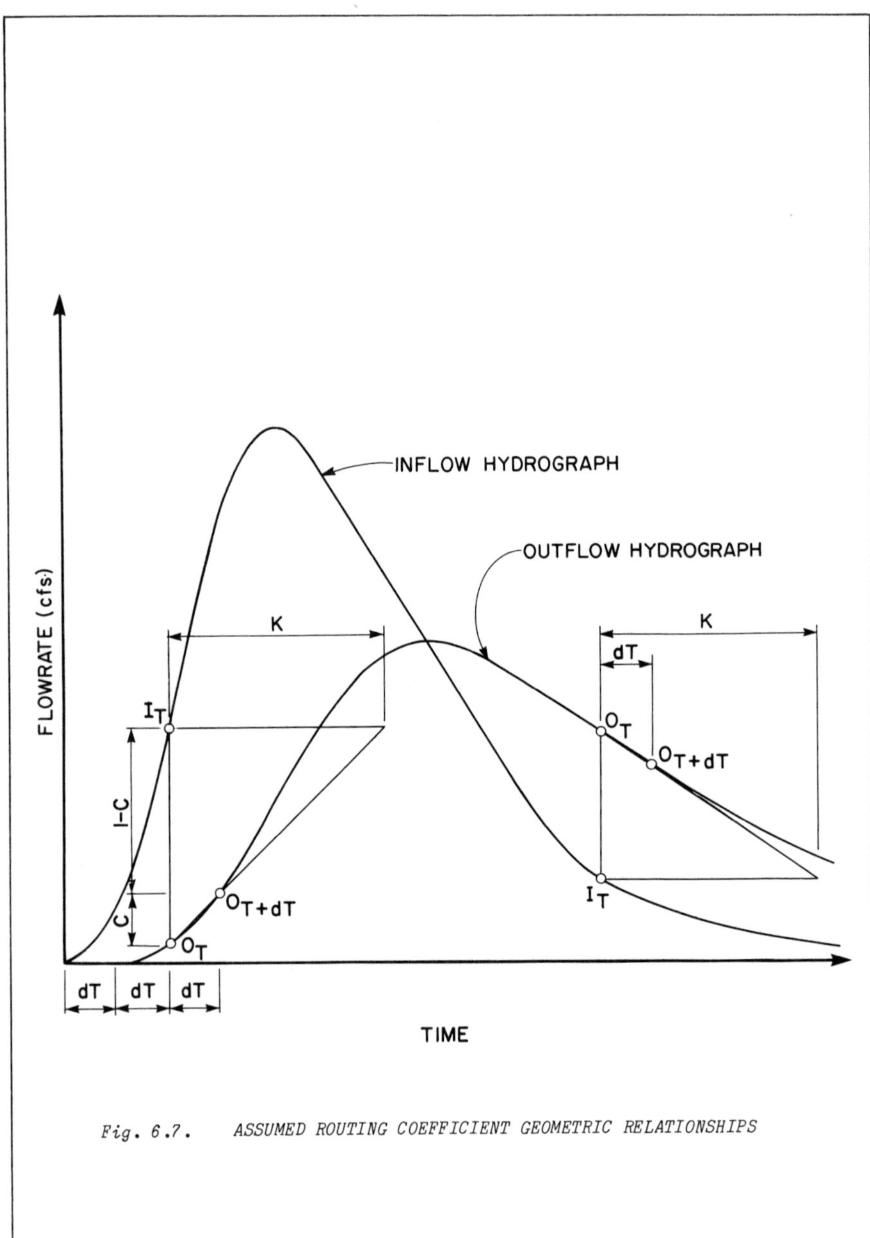

Fig. 6.7. ASSUMED ROUTING COEFFICIENT GEOMETRIC RELATIONSHIPS

table, the average flowrate in excess of the 50 percent peak flowrate value is 767.4 cfs.

Using Manning's equation, the normal depth flow velocity is calculated as

$$V = (1.486 R_h^{0.67} S_0^{0.50})/n \tag{6.13}$$

where R_h is the hydraulic radius in feet, and S_0 is the channel slope in feet per foot. For the example problem, V is 13.5 fps (feet per second) and the default routing coefficient from Eq. (6.8) is C = 0.89. From Eqs. (6.11) and (6.12),

$$K = (3000\ \text{ft})/(13.5\ \text{fps})(3600\text{sec/hr}) = 0.062\ \text{hour}$$

$$dT = (0.89)(0.062) = 0.055\ \text{hour}$$

From Eq. (6.9), C* is estimated for a timestep of dT* equal to 3 minutes by:

$$C^* = 1 - (1-0.89)^E = 0.874 \tag{6.14}$$

where E = (0.050 + (0.5)(0.055))/((1.5)(0.055)) = 1.9394. Thus the appropriate convex method routing approximation statement is

$$O_{T+dT} = (1-C^*)\, O_{T+dT-dT^*} + C^* I_T \tag{6.15}$$

where for the example problem

$$O_{T+dT} = (0.126)\, O_{T+dT-dT^*} + (0.874)\, I_T \tag{6.16}$$

The tabulation of Table 6.4 follows the presentation used in Mockus (1972, p.17.64).

6.6. The Modified Att-Kin Method

The Modified Att-Kin method (MAK) is an alternative to the Convex method. The MAK method is based on the same general routing equation (Eq. 6.5) but provides for translation of the hydrograph beyond the translation effect provided by Eq. (6.5). The MAK method uses a routing coefficient C which is:

$$C = 2dT/(2K + dT) \tag{6.17a}$$

The value K is assumed to be given by:

$$K = L/mV \tag{6.17b}$$

in which L is the channel reach length, V is the wave velocity, and m is the coefficient of the rating curve. The wave velocity can be computed using Manning's Equation.

TABLE 6.4
CONVEX ROUTING EXAMPLE PROBLEM SOLUTION

Inflow Time (minutes)	Inflow (cfs)	Outflow (cfs)	Outflow Time (minutes)
0	0.0	0.0	
3	0.8	0.0[1]	3.3
6	0.9	0.7	6.3
9	40.5	0.9	9.3
12	202.7	35.5	12.3
15	445.1	181.6	15.3
18	602.9	411.9	18.3
21	653.7	578.8	21.3
24	600.9	644.3	24.3
27	608.0	606.4	27.3
30	917.1	607.8	30.3
33	1186.7	878.1	33.3
36	1001.2	1147.8	36.3
39	763.6	1019.7	39.3
42	714.9	795.9	42.3
etc.	etc.	etc.	etc.

Note 1: Outflow hydrograph initiates at time
$dT = CK = (0.89)(0.62 \text{ hrs}) = 0.055 \text{ hrs} = 3.3$ minutes

After deriving the first estimate of the downstream hydrograph, it is necessary to check whether or not translation is necessary. The hydrograph computed with Eq. (6.5) is translated when the kinematic travel time (dt_p) is greater than dt_{ps} where dt_{ps} is the time difference between the upstream hydrograph I and the hydrograph computed with Eq. (6.5):

$$dt_{ps} = t_{po} - t_{pi} \tag{6.18}$$

in which t_{po} and t_{pi} are the time to peaks of the downstream (outflow) and upstream (inflow) hydrographs, respectively. The kinematic travel time is given by:

$$dt_p = (S_{po}/q_{po})\{[(q_I/q_{po})^{1/m}-1]/[(q_I/q_{po})-1]\}/3600 \qquad (6.19)$$

in which q_{po} is the peak discharge of the downstream hydrograph, q_I is the peak discharge of the upstream hydrograph, and S_{po} is given by:

$$S_{po} = (q_{po}/K)^{1/m} \qquad (6.20)$$

and

$$K = x/L^m \qquad (6.21)$$

If $dt_p > dt_{ps}$, then the storage-routed hydrograph from Eq. (6.5) is translated by an amount $(dt_p - dt_{ps})$.

This procedure can be summarized by the following steps:

(1). From cross-section information evaluate m and x;

(2). Compute K and then C;

(3). Use the routing equation (Eq. 6.5) to route the upstream hydrograph;

(4). Compute dt_{ps} from Eq. (6.18);

(5). Compute the kinematic travel time dt_p (Eq. 6.19); and

(6). If $dt_p > dt_{ps}$, then translate the computed downstream hydrograph.

6.7. Muskingum Channel Routing

The Muskingum channel routing method is another alternative to the Convex routing method. It is somewhat more flexible than the Convex method since the value of O_{T+dT} depends on I_{T+dT}, I_T, and O_T, and not just I_T and O_T. As an alternative to Eq. 6.5, the Muskingum method uses the following equation for routing:

$$O_{T+dT} = C_o I_{T+dT} + C_1 I_T + C_2 O_T \tag{6.22}$$

where the coefficients are given by:

$$C_o = -\ (Kx - 0.5dT)/(K - Kx + 0.5dT) \tag{6.23a}$$

$$C_1 = (Kx + 0.5dT)/(K - Kx + 0.5dT) \tag{6.23b}$$

$$C_2 = (K - Kx - 0.5dT)/(K - Kx + 0.5dT) \tag{6.23c}$$

The sum of the three coefficients C_o, C_1, and C_2 equals 1.0. In Eqs. (6.22) and (6.23), K and x are the routing parameters. While K and x can be estimated from hydrographs measured at the upstream and downstream channel sections (see Chapter 7),the value of K is taken to be equal to the reach travel time. Methods of assigning a value of x are less respected. Some suggest that a value for x of 0.2 is reasonable, while others suggest 0.4. As the value of x decreases from the upper limit of 0.5, the peak of the downstream hydrograph becomes more attenuated. To ensure nonnegative ordinates for O_{T+dT} the following limits should be respected:

$$x \leq dT/2K \leq (1 - x);\ x \leq 0.5 \tag{6.24}$$

Example 6.3. Muskingum Channel Routing

The data for Example 6.2 can be used to illustrate the use of the Muskingum method. For a velocity of 13.5 fps, the travel time which was computed to be 0.062 hours, was set equal to the routing parameter K. The other routing parameter, x, was assumed to be 0.2.

A time interval of 3-minutes was chosen which satisfied the inequality of Eq. 6.24. The three coefficients C_o, C_1, and C_2, are 0.169, 0.501, and 0.329, respectively. The results of the Muskingum routing are given in Table 6.5.

TABLE 6.5
MUSKINGUM ROUTING EXAMPLE PROBLEM SOLUTION

Time (minutes)	Inflow (cfs)	Outflow (cfs)
0	0.0	0.0
3	0.8	0.0
6	0.9	0.8
9	40.5	7.6
12	202.7	57.1
15	445.1	195.7
18	602.9	389.7
21	653.9	541.3
24	600.9	607.7
27	608.	604.4
30	917.10	659.1
33	1186.7	877.7
36	1001.2	1053.6
39	763.6	978.3
42	714.9	826.1
etc.	etc.	etc.

6.8. A Pipeflow Routing Model

Similar to the Convex routing approximation procedure, the pipeflow routing model develops an outflow hydrograph for a reach of pipe given an inflow hydrograph and appropriate pipe section data. In the pipeflow model, however, a limiting value of outflow is assumed whereby all inflows greater than this pipe capacity are temporarily stored at the upstream endpoint of the pipe. The stored floodwaters subsequently drain into the pipe at a rate equal to the pipe capacity. This modeling approach approximates the ponding of floodwaters in the urban street system low points where, generally, a significant volume of storage is available with a small change in flooding depths. Similar to the other channel routing methods, backwater effects are not included in the modeling approach.

The pipeflow routing process is modeled by calculating a normal depth flow velocity for each 3 minute unit period runoff value from the inflow hydrograph, and translating the unit runoff forward in storm time by the appropriate time increment. The computer program uses the pipe capacity flow rate corresponding to a flow depth of 0.82 of the pipe diameter. In order to simplify the normal depth calculations, the flow velocity is calculated as a tabulated function of the ratio (unit inflow)/(pipe capacity). Figure 6.8 shows the salient features of the pipeflow modeling approach used in the computer program.

6.9. Hydrograph Translation

This function moves the runoff hydrograph forward in storm time by a specified time increment. The method used is to simply move each 3-minute unit runoff value to a future unit period according to the specified time increment, (Fig. 6.9). The Hydrograph Translation function may be used as a channel routing model for cases where it is assumed that no attenuation of the peak flowrate occurs during the unsteady flow routing process.

6.10. Kinematic Wave Routing Approximation

The kinematic wave routing approximation utilizes the user entered channel geometric and friction factor data to calculate normal depth flow hydraulic information for each 3-minute unit interval in the runoff hydrograph. For each unit flow rate, the normal depth is computed and the corresponding flow velocity is calculated. Using this flow velocity, a translation timestep is computed and the unit interval is moved forward in model time accordingly. Because the larger flow rates have a faster normal depth flow velocity than smaller flow rates, this flow routing technique tends to develop a kinematic "shock wave" whereby the rising limb of the runoff hydrograph becomes steeper due to the routing method (see Fig. 6.10).

6.11. Subroutine FLOWBY

The relationship between the captured flows and the stream flow into a flowby basin is based on the inflow and outflow relationship shown in Figure 6.11. The simple cutoff inflow-outflow relationship (Curve 1, Fig. 6.11) can be replaced by a variable inflow-outflow relationship (Curve II, Fig. 6.11) in this subroutine. Dead storage and basin infiltration included in the FLOWBY basin model. The basin percolation equation is based on the Horton infiltration equation, i.e.,

$$f = fc + (fo - fc)\ \exp(-kt)$$

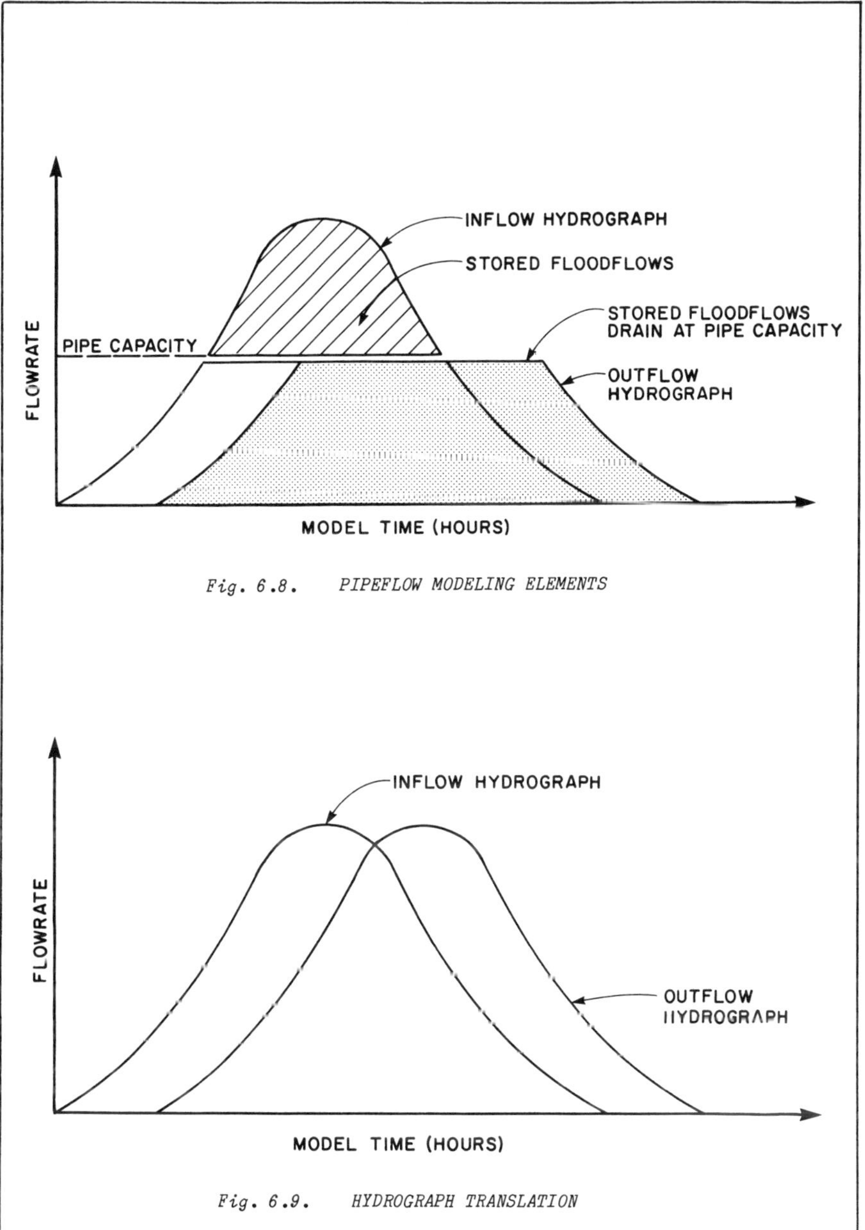

Fig. 6.8. PIPEFLOW MODELING ELEMENTS

Fig. 6.9. HYDROGRAPH TRANSLATION

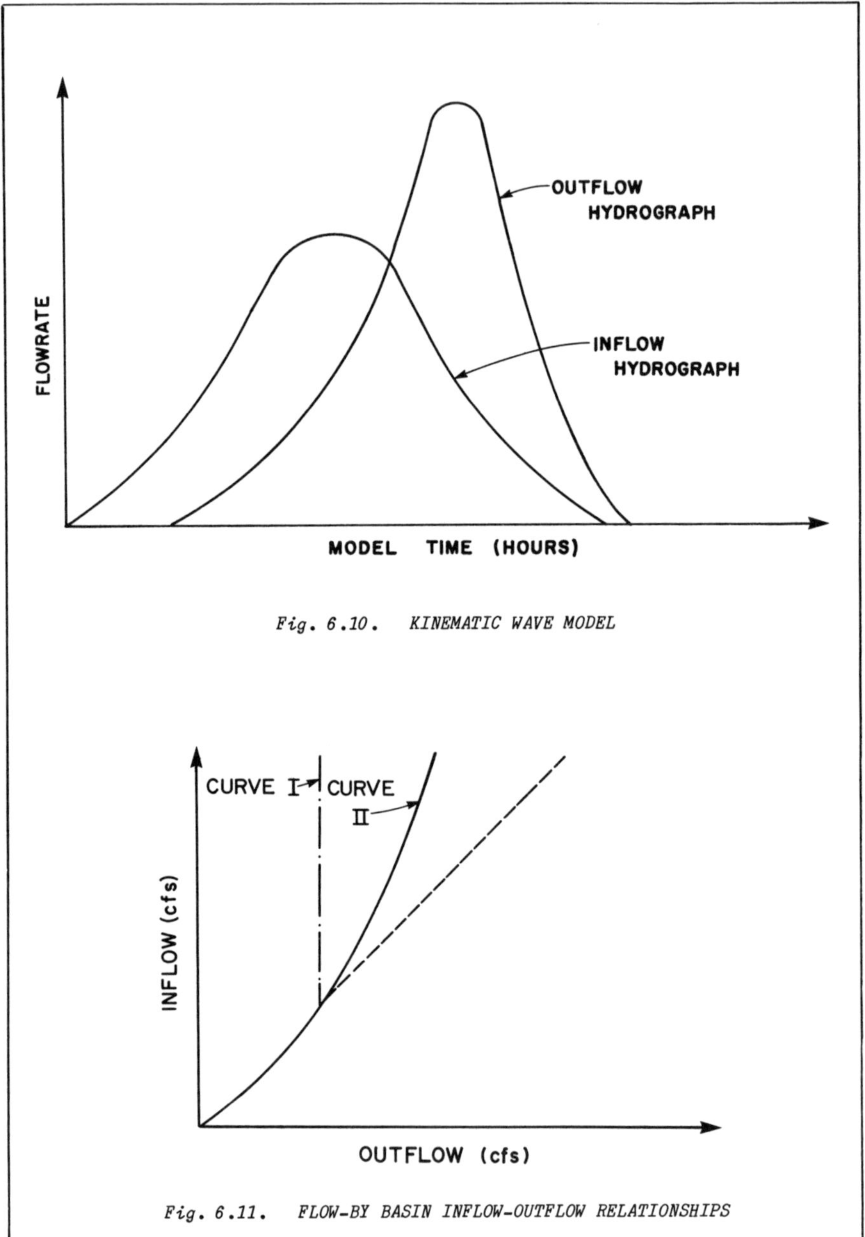

Fig. 6.10. KINEMATIC WAVE MODEL

Fig. 6.11. FLOW-BY BASIN INFLOW-OUTFLOW RELATIONSHIPS

where

f = infiltration rate (cfs) at time t (hour)

fc = final infiltration rate (cfs)

fo = initial infiltration rate (cfs)

k = exponential coefficient

t = time in hour

6.11.1. Application: Flow-by Basin

The runoff hydrograph of Fig. 3.6 is used as the inflow hydrograph for the flow-by basin routing model. Figure 6.12 illustrates the inflow/outflow hydrgraphs for the flow-by structure. Table 6.6 shows the inflow versus outflow relationship assumed for this flow-by structure.

TABLE 6.6
EXAMPLE INFLOW VS. OUTFLOW

INFLOW RATE (cfs)	OUTFLOW RATE (cfs)
0.	0.
1000.	1000.
1500.	1400.
2000.	1800.
2500.	2300.
2700.	2500.
4000.	2500.

The excess flow can either be stored at an adjacent detention basin, or be added to another stream. In this example, the total inflow runoff volume is 2208 acre-feet, the adjacent basin storage requirements is determined to be 200.66 acre-feet excluding infiltration, and 200.58 acre-feet for a basin with infiltration. The Horton infiltration equation used in this example is I(cfs) = (0.10) + (0.15 - 0.10) exp[-.1*TIME(hour)].

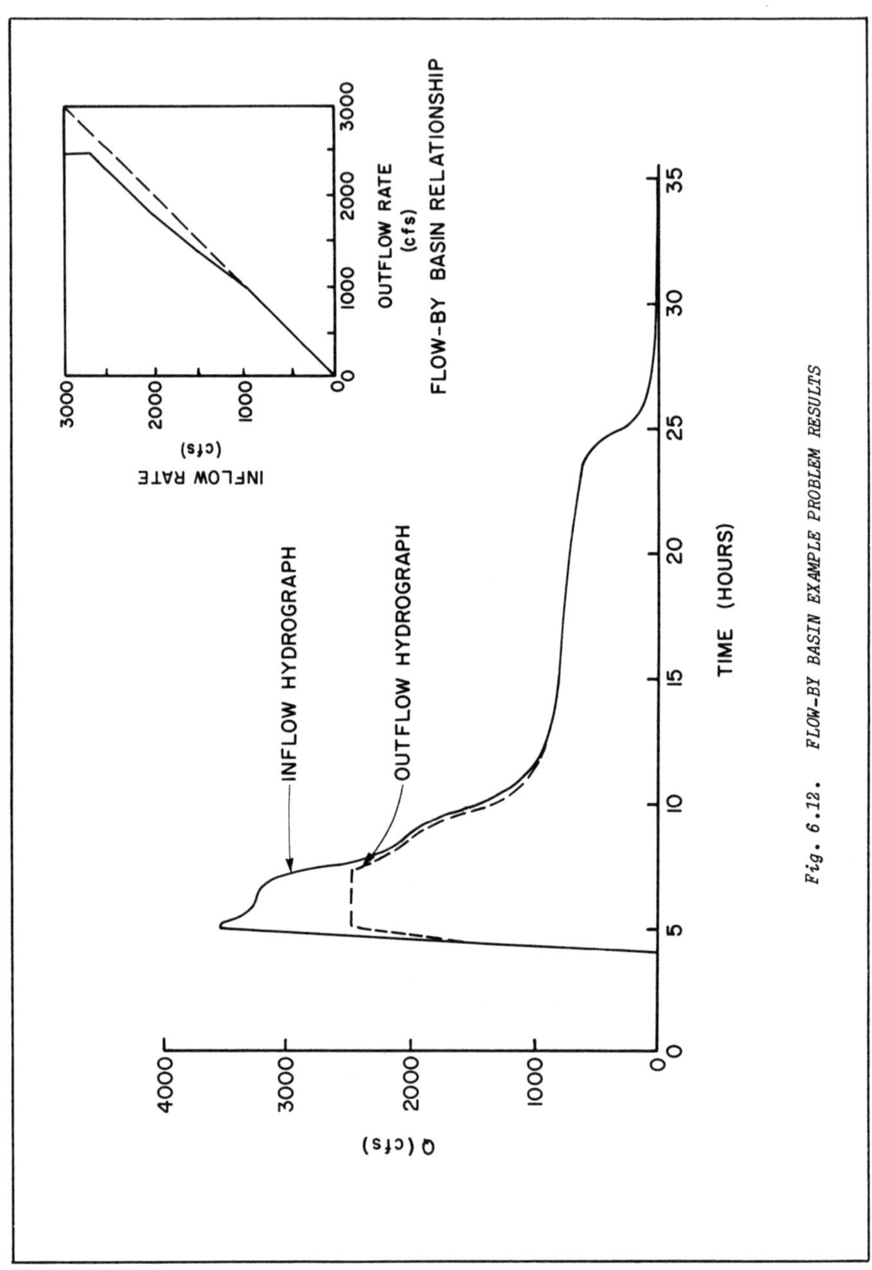

Fig. 6.12. FLOW-BY BASIN EXAMPLE PROBLEM RESULTS

6.11.2. Data Entry Sequence

The excess flow from the flow-by structure can be either stored in an adjacent detention basin, or be added to another stream. The required flow-by basin model data is as follows:

```
*** FLOWBY BASIN ***
  ENTER THE STREAM NUMBER THAT ENTERS THE FLOWBY BASIN
  (ALLOWABLE VALUE IS BETWEEN [1] AND [5])
=====> 'NA'
  ENTER NODE NUMBER
=====> 'INODE'

  ENTER THE STREAM NUMBER THAT STORES THE EXCESS FLOW
  (ALLOWABLE VALUE IS BETWEEN [1] AND [5])
=====> 'NB'
  ENTER (INFLOW, OUTFLOW) INFORMATION:
  THE FIRST SET OF DATA IS DEFINED TO BE (0,0)
  ENTER THE NUMBER OF DATA SETS (BETWEEN [3] AND [20])
=====> 'NSET'

  ENTER INFLOW RATE (CFS) > 'GS(I-1)' FOR SET #'I'
=====> 'QS(I)'
  ENTER OUTFLOW RATE BETWEEN ['QF(I)'] AND ['QS(I)'] CFS
=====> 'QF(I)'
```

If (NA.EQ.NB), then

```
  ENTER A [1] TO INCUDE BASIN INFILTRATION
=====> 'NIF'
```

If (NIF.EQ.1), then

```
  THE INFILTRATION IS APPROXIMATED BY:
   QOUT = QIN + (QO - QIN)*EXP(-k*TIME)
  ENTER QIN IN CFS
=====> 'QIN'
  ENTER QO IN CFS
=====> 'QO'
  ENTER k IN 1/HOUR
=====> 'COE'
  SELECT PRINT/VIEW OPTIONS:
  ENTER [1] = WHOLE HYDROGRAPH
        [2] = PARTIAL HYDROGRAPH
=====> 'KK'
```

If (KK.EQ.2), then

```
  ENTER MODEL TIME (HOURS) FOR BEGINNING OF RESULTS
=====> 'TIME1'
  ENTER MODEL TIME (HOURS) GREATER THAN ['TIME12'] FOR END
  OF RESULTS
=====> 'TIME2'
```

After the hydrograph in shown, then

```
  ENTER A [1] TO ACCEPT THE PROCESS
=====> 'OPT'
```

IF (OPT.NE.1), then the previous flow process is ignored.
IF (OPT.EQ.1), then the main menu will appear on the CRT screen.

6.12. Subroutine FLOWTR

The modified Puls method is used for the flow-through basin routing approximation. Dead storage and basin infiltration are also included in this subroutine. The infiltration equation is the same Horton modeling technique as that described in subroutine FLOWBY.

6.12.1. Application: Flow-through Basin

The runoff hydrograph (from the unit hydrograph model) is routed through a detention basin with the depth-storage-outflow relationships shown in Table 6.7.

TABLE 6.7
EXAMPLE BASIN DATA

Depth (Feet)	Basin Storage (Acre-Feet)	Outflow Rate (cfs)
0	0	0
1	50	100
2	100	300
3	150	500
4	200	800
5	250	1100
6	300	1500
7	350	1900
8	400	2300
9	450	2700
10	500	3200

Figure 6.13 illustrates the inflow and outflow hydrographs for the detention basin example problem. For demonstration purposes, the total outflow volume is reduced to 2146 Acre-feet for an assumed basin infiltration rate of

Infiltration (cfs) = 10 + (15 - 10) exp [-1.*TIME (hours)]

6.12.2. Data Entry Sequence

```
=====================================================
STORAGE HYDROGRAPH ROUTING IN A CONTROLLED RESERVOIR
=====================================================
  ENTER STREAM NUMBER (BETWEEN [1] AND [5]) OF THE
  INFLOW HYDROGRAPH
=====> 'NN'
*** ENTER THE UPSTREAM NODE NUMBER
=====> 'IUP'
*** ENTER THE DOWNSTREAM NODE NUMBER
=====> 'IDDWN'
  ENTER BASIN DEPTH-VERSUS-STORAGE AND-DISCHARGE INFORMATION:
  ENTER TOTAL NUMBER OF DEPTH-VERSUS-STORAGE DATA ENTRIES:
  (ALLOWABLE VALUES ARE BETWEEN [3] AND [20])
=====> 'NUMD'
  ENTER DEPTH-VERSUS-STORAGE AND DEPTH-VERSUS-DISCHARGE
  INFORMATION
 *FOR DEPTH-VERSUS-STORAGE DATA ENTRY #1
 *BOTTOM OF BASIN ASSUMED AS DEPTH (FEET) = 0.
  ENTER BASIN STORAGE (ACRE-FEET) AT DEPTH (FEET) = 0:
  (ALLOWABLE VALUES ARE BETWEEN [0] AND [100000])
=====> 'S(1)'
 *BASIN OUTFLOW AT DEPTH = 0 FOOT ASSUMED AS 0.0 CFS
  ENTER BASIN DEPTH (FEET) FOR DATA ENTRY #'K'
  (ALLOWABLE VALUES ARE BETWEEN ['DX'] AND [250]
  NOTE: ENTER [-1] TO RESTART BASIN INFORMATION SEQUENCE
=====> 'D(K)'
  ENTER BASIN STORAGE (ACRE-FEET) AT DEPTH (FEET) = 'D(K)'
  (ALLOWABLE VALUES ARE BETWEEN ['SX'] AND [100000])
=====> 'S(K)'
  ENTER BASIN OUTFLOW (CFS) AT DEPTH (FEET) = 'D(K)':
  (ALLOWABLE VALUES ARE BETWEEN ['OX'] AND [100000])
=====> 'O(K)'
```

```
NOTE:  DX = D(K-1) + 0.005
       SX = S(K-1)*1.0001
       OX = O(K-1)*1.0001
```

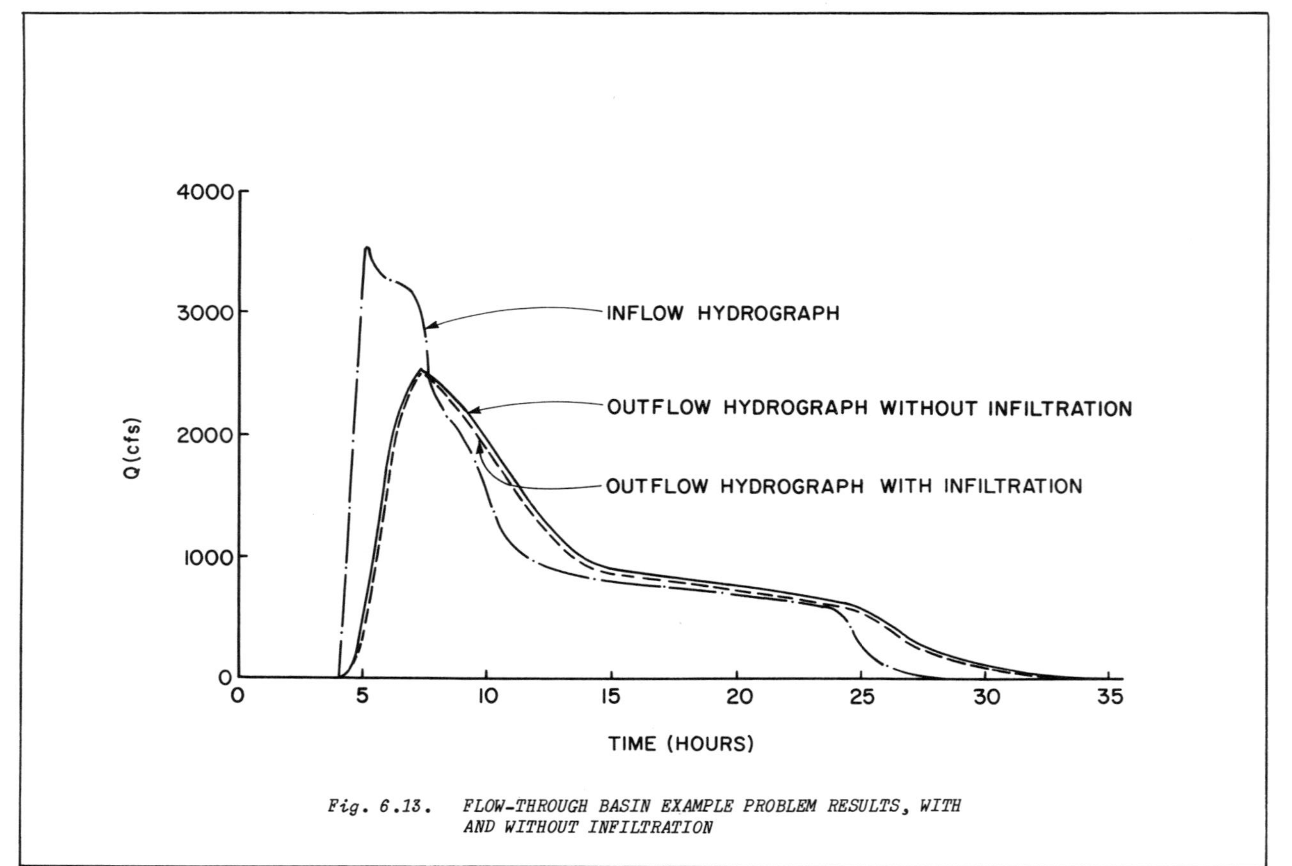

Fig. 6.13. FLOW-THROUGH BASIN EXAMPLE PROBLEM RESULTS, WITH AND WITHOUT INFILTRATION

```
  ENTER ASSUMED INITIAL DEPTH (FEET) OF WATER IN BASIN AT
  BEGINNING OF INFLOW HYDROGRAPH INTO BASIN:
  (ALLOWABLE VALUES ARE BETWEEN [0] AND ['D(NUMD)'])
=====> 'DO'
  ENTER MAXIMUM NUMBER OF OUTFLOW HYDROGRAPH INTERVALS
  (ALLOWABLE VALUES ARE BETWEEN ['NUMBER'] AND [1000])
=====> 'LIMIT'
  ENTER DEAD STORAGE (ACRE-FEET) FOR RESERVOIR
=====> 'QD'
  ENTER [1] TO INCLUDE THE INFILTRATION SUBMODEL
=====> 'KKK'
```

If (KKK.EQ.1), then

```
  THE INFILTRATION RATE IS APPROXIMATED BY
   QUOT = QIN + (QO-QIN)*EXP(-k*TIME)
  ENTER QIN IN CFS
=====> 'QIN'
  ENTER QO IN CFS
=====> 'QO'
  ENTER COEFFICIENT k IN 1/HOUR
=====> 'COE'
  SELECT PRINT/VIEW OPTIONS:
  ENTER [1] = WHOLE HYDROGRAPH
        [2] = PARTIAL HYDROGRAPH
=====> 'KK'
```

If (KK.EQ.2), then

```
  ENTER MODEL TIME (HOURS) FOR BEGINNING OF RESULTS
=====> 'TIME1'
  ENTER MODEL TIME (HOURS) GREATER THAN ['TIME12'] FOR
  END OF RESULTS
=====> 'TIME2'
```

After the hydrograph is displayed, then

```
  ENTER A [1] TO ACCEPT THE PROCESS
=====> 'OPT'
```

If (OPT.NE.1), then the previous flow process is ignored.
If (OPT.EQ.1), then the main menu will appear on the CRT screen.

6.13. Subroutine RCHAN

This subroutine includes five channel routing models: (1) convex routing, (2) pipeflow routing, (3) translation, (4) kinematic-wave routing, and (5) Muskingum routing.

6.13.1. Convex Routing Model

The convex routing technique is based on the following routing equation:

$$O_{t+dt} = C\ I_t + (1-C)\ O_t$$

where

O_{t+dt} = outflow rate (cfs) at time t+dt hours
I_t = inflow rate (cfs) at time t hours
O_t = outflow rate (cfs) at time t hours
C = routing coefficient

The routing coefficient C can be approximated by:

$$C = V/(V + 1.7)$$

in which V is a steady flow velocity (fps).

The user can either specify the steady flow velocity or request the computer to estimate a mean velocity from a specified percent of peak flowrate by the user. The Horton equation is used as the infiltration equation in the channel reach.

6.13.2. Pipeflow Routing Model

The pipeflow routing process is modeled by calculating a normal depth flow velocity for each 3-minute period runoff value from the inflow hydrograph, and translating the unit runoff forward in storm time by the appropriate time increment. A limiting value of outflow is assumed whereby all inflows greater than this pipe capacity are temporarily stored at the upstream endpoint of the pipe. The limiting value corresponds to the pipe capacity at a flow depth of 0.82 of the pipe diameter.

6.13.3. Hydrograph Translation Model

The hydrograph translation model moves the hydrograph in stream A forward in storm time by a specified time increment. This time increment can be estimated either by a user specified mean velocity or by the computer as described in the convex routing model. The Horton infiltration equation is also available.

6.13.4. Kinematic-Wave Routing Model

The kinematic-wave routing model routes the inflow hydrograph downstream by using the following technique:

(1). Each unit interval of the inflow hydrograph is translated downstream using a travel time based on the normal depth corresponding to the unit flow rate, and
(2). The new outflow hydrgraph is the summation of all the translated unit runoffs.

6.13.5. Muskingum Routing Model

The Muskingum method is based on the routing equation:

$$O(t+dt) = C_0 I(t+dt) + C_1 I(t) + C_2 O(t)$$

in which I(t) and O(t) are the inflow rate (cfs) and outflow rate (cfs) at time t (hours), respectively, dt is the modeling timestep (hours), and the routing coefficients are defined by

$$C_0 = (Kx - 0.5dt)/(K - Kx + 0.5dt)$$

$$C_1 = (Kx + 0.5dt)/(K - Kx + 0.5dt)$$

$$C_2 = (K - Kx - 0.5dt)/(K - Kx + 0.5dt)$$

where $C_0 + C_1 + C_2 = 1$.

The translation parameter, K, can be estimated by using the travel time through the reach, and a value of 0.2 is often used for the storage parameter x. A value of 0.5 for x is usually considered to be the upper limit of rationality; this value suggests equal weighting of inflow I(t) and outflow O(t), and it appears as primarily translation of the hydrograph with little attenuation. A value of 0 for x is the practical lower limit and suggests that the inflow has little effect; this would reflect reservoir storage type effects in which attenuation would be dominant.

6.13.6. Application: Channel Routing Modeling Techniques

The outflow hydrograph (Fig. 6.12) from the flow-by basin example problem is used as the inflow hydrograph for each of the five channel routing models. The five outlfow hdyrographs are shown in Fig. 6.14. Table 6.8 shows the routing coefficients for the five channel routing models.

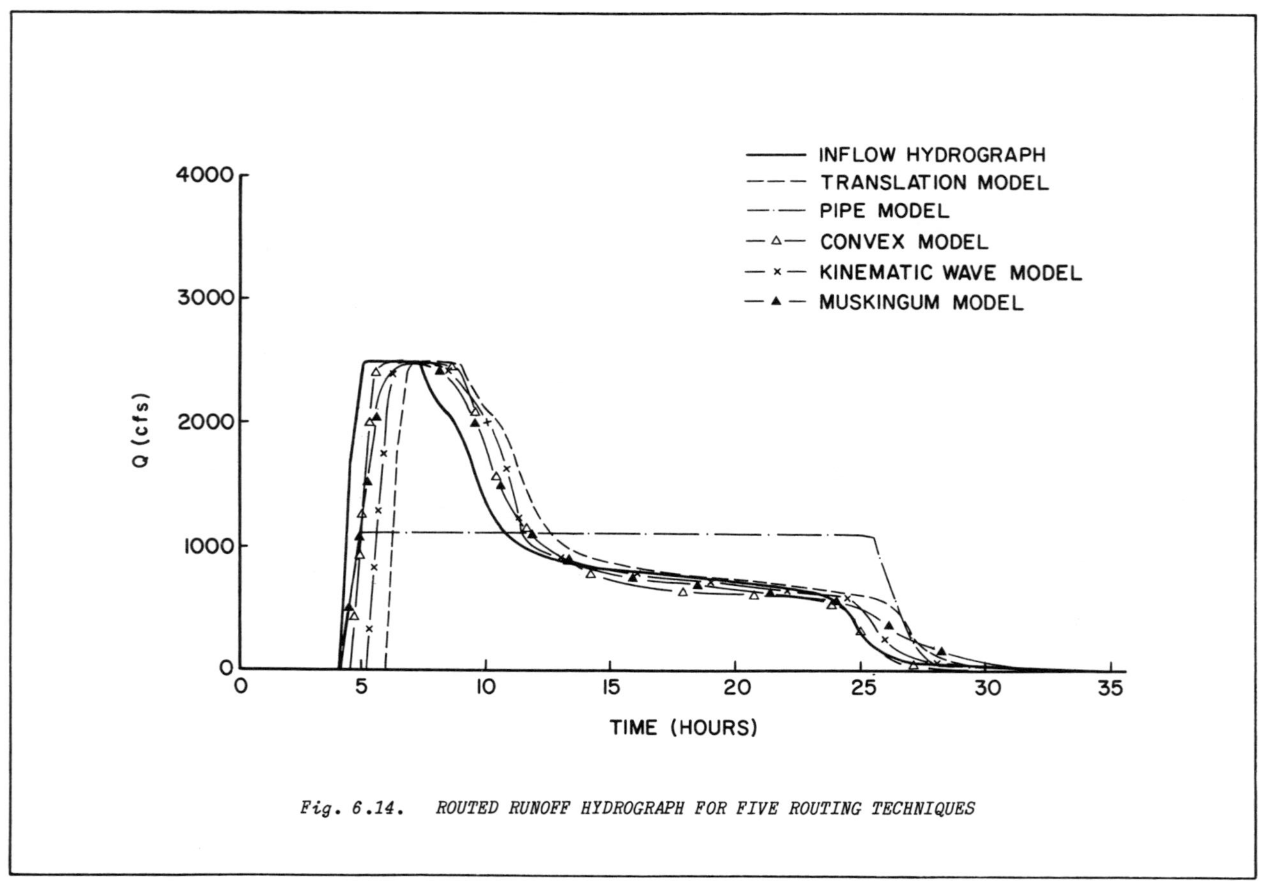

Fig. 6.14. ROUTED RUNOFF HYDROGRAPH FOR FIVE ROUTING TECHNIQUES

TABLE 6.8
EXAMPLE ROUTING PROBLEM INFORMATION

Model	Channel Geometry	Routing Coefficients
Translation	--	2 hour offset in time
Convex	Channel information: Base = 30' Side slope (vertical/ horizontal) = 2 Manning's n = 0.015 Length = 10,000 ft. Difference in Elevation = 1 ft.	C = 0.686
Pipe	Diameter = 18 ft. Manning's n = 0.013 Length = 5,000 ft. Difference in Elevation = 1 ft.	--
Muskingum	--	x = 0.2 K = 2500
Kinematic Wave	Same as Convex	--

6.13.7. Data Entry Sequence: Translation

```
  ENTER THE STREAM NUMBER OF THE HYDROGRAPH TO BE TRANSLATED
  FORWARD IN TIME (HOURS)
  (ALLOWABLE VALUE IS BETRWEEN [1] AND [5])
=====> 'NA'
  ENTER UPSTREAM NODE NUMBER
=====> 'NUP'
  ENTER DOWNSTREAM NODE NUMBER
=====> 'NDOWN'
  ENTER A [1] = USER SPECIFIED TRANSLATION TIME (HOURS)
     OR A [2] = COMPUTER ESTIMATED TRANSLATION TIME
=====> 'KODE'
```

If (KODE.EQ.1), then

```
  ENTER TRANSLATION TIME (HOURS)
=====> 'DELT'
```

If (KODE.EQ.2), then

```
  ENTER CHANNEL INFORMATION:
  BASE WIDTH (FEET)
=====> 'BB'
  ENTER CHANNEL SIDE SLOPE Z(Z/1 = HORIZONTAL/VERTICAL)
=====> 'Z'
  ENTER UPSTREAM ELEVATION (FEET)
=====> 'E1'
  ENTER DOWNSTREAM ELEVATION < 'E1' FEET
=====> 'E2'
  ENTER CHANNEL LENGTH (FEET)
=====> 'XL'
  ENTER MANNING'S FACTOR
=====> 'XN'
  ENTER PERCENT (DECIMAL) OF PEAK Q TO BE USED FOR WEIGHTING
  CALCULATIONS
  (ALLOWABLE VALUE IS BETWEEN [0.1] AND [1.0])
=====> 'XP1'
  ENTER PERCENTAGE OF WEIGHTING TO BE USED FOR NORMAL DEPTH
  CALCULATIONS (ALLOWABLE VALUE IS BETWEEN ['XP11'] AND [1.0])
=====> 'XP2'
```

NOTE: XP11 = XP1 + 0.01

```
  ENTER A [1] TO INCLUDE INFILTRATION THROUGH CHANNEL BED
=====> 'KINF'
```

If (KINF.EQ.1), then

```
  THE INFILTRATION RATE IS APPROXIMATED BY
  QOUT = QIN + (QO - QIN)*EXP(-k*TIME)
  ENTER QIN (CFS)
=====> 'QIN'
  ENTER QO (CFS)
=====> 'QO'
  ENTER COEFFICIENT k (1/HOUR)
=====> 'COE'
  SELECT PRINT/VIEW OPTIONS:
  ENTER [1] = WHOLE HYDROGRAPH
        [2] = PARTIAL HYDROGRAPH
=====> 'KK'
```

If (KK.EQ.2), then

```
  ENTER MODEL TIME (HOURS) FOR BEGINNING OF RESULTS
=====> 'TIME1'
  ENTER MODEL TIME (HOURS) GREATER THAN ['TIME12']
  FOR END OF RESULTS
=====> 'TIME2'
```

After the hydrograph is displayed, an accept or reject statement will be shown on the CRT screen, as discussed previously.

6.13.8. Data Entry Sequence: Convex Routing

```
  ENTER STREAM NUMBER
  (ALLOWABLE VALUE IS BETWEEN [1] AND [5])
=====> 'NA'
  ENTER THE UPSTREAM NODE NUMBER
=====> 'NUP'
  ENTER THE DOWNSTREAM NODE NUMBER
=====> 'NDOWN'
  ENTER A [1] TO USE COMPUTER ESTIMATED ROUTING COEFFICIENT
  OR A [2] TO ENTER USER SPECIFIED ROUTING COEFFICIENT
=====> 'KOPT'
```

If (KOPT.EQ.1), then

```
  ENTER CHANNEL INFORMATION:  BASE WIDTH (FEET)
=====> 'BB'
  ENTER CHANNEL SIDE SLOPE Z(Z/1 = HORIZONTAL/VERTICAL)
=====> 'Z'
  ENTER UPSTREAM ELEVATION (FEET)
=====> 'E1'
  ENTER DOWNSTREAM ELEVATION < 'E1' FEET
=====> 'E2'
  ENTER CHANNEL LENGTH (FEET)
=====> 'XL'
  ENTER MANNING'S FACTOR
=====> 'X1'
  ENTER A [1] TO INCLUDE THE INFILTRATION THROUGH CHANNEL BED
=====> 'KINF'
```

If (KINF.EQ.1), then

```
  THE INFILTRATION RATE IS APPROXIMATED BY
  QOUT = QIN + (QO - QIN)*EXP(-k*TIME)
  ENTER QIN (CFS)
=====> 'QIN'
  ENTER QO (CFS)
=====> 'QO'
  ENTER COEFFICIENT k (1/HOUR)
=====> 'COE'
  SELECT PRINT/VIEW OPTIONS:
  ENTER [1] = WHOLE HYDROGRAPH
        [2] = PARTIAL HYDROGRAPH
=====> 'KK'
```

If (KK.EQ.2), then

```
  ENTER MODEL TIME (HOURS) FOR BEGINNING OF RESULTS
=====> 'TIME1'
  ENTER MODEL TIME (HOURS) GREATER THAN ['TIME12']
  FOR END OF RESULTS
=====> 'TIME2'
```

After the hydrograph is displayed, an accept or reject statement will be shown on the CRT screen as discussed previously.

6.13.9. Data Entry Sequence: Pipe Routing

```
  ENTER STREAM NUMBER
  (ALLOWABLE VALUE IS BETWEEN [1] AND [5])
=====> 'NA'
  ENTER THE UPSTREAM NODE NUMBER
=====> 'NUP'
  ENTER THE DOWNSTREAM NODE NUMBER
=====> 'NDOWN'
  ENTER PIPE DIAMETER (FEET)
=====> 'XD'
  ENTER UPSTREAM ELEVATION
=====> 'E1'
  ENTER DOWNSTREAM ELEVATION < 'E1' FEET
=====> 'E2'
  ENTER CHANNEL LENGTH (FEET)
=====> 'XL'
  ENTER MANNING'S FACTOR
=====> 'XN'
  SELECT PRINT/VIEW OPTIONS:
  ENTER [1] = WHOLE HYDROGRAPH
        [2] = PARTIAL HYDROGRAPH
=====> 'KK'
```

If (KK.EQ.2), then

```
  ENTER MODEL TIME (HOURS) FOR BEGINNING OF RESULTS
=====> 'TIME1'
  ENTER MODEL TIME (HOURS) GREATER THAN ['TIME1 '] FOR
  END OF RESULTS
=====> 'TIME2'
```

After the computed hydrograph has been displayed, an accept or reject statement will be shown on the CRT screen as previously discussed.

6.13.10. Data Entry Sequence: Kinematic Wave

```
  ENTER STREAM NUMBER OF THE HYDROGRAPH TO BE ROUTED DOWNSTREAM
  BY THE KINEMATIC WAVE METHOD
  (ALLOWABLE VALUE IS BETWEEN [1] AND [5])
=====> 'NA'
  ENTER UPSTREAM NODE NUMBER
=====> 'NUP'
  ENTER DOWNSTREAM NODE NUMBER
=====> 'NDOWN'

  ENTER CHANNEL INFORMATION
  BASE WIDTH (FEET)
=====> 'BB'
```

```
  ENTER CHANNEL SIDE SLOPE Z(Z/1 = HORIZONTAL/VERTICAL)
=====> 'Z'
  ENTER UPSTREAM ELEVATION (FEET)
=====> 'E1'
  ENTER DOWNSTREAM ELEVATION < 'E1' FEET
=====> 'E2'
  ENTER CHANNEL LENGTH (FEET)
=====> 'XL'
  ENTER MANNING'S FACTOR
=====> 'XN'
  SELECT PRINT/VIEW OPTIONS:
  ENTER [1] = WHOLE HYDROGRAPH
        [2] = PARTIAL HYDROGRAPH
=====> 'KK'
```

If (KK.EQ.2), then

```
  ENTER MODEL TIME (HOURS) FOR BEGINNING OF RESULTS
=====> 'TIME1'
  ENTER MODEL TIME (HOURS) GREATER THAN ['TIME1 '] FOR
  END OF RESULTS
=====> 'TIME2'
```

After the hydrograph is displayed, an accept or reject statement will be shown on the CRT screen.

6.13.11. Data Entry Sequence: Muskingum Routing

```
  *** MUSKINGUM ROUTING ***
  ENTER STREAM NUMBER
  (ALLOWABLE VALUE IS BETWEEN [1] AND [5])
=====> 'NA'


  ENTER A [1] TO SKIP THE INTRODUCTION TEXT OR
        A [2] TO REVIEW THE INTRODUCTION TEXT
=====> '1'

  ENTER STORAGE PARAMETER X (0<X<.5):
=====> 'XX'
  ENTER TRANSLATION PARAMETER K IN SECONDS
=====>'XK'
  A SUGGESTED VALUE FOR THE COMPUTATION TIMESTEP, dt, IS BETWEEN
  [TI] AND [TT] SECONDS.
  SELECT dT VALUE OPTION TO BE USED:

      OPTION                 dt(min.)               dt(sec.)

       [1]                    1.                     60.
       [2]                    2.                    120.
```

```
       [3]                 3.                180.
       [4]                 4.5               270.
       [5]                 6.                360.
       [6]                 9.                540.
       [7]                12.                720.
       [8]                15.                900.
       [9]                30.               1800.
      [10]                60.               3600.

  ENTER THE TIMESTEP OPTION
=====>'KOPT'
```

Note: TI is the value that will ensure positive values for routing coefficients.
TT is the minimum value between K and one-fifth the time-to-peak.

```
  ENTER THE UPSTREAM NODE NUMBER
=====>'NUP'
  ENTER THE DOWNSTREAM NODE NUMBER
=====>'NDOWN'
  ENTER A [1] TO INCLUDE THE INFILTRATION THROUGH CHANNEL BED
=====>'KZNF'
```

The submodel for the infiltration rate, the output option, and the accept/reject options, are as discussed previously.

6.14. PROGRAM 6.1

```
C
C----------------------------------------------------------------------
C     THIS SUBROUTINE USES A THREE-MINUTE UNIT EXPLICIT MODEL
C     TO SIMULATE A FLOW-BY BASIN.
C----------------------------------------------------------------------
C
      SUBROUTINE FLOWBY(KRUN,KXY,KRD,KWT)
C
C  VARIABLES:
C        NA=STREAM NUMBER ENTERING THE SYSTEM
C        NB=STREAM NUMBER FOR FLOW EXCESS
C
C  MEMORY ALLOCATION
C
      COMMON/BLK 12/ STREAM(5,1000)
      COMMON/BLK 13/ NUMB(5)
      DIMENSION QS(20),QF(20)
      COMMON/BLK 16/ A(3000),B(3000)
C
      DO 100 I=1,20
      QS(I)=0.
      QF(I)=0.
 100  CONTINUE
      DO 105 I=1,1000
      A(I)=0.
      B(I)=0.
 105  CONTINUE
      DELT=3.
C
C  READ IN STREAM DATA
C
 110  IF(KRUN.NE.1)GO TO 107
      WRITE(KWT,501)
 501  FORMAT(24(/),'    *** FLOW-BY BASIN ***',/,'  ENTER THE STREAM',
     C ' NUMBER THAT ENTER THE FLOW-BY BASIN',/,'  (ALLOWABLE VALUE',
     C ' IS BETWEEN [1] AND [5])')
      WRITE(KWT,99)
 99   FORMAT('  ===== > ',$)
 107  READ(KRD,*)NA
      IF(NA.LT.1 .OR. NA.GT.5)GO TO 110
      NUMBER=NUMB(NA)
      IF(NUMBER.NE.0)GO TO 115
 101  WRITE(KWT,512)NA
 512  FORMAT(/,'  *** STREAM NUMBER ',I2,' IS EMPTY ***',/,
     1 '  ENTER A [1] TO CONTINUE')
      WRITE(KWT,99)
      READ(KRD,*)KX
      IF(KX.NE.1)GO TO 101
```

```
      RETURN
C
 115  IF(KRUN.NE.1)GO TO 117
      WRITE(KWT,513)
 513  FORMAT(/,2X,'ENTER NODE NUMBER')
      WRITE(KWT,99)
 117  READ(KRD,*)INODE
      IF(KRUN.NE.1)GO TO 127
      WRITE(KWT,502)
 502  FORMAT(/,'  ENTER THE STREAM NUMBER THAT STORES THE',
     C ' EXCESS FLOW',/,'  (ALLOWABLE VALUE IS BETWEEN [1] AND [5])')
      WRITE(KWT,99)
 127  READ(KRD,*)NB
      IF(NB.LT.1 .AND. NB.GT.5)GO TO 115
 120  IF(KRUN.NE.1)GO TO 137
      WRITE(KWT,503)
 503  FORMAT(/,'  ENTER (INFLOW,OUTFLOW) INFORMATION :',/,
     C '    THE FIRST SET OF DATA IS DEFINED TO BE (0.,0.)',/,
     C '    ENTER THE NUMBER OF DATA SETS',/,
     C '    (ALLOWABLE VALUE IS BETWEEN [3] AND [20])')
      WRITE(KWT,99)
 137  READ(KRD,*)NSET
      IF(NSET.LT.3 .AND. NSET.GT.20)GO TO 120
      DO 125 I=2,NSET
      IF(KRUN.NE.1)GO TO 147
 138  WRITE(KWT,504)QS(I-1),I
 504  FORMAT(/,'  ENTER INFLOW RATE (CFS) GREATER THAN [',F10.2,
     C '] FOR SET # ',I2)
      WRITE(KWT,99)
 147  READ(KRD,*)QS(I)
      IF(QS(I).LT.QS(I-1))GO TO 138
 130  IF(KRUN.NE.1)GO TO 157
      WRITE(KWT,506)QF(I-1),QS(I)
 506  FORMAT(/,'  ENTER OUTFLOW RATE BETWEEN ',
     C '[',F10.2,'] AND [',F10.2,'] CFS')
      WRITE(KWT,99)
 157  READ(KRD,*)QF(I)
      IF(QF(I).LT.QF(I-1).OR.QF(I).GT.QS(I))GO TO 130
 125  CONTINUE
      NIF=0
      IF(NA.NE.NB)GO TO 145
      IF(KRUN.NE.1)GO TO 167
      WRITE(KWT,507)
 507  FORMAT(/,'  ENTER A [1] TO INCLUDE BASIN INFILTRATION')
      WRITE(KWT,99)
 167  READ(KRD,*)NIF
      IF(NIF.NE.1)GO TO 145
      IF(KRUN.NE.1)GO TO 177
 168  WRITE(KWT,508)
 508  FORMAT(/,'  THE INFILTRATION RATE IS APPROXIMATED BY :',/,
     C '    QOUT = QIN + (QO-QIN)*EXP(-k*TIME)',//,'  ENTER',
     C ' QIN IN CFS')
      WRITE(KWT,99)
 177  READ(KRD,*)QIN
      IF(QIN.LT.0.)GO TO 168
```

```
      IF(KRUN.NE.1)GO TO 187
  178 WRITE(KWT,509)
  509 FORMAT('  ENTER QO IN CFS')
      WRITE(KWT,99)
  187 READ(KRD,*)QO
      IF(QO.LT.QIN)GO TO 178
      IF(KRUN.NE.1)GO TO 197
  188 WRITE(KWT,511)
  511 FORMAT('  ENTER k IN 1/HOUR')
      WRITE(KWT,99)
  197 READ(KRD,*)COE
      IF(COE.LT.0.)GO TO 188
  145 CONTINUE
  155 IF(KRUN.NE.1)GO TO 267
      WRITE(KWT,533)
  533 FORMAT('  *** SELECT PRINT/VIEW OPTIONS: ***'/,
     C '  ENTER A [1] = WHOLE HYDROGRAPH',/,5X,
     C 'OR A [2] = PARTIAL HYDROGRAPH')
      WRITE(KWT,99)
  267 READ(KRD,*)KK
      IF(KK.EQ.1)GO TO 160
      IF(KK.EQ.2)GO TO 170
      GO TO 155
  160 TIME1=0.
      TIME2=100.
      GO TO 180
  170 IF(KRUN.NE.1)GO TO 277
      WRITE(KWT,534)
  534 FORMAT(/,'  ENTER MODEL TIME(HOURS) FOR BEGINNING OF',
     C ' RESULTS')
      WRITE(KWT,99)
  277 READ(KRD,*)TIME1
      TIME12=TIME1+.05
      IF(KRUN.NE.1)GO TO 227
      WRITE(KWT,536)TIME12
  536 FORMAT(/,'  ENTER MODEL TIME(HOURS) GREATER THAN [ ',F8.3,
     C ' ] FOR END OF RESULTS')
      WRITE(KWT,99)
  227 READ(KRD,*)TIME2
      IF(TIME2 .LE. TIME12)GO TO 170
  180 CONTINUE
C
      CALL MREAD(NA,A)
      IF(NB.NE.NA)CALL MREAD(NB,B)
C
C  GRAPHICS
C
C  INITIALIZE VARIABLES
C
      TIME=0.
      STORE=0.
C
C  WRITE MODEL DATA
C
      WRITE(KWT,201)INODE,NA
```

```
 201  FORMAT(//,5X,'*** FLOW PROCESS AT NODE',I4,/,
     C 10X,'MODEL STREAM NUMBER',I2,' FLOWING PAST A',
     C ' FLOW-BY STRUCTURE:')
      IF(NB.EQ.NA)WRITE(KWT,202)
 202  FORMAT(10X,'FLOW EXCESS IS ASSUMED TO BE PERMANENTLY STORED.'/)
      IF(NB.NE.NA)WRITE(KWT,203)NB
 203  FORMAT(10X,'FLOW EXCESS IS ASSUMED TO BE ADDED TO STREAM '
     C ,'NUMBER',I2,//)
      IF(NB.EQ.NA)WRITE(KWT,221)NA,NA
 221  FORMAT(32X,'INFLOW',/,30X,'(STREAM',I2,')',/,
     C 3(35X,'|',/),21X,' ---------     |',/,
     C 21X,'|',9X,'|',3X,'|',/,21X,'|',9X,'|<--* <=flow-by structure',/,
     C 21X,'|  basin   |'/,
     C 21X,'| storage |    |',/,21X,' ---------     |',/,
     C 2(35X,'|',/),35X,'V',/,30X,' STREAM',I2,/,32X,'FLOW-BY',///)
      IF(NB.NE.NA)WRITE(KWT,223)NB,NA,NB,NA
 223  FORMAT(20X,'    INFLOW                INFLOW',/,
     C 20X,'| (STREAM',I2,')          (STREAM',I2,')',/,
     C 27X,'|',19X,'|',/,27X,'|',19X,'|',/,
     C 27X,'|',4X,'flow excess',4X,'|',/,
     C 27X,'|<------------------* <=flow-by structure',/,
     C 27X,'|',19X,'|'/,
     C 27X,'|',19X,'|',/,27X,'|',19X,'|',/,27X,'|',19X,'|',/,
     C 27X,'V',19X,'V',/,24X,'STREAM',I2,12X,'STREAM',I2,/,
     C 20X,'+ FLOW EXCESS',12X,'FLOW-BY',//)
      WRITE(KWT,211)
 211  FORMAT(/,5X,'INFLOW-VS.-OUTFLOW INFORMATION :',/,
     C 10X,'INFLOW (CFS)     OUTFLOW(CFS)')
      WRITE(KWT,212)(QS(I),QF(I),I=1,NSET)
 212  FORMAT(12X,F9.3,8X,F9.3)
      IF(NIF.EQ.1)WRITE(KWT,214)QO,QIN,QO,COE
 214  FORMAT(/,'  BASIN INFILTRATION RATE IS APPROXIMATED BY :',/,
     C '   QOUT (CFS) = ',F6.2,' + ( ',F6.2,' - ',F6.2,' ) * '
     C 'EXP ( -1. * TIME(HOURS) * ',F5.3)
      IF(NB.EQ.NA)WRITE(KWT,208)NA,NA
 208  FORMAT(//,10X,'FLOW-BY BASIN MODELING RESULTS:',//,
     C 10X,'  MODEL  STREAM',I2,'  STREAM',I2,'      BASIN',/,
     C 10X,'   TIME   INFLOW    FLOW-BY       VOLUME',/,
     C 10X,'  (HRS)    (CFS)     (CFS)        (AF)')
      IF(NB.NE.NA)WRITE(KWT,206)NB,NA,NB,NA
 206  FORMAT(//,10X,'FLOW-BY BASIN MODELING RESULTS:',//,
     C 10X,' MODEL         INFLOW     INFLOW    OUTFLOW     FLOW-BY',/,
     C 10X,'  TIME   ',4(' (STREAM',I2,')'),/,
     C 10X,' (HRS)  ',4X,4('   (CFS)  '))
C
C  MODEL FLOW-BY EFFECTS
C
 237  QOUT=0.
      IF(NB.NE.NA)GOTO 200
C
C  MODEL DEAD STORAGE
C
      DO 210 I=1,NUMBER
      TIME=TIME+0.05
      Z=A(I)
```

```
      IF(NIF.EQ.1)QOUT=QIN+(QO-QIN)*EXP(-1.*COE*TIME)
      CALL SEE1(Z,QS,Q1,Q2,I1,I2,NSET,KRUN,TIME1,TIME2,TIME,KWT)
      QCAP=QF(I1)+(Z-Q1)/(Q2-Q1)*(QF(I2)-QF(I1))
      X=Z-QCAP-QOUT
      IF(X.LT.0.)X=0.
      STORE=STORE+X/242.
      IF(STORE.LT.0.)STORE=0.
      A(I)=QCAP
      IF(TIME.LT.TIME1 .OR. TIME.GT.TIME2)GO TO 210
      WRITE(KWT,207)TIME,Z,A(I),STORE
 207  FORMAT(10X,F7.3,F9.1,F10.1,F13.3)
 210  CONTINUE
      GOTO 300
 200  DO 220 I=1,NUMBER
      TIME=TIME+0.05
      Z=A(I)
      ZB=B(I)
      CALL SEE1(Z,QS,Q1,Q2,I1,I2,NSET,KRUN,TIME1,TIME2,TIME,KWT)
      QCAP=QF(I1)+(Z-Q1)/(Q2-Q1)*(QF(I2)-QF(I1))
      X=Z-QCAP
      IF(X)220,250,250
 250  B(I)=B(I)+X
      A(I)=QCAP
 230  IF(TIME.LT.TIME1 .OR. TIME.GT.TIME2)GO TO 220
      WRITE(KWT,209)TIME,ZB,Z,B(I),A(I)
 209  FORMAT(10X,F7.3,3X,4F10.1)
 220  CONTINUE
 300  CONTINUE
      IF(KRUN.NE.1)GO TO 247
 310  WRITE(KWT,537)
 537  FORMAT(/,2X,'ENTER A [1] TO ACCEPT THE PROCESS',/,
     C 5X,'OR A [2] TO REJECT THE PROCESS')
      WRITE(KWT,99)
      READ(KRD,*)OPT
      IF(OPT.LT.1 .OR. OPT.GT.2)GO TO 310
      IF(OPT.EQ.2)RETURN
C
      WRITE(7,*)KXY
      WRITE(7,*)NA
      WRITE(7,*)INODE
      WRITE(7,*)NB
      WRITE(7,*)NSET
      DO 260 I=2,NSET
      WRITE(7,*)QS(I)
      WRITE(7,*)QF(I)
 260  CONTINUE
      WRITE(7,*)NIF
      IF(NIF.EQ.1)WRITE(7,*)QIN
      IF(NIF.EQ.1)WRITE(7,*)QO
      IF(NIF.EQ.1)WRITE(7,*)COE
      WRITE(7,*)KK
      IF(KK.EQ.2)WRITE(7,*)TIME1
      IF(KK.EQ.2)WRITE(7,*)TIME2
C
 247  CALL MWRITE(NA,A)
```

```
      IF(NB.NE.NA)CALL MWRITE(NB,B)
      IF(NB.NE.NA)NUMB(NB)=NUMBER
C
      RETURN
      END
C
C-----------------------------------------------------------------
C     THIS SUBROUTINE DETERMINES THE BASIN LIMITATION
C-----------------------------------------------------------------
C
      SUBROUTINE SEE1(X,Y,D1,D2,I1,I2,NB,KRUN,TIME1,TIME2,TIME,KWT)
C
      DIMENSION Y(20)
C
      DO 100 K=2,NB
      IF(X.LT.Y(K))GOTO 200
 100  CONTINUE
      IF(TIME.LT.TIME1 .OR. TIME.GT.TIME2)GO TO 110
      WRITE(KWT,201)
 201  FORMAT(3X,'NOTE: BASIN LIMITATIONS ARE EXCEEDED.',/,
     C 24X,'COMPUTED VALUES ARE EXTRAPOLATIONS')
 110  K=NB
 200  I1=K-1
      I2=K
      D1=Y(I1)
      D2=Y(I2)
C
      RETURN
      END
C
C-----------------------------------------------------------------
C     THIS SUBROUTINE DETERMINES RESERVOIR-ROUTED HYDROGRAPH
C-----------------------------------------------------------------
C
      SUBROUTINE FLOWTR(KRUN,KXY,KRD,KWT)
C
      DIMENSION S(20),O(20),D(20)
      COMMON/BLK 12/ STREAM(5,1000)
      COMMON/BLK 13/ NUMB(5)
      COMMON/BLK 16/ A(3000),B(3000)
C
      DO 10 I=1,20
      S(I)=0.
      O(I)=0.
      D(I)=0.
 10   CONTINUE
      DO 20 K=1,1000
 20   A(K)=0.
      DELT=3.
C
C  DATA INPUT
C
      IF(KRUN.NE.1)GO TO 107
      WRITE(KWT,501)
 501  FORMAT(24(/),76('='),/,'   STORAGE HYDROGRAPH ROUTING',
```

```
     1' IN A CONTROLLED RESERVOIR')
      WRITE(KWT,505)
 505  FORMAT(4('==================='))
      WRITE(KWT,504)
 504  FORMAT(//)
      WRITE(KWT,508)
 508  FORMAT(4(':::::::::::::::::::'))
      WRITE(KWT,502)
 502  FORMAT('  ENTER RESERVOIR INFLOW HYDROGRAPH INFORMATION:')
      WRITE(KWT,506)
 506  FORMAT(4('-------------------'))
 30   WRITE(KWT,513)
 513  FORMAT(/,'  ENTER STREAM NUMBER OF THE INFLOW HYDROGRAPH',/
     1 ,'  (ALLOWABLE VALUE IS BETWEEN [1] AND [5])')
      WRITE(KWT,99)
 99   FORMAT('  =====> ',$)
 107  READ(KRD,*)NN
      IF(NN.LT.1 .OR. NN.GT.5)GO TO 30
      NUMBER=NUMB(NN)
      IF(NUMBER.NE.0)GO TO 40
 109  WRITE(KWT,516)NN
 516  FORMAT(/,'  *** STREAM NUMBER ',I2,' IS EMPTY ***',/,
     1 '  ENTER A [1] TO CONTINUE')
      READ(KRD,*)KX
      IF(KX.NE.1)GO TO 109
      RETURN
C
 40   IF(KRUN.NE.1)GO TO 117
      WRITE(KWT,529)
 529  FORMAT(/,2X,'*** ENTER THE UPSTREAM NODE NUMBER')
      WRITE(KWT,99)
 117  READ(KRD,*)IUP
      IF(KRUN.NE.1)GO TO 127
      WRITE(KWT,511)
 511  FORMAT(/,2X,'*** ENTER THE DOWNSTREAM NODE NUMBER')
      WRITE(KWT,99)
 127  READ(KRD,*)IDOWN
      CALL MREAD(NN,A)
 50   IF(KRUN.NE.1)GO TO 137
      WRITE(KWT,504)
      WRITE(KWT,505)
      WRITE(KWT,512)
 512  FORMAT('  ENTER BASIN DEPTH-VERSUS-STORAGE AND-DISCHARGE '
     1 ,'INFORMATION:')
      WRITE(KWT,506)
 70   WRITE(KWT,504)
      WRITE(KWT,514)
 514  FORMAT('  ENTER TOTAL NUMBER OF DEPTH-VS.-STORAGE DATA ENTRIES:'
     1 /,'  (ALLOWABLE VALUES ARE BETWEEN [3] AND [20])')
      WRITE(KWT,99)
 137  READ(KRD,*)NUMD
      IF(NUMD.GE.3.AND.NUMD.LE.20)GOTO 80
      IF(KRUN.EQ.1)WRITE(KWT,503)
 503  FORMAT(2X,'*** INFORMATION IS UNACCEPTABLE:  TRY AGAIN.')
      GOTO 70
```

```
80    IF(KRUN.NE.1)GO TO 147
      WRITE(KWT,504)
      WRITE(KWT,508)
      WRITE(KWT,515)
515   FORMAT(/,'  ENTER DEPTH-VS.-STORAGE AND DEPTH-VS.-DISCHARGE'
     1 ,' INFORMATION:')
      WRITE(KWT,508)
90    WRITE(KWT,504)
      WRITE(KWT,517)
517   FORMAT('   *FOR DEPTH-VS.-STORAGE DATA ENTRY #1,'
     1 /,'   *BOTTOM OF BASIN ASSUMED AS DEPTH(FEET) =  0.'
     1 /,'   ENTER BASIN STORAGE(ACRE-FEET) AT DEPTH(FEET) =  0. :'
     1 /,'   (ALLOWABLE VALUES ARE BETWEEN [0] AND [100000])')
      WRITE(KWT,99)
147   READ(KRD,*)S(1)
      IF(S(1).GE.0..AND.S(1).LE.100000.)GOTO 100
      IF(KRUN.EQ.1)WRITE(KWT,503)
      GOTO 90
100   IF(KRUN.EQ.1)WRITE(KWT,518)
518   FORMAT('  *BASIN OUTFLOW AT DEPTH = 0 FOOT ASSUMED AS  0.0 CFS')
      J=1
      O(1)=0.
      D(1)=0.
200   K=J+1
      IF(K.GT.NUMD)GOTO 210
      IF(KRUN.EQ.1)WRITE(KWT,504)
      DX=D(J)+.05
      IF(KRUN.NE.1)GO TO 157
      WRITE(KWT,528)K,DX
528   FORMAT(3X,'ENTER BASIN DEPTH(FEET) FOR DATA ENTRY #',I3,':',/,
     C 3X,'(ALLOWABLE VALUES ARE BETWEEN [',F6.2,'] AND [200])',/,
     C 3X,'NOTE: ENTER [-1] TO RESTART BASIN INFORMATION SEQUENCE')
      WRITE(KWT,99)
157   READ(KRD,*)D(K)
      IF(D(K).EQ.-1.)GOTO 50
      IF(D(K).GE.DX.AND.D(K).LE.200.)GOTO 220
      IF(KRUN.EQ.1)WRITE(KWT,503)
      GOTO 200
220   SX=S(J)*1.0001
      IF(KRUN.NE.1)GO TO 167
      WRITE(KWT,519)D(K),SX
519   FORMAT(3X,'ENTER BASIN STORAGE(ACRE-FEET) AT DEPTH(FEET) = ',
     C F7.2,' :',/,3X,'(ALLOWABLE VALUES ARE BETWEEN [',F12.5,
     C '] AND [100000])')
      WRITE(KWT,99)
167   READ(KRD,*)S(K)
      IF(S(K).GE.SX.AND.S(K).LE.100000.)GOTO 230
      WRITE(KWT,503)
      GOTO 220
230   OX=O(J)*1.0001
      IF(KRUN.NE.1)GO TO 177
      WRITE(KWT,520)D(K),OX
520   FORMAT(3X,'ENTER BASIN OUTFLOW(CFS) AT DEPTH(FEET) = ',F7.2,' :',
     C /,3X,'(ALLOWABLE VALUES ARE BETWEEN [',F12.5,'] AND [100000])')
      WRITE(KWT,99)
```

```
177  READ(KRD,*)O(K)
     IF(O(K).GE.OX.AND.O(K).LE.100000.)GOTO 240
     WRITE(KWT,503)
     GOTO 230
240  J=K
     GOTO 200
210  CONTINUE
250  IF(KRUN.NE.1)GO TO 187
     WRITE(KWT,508)
     WRITE(KWT,504)
     WRITE(KWT,521)
521  FORMAT('  ENTER ASSUMED INITIAL DEPTH(FEET) OF WATER IN BASIN'
    1 ,/,'  AT BEGINNING OF INFLOW HYDROGRAPH INTO BASIN:')
     WRITE(KWT,531)D(NUMD)
531  FORMAT('  (ALLOWABLE VALUES ARE BETWEEN [0] AND [',F7.2,'])')
     WRITE(KWT,99)
187  READ(KRD,*)DO
     IF(DO.GE.0..AND.DO.LE.D(NUMD))GOTO 260
     IF(KRUN.EQ.1)WRITE(KWT,503)
     GOTO 250
260  CONTINUE
280  CONTINUE
     IF(KRUN.NE.1)GO TO 207
     WRITE(KWT,508)
     WRITE(KWT,523)
523  FORMAT('  ENTER DEAD STORAGE (ACRE-FEET) FOR RESERVOIR')
     WRITE(KWT,99)
207  READ(KRD,*)QD
     IF(KRUN.NE.1)GO TO 317
     WRITE(KWT,524)
524  FORMAT(/,'  ENTER [1] TO INCLUDE THE INFILTRATION SUBMODEL')
     WRITE(KWT,99)
317  READ(KRD,*)KKK
     IF(KKK.NE.1)GO TO 655
     IF(KRUN.NE.1)GO TO 227
318  WRITE(KWT,525)
525  FORMAT('  THE INFILTRATION RATE IS APPROXIMATED BY :',
    1 /,'   QOUT = QIN + (QO-QIN)*EXP(-k*TIME) ',//,
    2 '  ENTER QIN IN CFS')
     WRITE(KWT,99)
227  READ(KRD,*)QIN
     IF(QIN.LT.0.)GO TO 318
     IF(KRUN.NE.1)GO TO 237
328  WRITE(KWT,526)
526  FORMAT('  ENTER QO IN CFS')
     WRITE(KWT,99)
237  READ(KRD,*)QO
     IF(QO.LT.QIN)GO TO 328
     IF(KRUN.NE.1)GO TO 247
428  WRITE(KWT,527)
527  FORMAT('  ENTER COEFFICIENT k IN 1/HOUR')
     WRITE(KWT,99)
247  READ(KRD,*)COE
     IF(COE.LT.0.)GO TO 428
655  IF(KRUN.NE.1)GO TO 607
```

```
      WRITE(KWT,601)
 601  FORMAT(/,'  *** SELECT PRINT/VIEW OPTIONS: ***',/,
     C 2X,'ENTER A [1] = WHOLE HYDROGRAPH',/,5X,
     C 'OR A [2] = PARTIAL HYDROGRAPH')
      WRITE(KWT,99)
 607  READ(KRD,*)KK
      IF(KK.EQ.1)GO TO 660
      IF(KK.EQ.2)GO TO 670
      GO TO 655
 660  TIME1=0.
      TIME2=100.
      GO TO 680
 670  IF(KRUN.NE.1)GO TO 617
      WRITE(KWT,602)
 602  FORMAT(/,'  ENTER MODEL TIME(HOURS) FOR BEGINNING OF',
     C ' RESULTS')
      WRITE(KWT,99)
 617  READ(KRD,*)TIME1
      TIME12=TIME1+.05
      IF(KRUN.NE.1)GO TO 627
      WRITE(KWT,603)TIME12
 603  FORMAT(/,'  ENTER MODEL TIME(HOURS) GREATER THAN [ ',F8.3,
     C ' ] FOR END OF RESULTS')
      WRITE(KWT,99)
 627  READ(KRD,*)TIME2
      IF(TIME2 .LE. TIME12)GO TO 670
 680  CONTINUE
C
C  DATA OUTPUT
C
      WRITE(KWT,507)
 507  FORMAT(4('******************'))
      WRITE(KWT,211)IUP,IDOWN
 211  FORMAT(/,5X,'*** FLOW PROCESS FROM NODE',I4,' TO NODE',I4,
     C /,5X,'STORAGE BASIN HYDROGRAPH ROUTING MODEL')
      WRITE(KWT,505)
      IF(KRUN.EQ.1)GO TO 300
      WRITE(KWT,212)
 212  FORMAT(5X,'ENTERED INFORMATION:')
      WRITE(KWT,506)
      WRITE(KWT,213)NN,DELT,DO
 213  FORMAT(5X,'USE STREAM #',I2,' AS THE INFLOW HYDROGRAPH ',/,
     C 5X,'CONSTANT HYDROGRAPH TIME UNIT(MINUTES) = ',F8.3,/,
     C 5X,'ASSUMED INITIAL DEPTH(FEET) IN STORAGE BASIN = ',F7.2)
      WRITE(KWT,505)
      WRITE(KWT,216)
 216  FORMAT(5X,'DEPTH-VS.-STORAGE AND DEPTH-VS.-DISCHARGE ',
     C 'INFORMATION:')
      WRITE(KWT,506)
      WRITE(KWT,217)NUMD
 217  FORMAT(5X,'TOTAL NUMBER OF BASIN DEPTH INFORMATION ENTRIES = ',
     C I3)
      WRITE(KWT,506)
      WRITE(KWT,218)
 218  FORMAT(5X,2('*BASIN-DEPTH  STORAGE    OUTFLOW  *'),/,5X,
```

```
     C 2('*   (FEET)  (ACRE-FEET)   (CFS)   *'))
      WRITE(KWT,219)(D(J),S(J),O(J),J=1,NUMD)
 219  FORMAT(5X,'*',3F11.3,'**',3F11.3,'*')
      WRITE(KWT,507)
C
C  RESERVOIR ROUTING
C
 300  CONTINUE
      CALL BASIN(NUMBER,NUMD,DELT,DO,S,D,O,QO,QIN,COE,QD,
     C KKK,KRUN,TIME1,TIME2,KWT)
      IF(KRUN.NE.1)GO TO 257
 310  WRITE(KWT,534)
 534  FORMAT(/,2X,'ENTER A [1] TO ACCEPT THE PROCESS',/,
     C 5X,'OR A [2] TO REJECT THE PROCESS')
      WRITE(KWT,99)
      READ(KRD,*)KOP
      IF(KOP.LT.1 .OR. KOP.GT.2)GO TO 310
      IF(KOP.EQ.2)RETURN
C
      WRITE(7,*)KXY
      WRITE(7,*)NN
      WRITE(7,*)IUP
      WRITE(7,*)IDOWN
      WRITE(7,*)NUMD
      WRITE(7,*)S(1)
      DO 330 K=2,NUMD
      WRITE(7,*)D(K)
      WRITE(7,*)S(K)
      WRITE(7,*)O(K)
 330  CONTINUE
      WRITE(7,*)DO
      WRITE(7,*)QD
      WRITE(7,*)KKK
      IF(KKK.EQ.1)WRITE(7,*)QIN
      IF(KKK.EQ.1)WRITE(7,*)QO
      IF(KKK.EQ.1)WRITE(7,*)COE
      WRITE(7,*)KK
      IF(KK.EQ.2)WRITE(7,*)TIME1
      IF(KK.EQ.2)WRITE(7,*)TIME2
C
 257  CALL MWRITE(NN,A)
      RETURN
      END
C
C------------------------------------------------------------------
C     THIS SUBROUTINE ROUTS HYDROGRAPH THRU BASIN
C------------------------------------------------------------------
C
      SUBROUTINE BASIN(NUMBER,NUMD,DELT,DO,S,D,O,QO,QIN,
     C COE,QD,INF,KRUN,TIME1,TIME2,KWT)
      DIMENSION AA(20),BB(20),S(20),O(20),D(20)
      COMMON/BLK 16/ A(3000),B(3000)
C
      INTEGER BLANK,DOT,CROSS,DASH,LINE(33)
      DATA BLANK,DOT,CROSS,DASH,LINE/' ','.','I','O',33*' '/
```

```
C
      DO 90 I=1,1000
 90   B(I)=0.
C
C  CONSTRUCT S+-O/2DELT CURVES
C
C   USE AA-VECTOR TO STORE (S)-(O)*DELT/2
C   USE BB-VECTOR TO STORE (S)+(O)*DELT/2
C
      CON=60./43560.*DELT/2.
      DO 100 K=1,NUMD
      AA(K)=S(K)-O(K)*CON
 100  BB(K)=S(K)+O(K)*CON
C
C  GET INITIAL VALUES
C
      CALL SEE(DO,D,D1,D2,I1,I2,NUMD,KRUN,KWT)
      O2=0.
      S2=0.
      S1=S(I1)+(DO-D1)/(D2-D1)*(S(I2)-S(I1))
      O1=O(I1)+(DO-D1)/(D2-D1)*(O(I2)-O(I1))
      IF(S1.GT.0.)S1=S1+QD
      QA=QD
      WRITE(KWT,202)
 202  FORMAT(4('------------------'))
      WRITE(KWT,203)DO,S1,O1,QD
 203  FORMAT(5X,'INITIAL BASIN DEPTH(FEET) = ',F7.2,/,
     C 5X,'INITIAL BASIN STORAGE(ACRE-FEET) = ',F10.2,/,
     C 5X,'INITIAL BASIN OUTFLOW(CFS) = ',F10.2,/,
     C 5X,'DEAD STORAGE(ACRE-FEET) = ',F10.2)
      WRITE(KWT,202)
      IF(INF.EQ.1)WRITE(KWT,204)QIN,QO,QIN,COE
 204  FORMAT(/,5X,'BASIN INFILTRATION RATE IS APPROXIMATED BY :',/,
     C 2X,'QOUT (CFS) = ',F6.2,' + (',F6.2,' - ',F6.2,') * EXP ( ',
     C '-1. * TIME(HOURS) *',F6.3,')',/)
C
C  SET AA(1)=BB(1)
C
 155  AA(1)=BB(1)
      WRITE(KWT,202)
      WRITE(KWT,201)
 201  FORMAT(5X,'BASIN STORAGE, OUTFLOW AND DEPTH ROUTING VALUES:',/,
     C 5X,' INTERVAL   DEPTH   (S-O*DT/2)   (S+O*DT/2)',/,5X,
     C '  NUMBER    (FEET)   (ACRE-FEET)  (ACRE-FEET)')
      WRITE(KWT,222)(K,D(K),AA(K),BB(K),K=1,NUMD)
 222  FORMAT(5X,I7,3X,F6.2,2X,F12.5,1X,F12.5)
      WRITE(KWT,223)
 223  FORMAT(5X,'WHERE S=STORAGE(AF);O=OUTFLOW(AF/MIN.);DT=UNIT',
     C ' INTERVAL(MIN.)')
      WRITE(KWT,202)
 137  CON=CON*2.
      IF(S1.GT.0.)S1=S1-QD
      ATEMP=S1-O1*CON/2.
 147  K=0
      DEPTH1=DO
```

```
C
C  FIND XMAX
C
      XMAX=0.
      DO 150 K=1,NUMBER
      IF(A(K).GT.XMAX)XMAX=A(K)
 150  CONTINUE
      IF(XMAX.GT.0.)F=32./XMAX
      TO=0.
      T1=XMAX/4.
      T2=2.*T1
      T3=3.*T1
      T4=4.*T1
      WRITE(KWT,202)
      WRITE(KWT,207)
 207  FORMAT(//,10X,'*UNIT-HYDROGRAPH STORAGE-BASIN ROUTING*',/,
     C 5X,'NOTE: COMPUTED BASIN DEPTH, OUTFLOW, AND STORAGE ',
     C 'QUANTITIES',/,11X,'OCCUR AT THE GIVEN TIME. BASIN INFLOW ',
     C 'VALUES REPRESENT THE',/,11X,'AVERAGE INFLOW DURING THE ',
     C 'RECENT HYDROGRAPH UNIT INTERVAL.'/,5X,'GRAPH NOTATION: ',
     C '"I"=MEAN UNIT INFLOW; "O"=OUTFLOW AT GIVEN TIME')
      WRITE(KWT,202)
      WRITE(KWT,208)TO,T1,T2,T3,T4
 208  FORMAT(6X,'TIME',3X,'INFLOW',2X,'OUTFLOW',4X,'STORAGE',/
     C 5X,'(HOURS)',1X,'(CFS)',4X,'(CFS)',5X,' ACRE-FT',1X,F2.1,4F8.0)
      K=0
      IF(QD.EQ.0.)QRE=0.
      IF(QD.NE.0.)QRE=QD-S1
      IF(QRE.LT.0.)QRE=0.
      QA=QD-QRE
      S2=S1
 200  K=K+1
      IF(K .GT. 1000)GOTO 550
      XK=K
      TIME=XK*DELT/60.
      QQ=0.
      IF(K.LT.NUMBER)QQ=0.5*CON*(A(K)+A(K+1))
      IF(K.EQ.NUMBER)QQ=0.5*CON*A(K)
      IF(K.GT.NUMBER)A(K)=0.
      IF(INF.NE.1 .AND. QD.NE.0.)GO TO 210
      IF(INF.EQ.1 .AND. QD.EQ.0.)GO TO 240
      GO TO 220
 210  IF(QRE.EQ.0.)GO TO 220
      IF(QRE.LT.QQ)GO TO 230
      QRE=QRE-QQ
      QQ=0.
      O2=0.
      S2=0.
      DEPTH2=0.
      GO TO 400
 230  QQ=QQ-QRE
      QRE=0.
 220  TEMP=QQ+ATEMP
      IF(INF.NE.1)GO TO 300
C
```

```
C  RESERVOIR INFILTRATION
C
 240  CONTINUE
      QOUT=QIN+(QO-QIN)*EXP(-1.*TIME*COE)
      QQ=QQ-QOUT*CON
      TEMP=QQ+ATEMP
      IF(QQ.LE.0.)GO TO 310
      IF(QA.EQ.QD)GO TO 300
      IF(QQ.GT.QD-QA)GO TO 320
      IF(QQ.LE.QD-QA)GO TO 330
C
C  INFLOW INTO RESERVOIR AND DEAD STORAGE IS FULL
C
 320  QQ=QQ-(QD-QA)
      QA=QD
      TEMP=QQ+ATEMP
      GO TO 300
C
C  INFLOW INTO RESERVOIR AND DEAD STORAGE IS NOT FULL
C
 330  QA=QA+QQ
      S2=0.
      02=0.
      DEPTH2=0.
      GO TO 400
C
C  DISCHARGE FROM RESERVOIR
C
 310  IF(TEMP.GT.0.)GO TO 300
      IF(S22.GT.QA)GO TO 340
C
C  DISCHARGE FROM DEAD STORAGE
C
      S2=0.
      02=0.
      DEPTH2=0.
      IF(QA.EQ.0.)GO TO 400
      IF(QA.GT.ABS(QQ))QA=QA+QQ
      IF(QA.LE.ABS(QQ))QA=0.
      GO TO 400
C
C  DISCHARGE FROM BOTH LIVE AND DEAD STORAGES
C
 340  QA=S22+QQ
      S2=0.
      02=0.
      DEPTH2=0.
      GO TO 400
 300  CONTINUE
      CALL SEE(TEMP,BB,B1,B2,I1,I2,NUMD,KRUN,KWT)
      DEPTH2=D(I1)+(TEMP-B1)/(B2-B1)*(D(I2)-D(I1))
      S2=S(I1)+(DEPTH2-D(I1))/(D(I2)-D(I1))*(S(I2)-S(I1))
      02=0(I1)+(DEPTH2-D(I1))/(D(I2)-D(I1))*(0(I2)-0(I1))
 400  ATEMP=S2-02*CON/2.
      IF(ATEMP.LT.0.)ATEMP=0.
```

```
      TEST=F*A(K)
      TEST1=F*02
      LINE(1)=DOT
      LINE(9)=DOT
      LINE(17)=DOT
      LINE(25)=DOT
      LINE(33)=DOT
      JJ=TEST+1
      LINE(JJ)=CROSS
      JJJ=TEST1+1
      LINE(JJJ)=DASH
      IF(INF.NE.1 .AND. QD.NE.0.)GO TO 430
      GO TO 410
 430  IF(QRE.EQ.0.)S22=S2+QD
      IF(QRE.NE.0.)S22=QD-QRE
      GO TO 420
 410  IF(S2.EQ.0.)S22=QA
      IF(S2.NE.0.)S22=S2+QD
 420  IF(TIME.LT.TIME1 .OR. TIME.GT.TIME2)GO TO 500
      WRITE(KWT,209)TIME,A(K),02,S22,LINE
 209  FORMAT(5X,F7.2,2F9.1,F12.3,1X,33A1)
      WRITE(KWT,211)DEPTH2
 211  FORMAT(10X,'[BASIN DEPTH(FEET) = ',F6.1,']')
 500  B(K)=02
      LINE(JJ)=BLANK
      LINE(JJJ)=BLANK
      DEPTH1=DEPTH2
      GOTO 200
 550  WRITE(KWT,202)
      IF(KRUN.EQ.1)WRITE(KWT,202)
C
      DO 600 I=1,1000
 600  A(I)=B(I)
C
      RETURN
      END
C
C------------------------------------------------------------------------
C     THIS SUBROUTINE DETERMINES THE BASIN LIMITATION
C------------------------------------------------------------------------
C
      SUBROUTINE SEE(X,Y,D1,D2,I1,I2,NB,KRUN,KWT)
C
      DIMENSION Y(20)
C
      DO 100 K=2,NB
      IF(X.LT.Y(K))GOTO 200
 100  CONTINUE
      WRITE(KWT,201)
 201  FORMAT(3X,'NOTE: BASIN LIMITATIONS ARE EXCEEDED.',/,
     C 24X,'COMPUTED VALUES ARE EXTRAPOLATIONS')
      K=NB
 200  I1=K-1
      I2=K
      D1=Y(I1)
```

```
      D2=Y(I2)
C
      RETURN
      END
```

6.15. PROGRAM 6.2

```
C
C-----------------------------------------------------------------------
C     THIS SUBROUTINE MODELS CHANNEL ROUTING BY THE SIMPLER CONVEX METHOD
C     WHERE A CSTAR VALUE AS ESTIMATED DUE TO IRREGULAR DELT VALUES
C-----------------------------------------------------------------------
C
      SUBROUTINE CONVEX(KRUN,KXY,KRD,KWT)
C
      COMMON/BLK 12/ STREAM(5,1000)
      COMMON/BLK 13/ NUMB(5)
      COMMON/BLK 16/ A(3000),B(3000)
C
      DO 100 I=1,1000
      A(I)=0.
      B(I)=0.
 100  CONTINUE
C
C  INITIALIZE VARIABLES
C
      TIME=0.
      C=0.
      VO=0.
C
C  INPUT DATA
C
 105  IF(KRUN.NE.1)GO TO 107
      WRITE(KWT,501)
 501  FORMAT(24(/),'  *** CONVEX ROUTING ***',/,
     C 2X,'ENTER STREAM NUMBER',/,2X,'(ALLOWABLE VALUE'
     C ,' IS BETWEEN [1] AND [5])')
      WRITE(KWT,99)
 99   FORMAT('  =====> ',$)
 107  READ(KRD,*)NA
      IF(NA.LT.1 .OR. NA.GT.5)GO TO 105
      IF(NUMB(NA).NE.0)GO TO 120
 108  IF(KRUN.EQ.1)WRITE(KWT,518)NA
 518  FORMAT(/,'  *** STREAM NUMBER ',I2,' IS EMPTY ***',/,
     C '  ENTER A [1] TO CONTINUE')
      WRITE(KWT,99)
      READ(KRD,*)KX
      IF(KX.NE.1)GO TO 108
      RETURN
C
 120  IF(KRUN.NE.1)GO TO 117
      WRITE(KWT,519)
 519  FORMAT(/,2X,'ENTER THE UPSTREAM NODE NUMBER ')
      WRITE(KWT,99)
 117  READ(KRD,*)NUP
      IF(KRUN.NE.1)GO TO 127
      WRITE(KWT,521)
 521  FORMAT(/,2X,'ENTER THE DOWNSTREAM NODE NUMBER')
      WRITE(KWT,99)
 127  READ(KRD,*)NDOWN
      IF(KRUN.NE.1)GO TO 137
```

```
      WRITE(KWT,502)
502   FORMAT(/,2X,'ENTER A [1] TO USE COMPUTER ESTIMATED',
     C ' ROUTING COEFFICIENT',/,5X,'OR A [2] TO ENTER USER',
     C ' SPECIFIED ROUTING COEFFICIENT')
      WRITE(KWT,99)
137   READ(KRD,*)KOPT
      IF(KOPT.EQ.1)GO TO 115
      IF(KOPT.EQ.2)GO TO 110
      GO TO 120
110   IF(KRUN.NE.1)GO TO 147
138   WRITE(KWT,503)
503   FORMAT(/,2X,'ENTER USER SPECIFIED CHANNEL ROUTING ',
     C 'COEFICIENT "C"')
      WRITE(KWT,99)
147   READ(KRD,*)C
      IF(C.LE.0. .OR. C.GT.1.)GO TO 138
      IF(KRUN.NE.1)GO TO 157
148   WRITE(KWT,504)
504   FORMAT(/,2X,'ENTER USER SPECIFIED CHANNEL AVERAGED',
     C ' VELOCITY (FPS)')
      WRITE(KWT,99)
157   READ(KRD,*)VO
      IF(VO.LE.0.)GO TO 148
115   IF(KRUN.NE.1)GO TO 167
158   WRITE(KWT,506)
506   FORMAT(/,2X,'ENTER CHANNEL INFORMATION :',/,2X,'BASE',
     C ' WIDTH (FEET)')
      WRITE(KWT,99)
167   READ(KRD,*)BB
      IF(BB.LE.0.)GO TO 158
      IF(KRUN.NE.1)GO TO 177
168   WRITE(KWT,507)
507   FORMAT(/,2X,'ENTER CHANNEL SIDE SLOPE Z (Z/1 = ',
     C 'HORIZONTAL/VERTICAL)')
      WRITE(KWT,99)
177   READ(KRD,*)Z
      IF(Z.LE.0.)GO TO 168
      IF(KRUN.NE.1)GO TO 187
      WRITE(KWT,508)
508   FORMAT(/,2X,'ENTER UPSTREAM ELEVATION (FEET)')
      WRITE(KWT,99)
187   READ(KRD,*)E1
125   IF(KRUN.NE.1)GO TO 197
      WRITE(KWT,509)E1
509   FORMAT(/,2X,'ENTER DOWNSTREAM ELEVATION LESS THAN',F10.2,' FEET')
      WRITE(KWT,99)
197   READ(KRD,*)E2
      IF(E2.GE.E1)GO TO 125
      IF(KRUN.NE.1)GO TO 207
198   WRITE(KWT,511)
511   FORMAT(/,2X,'ENTER CHANNEL LENGTH (FEET)')
      WRITE(KWT,99)
207   READ(KRD,*)XL
      IF(XL.LE.0.)GO TO 198
      IF(KRUN.NE.1)GO TO 217
```

```
218  WRITE(KWT,512)
512  FORMAT(/,2X,'ENTER MANNINGS FACTOR')
     WRITE(KWT,99)
217  READ(KRD,*)XN
     IF(XN.LE.0.)GO TO 218
     IF(KRUN.NE.1)GO TO 227
     WRITE(KWT,513)
513  FORMAT(/,2X,'ENTER A [1] TO INCLUDE THE INFILTRATION',
    C ' THROUGH CHANNEL BED')
     WRITE(KWT,99)
227  READ(KRD,*)KINF
     IF(KINF.NE.1)GO TO 130
     IF(KRUN.NE.1)GO TO 237
228  WRITE(KWT,514)
514  FORMAT(/,2X,'THE INFILTRATION RATE IS APPROXIMATED BY',/,
    C 3X,'QOUT = QIN + (QO-QIN)*EXP(-K*TIME)',//,2X,'ENTER',
    C ' QIN (CFS)')
     WRITE(KWT,99)
237  READ(KRD,*)QIN
     IF(QIN.LE.0.)GO TO 228
     IF(KRUN.NE.1)GO TO 247
238  WRITE(KWT,516)
516  FORMAT(/,2X,'ENTER QO (CFS)')
     WRITE(KWT,99)
247  READ(KRD,*)QO
     IF(QO.LT.QIN)GO TO 238
     IF(KRUN.NE.1)GO TO 257
248  WRITE(KWT,517)
517  FORMAT(/,2X,'ENTER COEFFICIENT K (1/HOUR)')
     WRITE(KWT,99)
257  READ(KRD,*)COE
     IF(C.LE.0.)GO TO 248
130  CONTINUE
155  IF(KRUN.NE.1)GO TO 267
     WRITE(KWT,533)
533  FORMAT(/,'  *** SELECT PRINT/VIEW OPTIONS: ***',/,
    C 2X,'ENTER A [1] = WHOLE HYDROGRAPH',/,5X,
    C 'OR A [2] = PARTIAL HYDROGRAPH')
     WRITE(KWT,99)
267  READ(KRD,*)KK
     IF(KK.EQ.1)GO TO 160
     IF(KK.EQ.2)GO TO 170
     GO TO 155
160  TIME1=0.
     TIME2=100.
     GO TO 180
170  IF(KRUN.NE.1)GO TO 277
     WRITE(KWT,534)
534  FORMAT(/,'  ENTER MODEL TIME(HOURS) FOR BEGINNING OF',
    C ' RESULTS')
     WRITE(KWT,99)
277  READ(KRD,*)TIME1
     TIME12=TIME1+.05
     IF(KRUN.NE.1)GO TO 287
     WRITE(KWT,536)TIME12
```

```
 536  FORMAT(/,'  ENTER MODEL TIME(HOURS) GREATER THAN [ ',F8.3,
     C ' ] FOR END OF RESULTS')
      WRITE(KWT,99)
 287  READ(KRD,*)TIME2
      IF(TIME2 .LE. TIME12)GO TO 170
 180  CONTINUE
C
      WRITE(KWT,201)NUP,NDOWN,NA
 201  FORMAT(//,5X,'*** FLOW PROCESS FROM NODE',I4,' TO NODE',I4,/,
     C 10X,'MODEL CHANNEL ROUTING BY THE CONVEX METHOD '
     C 'WHERE',/,10X,
     C 'A MODIFIED C-ROUTING COEFFICIENT IS ESTIMATED IN ORDER',/,
     C 10X,'TO ROUT THE STREAM',I2,' INFLOW HYDROGRAPH BY 3-MINUTE'
     C ,/,10X,'INTERVALS(reference: the National Engineering Hand',
     C 'book',/10X,'Hydrology, Chapter 17, page 17-52, August,1972,'
     C ,/,10X,'U.S. Department of Commerce).',/)
      IF(C.GT.0.)WRITE(KWT,203)C
 203  FORMAT(10X,'USER-SPECIFIED CHANNEL ROUTING COEFFICIENT = ',
     C F5.3,/)
      IF(VO.GT.0.)WRITE(KWT,204)VO
 204  FORMAT(10X,'USER-SPECIFIED CHANNEL AVERAGE VELOCITY(FPS) = ',
     C F7.3,/)
      WRITE(KWT,205)BB,Z,E1,E2,XL,XN
 205  FORMAT(10X,'ASSUMED REGULAR CHANNEL INFORMATION:',/,
     C 12X,'BASEWIDTH(FT) = ',F7.2,5X,'CHANNEL Z = ',F5.2,/,
     C 12X,'UPSTREAM ELEVATION = ',F10.2,/,
     C 12X,'DOWNSTREAM ELEVATION = ',F10.2,/,12X,'CHANNEL ',
     C 'LENGTH(FT) = ',F10.2,5X,'MANNINGS FACTOR = ',F5.3,/)
      IF(KINF.EQ.1.)WRITE(KWT,206)QIN,QO,QIN,COE
 206  FORMAT(/,5X,'CHANNEL INFILTRATION RATE IS APPROXIMATED BY',
     C /,7X,'QOUT(CFS) = ',F6.2,' + (',F6.2,' - ',F6.2,') * EXP ('
     C , ' -1. * TIME(HOUR) * ',F6.3,')',/)
C
 297  CALL MREAD(NA,A)
      ISUM=0
      QSUM=0.
      NUMBER=NUMB(NA)
C
C  FIND QMAX OF STREAM
C
      QMAX=0.
      DO 200 I=1,NUMBER
      IF(A(I).GT.QMAX)QMAX=A(I)
 200  CONTINUE
      F=QMAX/2.
      IF(C.GT.0.)GOTO 210
C
C  FIND SUM OF Q50
C
      DO 220 I=1,NUMBER
      IF(A(I).LT.F)GOTO 220
      QSUM=QSUM+A(I)
      ISUM=ISUM+1
 220  CONTINUE
      X=ISUM
```

```
      QAVG=QSUM/X
      CALL TRAPV(QAVG,BB,Z,E1,E2,XL,XN,V)
      C=V/(V+1.7)
      WRITE(KWT,209)QMAX,QAVG,QAVG,V,C
 209  FORMAT(/,10X,'CHANNEL ROUTING COEFFICIENT ESTIMATED:',/,
     C 12X,'MAXIMUM INFLOW(CFS) = ',F8.2,/,12X,'AVERAGE',
     C ' FLOWRATE IN EXCESS OF 50% MAXIMUM INFLOW = ',F8.2,/,12X,
     C 'CHANNEL NORMAL VELOCITY FOR Q = ',F8.2,' CFS = ',F6.2,' FPS'
     C ,/,12X,'ESTIMATED CHANNEL ROUTING COEFFICIENT = ',F5.3,/)
 210  IF(VO.GT.0.)V=VO
      TT=XL/3600./V
      DELT=TT*C
      EX=(.05+.5*DELT)/(1.5*DELT)
      CSTAR=1.-(1.-C)**EX
      CIN=CSTAR
      COUT=1.-CSTAR
      X=DELT*20.
      NUM=X
      IF((NUMBER+NUM+1).LE.1000)GOTO 230
      NUMBER=999-NUM
 230  CONTINUE
      Y=NUM
      DA=X-Y
      DB=1.-DA
      WRITE(KWT,211)CSTAR
 211  FORMAT(10X,'MODIFIED CHANNEL ROUTING COEFFICIENT FOR 3-MINUTE',/,
     C 10X,'UNIT INTERVALS IS CSTAR = ',F5.3,/)
C
C  ROUTING LOOP AND FIND NEW QMAX
C
      WRITE(KWT,208)NA,NA
 208  FORMAT(/,10X,'CONVEX METHOD CHANNEL ROUTING RESULTS:',//,
     C 10X,' MODEL        INFLOW     OUTFLOW',/,
     C 10X,'  TIME      (STREAM',I2,')  (STREAM',I2,')',/,
     C 10X,' (HRS)       (CFS)       (CFS)')
      QMAX=0.
      QOUT=0.
      QQOUT=0.
      QQIN=A(1)
      DO 240 I=1,NUMBER
      TIME=TIME+.05
      IF(KINF.EQ.1)QOUT=QIN+(QO-QIN)*EXP(-1.*TIME*COE)
      QQIN=QQIN-QOUT
      IF(QQIN.LT.0.)QQIN=0.
      II=NUM+I+1
      QQOUT=QQOUT*COUT+QQIN*CIN
      B(II)=B(II)+DA*QQOUT
      II=II-1
      B(II)=B(II)+DB*QQOUT
      IF(TIME.LT.TIME1.OR.TIME.GT.TIME2)GOTO 240
      WRITE(KWT,213)TIME,A(I),B(I)
 213  FORMAT(10X,F7.3,1X,2F12.1)
 240  QQIN=A(I+1)
C
      IF(KRUN.NE.1)GO TO 307
```

```
 310  WRITE(KWT,214)
 214  FORMAT(/,2X,'ENTER A [1] TO ACCEPT THE PROCESS'
     C ,/,5X,'OR A [2] TO REJECT THE PROCESS')
      WRITE(KWT,99)
      READ(KRD,*)KOPT
      IF(KOPT.LT.1 .OR. KOPT.GT.2)GO TO 310
      IF(KOPT.EQ.2)RETURN
C
      WRITE(7,*)KXY
      WRITE(7,*)NA
      WRITE(7,*)NUP
      WRITE(7,*)NDOWN
      WRITE(7,*)KOPT
      IF(KOPT.EQ.2)WRITE(7,*)C
      IF(KOPT.EQ.2)WRITE(7,*)VO
      WRITE(7,*)BB
      WRITE(7,*)Z
      WRITE(7,*)E1
      WRITE(7,*)E2
      WRITE(7,*)XL
      WRITE(7,*)XN
      WRITE(7,*)KINF
      IF(KINF.EQ.1)WRITE(7,*)QIN
      IF(KINF.EQ.1)WRITE(7,*)QO
      IF(KINF.EQ.1)WRITE(7,*)COE
      WRITE(7,*)KK
      IF(KK.EQ.2)WRITE(7,*)TIME1
      IF(KK.EQ.2)WRITE(7,*)TIME2
C
 307  DO 250 I=1,1000
 250  A(I)=B(I)
      CALL MWRITE(NA,A)
      NUMB(NA)=NUMBER
C
      RETURN
      END
C
C---------------------------------------------------------------------
C     THIS SUBROUTINE PERFORMS NORMAL DEPTH ROUTING
C---------------------------------------------------------------------
C
      SUBROUTINE PIPE(KRUN,KXY,KRD,KWT)
C
      COMMON/BLK 12/ STREAM(5,1000)
      COMMON/BLK 13/ NUMB(5)
      DIMENSION F(21,2)
      COMMON/BLK 16/ A(3000),B(3000)
C
      DATA F/0.,.05,.1,.15,.2,.25,.3,.35,.4,.45,.5,.55,.6,.65,.7,
     C .75,.8,.85,.9,.95,1.,0.,.52,.63,.715,.78,.832,.88,.912,
     C .945,.97,1.,1.025,1.045,1.06,1.08,1.095,1.11,1.12,1.13,
     C 1.136,1.14/
C
      DO 100 I=1,1000
      A(I)=0.
```

```
      B(I)=0.
 100  CONTINUE
C
 105  IF(KRUN.NE.1)GO TO 107
      WRITE(KWT,501)
 501  FORMAT(24(/),'  *** PIPE ROUTING ***',/,
     C 2X,'ENTER STREAM NUMBER',/,2X,'(ALLOWABLE VALUE'
     C ,' IS BETWEEN [1] AND [5])')
      WRITE(KWT,99)
 99   FORMAT('  =====> ',$)
 107  READ(KRD,*)NA
      IF(NA.LT.1 .OR. NA.GT.5)GO TO 105
      IF(NUMB(NA).NE.0)GO TO 130
 108  WRITE(KWT,513)NA
 513  FORMAT(/,'  *** STREAM NUMBER ',I2,' IS EMPTY ***'
     C ,/,'  ENTER A [1] TO CONTINUE')
      WRITE(KWT,99)
      READ(KRD,*)KX
      IF(KX.NE.1)GO TO 108
      RETURN
C
 130  IF(KRUN.NE.1)GO TO 117
      WRITE(KWT,502)
 502  FORMAT(/,2X,'ENTER THE UPSTREAM NODE NUMBER')
      WRITE(KWT,99)
 117  READ(KRD,*)NUP
      IF(KRUN.NE.1)GO TO 127
      WRITE(KWT,503)
 503  FORMAT(/,2X,'ENTER THE DOWNSTREAM NODE NUMBER')
      WRITE(KWT,99)
 127  READ(KRD,*)NDOWN
      IF(KRUN.NE.1)GO TO 137
 128  WRITE(KWT,507)
 507  FORMAT(/,2X,'ENTER PIPE DIAMETER (FEET)')
      WRITE(KWT,99)
 137  READ(KRD,*)D
      IF(D.LE.0.)GO TO 128
      IF(KRUN.NE.1)GO TO 147
      WRITE(KWT,508)
 508  FORMAT(/,2X,'ENTER UPSTREAM ELEVATION (FEET)')
      WRITE(KWT,99)
 147  READ(KRD,*)E1
 125  IF(KRUN.NE.1)GO TO 157
      WRITE(KWT,509)E1
 509  FORMAT(/,2X,'ENTER DOWNSTREAM ELEVATION LESS THAN',F10.2,' FEET')
      WRITE(KWT,99)
 157  READ(KRD,*)E2
      IF(E2.GT.E1)GO TO 125
      IF(KRUN.NE.1)GO TO 167
 158  WRITE(KWT,511)
 511  FORMAT(/,2X,'ENTER CHANNEL LENGTH (FEET)')
      WRITE(KWT,99)
 167  READ(KRD,*)XL
      IF(XL.LE.0.)GO TO 158
      IF(KRUN.NE.1)GO TO 177
```

```
 168  WRITE(KWT,512)
 512  FORMAT(/,2X,'ENTER MANNINGS FACTOR')
      WRITE(KWT,99)
 177  READ(KRD,*)XN
      IF(XN.LE.0.)GO TO 168
 155  IF(KRUN.NE.1)GO TO 187
      WRITE(KWT,533)
 533  FORMAT(/,'  *** SELECT PRINT/VIEW OPTIONS: ***',/,
     C 2X,'ENTER A [1] = WHOLE HYDROGRAPH',/,5X,
     C 'OR A [2] = PARTIAL HYDROGRAPH')
      WRITE(KWT,99)
 187  READ(KRD,*)KK
      IF(KK.EQ.1)GO TO 160
      IF(KK.EQ.2)GO TO 170
      GO TO 155
 160  TIME1=0.
      TIME2=100.
      GO TO 180
 170  IF(KRUN.NE.1)GO TO 197
      WRITE(KWT,534)
 534  FORMAT(/,'  ENTER MODEL TIME(HOURS) FOR BEGINNING OF',
     C ' RESULTS')
      WRITE(KWT,99)
 197  READ(KRD,*)TIME1
      TIME12=TIME1+.05
      IF(KRUN.NE.1)GO TO 207
      WRITE(KWT,536)TIME12
 536  FORMAT(/,'  ENTER MODEL TIME(HOURS) GREATER THAN [ ',F8.3,
     C ' ] FOR END OF RESULTS')
      WRITE(KWT,99)
 207  READ(KRD,*)TIME2
      IF(TIME2 .LE. TIME12)GO TO 170
 180  CONTINUE
C
      WRITE(KWT,201)NUP,NDOWN,NA
 201  FORMAT(//,5X,'*** FLOW PROCESS FROM NODE',I4,' TO NODE',I4,
     C /,10X,'MODEL PIPEFLOW ROUTING OF STREAM',I2,' WHERE',
     C /,10X,'STORAGE EFFECTS ARE NEGLECTED WITHIN THE PIPE, FLOW',/,
     C 10X,'VELOCITIES ARE ESTIMATED BY ASSUMING STEADY FLOW FOR',/,
     C 10X,'EACH UNIT INTERVAL(NORMAL DEPTH), AND FLOWS IN EXCESS',/,
     C 10X,'OF (.82)(DIAMETER) ARE PONDED AT THE UPSTREAM INLET:',//)
      WRITE(KWT,206)XL,XN,E1,E2,D
 206  FORMAT(10X,'PIPELENGTH(FT) = ',F10.2,5X,'MANNINGS FACTOR = ',
     C F5.3,/,10X,'UPSTREAM ELEVATION(FT) = ',F10.2,5X,/,
     C 10X,'DOWNSTREAM ELEVATION(FT) = ',F10.2,
     C /,10X,'PIPE DIAMETER(FT) = ',F6.2,/)
C
C  INITIALIZE VARIABLES
C
 217  DO 200 I=1,1000
 200  B(I)=0.
      CALL MREAD(NA,A)
      NUMBER=NUMB(NA)
      TIME=0.
      STORE=0.
```

```
      S=SQRT((E1-E2)/XL)
      QCAP=35.628/XN*.013*S*D**2.6667
      VCAP=QCAP/.7854/D/D
      XA=XL/180.
      XB=180./43560.
      WRITE(KWT,202)
 202  FORMAT(//,10X,'NORMAL DEPTH VELOCITY PIPE ROUTING RESULTS:',/,
     C 10X,'  TIME      INFLOW   VELOCITY   OUTFLOW    UPSTREAM',/,
     C 10X,' (HRS)       (CFS)     (FPS)      (CFS)    PONDING(AF)')
C
C  MAIN LOOP
C
      DO 210 I=1,1000
      IF(I.GT.NUMBER.AND.STORE.LT..001)GOTO 300
      Q=A(I)
      IF(Q.LT.QCAP.AND.STORE.LE.0.)GOTO 220
C
C  PIPE IS UNDER PRESSURE
C
      V=VCAP
      STORE=STORE+(Q-QCAP)*XB
      Q=QCAP
      IF(STORE.GE.0.)GOTO 230
      Q=QCAP+STORE*145.2
      STORE=0.
C
C  OPEN FLOW
C
 220  QQ=Q/QCAP
      INDEX=QQ*20.+1.
      V=(QQ-F(INDEX,1))/.05*(F(INDEX+1,2)-F(INDEX,2))+F(INDEX,2)
      V=V*VCAP
 230  IF(V.GT.0.)XNUM=XA/V
      IF(V.LE.0.)XNUM=0.
      NUM=XNUM
      ZNUM=NUM
      DA=XNUM-ZNUM
      DB=1.-DA
      II=I+NUM+1
      IF(II.GT.1000)GOTO 240
      B(II)=B(II)+DA*Q
      II=I+NUM
      B(II)=B(II)+DB*Q
 240  TIME=TIME+.05
      IF(TIME.LT.TIME1 .OR. TIME.GT.TIME2)GO TO 210
      WRITE(KWT,203)TIME,A(I),V,B(I),STORE
 203  FORMAT(10X,F7.3,3F10.1,F13.3)
 210  CONTINUE
 300  IF(KRUN.NE.1)GO TO 227
 310  WRITE(KWT,537)
 537  FORMAT(/,2X,'ENTER A [1] TO ACCEPT THE PROCESS',/,
     C 5X,'OR A [2] TO REJECT THE PROCESS')
      WRITE(KWT,99)
      READ(KRD,*)KOPT
      IF(KOPT.LT.1 .OR. KOPT.GT.2)GO TO 310
```

```
      IF(KOPT.EQ.2)RETURN
C
      WRITE(7,*)KXY
      WRITE(7,*)NA
      WRITE(7,*)NUP
      WRITE(7,*)NDOWN
      WRITE(7,*)D
      WRITE(7,*)E1
      WRITE(7,*)E2
      WRITE(7,*)XL
      WRITE(7,*)XN
      WRITE(7,*)KK
      IF(KK.EQ.2)WRITE(7,*)TIME1
      IF(KK.EQ.2)WRITE(7,*)TIME2
C
 227  NUMBER=I
      DO 320 I=1,1000
 320  A(I)=B(I)
      CALL MWRITE(NA,A)
      NUMB(NA)=NUMBER
C
      RETURN
      END
C
C-----------------------------------------------------------------------
C     THIS SUBROUTINE TRANSLATES A STREAM FORWARD IN TIME
C     BY DELT HOURS
C-----------------------------------------------------------------------
C
      SUBROUTINE MOVE(KRUN,KXY,KRD,KWT)
C
      COMMON/BLK 12/ STREAM(5,1000)
      COMMON/BLK 13/ NUMB(5)
      COMMON/BLK 16/ A(3000),B(3000)
C
      DO 100 I=1,1000
      A(I)=0.
      B(I)=0.
 100  CONTINUE
C
      IF(KRUN.NE.1)GO TO 107
      WRITE(KWT,532)
 532  FORMAT(24(/),8X,'*** TRANSLATION MODEL ***',//,
     C 5X,'The translation flow routing techniqe simple ',/,
     C 3X,'moves incremental unit volumes forward in time'
     C ,/,3X,'by a specified translation timestep.'/,76('-'),//)
 105  WRITE(KWT,501)
 501  FORMAT(/,2X,'ENTER THE STREAM NUMBER OF THE HYDROGRAPH TO',
     C /,2X,'BE TRANSLATED FORWARD IN TIME (HOURS)',/,2X,'(ALLOWABLE',
     C ' VALUE IS BETWEEN [1] AND [5])')
      WRITE(KWT,99)
 99   FORMAT('  =====> ',$)
```

```
 107  READ(KRD,*)NA
      IF(NA.LT.1 .OR. NA.GT.5)GO TO 105
      IF(NUMB(NA).NE.0)GO TO 120
 108  WRITE(KWT,521)NA
 521  FORMAT(/,'  *** STREAM NUMBER ',I2,' IS EMPTY ***',
     C /,'  ENTER A [1] TO CONTINUE')
      WRITE(KWT,99)
      READ(KRD,*)KX
      IF(KX.NE.1)GO TO 108
      RETURN
C
 120  IF(KRUN.NE.1)GO TO 117
      WRITE(KWT,522)
 522  FORMAT(/,2X,'ENTER UPSTREAM NODE NUMBER')
      WRITE(KWT,99)
 117  READ(KRD,*)NUP
      IF(KRUN.NE.1)GO TO 127
      WRITE(KWT,523)
 523  FORMAT(/,2X,'ENTER DOWNSTREAM NODE NUMBER')
      WRITE(KWT,99)
 127  READ(KRD,*)NDOWN
      IF(KRUN.NE.1)GO TO 137
      WRITE(KWT,503)
 503  FORMAT(/,2X,'ENTER A [1] = USER SPECIFIED TRANSLATION TIME ',
     C '(HOURS)',/,5X,'OR A [2] = COMPUTER ESTIMATED TRANSLATION TIME')
      WRITE(KWT,99)
 137  READ(KRD,*)KODE
      IF(KODE.EQ.1)GO TO 110
      IF(KODE.EQ.2)GO TO 115
      GO TO 120
 115  IF(KRUN.NE.1)GO TO 147
 118  WRITE(KWT,506)
 506  FORMAT(/,2X,'ENTER CHANNEL INFORMATION :',/,2X,'BASE',
     C ' WIDTH (FEET)')
      WRITE(KWT,99)
 147  READ(KRD,*)BB
      IF(BB.LE.0.)GO TO 118
      IF(KRUN.NE.1)GO TO 157
 148  WRITE(KWT,507)
 507  FORMAT(/,2X,'ENTER CHANNEL SIDE SLOPE Z (Z/1 = ',
     C 'HORIZONTAL/VERTICAL)')
      WRITE(KWT,99)
 157  READ(KRD,*)Z
      IF(Z.LE.0.)GO TO 148
      IF(KRUN.NE.1)GO TO 167
      WRITE(KWT,508)
 508  FORMAT(/,2X,'ENTER UPSTREAM ELEVATION (FEET)')
      WRITE(KWT,99)
 167  READ(KRD,*)E1
 125  IF(KRUN.NE.1)GO TO 177
      WRITE(KWT,509)E1
 509  FORMAT(/,2X,'ENTER DOWNSTREAM ELEVATION LESS THAN',F10.2,' FEET')
      WRITE(KWT,99)
 177  READ(KRD,*)E2
```

```
      IF(E2.GT.E1)GO TO 125
      IF(KRUN.NE.1)GO TO 187
178   WRITE(KWT,511)
511   FORMAT(/,2X,'ENTER CHANNEL LENGTH (FEET)')
      WRITE(KWT,99)
187   READ(KRD,*)XL
      IF(XL.LE.0.)GO TO 178
      IF(KRUN.NE.1)GO TO 197
188   WRITE(KWT,512)
512   FORMAT(/,2X,'ENTER MANNINGS FACTOR')
      WRITE(KWT,99)
197   READ(KRD,*)XN
      IF(XN.LE.0.)GO TO 188
135   IF(KRUN.NE.1)GO TO 207
      WRITE(KWT,518)
518   FORMAT(/,2X,'ENTER PERCENT (DECIMAL) OF PEAK Q TO BE USED ',
     C 'FOR WEIGHTING CALCULATIONS',/,2X,'(ALLOWABLE VALUE IS ',
     C 'BETWEEN [0.1] AND [1.0])')
      WRITE(KWT,99)
207   READ(KRD,*)XP1
      IF(XP1.LT.0.1 .OR. XP1.GT.1.)GO TO 135
      IF(XP1.EQ.1)XP2=1.
      IF(XP1.EQ.1.)GO TO 140
      XP11=XP1+.01
145   IF(KRUN.NE.1)GO TO 217
      WRITE(KWT,519)XP11
519   FORMAT(/,2X,'ENTER PERCENTAGE OF WEIGHTING TO BE',/,
     C '  USED FOR NORMAL DEPTH CALCULATIONS',/,2X,'(ALLOWABLE',
     C ' VALUE IS BETWEEN [',F6.3,'] AND [1.0])')
      WRITE(KWT,99)
217   READ(KRD,*)XP2
      IF(XP2.LT.XP11 .OR. XP2.GT.1.)GO TO 145
      GO TO 140
110   IF(KRUN.NE.1)GO TO 227
198   WRITE(KWT,502)
502   FORMAT(/,2X,'ENTER THE TRANSLATION TIME IN HOUR')
      WRITE(KWT,99)
227   READ(KRD,*)DELT
      IF(DELT.LT.0.)GO TO 198
140   IF(KRUN.NE.1)GO TO 237
      WRITE(KWT,513)
513   FORMAT(/,2X,'ENTER A [1] TO INCLUDE INFILTRATION',
     C ' THROUGH CHANNEL BED')
      WRITE(KWT,99)
237   READ(KRD,*)KINF
      IF(KINF.NE.1)GO TO 130
      IF(KRUN.NE.1)GO TO 247
238   WRITE(KWT,514)
514   FORMAT(/,2X,'THE INFILTRATION RATE IS APPROXIMATED BY',/,
     C 3X,'QOUT = QIN + (QO-QIN)*EXP(-K*TIME)',//,2X,'ENTER',
     C ' QIN (CFS)')
      WRITE(KWT,99)
247   READ(KRD,*)QIN
      IF(QIN.LE.0.)GO TO 238
      IF(KRUN.NE.1)GO TO 257
```

```
 248  WRITE(KWT,516)
 516  FORMAT(/,2X,'ENTER QO (CFS)')
      WRITE(KWT,99)
 257  READ(KRD,*)QO
      IF(QO.LT.QIN)GO TO 248
      IF(KRUN.NE.1)GO TO 267
 258  WRITE(KWT,517)
 517  FORMAT(/,2X,'ENTER COEFFICIENT K (1/HOUR)')
      WRITE(KWT,99)
 267  READ(KRD,*)COE
      IF(COE.LE.0.)GO TO 258
 130  CONTINUE
 155  IF(KRUN.NE.1)GO TO 277
      WRITE(KWT,533)
 533  FORMAT(/,'  *** SELECT PRINT/VIEW OPTIONS: ***',/,
     C 2X,'ENTER A [1] = WHOLE HYDROGRAPH',/,5X,
     C 'OR A [2] - PARTIAL HYDROGRAPH')
      WRITE(KWT,99)
 277  READ(KRD,*)KK
      IF(KK.EQ.1)GO TO 160
      IF(KK.EQ.2)GO TO 170
      GO TO 155
 160  TIME1=0.
      TIME2=100.
      GO TO 180
 170  IF(KRUN.NE.1)GO TO 287
      WRITE(KWT,534)
 534  FORMAT(/,'  ENTER MODEL TIME(HOURS) FOR BEGINNING OF',
     C ' RESULTS')
      WRITE(KWT,99)
 287  READ(KRD,*)TIME1
      TIME12=TIME1+.05
      IF(KRUN.NE.1)GO TO 297
      WRITE(KWT,536)TIME12
 536  FORMAT(/,'  ENTER MODEL TIME(HOURS) GREATER THAN [ ',F8.3,
     C ' ] FOR END OF RESULTS')
      WRITE(KWT,99)
 297  READ(KRD,*)TIME2
      IF(TIME2 .LE. TIME12)GO TO 170
 180  CONTINUE
C
      CALL MREAD(NA,A)
      NUMBER=NUMB(NA)
      IF(KODE.EQ.1)GO TO 200
C
C  ESTIMATION OF TRANSLATION TIME
C
      ISUM=0
      QSUM=0.
C
C  FIND QMAX OF STREAM
C
      QMAX=0.
      DO 210 I=1,NUMBER
      IF(A(I).GT.QMAX)QMAX=A(I)
```

```
 210  CONTINUE
      F=QMAX*XP1
C
C  FIND SUM OF Q(XP1)
C
      DO 220 I=1,NUMBER
      IF(A(I).LT.F)GOTO 220
      QSUM=QSUM+A(I)
      ISUM=ISUM+1
 220  CONTINUE
      X=ISUM
      QAVG=XP2*QSUM/X
      CALL TRAPV(QAVG,BB,Z,E1,E2,XL,XN,V)
      DELT=XL/V/3600.
C
 200  WRITE(KWT,201)NUP,NDOWN,NA,DELT
 201  FORMAT(/,76('-'),/,5X,'*** FLOW PROCESS FROM NODE',I4,
     C ' TO NODE',I4,/,10X,'TRANSLATE STREAM NUMBER',I2,' FORWARD',
     C ' IN TIME BY',F7.3,' HOURS:',//)
      IF(KODE.EQ.2)WRITE(KWT,202)BB,Z,E1,E2,XL,XN
 202  FORMAT(10X,'ASSUMED REGULAR CHANNEL INFORMATION:',/,
     C 12X,'BASEWIDTH(FT) = ',F7.2,5X,'CHANNEL Z = ',F5.2,/,
     C 12X,'UPSTREAM ELEVATION = ',F10.2,/,
     C 12X,'DOWNSTREAM ELEVATION = ',F10.2,/,12X,'CHANNEL ',
     C 'LENGTH(FT) = ',F10.2,5X,'MANNINGS FACTOR = ',F5.3,/)
      IF(KINF.EQ.1)WRITE(KWT,203)QIN,QO,QIN,COE
 203  FORMAT(/,5X,'CHANNEL INFILTRATION RATE IS APPROXIMATED BY',
     C /,7X,'QOUT(CFS) = ',F6.2,' + (',F6.2,' - ',F6.2,') * EXP ('
     C , ' -1. * TIME(HOUR) * ',F6.3,')',/)
      WRITE(KWT,204)NA,NA,DELT
 204  FORMAT(10X,' MODEL     STREAM',I2,5X,'  STREAM',I2,/,
     C 10X,'  TIME       (CFS)       TRANSLATED',F7.3,' HOURS')
C
      XM=DELT*20.
      M=IFIX(XM)
      TIME=0.
      QOUT=0.
      NUM1=NUMBER+M
      IF(NUM1.LT.1000)GO TO 250
      NUMBER=999-M
C
C  TRANSLATE HYDROGRAPH FORWARD
C
 250  XA=REAL(M)
      XA=XM-XA
      XB=1.-XA
      DO 260 I=1,NUMBER
      J=I+M+1
      JJ=I+M
      TIME=TIME+.05
      IF(KINF.EQ.1)QOUT=QIN+(QO-QIN)*EXP(-1.*TIME*COE)
      QQIN=A(I)-QOUT
      IF(QQIN.LT.0.)QQIN=0.
      B(J)=QQIN*XA
      B(JJ)=QQIN*XB+B(JJ)
```

```
      IF(TIME.LT.TIME1 .OR. TIME.GT.TIME2)GO TO 260
      WRITE(KWT,205)TIME,A(I),B(I)
 205  FORMAT(10X,F7.3,3X,F10.1,F10.1)
 260  CONTINUE
C
      IF(KRUN.NE.1)GO TO 317
 310  WRITE(KWT,206)
 206  FORMAT(/,2X,'ENTER A [1] TO ACCEPT THE PROCESS',/,
     C 5X,'OR A [2] REJECT THE PROCESS')
      WRITE(KWT,99)
      READ(KRD,*)KOPT
      IF(KOPT.LT.1 .OR. KOPT.GT.2)GO TO 310
      IF(KOPT.EQ.2)RETURN
C
      WRITE(7,*)KXY
      WRITE(7,*)NA
      WRITE(7,*)NUP
      WRITE(7,*)NDOWN
      WRITE(7,*)KODE
      IF(KODE.EQ.2)WRITE(7,*)BB
      IF(KODE.EQ.2)WRITE(7,*)Z
      IF(KODE.EQ.2)WRITE(7,*)E1
      IF(KODE.EQ.2)WRITE(7,*)E2
      IF(KODE.EQ.2)WRITE(7,*)XL
      IF(KODE.EQ.2)WRITE(7,*)XN
      IF(KODE.EQ.2)WRITE(7,*)XP1
      IF(KODE.EQ.2)WRITE(7,*)XP2
      IF(KODE.EQ.1)WRITE(7,*)DELT
      WRITE(7,*)KINF
      IF(KINF.EQ.1)WRITE(7,*)QIN
      IF(KINF.EQ.1)WRITE(7,*)QO
      IF(KINF.EQ.1)WRITE(7,*)COE
      WRITE(7,*)KK
      IF(KK.EQ.2)WRITE(7,*)TIME1
      IF(KK.EQ.2)WRITE(7,*)TIME2
C
 317  NUMB(NA)=J
C
      CALL MWRITE(NA,B)
C
      RETURN
      END
C
C-----------------------------------------------------------------------
C     THIS SUBRUOTINE ESTIMATES THE AVERAGED FLOW VELOCITY
C-----------------------------------------------------------------------
C
      SUBROUTINE TRAPV(Q,B,Z,E1,E2,XL,RN,V)
C
      FN(QQ,BB,DD,ZZ,SS)=1.-QQ*RN*(BB+2.*DD*SQRT(ZZ*ZZ+1.))**.66667/
     C (1.486*((BB+ZZ*DD)*DD)**1.6667*SQRT(SS))
C
      S=(E1-E2)/XL
      IF(B.LE.0.)B=.0001
      DMAX=1000.
```

```
      YMAX=DMAX
      YMIN=0.
      DO 100 I=1,17
      DN=.5*(YMIN+YMAX)
      F=FN(Q,B,DN,Z,S)
      IF(F)120,150,130
120   YMIN=DN
      GOTO 100
130   YMAX=DN
100   CONTINUE
150   CONTINUE
      TW=B+2.*Z*DN
      AREA=.5*(B+TW)*DN
      V=Q/AREA
      RETURN
      END
C
C-----------------------------------------------------------------------
C     THIS SUBROUTINE USES KINEMATIC WAVE METHOD
C     TO ROUT THE HYDROGRAPH DOWNSTREAM
C-----------------------------------------------------------------------
C
      SUBROUTINE KROUT(KRUN,KXY,KRD,KWT)
C
      COMMON/BLK 12/ STREAM(5,1000)
      COMMON/BLK 13/ NUMB(5)
      COMMON/BLK 16/ A(3000),B(3000)
C
      DO 100 I=1,1000
      A(I)=0.
      B(I)=0.
 100  CONTINUE
C
 103  IF(KRUN.NE.1)GO TO 107
      WRITE(KWT,519)
 519  FORMAT(24(/),10X,'*** KINEMATIC WAVE METHOD ***',/,
     C 5X,'The kinematic flow routing technique moves each',/,
     C 3X,'unit period flowrate (3-minute periods) forward in',/,
     C 3X,'time by pure translation based on a travel time',/,
     C 3X,'computed using normal depth flow velocities for',/,
     C 3X,'each unit period',/,76('-'),/,'  ENTER A [1] TO',
     C ' CONTINUE')
      WRITE(KWT,99)
 99   FORMAT('  =====> ',$)
      READ(KRD,*)II
      IF(II.NE.1)GO TO 103
 105  WRITE(KWT,501)
 501  FORMAT(/,2X,'ENTER THE STREAM NUMBER OF THE HYDROGRAPH TO',
     C /,2X,'BE ROUTED DOWNSTREAM BY THE KINEMATIC WAVE METHOD',/,2X,
     C '(ALLOWABLE VALUE IS BETWEEN [1] AND [5])')
      WRITE(KWT,99)
 107  READ(KRD,*)NA
      IF(NA.LT.1 .OR. NA.GT.5)GO TO 105
      IF(NUMB(NA).NE.0)GO TO 120
 108  WRITE(KWT,502)NA
```

```
 502  FORMAT(/,'  *** STREAM NUMBER ',I2,' IS EMPTY ***',
     C /,'  ENTER A [1] TO CONTINUE')
      WRITE(KWT,99)
      READ(KRD,*)KX
      IF(KX.NE.1)GO TO 108
      RETURN
C
 120  IF(KRUN.NE.1)GO TO 117
      WRITE(KWT,503)
 503  FORMAT(/,2X,'ENTER UPSTREAM NODE NUMBER')
      WRITE(KWT,99)
 117  READ(KRD,*)NUP
      IF(KRUN.NE.1)GO TO 127
      WRITE(KWT,504)
 504  FORMAT(/,2X,'ENTER DOWNSTREAM NODE NUMBER')
      WRITE(KWT,99)
 127  READ(KRD,*)NDOWN
 115  IF(KRUN.NE.1)GO TO 147
 118  WRITE(KWT,505)
 505  FORMAT(/,2X,'ENTER CHANNEL INFORMATION :',/,2X,'BASE',
     C ' WIDTH (FEET)')
      WRITE(KWT,99)
 147  READ(KRD,*)BB
      IF(BB.LE.0.)GO TO 118
      IF(KRUN.NE.1)GO TO 157
 148  WRITE(KWT,506)
 506  FORMAT(/,2X,'ENTER CHANNEL SIDE SLOPE Z (Z/1 = ',
     C 'HORIZONTAL/VERTICAL)')
      WRITE(KWT,99)
 157  READ(KRD,*)Z
      IF(Z.LE.0.)GO TO 148
      IF(KRUN.NE.1)GO TO 167
      WRITE(KWT,507)
 507  FORMAT(/,2X,'ENTER UPSTREAM ELEVATION (FEET)')
      WRITE(KWT,99)
 167  READ(KWT,*)E1
 125  IF(KRUN.NE.1)GO TO 177
      WRITE(KWT,508)E1
 508  FORMAT(/,2X,'ENTER DOWNSTREAM ELEVATION LESS THAN',F10.2,' FEET')
      WRITE(KWT,99)
 177  READ(KRD,*)E2
      IF(E2.GT.E1)GO TO 125
      IF(KRUN.NE.1)GO TO 187
 178  WRITE(KWT,509)
 509  FORMAT(/,2X,'ENTER CHANNEL LENGTH (FEET)')
      WRITE(KWT,99)
 187  READ(KRD,*)XL
      IF(XL.LE.0.)GO TO 178
      IF(KRUN.NE.1)GO TO 197
 188  WRITE(KWT,510)
 510  FORMAT(/,2X,'ENTER MANNINGS FACTOR')
      WRITE(KWT,99)
 197  READ(KRD,*)XN
      IF(XN.LE.0.)GO TO 188
 140  IF(KRUN.NE.1)GO TO 237
```

```
      WRITE(KWT,513)
513   FORMAT(/,2X,'ENTER A [1] TO INCLUDE INFILTRATION',
     C ' THROUGH CHANNEL BED')
      WRITE(KWT,99)
237   READ(KRD,*)KINF
      IF(KINF.NE.1)GO TO 130
      IF(KRUN.NE.1)GO TO 247
238   WRITE(KWT,514)
514   FORMAT(/,2X,'THE INFILTRATION RATE IS APPROXIMATED BY',/,
     C 3X,'QOUT = QIN + (QO-QIN)*EXP(-K*TIME)',//,2X,'ENTER',
     C ' QIN (CFS)')
      WRITE(KWT,99)
247   READ(KRD,*)QIN
      IF(QIN.LE.0.)GO TO 238
      IF(KRUN.NE.1)GO TO 257
248   WRITE(KWT,515)
515   FORMAT(/,2X,'ENTER QO (CFS)')
      WRITE(KWT,99)
257   READ(KRD,*)QO
      IF(QO.LT.QIN)GO TO 248
      IF(KRUN.NE.1)GO TO 267
258   WRITE(KWT,517)
517   FORMAT(/,2X,'ENTER COEFFICIENT K (1/HOUR)')
      WRITE(KWT,99)
267   READ(KRD,*)COE
      IF(COE.LE.0.)GO TO 258
130   CONTINUE
155   IF(KRUN.NE.1)GO TO 277
      WRITE(KWT,511)
511   FORMAT(/,'  *** SELECT PRINT/VIEW OPTIONS: ***',/,
     C 2X,'ENTER A [1] = WHOLE HYDROGRAPH',/,5X,
     C 'OR A [2] = PARTIAL HYDROGRAPH')
      WRITE(KWT,99)
277   READ(KRD,*)KK
      IF(KK.EQ.1)GO TO 160
      IF(KK.EQ.2)GO TO 170
      GO TO 155
160   TIME1=0.
      TIME2=100.
      GO TO 180
170   IF(KRUN.NE.1)GO TO 287
      WRITE(KWT,512)
512   FORMAT(/,'  ENTER MODEL TIME(HOURS) FOR BEGINNING OF',
     C ' RESULTS')
      WRITE(KWT,99)
287   READ(KRD,*)TIME1
      TIME12=TIME1+.05
      IF(KRUN.NE.1)GO TO 297
      WRITE(KWT,516)TIME12
516   FORMAT(/,'  ENTER MODEL TIME(HOURS) GREATER THAN [ ',F8.3,
     C ' ] FOR END OF RESULTS')
      WRITE(KWT,99)
297   READ(KRD,*)TIME2
      IF(TIME2 .LE. TIME12)GO TO 170
180   CONTINUE
```

```
C
      CALL MREAD(NA,A)
      NUMBER=NUMB(NA)
      WRITE(KWT,201)NUP,NDOWN,NA
 201  FORMAT(/,76('-'),/,5X,'*** FLOW PROCESS FROM NODE',I4,
     C ' TO NODE',I4,/,10X,'STREAM NUMBER',I2,' IS ROUTED',
     C ' BY KINEMATIC WAVE METHOD',//)
      WRITE(KWT,202)BB,Z,E1,E2,XL,XN
 202  FORMAT(10X,'ASSUMED REGULAR CHANNEL INFORMATION:',/,
     C 12X,'BASEWIDTH(FT) = ',F7.2,5X,'CHANNEL Z = ',F5.2,/,
     C 12X,'UPSTREAM ELEVATION = ',F10.2,/,
     C 12X,'DOWNSTREAM ELEVATION = ',F10.2,/,12X,'CHANNEL ',
     C 'LENGTH(FT) = ',F10.2,5X,'MANNINGS FACTOR = ',F5.3,/)
      IF(KINF.EQ.1)WRITE(KWT,203)QIN,QO,QIN,COE
 203  FORMAT(/,5X,'CHANNEL INFILTRATION RATE IS APPROXIMATED BY',
     C /,7X,'QOUT(CFS) = ',F6.2,' + (',F6.2,' - ',F6.2,') * EXP ('
     C , ' -1. * TIME(HOUR) * ',F6.3,')',/)
      WRITE(KWT,214)
 214  FORMAT(10X,' MODEL      INFLOW     OUTFLOW'/,
     C 10X,'  TIME       (CFS)      (CFS)')
C
C  ESTIMATION OF TRAVELING TIME
C
C  FIND QMAX OF STREAM
C
 307  QMAX=0.
      DO 210 I=1,NUMBER
      IF(A(I).GT.QMAX)QMAX=A(I)
 210  CONTINUE
      CALL TRAPV(QMAX,BB,Z,E1,E2,XL,XN,V)
      DELT=XL/V/3600.
      SO=(E1-E2)/XL
      CALL NORVEL(BB,Z,SO,XN,QMAX,V,1)
C
C  KINEMATIC WAVE METHOD
C
      JJ=0
      DO 250 I=1,NUMBER
      QXX=A(I)
      IF(QXX.LE.0.)GO TO 250
      CALL NORVEL(BB,Z,SO,XN,QXX,VEL,2)
      DELT=XL/VEL/3600.
      XX=DELT*20.
      DA=XX-IFIX(XX)
      DB=1.-DA
      J=I+IFIX(XX)+1
      IF(J.GT.JJ)JJ=J
      IF(J.GT.1000)GO TO 250
      B(J)=B(J)+QXX*DA
      B(J-1)=B(J-1)+QXX*DB
 250  CONTINUE
      IF(JJ.GT.1000)JJ=1000
      TIME=0.
      DO 270 I=1,JJ
      TIME=TIME+.05
```

```
      IF(KINF.EQ.1)B(I)=B(I)-(QIN+(QO-QIN)*EXP(-1.*COE*TIME))
      IF(TIME.LT.TIME1 .OR. TIME.GT.TIME2)GO TO 270
      WRITE(KWT,204)TIME,A(I),B(I)
 204  FORMAT(10X,F7.3,3X,F10.1,F10.1)
 270  CONTINUE
C
      IF(KRUN.NE.1)GO TO 317
 310  WRITE(KWT,205)
 205  FORMAT(/,2X,'ENTER A [1] TO ACCEPT THE PROCESS',/,
     C 5X,'OR A [2] TO REJECT THE PROCESS')
      WRITE(KWT,99)
      READ(KRD,*)KOPT
      IF(KOPT.LT.1 .OR. KOPT.GT.2)GO TO 310
      IF(KOPT.EQ.2)RETURN
C
      WRITE(7,*)KXY
      WRITE(7,*)NA
      WRITE(7,*)NUP
      WRITE(7,*)NDOWN
      WRITE(7,*)BB
      WRITE(7,*)Z
      WRITE(7,*)E1
      WRITE(7,*)E2
      WRITE(7,*)XL
      WRITE(7,*)XN
      WRITE(7,*)KINF
      IF(KINF.EQ.1)WRITE(7,*)QIN
      IF(KINF.EQ.1)WRITE(7,*)QO
      IF(KINF.EQ.1)WRITE(7,*)COE
      WRITE(7,*)KK
      IF(KK.EQ.2)WRITE(7,*)TIME1
      IF(KK.EQ.2)WRITE(7,*)TIME2
C
 317  NUMB(NA)=JJ
C
      CALL MWRITE(NA,B)
C
      RETURN
      END
C
C-----------------------------------------------------------------------
C     THIS SUBROUTINE PREPARES THE NORMAL DEPTH TABLE
C     FOR DISCHARGE AND AVERAGED VELOCITY
C     THEN INTERPOLATES THE AVERAGED VELOCITIES FOR
C     CORRESPONDING DISCHARGE
C-----------------------------------------------------------------------
C
      SUBROUTINE NORVEL(B,Z,SO,XN,QX,VEL,KODE)
C
      COMMON/BLK 14/ QNOR(21),VNOR(21)
C
      IF(KODE.EQ.2)GO TO 150
      A=QX/VEL
      IF(Z.EQ.0.)D=A/B
      IF(Z.NE.0.)D=(-B+SQRT(B*B+4.*A*Z))/(2.*Z)
```

```
      DELTD=D/20.
      D=0.
      VNOR(1)=0.
      QNOR(1)=0.
      DO 100 I=2,21
      D=D+DELTD
      A=Z*D*D+B*D
      R=A/(B+2.*D*SQRT(1.+Z*Z))
      VNOR(I)=1.486*SQRT(SO)*(R**.667)/XN
      QNOR(I)=A*VNOR(I)
 100  CONTINUE
      RETURN
C
C  INTERPOLATION OF AVERAGED VELOCITY
C
 150  DO 200 I=2,21
      IF(QX.GT.QNOR(I))GO TO 200
      VEL=VNOR(I-1)+(QX-QNOR(I-1))*(VNOR(I)-VNOR(I-1))/
     C    (QNOR(I)-QNOR(I-1))
      RETURN
 200  CONTINUE
      RETURN
      END
C
```

```
C
C-----------------------------------------------------------------
      SUBROUTINE MUSK(KRUN,KXY,KRD,KWT)
C-----------------------------------------------------------------
C     MUSKINGUM HYDROGRAPH ROUTING MODEL
C
      COMMON/BLK 12/ STREAM(5,1000)
      COMMON/BLK 13/ NUMB(5)
      COMMON/BLK 16/ A(3000),B(3000)
C
C  INPUT DATA
C
 105  IF(KRUN.NE.1)GO TO 107
      WRITE(KWT,501)
 501  FORMAT(24(/),'  *** MUSKINGUM ROUTING ***',/,
     C 2X,'ENTER THE STREAM NUMBER OF THE HYDROGRAPH TO BE',/,
     C '  ROUTED DOWNSTREAM BY THE MUSKIGUM METHOD',/
     C ,2X,'(ALLOWABLE VALUE IS BETWEEN [1] AND [5])')
      WRITE(KWT,99)
 99   FORMAT('  =====> ',$)
 107  READ(KRD,*)NA
      IF(NA.LT.1 .OR. NA.GT.5)GO TO 105
      IF(NUMB(NA).NE.0)GO TO 125
 108  IF(KRUN.EQ.1)WRITE(KWT,502)NA
 502  FORMAT(/,'  *** STREAM NUMBER ',I2,' IS EMPTY ***',/,
     C '  ENTER A [1] TO CONTINUE')
      WRITE(KWT,99)
      READ(KRD,*)KX
      IF(KX.NE.1)GO TO 108
      RETURN
C
 125  IF(KRUN.NE.1)GO TO 40
 10   WRITE(KWT,524)
 524  FORMAT(/,'  ENTER A [1] TO  SKIP  THE INTRODUCTION TEXT',/,
     C 5X,'OR A [2] TO REVIEW THE INTRODUCTION TEXT')
      WRITE(KWT,99)
      READ(KRD,*)II
      IF(II.EQ.1)GO TO 40
      IF(II.EQ.2)GO TO 20
      GO TO 10
 20   WRITE(KWT,505)
 505  FORMAT(24(/),'    ***  MUSKINGUM ROUTING MODEL ***',//,
     C '   The Muskingum Hydrograph routing technique is based',/,
     C ' on two parameters: X (a storage parameter) and K (a',/,
     C ' translation parameter).',//,
     C '   A general rule-of-thumb is that K can be estimated by',/,
     C ' the travel time through the reach and a value of 0.2 ',/,
     C ' can be used for X.  The parameter X is restricted to lie',/,
     C ' between 0.0 and 0.5.  A value of 0.5 suggests equal',/,
     C ' weighting of inflow and outflow, and it appears as',/,
     C ' primarily translation of the hydrograph with little',/,
     C ' attenuation.  A value of 0.0 suggests that the inflow',/,
     C ' has little effect; this would reflect reservoir storage',/,
     C ' type effects in which attenuation would dominant.')
 30   WRITE(KWT,91)
 91   FORMAT(76('-'),/,'  ENTER A [1] TO CONTINUE')
```

```
      WRITE(KWT,99)
      READ(KRD,*)KK
      IF(KK.NE.1)GO TO 30
      WRITE(KWT,506)
 506  FORMAT(24(/),
     C '   The translation parameter, K, can be estimate by',/,
     C '                     K = L/V',/,
     C ' where L is the channel length (feet) and V is a',/,
     C ' "representative" flow velocity (feet/sec.).',/,' Therefore,',
     C ' K has units of seconds.',/,
     C '   The parameters X and K are usually assumed related by',/,
     C '          X =< dT/(2K) =< (1-X) and X =< 0.5',/,
     C ' where dT is the computer modeling timestep in seconds.',/,
     C ' A mean value for dT then is dT = K, but the routing',/,
     C ' interval (dT) should be less than one-fifth the time-to-',
     C 'peak',/,' of the inflow hydrograph.',//,76('-'),/)
C
 40   DO 5 I=1,3000
      A(I)=0.
      B(I)=0.
 5    CONTINUE
      CALL MREAD(NA,A)
      NUMBER=NUMB(NA)
C
C  FIND STARTING TIME OF HYDROGRAPH
C
      DO 15 I=1,NUMBER
      IF(A(I).GT.0.)GO TO 25
 15   CONTINUE
 25   N1=I
      NX=0
      NP1=I-1
C
C  FIND PEAK Q AND TIME-TO-PEAK
C
      IMAX=0.
      QMAX=0.
      ASUM=0.
      DO 17 I=1,NUMBER
      IF(A(I).LT.QMAX)GO TO 13
      QMAX=A(I)
      IMAX=I
 13   ASUM=ASUM+A(I)
 17   CONTINUE
      IF(KRUN.EQ.2)GO TO 65
      WRITE(KWT,507)
 507  FORMAT(/,'  ENTER STORAGE PARAMETER X (0 < X < .5):')
      WRITE(KWT,99)
      READ(KRD,*)XX
      IF(XX.LT.0. .OR. XX.GT.0.5)GO TO 40
 50   WRITE(KWT,508)
 508  FORMAT(/,'  ENTER TRANSLATION PARAMETER K IN SECONDS')
      WRITE(KWT,99)
      READ(KRD,*)XK
      IF(XK.LT.0.)GO TO 50
```

```
      TT=32.*FLOAT(IMAX-NP1)
      TT=AMIN1(TT,XK)
      T1=2.*XK*XX
60    IF(T1.EQ.TT)TT=T1+60.
      IF(T1.LT.TT)WRITE(KWT,509)T1,TT
      IF(T1.GT.TT)WRITE(KWT,509)TT,T1
509   FORMAT(//,'  A SUGGESTED VALUE FOR THE COMPUTATION',
     C ' TIMESTEP, dT,',/,'  IS BETWEEN [',F8.2,'] AND [',F8.2,
     C '] SECONDS.',//,
     C '  SELECT dT VALUE OPTION TO BE USED:',/,
     C '   OPTION      dT(min.)      dT(sec.)',/,
     C '     [1]          1.            60.',/,
     C '     [2]          2.           120.',/,
     C '     [3]          3.           180.',/,
     C '     [4]          4.5          270.',/,
     C '     [5]          6.           360.',/,
     C '     [6]          9.           540.',/,
     C '     [7]         12.           720.',/,
     C '     [8]         15.           900.',/,
     C '     [9]         30.          1800.',/,
     C '    [10]         60.          3600.',/)
      WRITE(KWT,510)
510   FORMAT(/,'  ENTER THE TIMESTEP OPTION')
      WRITE(KWT,99)
      READ(KRD,*)KOPT
      IF(KOPT.LT.0 .OR. KOPT.GT.10)GO TO 60
      GO TO 67
65    READ(KRD,*)XX
      READ(KRD,*)XK
      READ(KRD,*)KOPT
67    IF(KOPT.EQ.1)DELT=60.
      IF(KOPT.EQ.2)DELT=120.
      IF(KOPT.EQ.3)DELT=180.
      IF(KOPT.EQ.4)DELT=270.
      IF(KOPT.EQ.5)DELT=360.
      IF(KOPT.EQ.6)DELT=540.
      IF(KOPT.EQ.7)DELT=720.
      IF(KOPT.EQ.8)DELT=900.
      IF(KOPT.EQ.9)DELT=1800.
      IF(KOPT.EQ.10)DELT=3600.
      IF(KRUN.NE.1)GO TO 117
      WRITE(KWT,503)
503   FORMAT(/,2X,'ENTER THE UPSTREAM NODE NUMBER ')
      WRITE(KWT,99)
117   READ(KRD,*)NUP
      IF(KRUN.NE.1)GO TO 127
      WRITE(KWT,504)
504   FORMAT(/,2X,'ENTER THE DOWNSTREAM NODE NUMBER')
      WRITE(KWT,99)
127   READ(KRD,*)NDOWN
      IF(KRUN.NE.1)GO TO 227
      WRITE(KWT,513)
513   FORMAT(/,2X,'ENTER A [1] TO INCLUDE THE INFILTRATION',
     C ' THROUGH CHANNEL BED')
      WRITE(KWT,99)
```

```
  227  READ(KRD,*)KINF
       IF(KINF.NE.1)GO TO 135
       IF(KRUN.NE.1)GO TO 237
  228  WRITE(KWT,514)
  514  FORMAT(/,2X,'THE INFILTRATION RATE IS APPROXIMATED BY',/,
      C 3X,'QOUT = QIN + (QO-QIN)*EXP(-K*TIME)',//,2X,'ENTER',
      C ' QIN (CFS)')
       WRITE(KWT,99)
  237  READ(KRD,*)QIN
       IF(QIN.LE.0.)GO TO 228
       IF(KRUN.NE.1)GO TO 247
  238  WRITE(KWT,515)
  515  FORMAT(/,2X,'ENTER QO (CFS)')
       WRITE(KWT,99)
  247  READ(KRD,*)QO
       IF(QO.LT.QIN)GO TO 238
       IF(KRUN.NE.1)GO TO 257
  248  WRITE(KWT,516)
  516  FORMAT(/,2X,'ENTER COEFFICIENT K (1/HOUR)')
       WRITE(KWT,99)
  257  READ(KRD,*)COE
       IF(C.LE.0.)GO TO 248
  135  CONTINUE
  155  IF(KRUN.NE.1)GO TO 267
       WRITE(KWT,517)
  517  FORMAT(/,'  *** SELECT PRINT/VIEW OPTIONS: ***',/,
      C 2X,'ENTER A [1] = WHOLE HYDROGRAPH',/,5X,
      C 'OR A [2] = PARTIAL HYDROGRAPH')
       WRITE(KWT,99)
  267  READ(KRD,*)KK
       IF(KK.EQ.1)GO TO 160
       IF(KK.EQ.2)GO TO 177
       GO TO 155
  160  TIME1=0.
       TIME2=100.
       GO TO 180
  177  IF(KRUN.NE.1)GO TO 277
       WRITE(KWT,518)
  518  FORMAT(/,'  ENTER MODEL TIME(HOURS) FOR BEGINNING OF',
      C ' RESULTS')
       WRITE(KWT,99)
  277  READ(KRD,*)TIME1
       TIME12=TIME1+.05
       IF(KRUN.NE.1)GO TO 287
       WRITE(KWT,519)TIME12
  519  FORMAT(/,'  ENTER MODEL TIME(HOURS) GREATER THAN [ ',F8.3,
      C ' ] FOR END OF RESULTS')
       WRITE(KWT,99)
  287  READ(KRD,*)TIME2
       IF(TIME2 .LE. TIME12)GO TO 177
  180  CONTINUE
C
C  CALCULATE ROUTING COEFFICIENTS CO, C1, AND C2
C
       DD=XK-(XK*XX)+0.5*DELT
```

```
      CO=-(XK*XX-0.5*DELT)/DD
      C1=(XK*XX+.5*DELT)/DD
      C2=1.-CO-C1
C
C  OUTPUT ROUTING PARAMETERS
C
  70  WRITE(KRT,511)NUP,NDOWN,NA,XX,XK,DELT
 511  FORMAT(//,'    *** MUSKINGUM ROUTING MODEL ***',//,
     C '  FLOW PROCESS FROM NODE',I4,' TO NODE',I4,/,
     C '  STREAM NUMBER [',I2,' ] IS ROUTED BY THE MUSKINGUM',
     C ' METHOD',/,'   ROUTING PARAMETERS:',/,7X,'X = ',F5.3,
     C /,7X,'K = ',F7.1,' SEC.',/,7X,'USER-SELECTED TIMESTEP = ',
     C F7.1,' SEC.')
      WRITE(KWT,512)CO,C1,C2
 512  FORMAT(/,'   THE ROUTING EQUATION IS:',
     C ' O2 = CO*I2 + C1*I1 + C2*O1',/,
     C 7X,'WITH CO = ',F5.3,' ;C1 = ',F5.3,' ;C2 = ',F5.3)
      IF(KINF.EQ.1)WRITE(KWT,203)QIN,QO,QIN,COE
 203  FORMAT(/,5X,'CHANNEL INFILTRATION RATE IS APPROXIMATED BY',
     C /,7X,'QOUT(CFS) = ',F6.2,' + (',F6.2,' - ',F6.2,') * EXP ('
     C , ' -1. * TIME(HOUR) * ',F6.3,')',/)
C
C  ADJUST HYDROGRAPH TO NEW TIME INTERVAL
C
      GO TO (100,110,120,130,140,140,140,140,140,140),KOPT
 100  K=NP1
      DO 200 I=N1,NUMBER
      DO 200 J=1,3
      K=K+1
      B(K)=A(I)
 200  CONTINUE
      GO TO 170
 110  K=0
      DO 210 I=N1,NUMBER,2
      K1=3*K+NP1
      B(K1+1)=A(I)
      B(K1+2)=(A(I)+A(I+1))/2.
      B(K1+3)=A(I+1)
      K=K+1
 210  CONTINUE
      K=K1+3
      GO TO 170
 120  DO 220 K=1,NUMBER
 220  B(K)=A(K)
      GO TO 170
 130  K=0
      DO 230 I=N1,NUMBER,3
      K1=2*K+NP1
      B(K1+1)=(3.*A(I)+1.5*A(I+1))/4.5
      B(K1+2)=(1.5*A(I+1)+3.*A(I+2))/4.5
      K=K+1
 230  CONTINUE
      K=K1+2
      GO TO 170
 140  K=NP1
```

```
      NX=IFIX(DELT/180.)
      DO 240 I=N1,NUMBER,NX
      K=K+1
      IJ=I-1
      DO 240 J=1,NX
      B(K)=B(K)+A(IJ+J)/FLOAT(NX)
 240  CONTINUE
 170  INTER=K
      DO 175 I=1,3000
 175  A(I)=0.
C
C  MUSKINGUM ROUTING MODEL
C
      IF(KINF.NE.1)GO TO 280
      TIME=0.
      DO 295 I=1,INTER
      TIME=TIME+DELT/3600.
      QOUT=QIN+(QO-QIN)*EXP(-1.*TIME*COE)
      B(I)=B(I)-QOUT
      IF(B(I).LT.0.)B(I)=0.
 295  CONTINUE
 280  DO 290 I=1,N1
 290  A(I)=0.
      A(N1+1)=B(N1)
      CSUM=B(N1)
      AASUM=B(N1)
      DO 300 I=N1+1,INTER
      AASUM=AASUM+B(I)
      A(I+1)=CO*B(I+1)+C1*B(I)+C2*A(I)
      IF(A(I+1).LT.0.)A(I+1)=0.
      B(I)=0.
      CSUM=CSUM+A(I+1)
 300  CONTINUE
      DQ=AASUM-CSUM
      IF(DQ.LE.0.)GO TO 309
      IF(DQ.LE.A(INTER+1))GO TO 305
      IN=IFIX(DQ/A(INTER+1))
      DO 307 J=1,IN
 307  A(INTER+1+J)=A(INTER+1)
      A(INTER+IN+2)=DQ-FLOAT(IN)*A(INTER+1)
      INTER=INTER+IN+2
      GO TO 309
 305  A(INTER+2)=DQ
      INTER=INTER+2
C
C  RETURN THE HYDROGRAPH TO 3-MINUTE UNIT INTERVAL
C
 309  GO TO (310,320,330,340,350,350,350,350,350,350),KOPT
 310  K=NP1
      DO 315 I=N1,INTER,3
      K=K+1
      B(K)=(A(I)+A(I+1)+A(I+2))/3.
 315  CONTINUE
      GO TO 380
 320  K=0
```

```
      DO 325 I=N1,INTER,3
      K1=2*K+NP1
      B(K1+1)=(2.*A(I)+A(I+1))/3.
      B(K1+2)=(A(I+1)+2.*A(I+2))/3.
      K=K+1
  325 CONTINUE
      K=K1+2
      GO TO 380
  330 DO 335 K=N1,INTER
  335 B(K)=A(K)
      GO TO 380
  340 K=0
      DO 345 I=N1,INTER,2
      K1=3*K+NP1
      B(K1+1)=A(I)
      B(K1+2)=(A(I)+A(I+1))/2.
      B(K1+3)=A(I+1)
      K=K+1
  345 CONTINUE
      K=K1+3
      GO TO 380
  350 K=0
      DO 355 I=N1,INTER
      K1=NX*K+NP1
      K=K+1
      DO 355 J=1,NX
      B(K1+J)=A(I)
  355 CONTINUE
      K=K1+NX
  380 TIME=0.
      WRITE(KWT,521)
  521 FORMAT(//,5X,'   TIME       INFLOW RATE     OUTFLOW',
     C ' RATE',/,5X,'  (HOURS)        (CFS)             (CFS)',/)
      DO 390 I=1,3000
  390 A(I)=0.
      CALL MREAD(NA,A)
C
C  CHECK MASS BALANCE
C
      CSUM=0.
      DO 405 I=1,K
      CSUM=CSUM+B(I)
      IF(CSUM.LT.ASUM)GO TO 405
      B(I)=CSUM-ASUM
      B(I+1)=0.
      CSUM=ASUM
  405 CONTINUE
      IF(CSUM.GE.ASUM)GO TO 407
      DQ=ASUM-CSUM
      IF(DQ.GE.B(K))GO TO 409
      B(K+1)=DQ
      K=K+1
      GO TO 407
  409 IN=IFIX(DQ/B(K))
      DO 411 I=1,IN
```

```
      K=K+1
      B(K)=B(K-1)
 411  CONTINUE
      B(K+1)=DQ-B(K-1)*FLOAT(IN)
      IF(B(K+1).LT.0.)B(K+1)=0.
      K=K+1
C
 407  DO 400 I=1,K
      TIME=TIME+.05
      IF(TIME.LT.TIME1)GO TO 400
      IF(TIME.GT.TIME2)GO TO 410
      WRITE(KWT,522)TIME,A(I),B(I)
 522  FORMAT(6X,F6.2,6X,F8.2,7X,F8.2)
 400  CONTINUE
 410  IF(KRUN.NE.1)GO TO 317
 370  WRITE(KWT,523)
 523  FORMAT(/,2X,'ENTER A [1] TO ACCEPT THE PROCESS'
     C ,/,5X,'OR A [2] TO REJECT THE PROCESS')
      WRITE(KWT,99)
      READ(KRD,*)KOP
      IF(KOP.LT.1 .OR. KOP.GT.2)GO TO 370
      IF(KOP.EQ.2)RETURN
C
      WRITE(7,*)KXY
      WRITE(7,*)NA
      WRITE(7,*)XX
      WRITE(7,*)XK
      WRITE(7,*)KOPT
      WRITE(7,*)NUP
      WRITE(7,*)NDOWN
      WRITE(7,*)KINF
      IF(KINF.NE.1)GO TO 430
      WRITE(7,*)QIN
      WRITE(7,*)QO
      WRITE(7,*)COE
 430  WRITE(7,*)KK
      IF(KK.EQ.2)WRITE(7,*)TIME1
      IF(KK.EQ.2)WRITE(7,*)TIME2
C
 317  DO 420 I=1,K
 420  A(I)=B(I)
      CALL MWRITE(NA,A)
      NUMB(NA)=K
C
      RETURN
      END
C
```

considerations in calibrating elements of design storm models

7

7.1. Introduction

To this point, the methods presented have been oriented towards design, rather than the analyses that would be involved in developing the design methods. The process that is necessary to develop design methods is sometimes referred to as calibration, which involves the evaluation of model coefficients by adjusting the coefficients so that the model reproduces measured hydrologic data as accurately as possible. A design model that is calibrated from local or regional data should provide more accurate designs than a model that has been calibrated for either another region or locality or locations where the importance of the physical processes is different. Designs that are obtained using a calibrated model will probably be more accurate and, thus, will enhance public safety, as well as reduce public costs. Therefore, every effort should be made to calibrate a design method prior to adopting the method for design. If a proposal has been made to adopt an existing design model, the model should be tested with local data prior to its adoption.

Each of the categories of design methods covered in previous chapters can be calibrated. The type of calibration undertaken will depend on the data that are available. Where measured times of concentration or hydrograph lag times are available, the coefficients of a prediction equation can be calibrated. Where annual maximum peak discharge data are available, peak discharge models can be calibrated for any exceedence probability using the flood frequency curve derived from the annual maximum data. Where measured rainfall hyetograph and runoff hydrograph models are available, the loss function and unit hydrograph models can be calibrated. Where point precipitation measurements of reasonable record lengths are available, local point intensity-duration-frequency relationships can be calibrated or existing relationships can be tested. Storm depth-area relationships can be calibrated from isohyetal patterns of measured storm event rainfall. Channel routing coefficients can be calibrated for stream reaches where hydrographs have been measured at both the upstream and downstream cross sections along a stream reach. In each of these calibration problems every attempt should be made to obtain as much data as possible and have data that will be representative of design conditions. Calibration of these design methods should lead to greater design accuracy.

Each of these calibration problems will be investigated here. Before discussing the specific procedures, the goals and objectives of calibration need to be reviewed. Also, since the calibration methods use similar criteria, the specific criteria that should be used in calibration are discussed in a general context. Examples will be provided to illustrate the calibration processes.

7.2. Goals and Objectives of Model Calibration

Models are developed for a number of reasons, which can be separated into two groups: (1) to develop a prediction equation, and (2) to explain the importance and effects of variables and processes. These two groups reflect different goals of modeling. If the only need is to obtain a predicted value of some criterion variable, such as a peak discharge, then a very simple model, such as the Rational method, may be acceptable. If the goal is to develop a model that can be used to measure the effects of the physical processes involved, then a more detailed and physically based model will be necessary. For example, if it is necessary to show the effect of urbanization on flood runoff rates, then a model that is sensitive to the level of urban development must be calibrated using data collected over a range of levels in urban development. In addition to affecting the complexity of the model, the two goals of modeling will require different levels of calibration. However, in spite of these differences, there are similarities in the objectives used in calibration.

The most fundamental objective of calibration is to eliminate errors of prediction. Recognizing that there is no way error can be eliminated entirely, the error of prediction can be divided into two parts, systematic error and nonsystematic error. The presence of systematic error in model predictions is often referred to as bias. Thus, one objective of calibration is to eliminate bias. The sum of the errors (i.e., the sum of the differences between the predicted and measured values of the criterion variable) is a measure of bias. A positive sum would suggest that the predicted values are, on the average, larger than the measured values, thus overprediction. Similarly, a negative sum would suggest a negative bias, or underprediction. However, even when the sum of the errors equals zero, as it would using regression analysis, it does not necessarily follow that the model is unbiased. The model will be unbiased only if the sum of the errors is essentially zero and there are no local biases. For example, if a model for predicting peak discharge were being calibrated and the sum of the errors equaled zero, the model might still be biased if all of the small (or large) discharges were either over predicted or under predicted. In summary, one criterion to use in calibrating a model is the degree to which the calibration process provides unbiased estimates. Failure of a model to provide unbiased estimates suggests that the model may not have been properly calibrated.

To illustrate the concept of bias, the data of Table 7.1, which are shown graphically in Fig. 7.1, represent the times of concentration predicted with models $\hat{T}c$ and values obtained from measured hydrologic data (Tc). The difference, ($\hat{T}c$ - Tc) is the error (e) of prediction. For the data shown in Fig. 7.1a the model consistently overpredicted the true value Tc, and so the errors are all positive. The sum of the errors equals 3.20 hours, which indicates a bias of 0.40 hours. For the eight watersheds of Fig. 7.1b, the model tends to underpredict, with seven of the eight errors being negative. The model shows a bias of -0.30 hours, which means that the average error is -0.30 hours. For the eight watersheds of Fig. 7.1c, the errors have a sum of zero, which suggests that the model is unbiased; that is, the model tends to overpredict as much as it does underpredict. While the overall bias is zero, it should be evident from Fig. 7.1c that the model is actually biased. For both small and large times of concentration the model tends to overpredict. For times of concentration between approximately 0.60 hours and 1.25 hours, the model tends to underpredict as indicated by the negative errors. Thus while the overall bias of the model is zero, the model appears to have some significant local biases. In the case of Fig. 7.1c a nonlinear relationship between $\hat{T}c$ and Tc is evident, which suggests that the assumed structure of the model is incorrect.

In addition to being unbiased, an objective of calibration should be to produce a precise model. Precision will be the term used here to reflect the degree of nonsystematic variation in the errors. Precision is a component of the overall accuracy of a model, with the accuracy composed of the precision and bias. Even when a model is unbiased, the predicted values of the criterion variable will not exactly equal the measured values. Plots of the measured and predicted values of a criterion variable will be characterized by a certain amount of random scatter. A precise model would be one that shows little scatter; and imprecise model would be characterized by a significant amount of scatter in the plot of the predicted and measured values of the criterion variable. An imprecise model may be the result of an incorrect model structure, the ignoring of important predictor variables or physical processes, or inadequate calibration.

The standard deviation of the errors is often used as a measure of the precision of a model. Where the model is unbiased, the standard deviation will also measure the accuracy of the model. A model that has a relatively small standard deviation is said to be precise, while a large standard deviation indicates an imprecise model. The standard deviations of the errors for the data of Table 7.1 suggest that the model used with Fig. 7.1a is more precise than the models for either Fig. 7.1b or Fig. 7.1c; however, the differences between the three standard deviations is not significant. Figures 7.1d and 7.1e show data with relatively large scatter. The model behind Fig. 7.1d is highly biased, with a mean error of 0.70 hours. While the model behind Fig.7.1e is unbiased, it has the same scatter, and therefore imprecision, as the model behind Fig. 7.1d. The model behind Fig. 7.1e is more accurate than the model behind Fig. 7.1d because the combination of the bias and precision is more favorable.

Whether one is interested in calibrating a model solely for prediction or both prediction and measuring of effects, the objective of model formulation and calibration is to develop an unbiased, precise model. A biased or imprecise model may not provide either acceptable predictions or realistic assessments of the effects of the variables or physical processes.

7.3. Criteria for Model Assessment

Having defined the goals of calibration, it is now necessary to translate the goals and objectives into specific criteria for calibration and assessment. Calibration criteria can be separated into three categories: rationality, graphical, and statistical. The task of model assessment should not be viewed as the process of selecting one of these three groups. In fact, criteria from each group should be used in calibrating and assessing a model.

The first step in data analysis should be to graph the data. Of course, the type of graphical analysis depends on the variables involved. For analyses involving a single random variable (i.e., univariate analyses), the data can be graphed in the form of either histograms or box-and-whisker plots. For example, an array of measured times of concentration, peak discharges, or loss rates could be graphically analyzed as a single random variable. A histogram gives a pictorial representation of data characteristics such as the central tendency, the scatter, the skewness, the presence of extreme events, and the uniformity in the distribution of the data. For example, Fig. 7.2 shows the distribution of times of concentration for a set of 39 urban watersheds. The histogram indicates a central tendency of about 1.5 hours, with most of the values less than 2 hours. While the Tc values ranged as high as 5 hours, 87 percent of the sample watersheds had Tc's less than two hours. Thus, any prediction equation developed with this data may only be accurate for Tc's less than 2 hours.

TABLE 7.1
EVALUATION OF BIAS AND PRECISION IN RESULTS OF HYDROLOGIC MODELS FOR ESTIMATING THE TIME OF CONCENTRATION

$\hat{T}_c$ (hrs)	T_c (hrs)	e (hrs)	$\hat{T}_c$ (hrs)	T_c (hrs)	e (hrs)	$\hat{T}_c$ (hrs)	T_c (hrs)	e (hrs)	$\hat{T}_c$ (hrs)	T_c (hrs)	e (hrs)	$\hat{T}_c$ (hrs)	T_c (hrs)	e (hrs)
0.44	0.18	0.26	0.50	0.68	-0.18	0.50	0.12	0.38	0.62	0.18	0.44	0.62	0.88	-0.26
0.68	0.32	0.36	0.75	1.00	-0.25	0.45	0.25	0.20	1.00	0.12	0.88	1.00	0.82	0.18
0.94	0.38	0.56	0.81	1.25	-0.44	0.68	0.82	-0.14	0.84	0.62	0.22	0.84	1.32	-0.48
0.88	0.62	0.26	1.32	1.18	0.14	0.62	1.00	-0.38	1.50	0.42	1.08	1.50	1.12	0.38
1.25	0.75	0.50	1.18	1.81	-0.63	0.94	1.12	-0.18	1.32	0.68	0.64	1.32	1.38	-0.06
1.50	0.82	0.68	1.30	1.62	-0.32	1.56	1.32	0.24	1.94	0.56	1.38	1.94	1.26	0.68
1.38	1.24	0.14	1.50	2.00	-0.50	1.62	1.44	0.18	1.92	1.12	0.80	1.92	1.82	0.10
1.68	1.24	0.44	1.62	1.84	-0.22	0.56	0.86	-0.30	1.52	1.36	0.16	1.52	2.06	-0.54
sum		3.20			-2.40			0.00			5.60			0.00
bias		0.40			-0.30			0.00			0.70			0.00
std.dev.		0.18			0.23			0.28			0.42			0.42

T_c = measured time of concentration
$\hat{T}_c$ = predicted estimate of T_c
e = error = $\hat{T}_c - T_c$
sum = sum of errors
bias = mean error
std. dev. = standard deviation of the errors

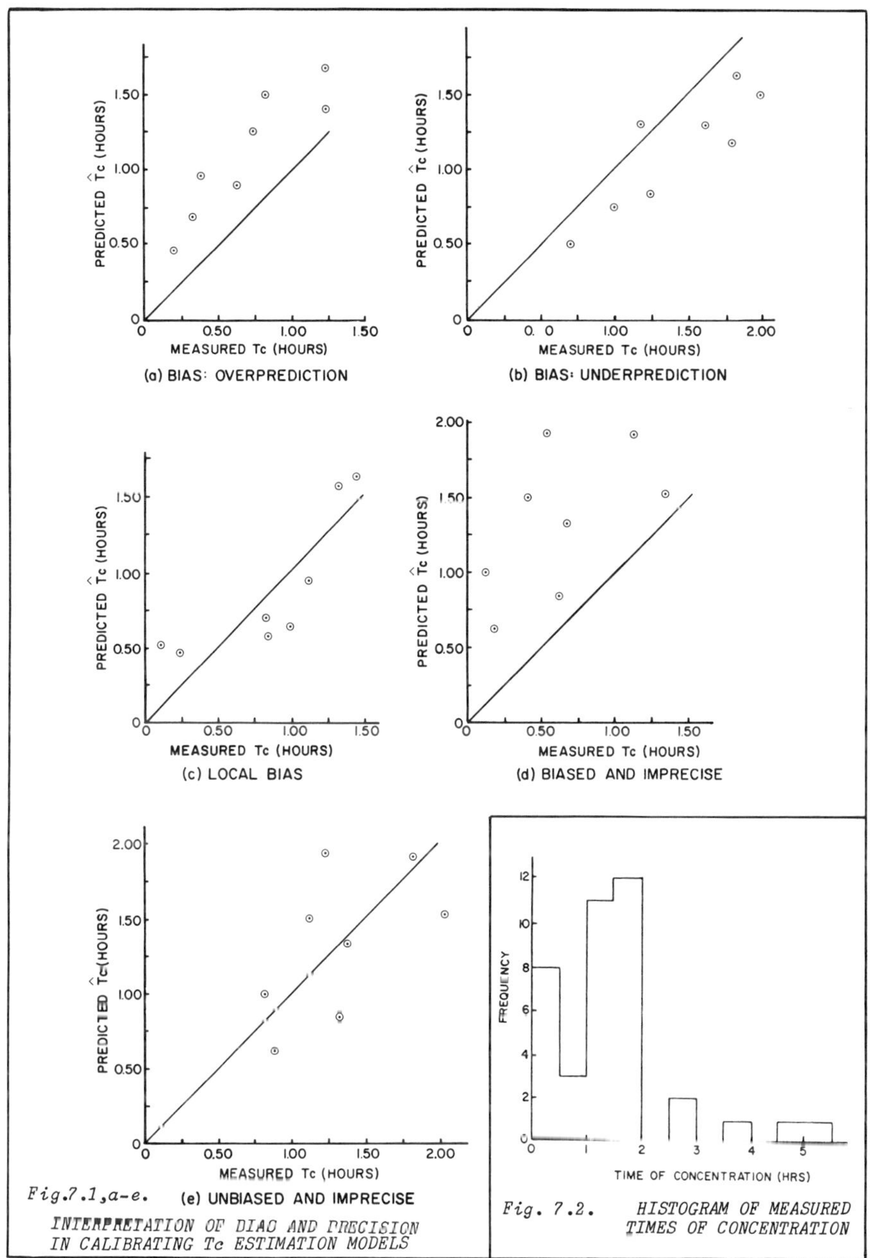

Fig.7.1,a-e. INTERPRETATION OF BIAS AND PRECISION IN CALIBRATING Tc ESTIMATION MODELS

Fig. 7.2. HISTOGRAM OF MEASURED TIMES OF CONCENTRATION

Where comparisons must be made for groups of data on a single random variable, different types of graphical analyses can be used. A box-and-whisker plot shows the following: the sample size, the mean, the minimum and maximum values, and the 10, 25, 50, 75, and 90 percentile values. Such information is useful for comparing data. Figure 7.3 shows box-and-whisker plots for the percent deviation of peak discharges estimated with three hydrologic models in five drainage area size groups. It is evident that the scatter for hydrologic model 2 is much greater than that for either model 1 or 5. The closeness of the mean for model 5 to zero suggests that it is unbiased. Model 2 appears to provide biased results, especially for the smaller drainage area sizes. The box-and-whisker plots, while presenting a single variable, provide a better means of comparison than the histogram.

Statistical indices, such as correlation coefficients and error sums of squares, can be used as criteria for assessing the goodness of fit of a calibrated model. The values of the criterion variable predicted with the calibrated model can be used to form a vector. Letting the vectors of predicted and measured values be denoted as Y and X, respectively, the correlation coefficient R can be determined by:

$$R = \frac{\sum_{i=1}^{n} X_i Y_i - \sum_{i=1}^{n} X_i \sum_{i=1}^{n} Y_i / n}{\left(\sum_{i=1}^{n} X_i^2 - \left(\sum_{i=1}^{n} X_i \right)^2 / n \right)^{0.5} \left(\sum_{i=1}^{n} Y_i^2 - \left(\sum_{i=1}^{n} Y_i \right)^2 / n \right)^{0.5}} \tag{7.1}$$

in which n is the number of elements in each vector. The value of the correlation coefficient varies from -1 to 1, with a value of zero indicating no relationship between the values of X and Y. An absolute value of 1 indicates a perfect fit, with the sign indicating whether the relationship is direct (+) or indirect (-). Figure 7.4 shows a plot of 100-year peak discharge estimates obtained from flood frequency analyses at a gaged location and estimates obtained from a hydrologic model; the data are given in Table 7.2. A correlation coefficient was computed for the data, with a value of 0.92. This suggests the model that provided the estimated values of Q_p provides reasonably accurate estimates of gage estimates. The square of R is a measure of the fraction of the variation in the gage estimates that could be explained by the hydrologic model; in this case R^2 equals 0.85, which indicates an explained variation of 85 percent.

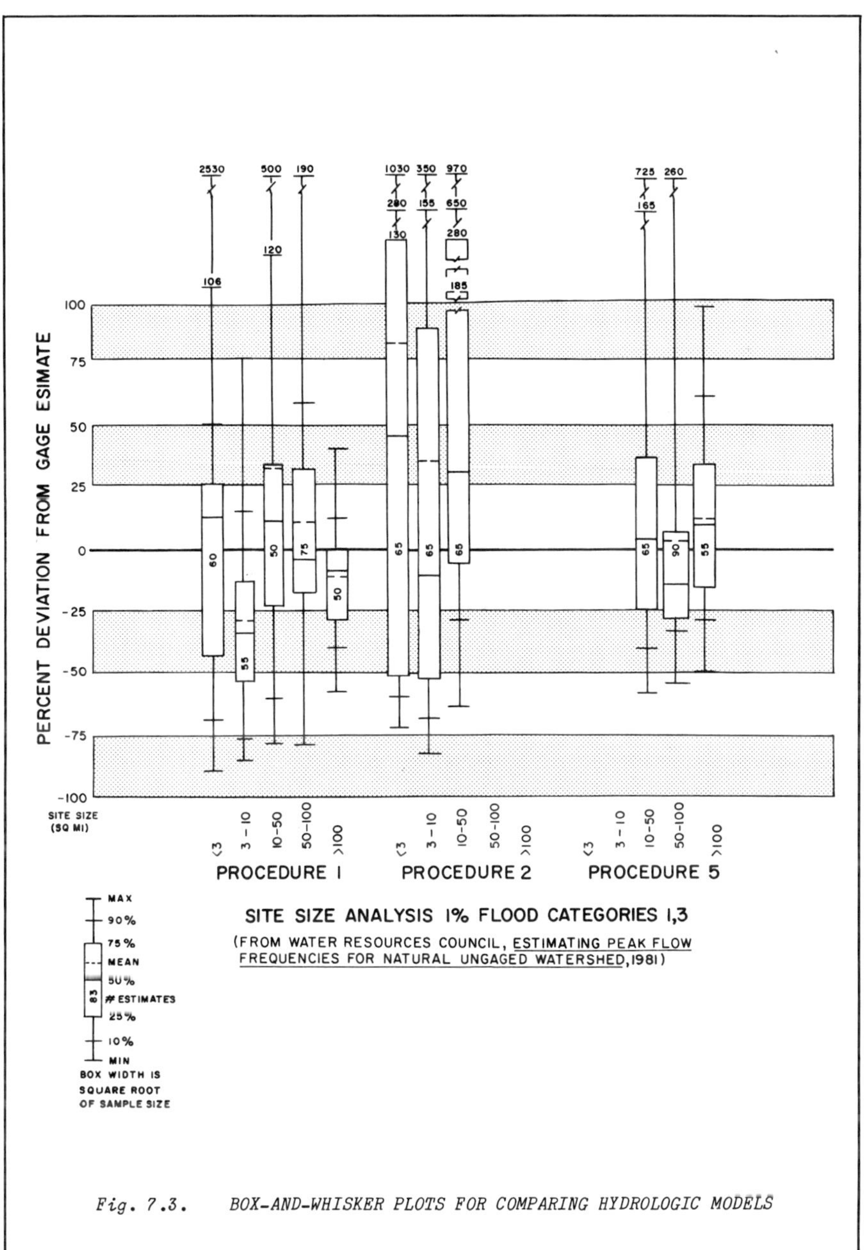

Fig. 7.3. BOX-AND-WHISKER PLOTS FOR COMPARING HYDROLOGIC MODELS

TABLE 7.2

PREDICTED 100-YEAR PEAK DISCHARGE AND CORRESPONDING LOG PEARSON TYPE III ESTIMATE

Predicted q_p(cfs)	Gage Estimate (cfs)
100	50
100	75
200	150
250	150
250	300
500	1050
700	750
800	850
850	400
850	500
950	550
950	800
1100	800
1400	650
1900	800
1850	1200
2250	1750
2800	2850
5500	2600
5200	4300
5800	5850

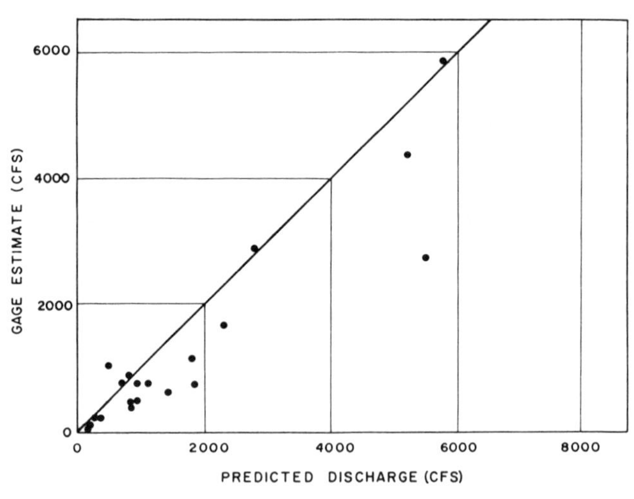

Fig. 7.4. PREDICTED AND MEASURED PEAK DISCHARGE ESTIMATES

The third criterion for assessing the quality of calibration is rationality. An irrational model is of little value either for prediction or measuring the importance of the model variables. Rationality can be evaluated by examining the signs and magnitudes of the fitted coefficients and by testing the sensitivity of the model output to changes in the variables. The concept of model rationality is discussed by McCuen (1985), and McCuen and Snyder (1986) provide a comprehensive discussion of model sensitivity.

7.4. Calibrating Unit Hydrographs and Loss Functions

For watersheds where storm event data are available, a unit hydrograph should be calibrated for the watershed. Where such data are not available for the watershed of interest but data are available for other watersheds in the homogeneous region, a regionalized unit hydrograph should be developed. The procedure for computing a site specific unit hydrograph is just one step of the procedure for developing a regionalized unit hydrograph. Specifically, a regionalized unit hydrograph is developed from an averaging of unit hydrographs developed for specific sites in the region. Since the development of a regionalized unit hydrograph just involves an averaging of site specific unit hydrographs, only the latter will be emphasized here.

The data required to compute a unit hydrograph for a given storm event includes: (1) the total rainfall hyetograph and (2) the storm runoff hydrograph. In addition to the data, two models will be required: (1) a baseflow model, if baseflow was present during the storm event, and (2) a model to represent the rainfall losses. The loss function may consist of two elements, one for modeling initial abstractions and one for other losses. In other analyses, a single loss function model may be used to represent the effects of both the initial abstraction and other losses.

The objective of the data analysis is to calibrate both the unit hydrograph and the loss function. Both of these time distributed functions are necessary for design. To meet this objective it is necessary to separate both the baseflow from the total runoff hydrograph and the losses from the rainfall hyetograph. The procedure is as follows:

1. separate baseflow from the total runoff hydrograph using an assumed baseflow model;
2. compute the direct runoff hydrograph by subtracting the baseflow from the total runoff hydrograph;
3. compute the volume of direct runoff and convert the distribution of direct runoff to a unit hydrograph that has a volume of 1 area-inch (i.e., a depth of 1 inch);

4. define the form of the loss function; and
5. fit the coefficients of the loss function so that the volume of rainfall excess equals the volume of direct runoff.

This five step procedure is very generalized, and many more elaborate procedures have been proposed, especially for use in analyzing complex storms.

The selected baseflow model will affect the shape of the direct runoff. Additionally, the selection of the baseflow can determine the rationality of the loss function and the unit hydrograph. Baseflow functions $q_b(t)$ that could be used are the constant function:

$$q_b(t) = c_o \tag{7.2}$$

and the straight-line function:

$$q_b(t) = b_o + b_1 t \tag{7.3}$$

in which c_o, b_o, and b_1 are fitting coefficients, t is time, and $q_b(t)$ is the baseflow at time t. If the values of c_o in Eq. 7.2 equals the discharge rate at the start of direct runoff, then direct runoff is assumed to continue until a discharge of c_o is reached; this may not be a realistic assumption. When using the straight-line method of Eq. (7.3), it is necessary to select both an initial point for baseflow separation and a time when direct runoff ends; two points are necessary in order to fit the two coefficients, b_o and b_1, of Eq. (7.3). The inflection point on the hydrograph recession is frequently assumed to be the end of direct runoff. Once the two points are selected, values for b_o and b_1 can be computed.

In addition to selecting a baseflow model, it is necessary to select the form of the loss function. A number of loss functions were discussed in Chapter 3. For the constant percentage loss function, the losses are assumed to be a fraction $\overline{Y}$ of the total rainfall volume P; thus, the ordinates of the loss function, f(t), equal the product of the fraction $\overline{Y}$ and the rainfall ordinate, where $\overline{Y}$ is given by:

$$\overline{Y} = 1 - (V_d/V_p) \tag{7.4}$$

in which V_d and V_p are the volumes of direct runoff and total rainfall, respectively.

While the constant percentage loss function is easy to compute, the phi-index method is probably a more rational model form; however, the value of phi is slightly more difficult to compute. The easiest solution is a numerically determined estimate. Specifically, an initial value can be assumed and used to compute the loss function and the distribution of rainfall excess. The initial estimate is corrected such that the volume of rainfall excess is made to equal the volume of direct runoff. The iterative procedures is continued until the difference between the volumes of direct runoff and rainfall excess are within some tolerance, say 0.5 percent.

The hydrograph separation and unit hydrograph estimation can be illustrated using the storm event shown in Fig. 7.5. The data are from the October 25, 1964, storm event on a 1.68 acre watershed, in Cherokee, Oklahoma. The runoff hydrograph and rainfall hyetograph are given in Tables 7.3 and 7.4. The rainfall is characterized by a single large burst for a duration of 15 minutes, with almost negligible rainfall for other times. The runoff hydrograph has a single peak that occurs approximately four minutes after the end of the large burst of rainfall.

While baseflow is not expected for such a small watershed, a baseflow model of the form of Eq. (7.3) was selected to eliminate the long tail of the measured runoff hydrograph. The extended tail may have resulted from the low intensity rain that occurred from 1920 hours to 2030 hours. The model of Eq. (7.3) was used to define a baseflow loss from 1914 hours to 2000 hours:

$$q_b(t) = 0.0204 + 0.00007826(t - 1914)$$

in which $q_b(t)$ is measured in inches/hour and t is the storm time in hours. The baseflow losses, which are given in Table 7.3, were subtracted from the total storm hydrograph to get the direct runoff hydrograph, which is also given in Table 7.3. The volume of direct runoff is:

$$V_{DRO} = 11.7349 \frac{\text{in}}{\text{hr}} \times 1 \text{ min} \times \frac{1 \text{ hr}}{60 \text{ min}} = 0.19558 \text{ inches}$$

Thus, the unit hydrograph in inches/hour is computed by multiplying each ordinate of the direct runoff hydrograph by the reciprocal of the volume. The unit hydrograph in cfs can be computed by multiplying by the drainage area and making the necessary adjustments for units. The duration of the unit hydrograph can be determined by subtracting initial abstraction losses and other losses from the rainfall hyetograph. All rainfall prior to the start of direct runoff (storm time 1914 hours) was considered to be initial abstraction losses. The phi index method can be used to separate the other losses from the

TABLE 7.3
CALCULATION OF SYNTHETIC UNIT HYDROGRAPH FROM MEASURED TOTAL RUNOFF HYDROGRAPH (q) FOR THE STORM EVENT OF OCTOBER 25, 1964, AT CHEROKEE, OKLAHOMA

time of day	q (in/hr)	q_b (in/hr)	$q-q_b$ (in/hr)	UH (in/hr)	UH (cfs)	Synthetic UH (cfs)
1912	0	0	0	0	0	
1913	0.0204	0.0204	0	0	0	
1914	0.0204	0.0204	0	0	0	
1915	0.0665	0.0205	0.0460	0.24	0.4	0
1916	0.0665	0.0206	0.0459	0.23	0.4	0
1917	0.122	0.0206	0.1014	0.51	0.9	0
1918	0.122	0.0207	0.1013	0.52	0.9	0
1919	0.285	0.0208	0.2642	1.35	2.3	0.3
1920	0.285	0.0209	0.2641	1.35	2.3	1.8
1921	0.524	0.0209	0.5031	2.57	4.4	6.0
1922	0.524	0.0210	0.5030	2.57	4.4	12.4
1923	1.51	0.0211	1.4889	7.61	12.9	18.4
1924	2.41	0.0202	2.3898	12.27	20.7	20.7
1925	1.39	0.0213	1.3687	6.99	11.9	18.7
1926	0.804	0.0213	0.7827	4.00	6.8	14.0
1927	0.636	0.0214	0.6146	3.14	5.3	9.0
1928	0.490	0.0215	0.4685	2.40	4.1	5.1
1929	0.365	0.0216	0.3434	1.76	3.0	2.6
1930	0.365	0.0217	0.3435	1.76	3.0	1.2
1931	0.237	0.0217	0.2153	1.10	1.9	0.5
1932	0.237	0.0218	0.2152	1.10	1.9	0.2
1933	0.237	0.0219	0.2151	1.10	1.9	0.1
1934	0.237	0.0220	0.2150	1.10	1.9	
1935	0.174	0.0220	0.1520	0.70	1.3	
1936	0.174	0.0221	0.1519	0.78	1.3	
1937	0.130	0.0222	0.1078	0.55	0.9	
1938	0.130	0.0223	0.1077	0.55	0.9	
1939	0.0990	0.0224	0.0766	0.39	0.7	
1940	0.0990	0.0224	0.0766	0.39	0.7	
1941	0.0990	0.0225	0.0765	0.39	0.7	
1942	0.0990	0.0226	0.0764	0.39	0.7	
1943	0.0990	0.0227	0.0763	0.39	0.7	
1944	0.0990	0.0227	0.0763	0.39	0.7	
1945	0.0852	0.0278	0.0624	0.32	0.5	
1946	0.0852	0.0229	0.0623	0.32	0.5	
1947	0.0852	0.0230	0.0622	0.32	0.5	
1948	0.0501	0.0231	0.0270	0.14	0.2	
1949	0.0360	0.0231	0.0129	0.07	0.1	
1950	0.0360	0.0232	0.0128	0.07	0.1	
1951	0.0360	0.0233	0.0127	0.07	0.1	
1952	0.0360	0.0234	0.0126	0.07	0.1	
1953	0.0240	0.0235	0.0005	0.00	0.0	
1954	0.0240	0.0235	0.0005	0.00	0.0	
1955	0.0240	0.0236	0.0004			
1956	0.0240	0.0237	0.0003			
1957	0.0240	0.0238	0.0002			
1958	0.0240	0.0238	0.0002			
1959	0.0240	0.0239	0.0001			
1960	0.0240	0.0240	0.0000			
			11.7349			

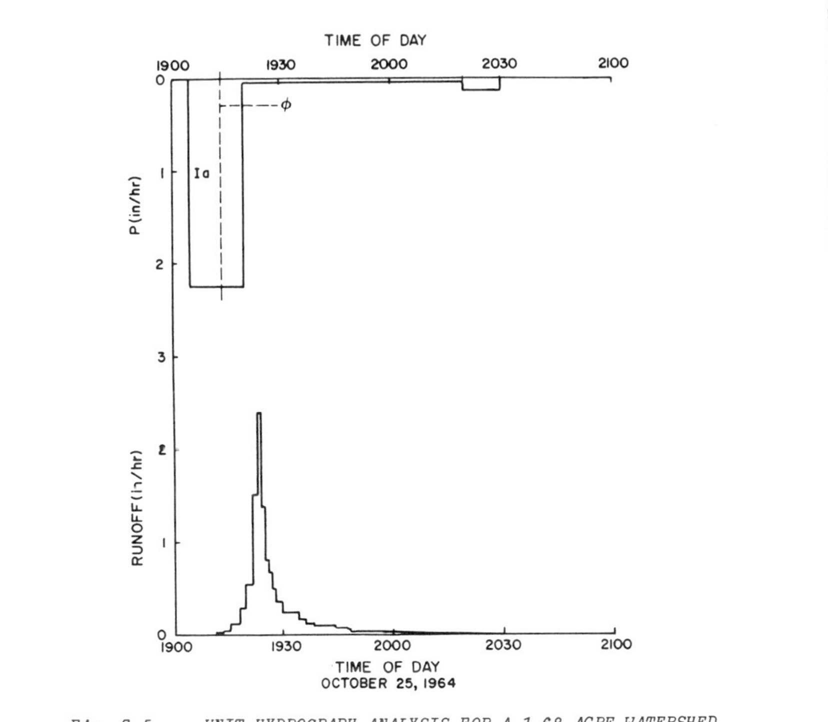

Fig. 7.5. *UNIT HYDROGRAPH ANALYSIS FOR A 1.68 ACRE WATERSHED*

TABLE 7.4

RAINFALL HYETOGRAPH AND LOSS FUNCTION ESTIMATES FOR THE STORM EVENT OF OCTOBER 25, 1964, AT CHEROKEE, OKLAHOMA

Time of Day(hrs)	Rainfall (in/hr)	Initial Abstraction (in/hr)	Losses (in/hr)	Rainfall Excess (in/hr)
1705				
1805	0.01	0.01	0	0
1905	0.00	0.00	0	0
1914	2.24	2.24	0	0
1920	2.24	0	0.28	1.96
2020	0.02	0	0.02	0
2030	0.12	0	0.12	0
2045	0.04	0	0.04	0

remaining rainfall. The high intensity rainfall covers a duration of six minutes (1914 to 1920), with a rate of 2.24 inches/hour. In order to have a volume of 0.19558 inches, the rainfall excess intensity would be:

$$i = (0.19558 \text{ in.}/6\text{-minutes})(60 \text{ min.}/\text{hr.}) = 1.9558 \text{ in.}/\text{hr.}$$

Thus, the phi index loss rate is (2.24-1.96) = 0.28 in./hr. The losses are shown in Fig. 7.5. Thus, the unit hydrograph has a duration of six minutes.

7.5. Least Squares Calibration of Unit Hydrographs

Having computed the rainfall excess and direct runoff, a number of methods have been proposed for obtaining the T-hour unit hydrograph. While the least squares method has some limitations, which will be discussed, it is a convenient method for deriving unit hydrograph ordinates from rainfall excess and direct runoff ordinates. It is easiest to introduce the least squares method using an example. The example will assume that the distribution of rainfall excess has two ordinates, p_1 and p_2, and the unit hydrograph has three ordinates, U_1, U_2, and U_3. Expressing the relationship between rainfall excess, the unit hydrograph, and the ordinates of the direct runoff hydrograph (r_i, i=1, 2,..., n) using the discrete form of the convolution integral provides the following:

$$\begin{aligned} p_1 U_1 &= r_1 \\ p_2 U_1 + p_1 U_2 &= r_2 \\ p_2 U_2 + p_1 U_3 &= r_3 \\ p_2 U_3 &= r_4 \end{aligned} \tag{7.5}$$

Equation 7.5 represents four equations and three unknowns, which are the unit hydrograph ordinates. There is also the constraint that the volume of direct runoff equals one area-inch. These equations can be expressed in matrix form as follows:

$$\begin{bmatrix} p_1 & 0 & 0 & 0 \\ p_2 & p_1 & 0 & 0 \\ 0 & p_2 & p_1 & 0 \\ 0 & 0 & p_2 & 0 \end{bmatrix} \begin{bmatrix} U_1 \\ U_2 \\ U_3 \\ 0 \end{bmatrix} = \begin{bmatrix} r_1 \\ r_2 \\ r_3 \\ r_4 \end{bmatrix} \tag{7.6}$$

which can be simplified to:

$$\underline{P}\,\underline{U} = \underline{R} \qquad (7.7)$$

in which $\underline{P}$ is a 4x4 matrix, and both $\underline{U}$ and $\underline{R}$ are vectors with four elements. Since the elements of $\underline{P}$ and $\underline{R}$ are known, while the elements of $\underline{U}$ are unknown, Eq. (7.7) can be transformed so that the knowns are on one side of the equation and the unknowns on the other side. Specifically, using concepts in matrix algebra provides the following:

$$\begin{aligned} \underline{P}\,\underline{U} &= \underline{R} \\ (\underline{P}^T\underline{P})\underline{U} &= \underline{P}^T\underline{R} \\ (\underline{P}^T\underline{P})^{-1}(\underline{P}^T\underline{P})\underline{U} &= (\underline{P}^T\underline{P})^{-1}\underline{P}^T\underline{R} \\ \underline{I}\,\underline{U} &= (\underline{P}^T\underline{P})^{-1}\underline{P}^T\underline{R} \\ \underline{U} &= (\underline{P}^T\underline{P})^{-1}\underline{P}^T\underline{P} \end{aligned} \qquad (7.8)$$

in which $\underline{P}^T$ is the transpose of $\underline{P}$, $\underline{P}^{-1}$ is the matrix inverse of $\underline{P}$, and $\underline{I}$ in the identity matrix. The unit hydrograph obtained from Eq. (7.8) will very likely need adjusting because of irregularities in the shape. Specifically, either negative ordinates or undulations in the ordinates of the unit hydrograph recession will probably occur. These irregularities can be eliminated through either manual smoothing or by fitting a synthetic functional form to the resulting moments. One possible method of calibrating a synthetic form is given in the next section. Of course, the solution of Eq. (7.8) must be obtained using a computer because of the length and complexity of the calculations.

To illustrate the use of Eq. (7.8) for deriving a unit hydrograph from the time distributions of rainfall excess (PE) and direct runoff (DRO), the simplest possible problem will be used. Consider the case where the rainfall excess consists of two ordinates (0.4, 0.8) and the unit hydrograph, which is unknown, has the ordinates (0.6, 0.4). From convolution, the direct runoff ordinates would be (0.24, 0.64, 0.32). Placing the rainfall excess and direct runoff in matrix form, and using Eq. (7.6),

$$\begin{bmatrix} 0.4 & 0 & 0 \\ 0.8 & 0.4 & 0 \\ 0 & 0.8 & 0 \end{bmatrix} \begin{bmatrix} U_1 \\ U_2 \\ 0 \end{bmatrix} = \begin{bmatrix} 0.24 \\ 0.64 \\ 0.32 \end{bmatrix}$$

The inverse matrix $(\underline{P}^T\underline{P})^{-1}$ of Eq. (7.8) is:

$$(\underline{P}^T\underline{P})^{-1} = \left(\begin{bmatrix} 0.4 & 0.8 & 0 \\ 0 & 0.4 & 0.8 \\ 0 & 0 & 0 \end{bmatrix} \begin{bmatrix} 0.4 & 0 & 0 \\ 0.8 & 0.4 & 0 \\ 0 & 0.8 & 0 \end{bmatrix} \right)^{-1} = \begin{bmatrix} 2.5/1.68 & -1/1.68 & 0 \\ -1/1.68 & 2.5/1.68 & 0 \\ 0 & 0 & 0 \end{bmatrix}$$

Post-multiplying this inverse by $\underline{P}^T$ yields:

$$(\underline{P}^T\underline{P})^{-1}\underline{P}^T = \begin{bmatrix} 2.5/1.68 & -1/1.68 & 0 \\ -1/1.68 & 2.5/1.68 & 0 \\ 0 & 0 & 0 \end{bmatrix} \begin{bmatrix} 0.4 & 0.8 & 0 \\ 0 & 0.4 & 0.8 \\ 0 & 0 & 0 \end{bmatrix} = \frac{1}{1.68} \begin{bmatrix} 1 & 1.60 & -0.8 \\ -0.4 & 0.2 & 2.0 \\ 0 & 0 & 0 \end{bmatrix}$$

Completing the matrix computations of Eq. (7.8) yields:

$$(\underline{P}^T\underline{P})^{-1}\underline{P}^T\underline{R} = \frac{1}{1.68} \begin{bmatrix} 1 & 1.60 & -0.8 \\ -0.4 & 0.2 & 2.0 \\ 0 & 0 & 0 \end{bmatrix} \begin{bmatrix} 0.24 \\ 0.64 \\ 0.32 \end{bmatrix} = \begin{bmatrix} 0.6 \\ 0.4 \\ 0 \end{bmatrix} = \begin{bmatrix} U_1 \\ U_2 \\ 0 \end{bmatrix}$$

The matrix computations yield the unit hydrograph that was used originally to derive the distribution of direct runoff by convolution with the rainfall excess.

7.6. Fitting Synthetic Unit Hydrographs

When unit hydrographs are derived from measured data, the ordinates must be smoothed to provide a reasonable shape. In order to provide for a systematically varying shape, it may be preferable to use a probability function and use the moments of the measured data unit hydrograph to fit the parameters of the probability function. Aron and White (1982) provided a method of estimating the parameters of a gamma probability distribution using the peak discharge and time to peak of a computed unit hydrograph. The gamma distribution has the form:

$$f(t;a,b) = \frac{t^a e^{-t/b}}{b^{a+1} \Gamma(a+1)} \tag{7.9}$$

in which "a" and "b" are called the shape and scale parameters, respectively; and $\Gamma(a)$ is the gamma function with argument a. The parameters a and b are obtained from the computed peak discharge q_p and the time to peak t_p of the unit hydrograph using the following steps:

1. Compute $f_a = \dfrac{q_p\ t_p}{A}$ (7.10)

   in which q_p has units of cfs, t_p has units of hours, and A is the drainage area in acres.

2. Find the value of "a" from the following:

$$a = 0.045 + 0.5f_a + 5.6f_a^2 + 0.3f_a^3 \tag{7.11}$$

3. Compute the parameter b from:

$$b = t_p/a \tag{7.12}$$

Given values of the parameters a and b, the ordinates of the gamma distribution unit hydrograph at time t can be computed from the following:

$$q(f) = q_p f^a\ e^{a(1-f)} \tag{7.13}$$

in which f is any fraction of t_p from 0 to $5t_p$.

Unit hydrographs computed from measured storm events on a single watershed can take on a variety of shapes, with some shapes not being indicative of general runoff characteristics of the watershed. The peak discharge and time to peak of such unit hydrographs can be used to fit a synthetic unit hydrograph, such as the gamma function of Eq. (7.9). Figure 7.6 shows the unit hydrograph for the October 19, 1960, storm event on Buffalo Bayou, which is a 58 square mile watershed in Houston, Texas. The computed unit hydrograph as a peak of 1456 cfs and a time to peak of 16 hours. From Eqs. (7.10), (7.11), and (7.12), f_a, a, and b are 0.62759, 2.6386, and 6.0638, respectively. Thus, the synthetic unit hydrograph Eq. (7.9) is:

$$f(t) = \frac{t^{2.6386} e^{-t/6.0638}}{704.848\,\Gamma(3.6386)}$$

Ordinates of the unit hydrograph can be computed using Eq. 7.13, which yields:

$$q(f) = q_p f^{2.6386} e^{2.6386(1-f)}$$

The ordinates are given in Table 7.5, and the synthetic unit hydrograph is shown in Fig. 7.6.

The 6-minute unit hydrograph of Fig. 7.5 and Table 7.3 was fit with the synthetic unit hydrograph of Eq. (7.9). For a peak discharge of 20.7 cfs, a time-to-peak of 9 minutes, or 0.15 hours, and an area of 1.68 acres, Eq. (7.10) yields a value of f_a of 1.8482, and values of a and b of 21.992 and 0.0068206, respectively. The resulting synthetic unit hydrograph is shown in Fig. 7.7.

7.7. Calibration of S-Hydrographs

Where S-hydrographs have been derived from computed unit hydrographs for different durations of rainfall excess, a generalized S-hydrograph can be calibrated from the individual event S-hydrographs. Three general approaches exist for calibrating S-hydrographs: functional form fitting, fitting by moments, and standardized form-free fitting. A function, such as a logistic function, can be fitted using least squares; this approach has the advantage that variations in the entire shape of the S-hydrograph influence the final form of the function. A watershed average S-hydrograph can be obtained after fitting the function to a number of storm events from the same watershed.

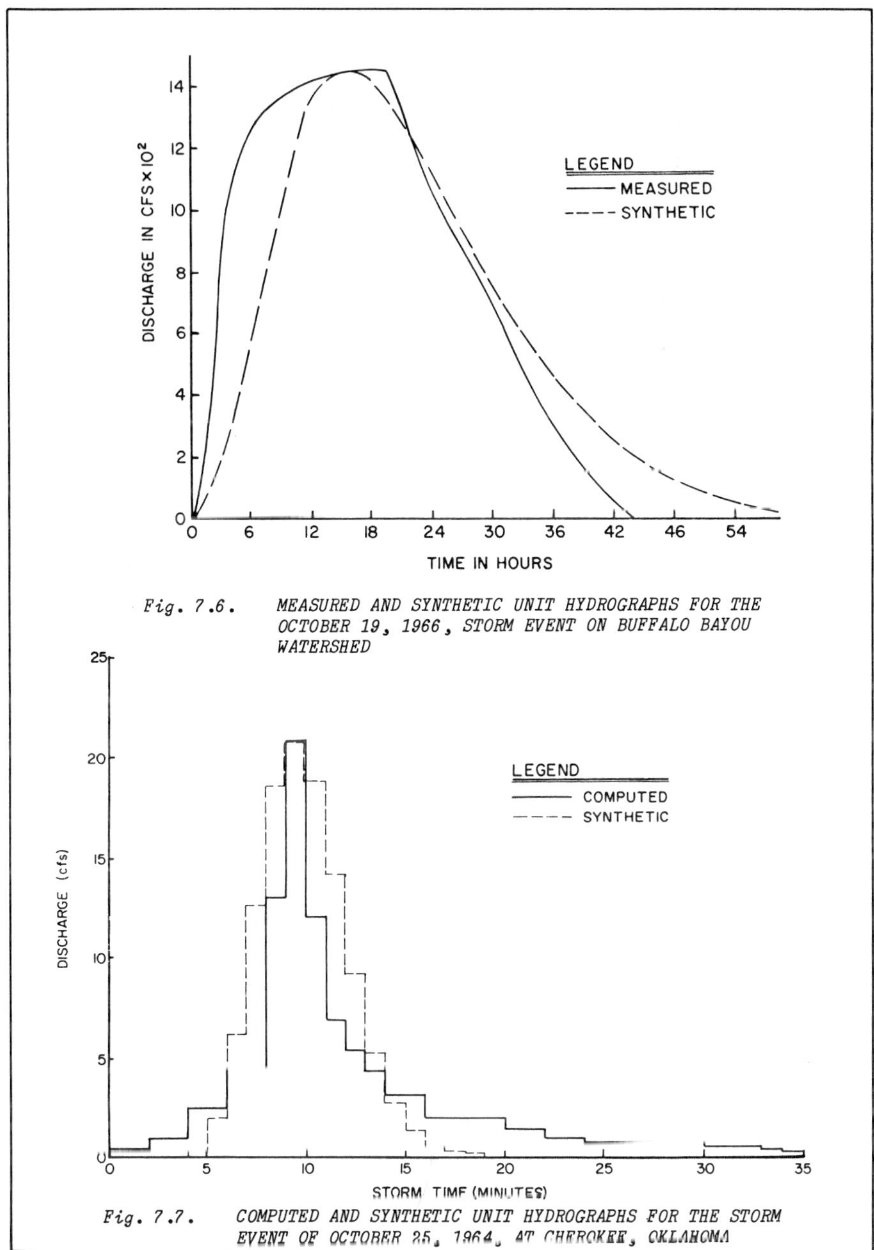

Fig. 7.6. MEASURED AND SYNTHETIC UNIT HYDROGRAPHS FOR THE OCTOBER 19, 1966, STORM EVENT ON BUFFALO BAYOU WATERSHED

Fig. 7.7. COMPUTED AND SYNTHETIC UNIT HYDROGRAPHS FOR THE STORM EVENT OF OCTOBER 25, 1964, AT CHEROKEE, OKLAHOMA

TABLE 7.5
ORDINATES OF THE MEASURED (q) AND SYNTHETIC ($\hat{q}$) UNIT HYDROGRAPHS AT BUFFALO BAYOU FOR THE STORM EVENT OF OCTOBER 19, 1960

Time (hours)	q (cfs)	f	$\hat{q}$ (cfs)
0	0	0	0
4	1050	0.25	272
8	1340	0.50	875
12	1400	0.75	1318
16	1456	1.00	1456
20	1400	1.25	1356
24	1060	1.50	1135
28	800	1.75	881
32	540	2.00	648
40	110	2.50	312
48	0	3.00	135
56	0	3.50	54
64	0	4.00	21
72	0	4.50	8
80	0	5.00	3

Moments estimation requires the first two moments of the S-hydrograph and expressions relating these sample moments to the moments of a known function. An approach to S-hydrograph fitting that does not require a specific function involves standardizing the individual S-hydrographs and subjectively constructing an average watershed S-hydrograph. For example, a storm event S-hydrograph can be standardized by scaling the discharge axis using the maximum or ultimate discharge and the time axis using the storm event lag time. Figure 7.8 shows the S-hydrograph for four storm events on a single watershed. The S-hydrographs were standardized using the ultimate discharge and the storm event lag time. The watershed average S-hydrograph was drawn such that it provided a reasonable fit to each of the individual storm event S-hydrographs.

Given a calibrated standardized S-hydrograph, an S-hydrograph for a design storm can be formed using a unit peak discharge computed with a hydrologic model for a selected return period and an estimate of the watershed lag time.

7.8. Calibration to Flood Frequency Discharges

Ideally, flood hydrograph data would be available when calibrating a time dependent model such as a unit hydrograph. The parameters of a time dependent model reflect both the timing and magnitude of flood runoff. Calibration with measured rainfall hyetograph and runoff hydrograph data enables model parameters that are sensitive to timing effects to be calibrated to reflect timing; similarly, the magnitude sensitive model parameters can be calibrated to reflect runoff magnitudes of the measured data.

Where sufficient hyetograph/hydrograph data are not available, time dependent models can be calibrated using intensity-duration-frequency rainfall information and peak discharge data. Based on the assumption that the T-year peak discharge results from the T-year rainfall, the parameters of the model can be varied in a way that the estimated peak discharges for each return period match the peak discharge estimates of the same return period obtained from flood frequency estimates. For such calibration, the model parameters will be adjusted so that the accuracy of flood magnitudes best reproduce the peak discharges obtained from the statistical frequency analyses. The effects of timing will not be independently assessed with such calibration and the accuracy of time dependent estimates will depend on the association between timing and magnitude effects. In spite of this weakness of this type of calibration, it is certainly preferable to use this method of calibration than not to calibrate at all.

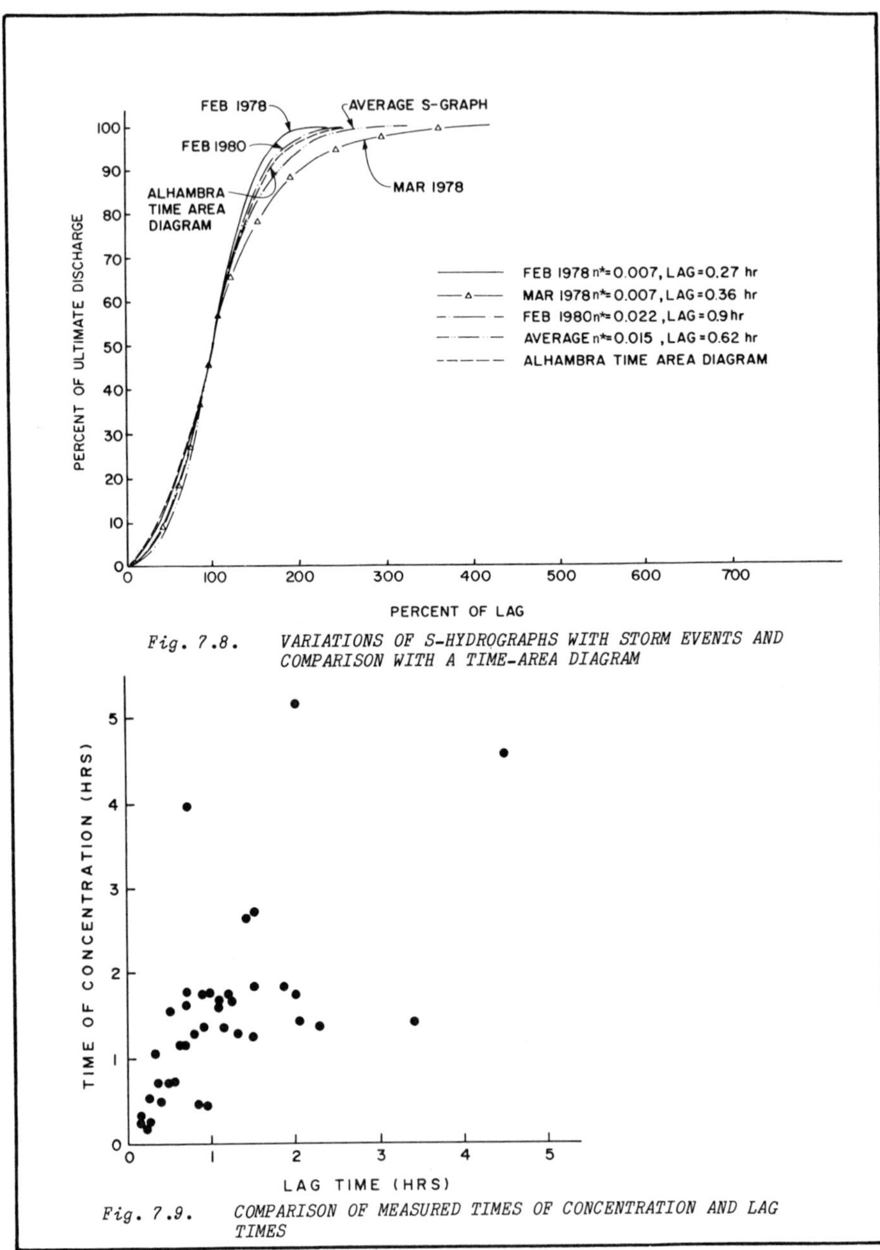

Fig. 7.8. VARIATIONS OF S-HYDROGRAPHS WITH STORM EVENTS AND COMPARISON WITH A TIME-AREA DIAGRAM

Fig. 7.9. COMPARISON OF MEASURED TIMES OF CONCENTRATION AND LAG TIMES

The input required to calibrate a time dependent model using flood frequency peak discharge estimates includes the intensity-duration-frequency rainfall information and peak discharge estimates obtained from flood frequency analyses for selected return periods. The rainfall information is used as input along with a design storm pattern (see Chapter 2) to define the design storm hydrograph. The model parameters are adjusted using subjective optimization (McCuen and Snyder, Chapter 8, 1986) until the peak discharges predicted with the model provide the best estimates of the flood frequency estimates.

7.9. Calibration of Time of Concentration Models

Of all the inputs to hydrologic design methods, the time of concentration is the input that combines the least accuracy with the highest importance. That is, estimates of Tc are very important to the accuracy of a design, yet it is widely recognized that methods used to estimate times of concentration are not regarded as being highly accurate indicators of travel times of flood runoff. This lack of accuracy should be viewed relative to other important inputs such as rainfall intensities and the drainage area.

Given the importance of times of concentration and the relative lack of accuracy, especially when transferring existing prediction equations between regions or localities, every effort should be made to calibrate an equation for predicting times of concentration. Variations in the importance of physical processes will effect travel times and thus the accuracy of Tc estimates made using existing equations. Factors that can significantly effect the ability to transfer a Tc model that was calibrated elsewhere include storage in swamps and wetlands, channels where transmission losses are significant, the presence of debris flow, and excessive control by pipe systems or culverts. Where such watershed conditions exist, every effort should be made to calibrate an equation for estimating times of concentration.

In order to calibrate a Tc model, two types of data are required. First, it is necessary to obtain estimates of Tc for a number of watersheds in a region. These must be obtained from calculations of watershed lags from measured rainfall hyetographs and runoff hydrographs. The watershed lag is proportional to the time of concentration (see Fig. 7.9). For each watershed a number of storm events should be analyzed and the travel time computed for each event. Then some measure of central tendency (i.e., mean or median) of the storm event times of concentration should be used to represent the watershed Tc. The Tc estimates for each watershed are used to form a vector, which represents one element of the Tc calibration process.

For each of the watersheds included in the study, watershed characteristics must be determined. The most frequently used variables are the slope, length, roughness coefficient, and a rainfall index, such as the 2-year, 2-hour rainfall intensity. For calibrating overland flow travel times, a surface roughness or land use index may serve as the variable. For watersheds where the travel times are dominated by channel conditions, the slope, length, and roughness characteristics should reflect average channel conditions of the principal flow path or an average type of channel. In some cases, a mixed set of watershed characteristics might be more appropriate. This matrix of watershed characteristics for each watershed for which Tc estimates are available from storm data represents the second type of data needed to calibrate a Tc model. Using an equation that was designed for estimating overland flow Tc's to predict a Tc for a watershed where channel flow dominates can lead to very inaccurate estimates of Tc. Thus, calibration would be warranted where a tested model is not available.

There are two definitions of the time of concentration: (1) the time required for a particle of water to travel from the hydraulically most distant point in the watershed to the outlet or design point; and (2) the time between the end of rainfall excess to the inflection point on the recession of the direct runoff hydrograph. There are variations on the second definition, such as the time between the centroids of rainfall excess and direct runoff. While these two definitions are intended to represent the same concept, they do not lead to identical values for Tc because the method of computing both are inexact. The first definition is usually applied using a method for estimating travel velocities, and the equations used to compute travel velocities are inexact themselves. The second definition requires measured rainfall hyetographs and runoff hydrographs, and in analyzing the measured data many simplifying assumptions must be made to separate the rainfall excess from the hyetograph and the direct runoff from the hydrograph. The application of these two definitions leads to two types of calibration of equations for estimating Tc.

To calibrate an equation for estimating Tc it is necessary to evaluate hyetographs and hydrographs. The rainfall excess is separated from the total rainfall, and the direct runoff is separated from the total runoff hydrograph. Depending on the definition of Tc used, the storm event Tc is computed. A number of watersheds should be included in the analyses, with the watersheds covering a range of drainage areas, slopes, and land use types. A matrix can be developed with the time of concentrations that were determined from the hyetograph-hydrograph analyses as the criterion or dependent variable and the values of the hydraulic length, the slope, and other watershed and/or rainfall characteristics as the predictor variables. The values of the time of concentration can then be regressed on the values of the predictor variables to develop a prediction equation.

Having compiled the data base, which consists of a vector of Tc estimates and a matrix of watershed characteristics, the data can be analyzed to determine the best model structure and the empirical coefficients that provide the best prediction model. While the power model form (i.e., log-log multiple regression form) is the most widely used to develop Tc models, the data should be plotted to ensure that the optimum model structure is obtained. The Tc estimates should be plotted against each of the watershed/channel characteristics, with both linear and log-log plots made. The plots should be examined for overall fit, local biases in the fit, the presence of extreme events, or the need for intercepts or shift parameters. The factors that will influence the best model structure are illustrated in Fig. 7.10.

Having decided on the best model structure from the graphical analysis, the model coefficients should be fit with the data. Simple structures can be fit using least squares regression (McCuen, 1985). More complex structures can be fit using a numerical method such as nonlinear least squares (McCuen and Snyder, 1986). Having fit the assumed model structure to the data base, the model should be examined for goodness-of-fit and rationality, with model rationality being the most important criterion. For example, the coefficients should suggest that the time of concentration decreases as either the length or roughness decreases, while an inverse relationship would be expected between Tc and slope. Goodness of fit can be assessed using any of the graphical or statistical measures discussed above.

The values of the Tc computed from the hyetograph-hydrograph analyses can also be used to develop a relationship between the measured values and estimates of Tc obtained using the velocity method. The velocity method uses existing equations to compute flow velocities that are used to compute a travel time for the watershed (T_t). The travel time and Tc are then related by:

$$Tc = KT_t \tag{7.14}$$

in which K is a proportionality constant. The value of K adjusts the computed travel times for the inaccuracy of the velocity equations used to compute T_t.

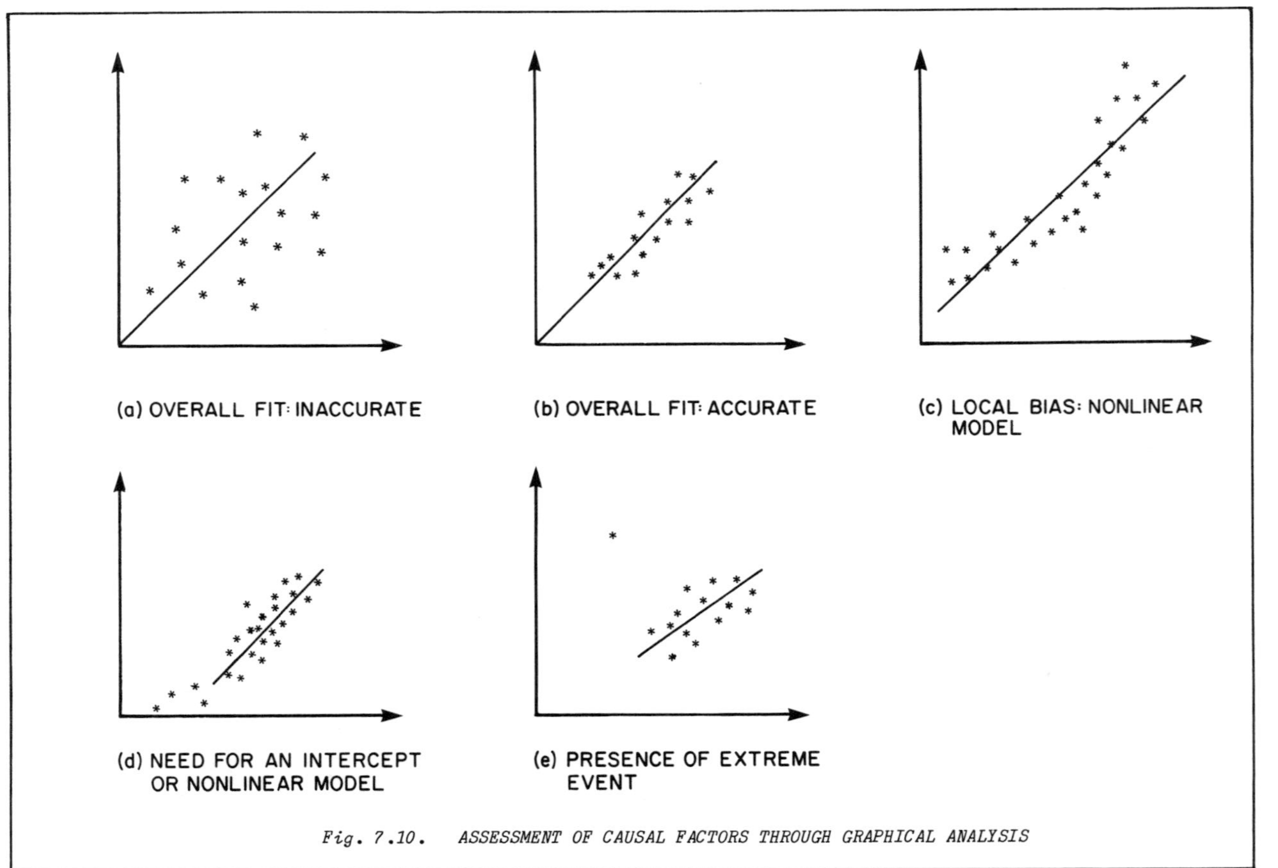

Fig. 7.10. ASSESSMENT OF CAUSAL FACTORS THROUGH GRAPHICAL ANALYSIS

7.10. Calibration of Channel Routing Parameters

In Chapter 6, methods of hydrologic channel routing were introduced: the Convex, the Modified Att-Kin, and the Muskingum methods. The Convex and Modified Att-Kin methods use the following routing equation:

$$O_2 = CI_1 + (1-C)\ O_1 \qquad (7.15)$$

in which C is the routing coefficient, I_1 is the inflow hydrograph ordinate at time t_i, and O_1 and O_2 are the outflow hydrograph ordinates at times t_i and t_{i+1}, respectively. The Muskingum method uses the following routing equation:

$$O_2 = C_0I_2 + C_1I_1 + C_2O_1 \qquad (7.16)$$

in which C_0, C_1, and C_2 are the Muskingum routing coefficients; and I_2 is the ordinate of the inflow hydrograph at time t_{i+1}. While the routing coefficient C of Eq. (7.15) can be calibrated directly, the coefficients of the Muskingum method are a function of two parameters, x and K. The routing parameters can be calibrated directly rather than calibrating the coefficients of Eq. (7.16).

7.11. Calibration of the Convex Method

To calibrate the routing coefficient C of Eq. (7.15) for the Convex Method requires ordinates of the flood hydrographs at the upstream and downstream cross sections. Since the value of C will depend on the characteristics of the stage-discharge relationship of the reach, hydrographs for several flood events should be obtained so that a representative value of C can be obtained by rearranging (7.15) to the form:

$$O_2 - O_1 = C(I_1 - O_1) \qquad (7.17)$$

A value of C is obtained by computing values for both (O_2-O_1) and (I_1-O_1) for all possible ordinates of the hydrographs and then plotting the pairs of points; the routing coefficient C is the slope of the line that provides the best fit to the data.

The upstream and downstream hydrographs for a channel are given in Table 7.6; the terms of Eq. (7.17) are also given. A graph of (O_2-O_1) versus (I_1-O_1) is shown in Fig. 7.11. While the construction of a line is somewhat subjective, a reasonable fit is provided by the line shown in Fig. 7.11. The slope of the line is 0.23; therefore, the routing equation is:

TABLE 7.6
EVALUATION OF THE ROUTING COEFFICIENT OF THE CONVEX ROUTING METHOD

Upstream Hydrograph (cfs)	Downstream Hydrograph (cfs)	$O_2 - O_1$ (cfs)	$I_1 - O_1$ (cfs)	$\hat{O}$ (cfs)
0	0	0	0	0
12	0	4	12	0
33	4	6	29	3
41	10	6	31	10
38	16	5	22	17
32	21	4	11	22
22	25	-1	-3	24
15	24	-2	-9	24
9	22	-3	-13	22
4	19	-4	-15	19
2	15	-4	-13	16
0	11	-	-	13

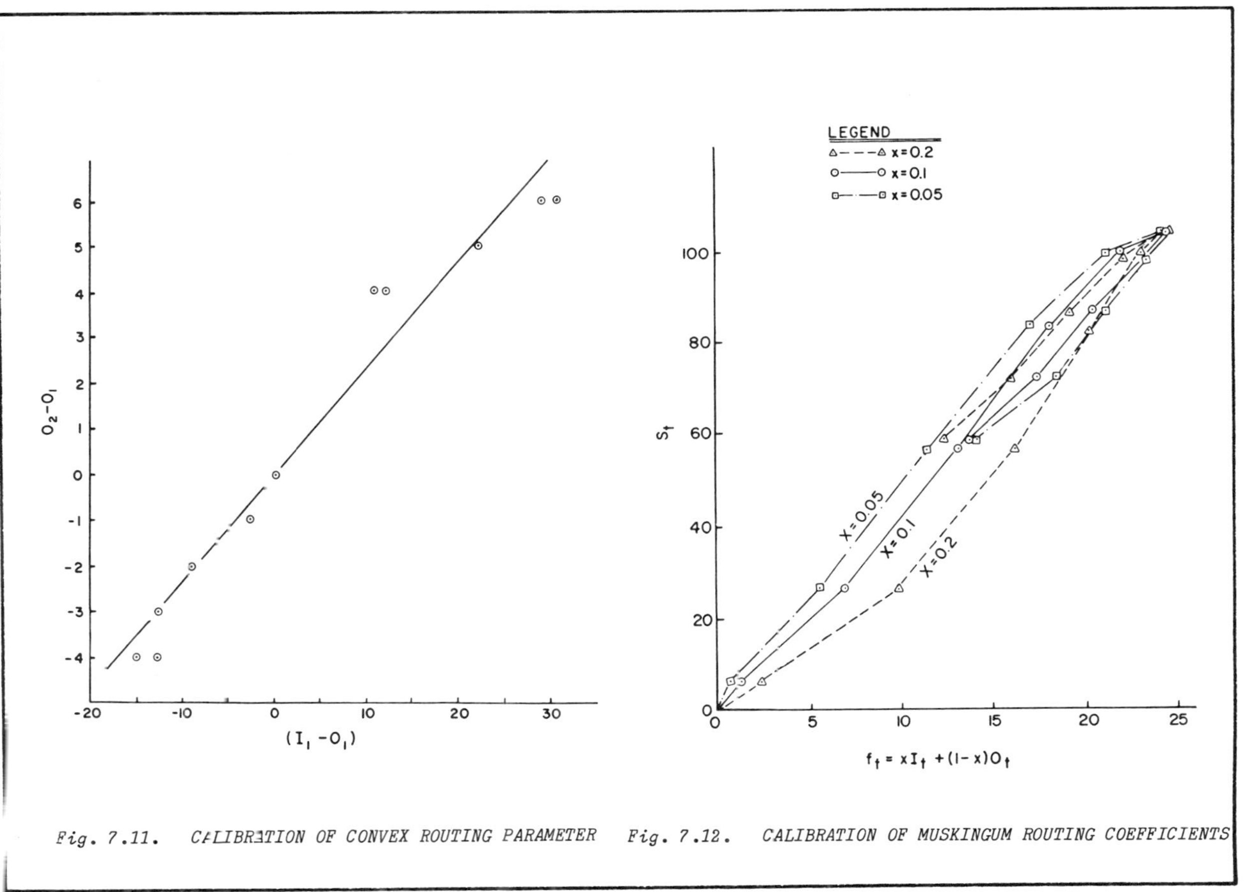

Fig. 7.11. CALIBRATION OF CONVEX ROUTING PARAMETER

Fig. 7.12. CALIBRATION OF MUSKINGUM ROUTING COEFFICIENTS

$$O_2 = 0.23I_1 + 0.77O_1 \qquad (7.18)$$

Equation (7.18) was used to compute the ordinates of the downstream hydrograph using the ordinates of the upstream hydrograph as input to Eq. (7.18), with the initial value of O_1 equal to zero. The computed ordinates of the downstream hydrograph $\hat{O}$ are given in Table 7.6. The graphical estimate of C provides a reasonably good fit for the downstream hydrograph.

7.12. Calibration of the Modified Att-Kin Method

While the Modified Att-Kin method uses the same routing Eq. (7.15) as the Convex method, the calibration is not as straightforward because of the translation element of the method. Where translation is not involved, calibration of the routing coefficient is the same as for the Convex method. However, when translation would be involved, then a subjective element to the calibration is necessary. An estimate of the routing coefficient can be made by making a number of trial translations, with each trial translation followed by the fitting method introduced in the previous section of the Convex method. The value of C that produces the best fit to the measured downstream hydrograph should be used for that channel reach.

7.13. Calibration of the Muskingum Method

The Muskingum method is a function of two parameters, K and x; the coefficients of Eq. (7.16) are a function of the two parameters and the routing time increment, dt. Values of K and x are obtained by assuming a value of the function f_t for each time ordinate:

$$f_t = xI_t + (1-x)\,O_t \qquad (7.19)$$

The values of f_t are plotted with the storage estimates obtained from

$$S_t = S_{t-1} + dt\,(I_a - O_a) \qquad (7.20)$$

in which I_a and O_a are the average inflow and average outflow for time periods t and (t-1), respectively. If the pairs of points, f_t versus S_t plot as a straight line, then the assumed value of x is correct and the slope of the line is the value of K. If the pairs of points do not form a straight line, then a new value of x should be tried until the data provide a straight line.

The hydrographs used with the Convex method can be used to illustrate the graphical calibration of the Muskingum coefficients. Values of f_t (Eq. 7-19) and S_t (Eq. 7-20) are given in Table 7.7 for values of x of 0.05, 0.1, and 0.2; Figure 7.12 shows plots of S_t versus f_x for the three values of x. The smallest departure from a straight line is for x equal to 0.10. The plot has a slope of approximately 4.4, so the graphical estimate of K is 4.4. More accurate estimates could be obtained by trying other values of x to minimize the departure of the plot from a straight line. Values of the discharge at the downstream cross section were computed using the following equation:

$$\hat{O}_t = 0.013\ I_t + 0.211\ I_{t-1} + 0.776\ O_{t-1} \qquad (7.21)$$

The values of $\hat{O}_t$ are given in Table 7.7. The largest absolute error is 2, which is 12.5 percent of the measured value.

TABLE 7.7

ESTIMATION OF THE COEFFICIENTS, K AND x, OF THE MUSKINGUM ROUTING METHOD

		S_t	f_t for x =			$\hat{O}$	
I_t	O_t	(Eq. 7.20)	0.2	0.1	0.05	(Eq. 7.21)	|e|
0	0	0				0	
12	0	6.0	2.4	1.2	0.6	0.2	0.2
33	4	26.5	9.8	6.9	5.5	3.1	0.9
41	10	56.5	16.2	13.1	11.5	11.5	1.5
38	16	83.0	20.4	18.2	17.1	18.0	2.0
32	21	99.5	23.2	22.1	21.6	22.4	1.4
22	25	103.5	24.4	24.7	24.8	24.4	0.6
15	24	97.5	22.2	23.1	23.6	23.8	0.2
9	22	86.5	19.4	20.7	21.4	21.8	0.2
4	19	72.0	16.0	17.5	18.2	18.8	0.2
2	15	58.0	12.4	13.7	14.4	15.5	0.5
0	11	46.0				12.4	1.4

complex watershed modeling

8

8.1. Link-Node Model

The design storm and unit hydrograph computer programs that were discussed in Chapters 2 and 5 will be used as the basis for a generalized watershed modeling computer program, RUNOFF. The elements of this program system are listed in Table 8.1. With these several submodels, a watershed can be modeled as a link-node system composed of individual subareas that contribute runoff at nodal points and are linked together by channel or pipe routing submodels. Analogous to the rational method master planning/design program system, the RUNOFF system is intended to be used interactively rather than in the batch mode. In this fashion, the watershed link-node model is generated correctly on the first pass with the user accepting or rejecting the intermediate results as the model schematic is constructed. The RUNOFF program includes several submodels that approximate the important effects of detention basin storage for both a flow-through and a flow-by (or hydrograph separation) detention basin processes. These storage approximations allow the design storm modeling approach to be used to study complex urban watershed flood control systems and enables the user to optimize the necessary flood control system elements.

As discussed in Chapter 1, each of the submodels has an associated uncertainty function that is a function of the several submodel parameter uncertainty functions. Consequently, watershed modeling with complex model is not always the ideal solution to analyzing a gaged or ungaged watershed. That is, the approximation results may not be any more reliable than the approximations produced by the lumped parameter unit hydrograph model of Chapter 5. However, in watersheds where channel storage or detention basin effects are considered significant, then use of a detention basin or channel routing model may considerably reduce the modeling uncertainty that would be generated by a lumped-parameter approach operating upon parameters (e.g., lag values, S-graphs, losses) averaged from watersheds that do not include these additional storage components.

8.2. A Watershed Computer Program

The design storm runoff hydrograph computer program of Chapter 5 may be used as a submodel for a generalized watershed link-node model. The subroutines listed in Table 8.1 may be connected by a main driver program that engages the selected submodel when called upon during the link-node model schematic generation process. Preceeding each FORTRAN computer program listing will be the data entry requirements and program variable definitions. This information will be presented on example CRT display text pages as prepared for design-interactive computer software (Hromadka et al, 1983b). The variable names are superimposed on each example CRT text page in order to associate the program variable to the appropriate CRT data entry prompt.

TABLE 8.1

WATERSHED LINK-NODE MODEL SUBMODELS

Program Number	Description
8.1	main driver program
8.2	flow-through detention basin model based on the modified Puls method
8.3	flow-by detention basin model (hydrograph separation)
8.4	link-node model functions of CLEAR, ADD, and SPLIT
8.5	channel flow routing model based on a convex routing approach
8.6	pipeflow routing model based on normal depth velocities for each unit runoff value less than the pipe capacity

8.3. Link-Node Model Functions

The RUNOFF program can be envisioned as a collection of several (five) working streams. The program user models the watershed by generating runoff hydrographs in subareas and adding the resulting runoff hydrographs to one of the streams. Channel and basin storage-routing effects are simulated by operating upon the contents of a particular stream with a specified routing submodel. To facilitate use of the working streams, three link-node functions are available in the watershed link-node model:

CLEAR. This function clears a working stream of all data.

ADD. This function adds the contents of one straem (A) to another stream (B) such that after this operation is completed, the contents of stream (A) remain unchanged and the contents of stream (B) becomes the sum of (A) + (B).

SPLIT. This function operates on a stream (A) by splitting (A) into two parts. Given a decimal ratio R, after the process is completed the contents of stream (A) becomes (1-R)(A), and R(A) is added to a specified receiving stream (B) so that (B) becomes (B)+(R)(A).

The RUNOFF program requires watershed information, rainfall patterns, soil loss rates, and representative watershed S-graphs. A synthetic unit hydrograph is generated, based upon the representative watershed S-graph and watershed lag time. Then the synthetic unit hydrograph is convoluted with effective rainfall to estimate the watershed outflow hydrograph. A specified hydrologic model is used to route the hydrograph downstream.

Figure 8.1 illustrates the flow chart of the RUNOFF program. Interactively, the user enter the proper input data via the request of the CRT. The RUNOFF program can be subdivided into five main modules. Table 8.2 illustrates the MAIN MENU OF THE RUNOFF model, and Table 8.3 illustrates the subroutines that are contained in each module of the RUNOFF model.

Before executing the RUNOFF model, the DBANK (data bank development) program should be executed to ensure sufficient data are available for the RUNOFF model. The DBANK program is an interactive program that guides the program user in storing watershed information through the CRT, and then stores the information in the data file "CONTR.DAT". The file CONTR.DAT can be revised, if needed; otherwise, the data are permanently stored. In this program, data is stored for up to four sets of depth-area adjustment curves, four sets of user specified 24-hour storm patterns, a single user specified 24-hour storm pattern with continuous depth-area adjustment, and ten sets of S-graphs. These data are retrievable within the RUNOFF program.

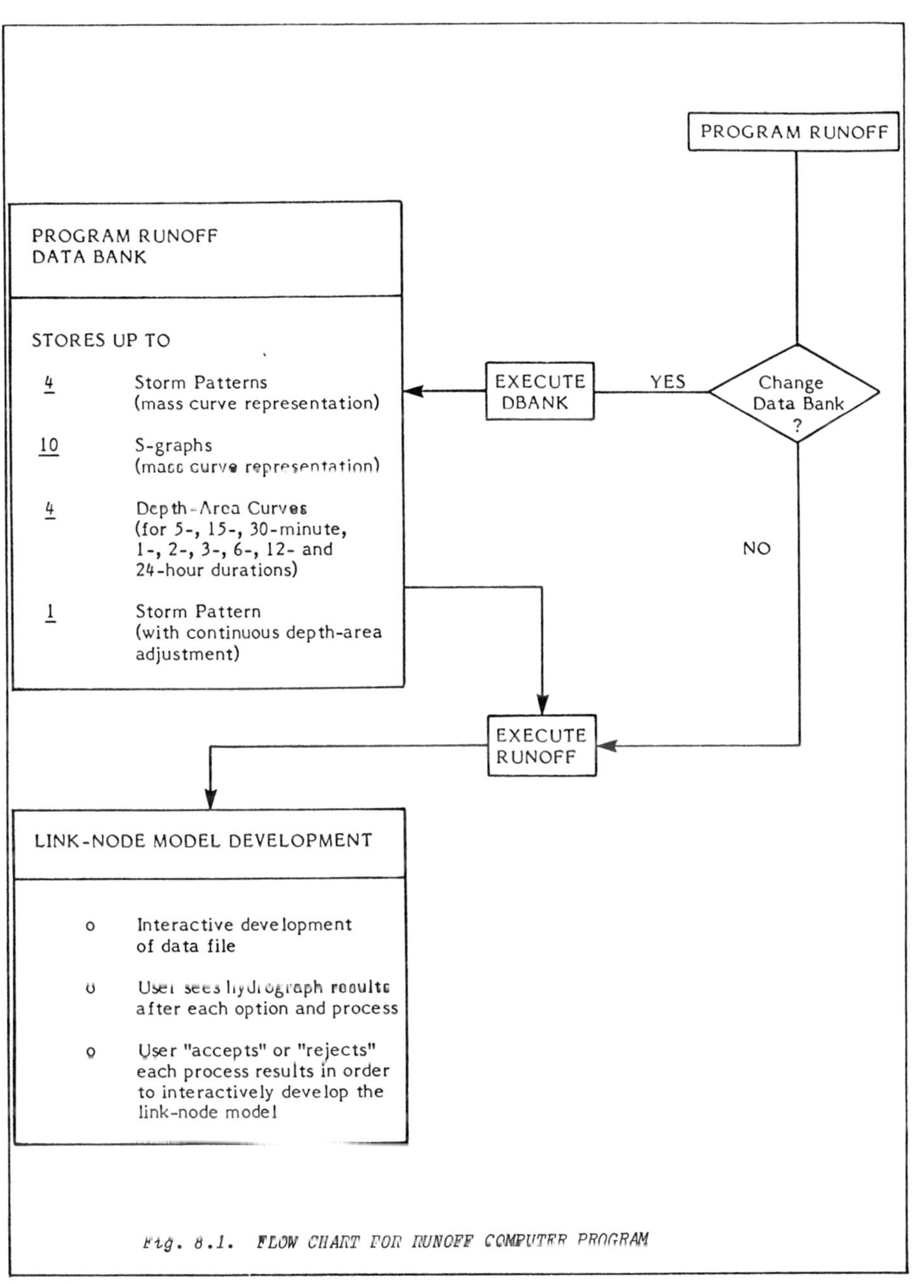

Fig. 8.1. FLOW CHART FOR RUNOFF COMPUTER PROGRAM

TABLE 8.2
MAIN MENU

Options	Description
Unit Hydrograph Model	
1	Synthetic Unit Hydrograph
Basin Routing Models	
2	Flow-by basin model
3	Flow-through basin model
Channel Routing Models	
4	Hydrograph translation model
5	Convex routing model
6	Pipe routing model
7	Kinematic routing model
8	Muskingum routing model
Utility Commands	
9	Add model
10	Split model
11	Clear model
12	View model
13	Print model
14	Exit program

TABLE 8.3
SUBROUTINES FOR 'RUNOFF' PROGRAM

Module	Subroutines
UNIHY	RAIN, SLOSS, SGRAPH, HGRAPH
RBASIN	FLOWBY, FLOWTR
RCHAN	CONVEX, PIPE, MOVE, KROUT, MUSK
UTIL	ADD, SPLIT, CLEAR

8.4. Data Bank Development

The data bank is a storehouse of several types of hydrologic information which are used in the RUNOFF program. The data bank includes: the depth-area curve sets, the user specified 24-hour storm patterns, the user specified storm pattern with continuous depth-area adjustment, and the S-graphs. This information is then permanently stored in the file CONTR.DAT until further revision is needed.

8.4.1. Depth-Area Adjustment Curve Sets

The 5-minute, 15-minute, 30-minute, 1-hour, 2-hour, 3-hour, 6-hour, 12-hour, and 24-hour depth-area curves shall be entered at 5-, 10-, 15-, 20-, 40-, 60-, 80-, 100-, 125-, and 150- square mile areas. Up to four sets of the depth-area adjustment curve sets can be stored in this subroutine.

8.4.2. User Specified 24 Hour Storm Pattern

A 24-hour design storm pattern is entered as a volume distribution curve. Each data pair consists of the time ordinate (in hours) and its corresponding mass value in decimal form (0.0 to 1.00). The last data pair shall pass through the 24-hour time ordinate and the mass value of 1.0 (100 percent of rainfall). Up to 50 data pairs are allowed, with a maximum of four sets of 24-hour storm patterns. These storm patterns provide a fixed time distribution of rainfall, according to a single 24-hour storm rainfall value.

8.4.3. User Specified Storm Pattern With Continuous Depth-Area Adjustment

A 24-hour design storm pattern with continuous depth-area adjustment can be generated by rearranging the intervals of a 24-hour nested design storm pattern with the peak defined at hour 0 (see Section 2.15). Then a depth-area adjustment curve set can be chosen from a previously entered set of depth-area adjustment curve sets for the 24-hour design storm pattern.

8.4.4. S-Graphs

Up to ten S-graphs can be stored for retrieval. Data are entered as ordered pairs (% volume, lag). The program prompts the user in 2% increments of total volume until 100% is reached.

8.4.5. Interactive Procedure

In this section, a detailed step-by-step user's instructions on the DBANK program are presented. By executing the "DBANK" program, the interactive procedures are initiated on the CRT with the following display:

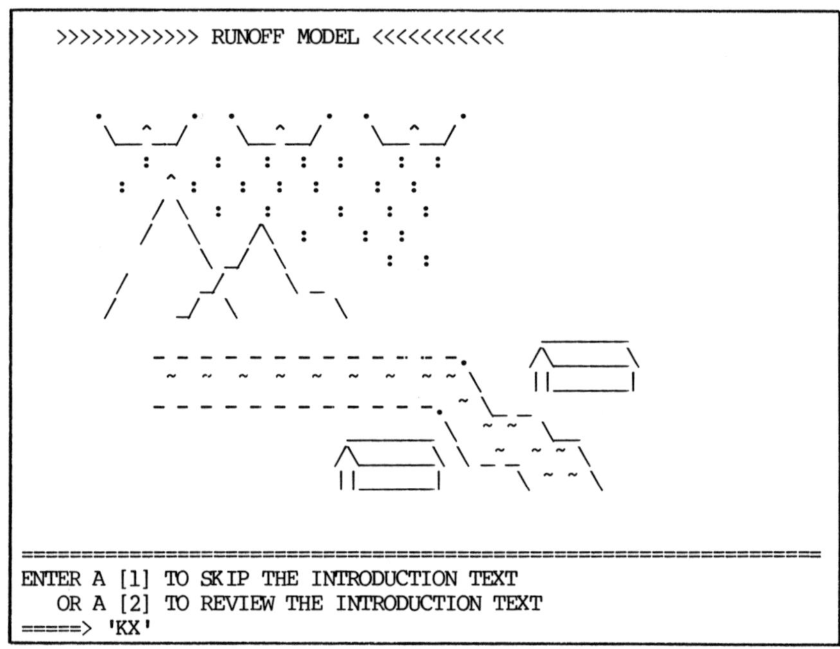

By entering a [2], the introduction text of the RUNOFF program will be displayed as follows:

>>>>>>>> PROGRAM RUNOFF >>>>>>>>

This is a computer-aided-design program in which the user interacts with the development of the link-node model. The user must first define a DATA BANK which stores various hydrologic parameters. (This DATA BANK will be stored permanently, unless changed at a later time.) Then the user proceeds to the link-node model development for a catchment.

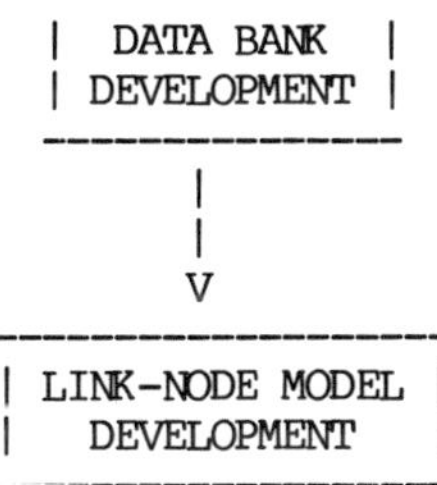

ENTER A [1] TO CONTINUE
=====> '1'

The link-node model is first conceptually developed on paper by:
(1) subdividing the catchment into subareas,
(2) linking the subareas by channel routing elements,
(3) identifying retention basin characteristics,
(4) developing catchment hydrologic parameters, and
(5) preparing a catchment hydrology map with node numbers assigned.

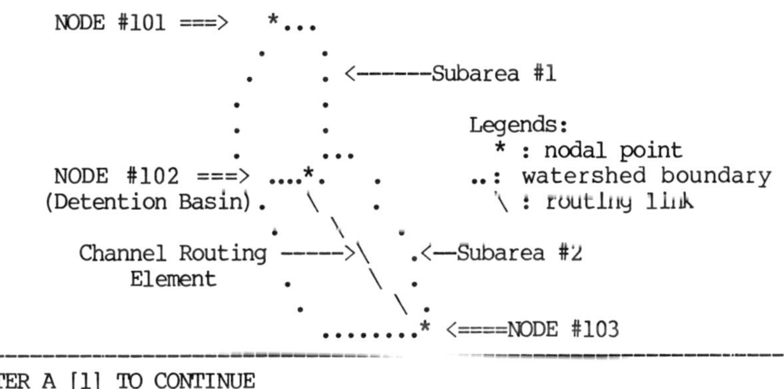

ENTER A [1] TO CONTINUE
=====> '1'

The link-node model is developed INTERACTIVELY because the user "accepts" or "rejects" all intermediate computational results as the model is constructed in the downstream flow direction. After using a certain process (such as channel routing, developing a subarea runoff hydrograph, etc.), the computed results are displayed for the user to "accept" (entering a [1]) or "reject" (entering a [2]). Rejecting intermediate results returns the program to the previous study point so that the user may make another attempt to solve this part of the problem. After building the link-node model, the data file is now available for other editing. The user is now in a BATCH MODE. Particulars of the program are contained in the book.

------------------------------END------------------------------

ENTER A [1] TO CONTINUE
=====> '1'

By entering a [1] the data bank information will be displayed as:

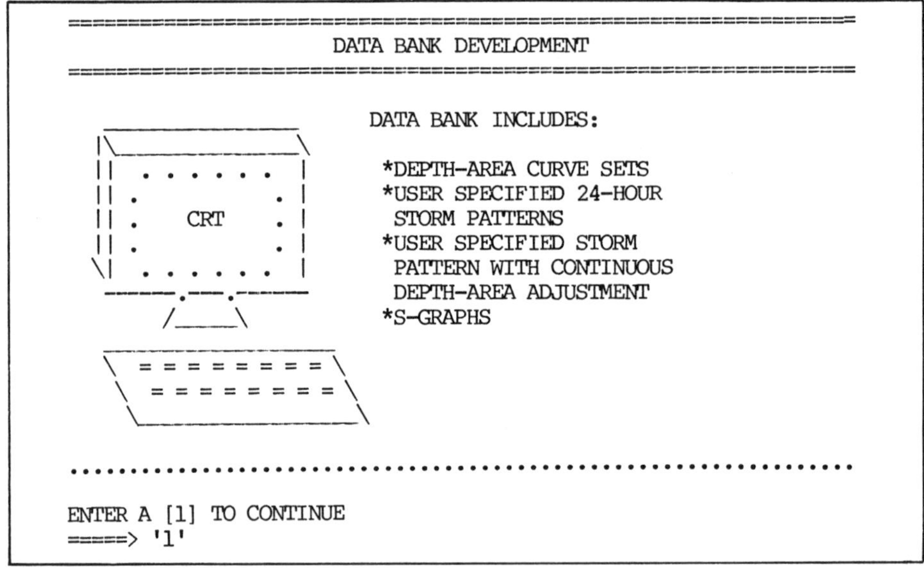

DATA BANK DEVELOPMENT

DATA BANK INCLUDES:

*DEPTH-AREA CURVE SETS
*USER SPECIFIED 24-HOUR STORM PATTERNS
*USER SPECIFIED STORM PATTERN WITH CONTINUOUS DEPTH-AREA ADJUSTMENT
*S-GRAPHS

CRT

ENTER A [1] TO CONTINUE
=====> '1'

```
                 *** PROGRAM DATA BANK DEVELOPMENT ***
-------------------------------------------------------------------

DATA BANK INCLUDES:
   .MAXIMUM OF [4] DEPTH-AREA CURVE SETS
   ([0] ARE DEFINED)
   .MAXIMUM OF [4] USER SPECIFIED 24-HOUR STORM PATTERNS
   ([0] ARE DEFINED)
   .[1] USER SPECIFIED STORM PATTERN WITH CONTINUOUS DEPTH-AREA
    ADJUSTMENT
   (CURRENTLY IS UNDEFINED)
   .MAXIMUM OF [10] S-GRAPHS
   ([0] ARE DEFINED)


SELECT PROGRAM INSTRUCTION NUMBER:
   [1] = ADD TO DATA BANK STORAGE
   [2] = DELETE DATA FROM STORAGE
   [3] = EXIT PROGRAM AND PROCEED TO LINK-NODE MODEL DEVELOPMENT
=====> 'KDATA'
```

By entering a [1], one of the four options can be chosen to modify the data bank.

```
SELECT A DATA BANK TO MODIFY:
 [1] = DEPTH-AREA CURVE SETS
 [2] = USER SPECIFIED 24-HOUR STORM PATTERNS
 [3] = USER SPECIFIED STORM PATTERN WITH CONTINUOUS DEPTH-AREA
       AJUSTMENT
 [4] = S-GRAPHS
=====> 'KMOD'
```

8.4.5.1. Depth-area curve sets

```
     Depth-area curve sets provide a reduction factor to area-
averaged point rainfall data for watershed size effects.  Each
curve set is composed of 9 curves, with a curve for the 5-, 15-,
30-minute, 1-, 2-, 3-, 6-, 12-, and 24-hour storm durations.  (Each
curve is  defined for ten catchment  area sizes  of 5-, 10-, 15-,
20-, 40-, 60-, 80-, 100-, 125-, and 150-square miles.)
     Date input are checked to guarantee that:
(1) all factors are between 0 and 1;
(2) all factors decrease with increasing catchment area size; and
(3) all factors decrease with decreasing precipitation duration.
-------------------------------------------------------------------

ENTER A [1] TO CONTINUE
=====> '1'
```

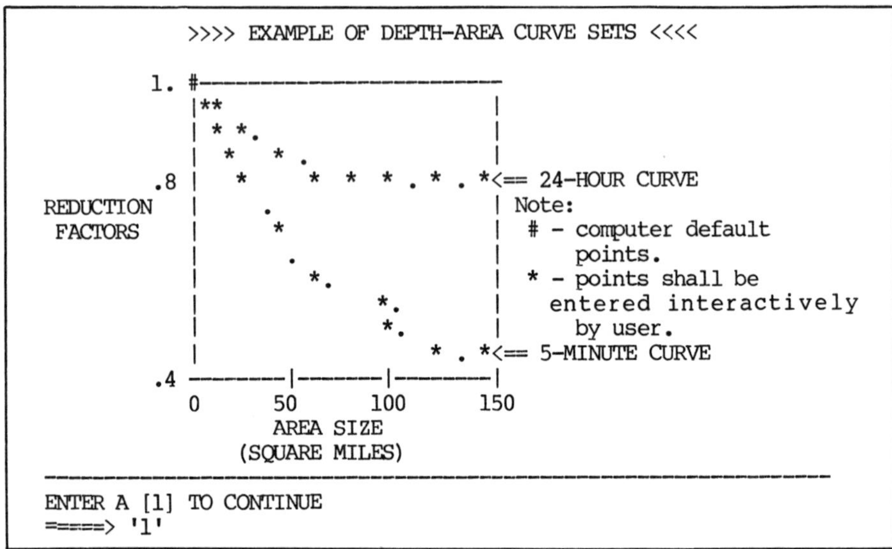

An example of a partial input data set is shown as follows:

```
*** DEPTH-AREA CURVE SETS ***
 (NOTE: MAXIMUM ALLOWABLE NUMBER OF SETS IS "4")

ENTER 5-MINUTE DEPTH-AREA CURVE DATA:
  ENTER THE 5-MINUTE ADJUSTMENT FACTORS CORRESPONDING TO
      5-, 10-, 15-, 20-, 40-, 60-, 80-, 100-,
      125-, AND 150-SQUARE MILES OF CATCHMENT AREA
 '.84 .75 .67 .62 .48 .395 .34 .3 .262 .24'

ENTER 15-MINUTE DEPTH-AREA CURVE DATA:
  ENTER THE 15-MINUTE ADJUSTMENT FACTORS CORRESPONDING TO
      5-, 10-, 15-, 20-, 40-, 60-, 80-, 100-,
      125-, AND 150-SQUARE MILES OF CATCHMENT AREA
 '.84 .75 .67 .63 .49 .41 .36 .32 .29 .27'

ENTER 30-MINUTE DEPTH-AREA CURVE DATA:
  ENTER THE 30-MINUTE ADJUSTMENT FACTORS CORRESPONDING TO
      5-, 10-, 15-, 20-, 40-, 60-, 80-, 100-,
      125-, AND 150-SQUARE MILES OF CATCHMENT AREA
 '.84 .75 .67 .64 .51 .44 .39 .357 .33 .31'

ENTER 1-HOUR DEPTH-AREA CURVE DATA:
  ENTER THE 1-HOUR ADJUSTMENT FACTORS CORRESPONDING TO
      5-, 10-, 15-, 20-, 40-, 60-, 80-, 100-,
      125-, AND 150-SQUARE MILES OF CATCHMENT AREA
 '.84 .75 .67 .64 .53 .47 .43 .4 .377 .362'
```

8.4.5.2. User specified 24-hour storm patterns

Up to FOUR 24-hour fixed storm patterns can be defined in the DATA BANK. Each pattern is a plot of storm time (hours) versus percent of total 24-hour storm rainfall. The program assumes 0% rainfall for 0-hour storm time as the first data pair. Then the user enters successively increasing storm [time (hours), % (24-hour total rainfall)] data pairs. The pattern must terminate on or before 24-hours in time.

ENTER A [1] TO CONTINUE
=====> '1'

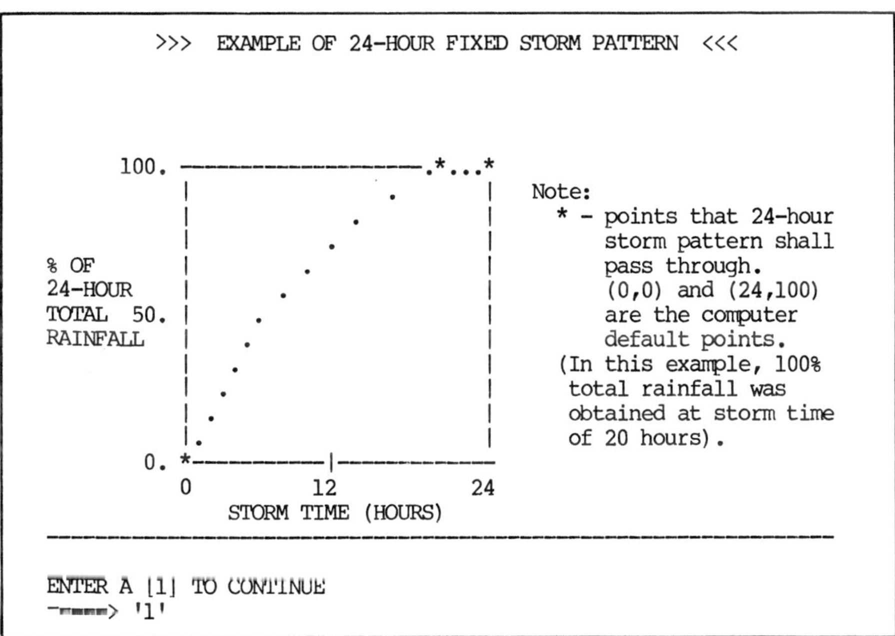

```
            >>>  EXAMPLE OF 24-HOUR FIXED STORM PATTERN  <<<


     100. -------------------.*...*
          |                .     |   Note:
          |              .       |     * - points that 24-hour
          |            .         |         storm pattern shall
% OF      |          .           |         pass through.
24-HOUR   |        .             |         (0,0) and (24,100)
TOTAL  50.|       .              |         are the computer
RAINFALL  |      .               |         default points.
          |     .                |     (In this example, 100%
          |    .                 |      total rainfall was
          |   .                  |      obtained at storm time
          |  .                   |      of 20 hours).
       0. *----------|-----------
          0          12          24
              STORM TIME (HOURS)
```

ENTER A [1] TO CONTINUE
=====> '1'

An example of a partial input data set is shown as follows:

```
*** USER SPECIFIED 24-HOUR STORM PATTERNS ***
  (NOTE: MAXIMUM ALLOWABLE NUMBER OF SETS IS "4")

ENTER THE # 1 24-HOUR STORM PATTERN AS TIME VS. PERCENT OF
24-HOUR RAINFALL
   NOTE: 1. 0 < TIME < 24  AND 0 < DECIMAL PERCENT < 1
         2. NO MORE THAN 50 DATA PAIRS PER SET ARE ALLOWED
         3. THE FIRST DATA PAIR IS DEFINED TO BE (0-HOUR,0-%)

ENTER THE TIME ORDINATE(HOURS) FOR DATA PAIR #  2
=====> '1'

ENTER DECIMAL PERCENT OF 24-HOUR RAINFALL FOR DATA PAIR #  2
=====> '.1'

ENTER THE TIME ORDINATE(HOURS) FOR DATA PAIR #  3
=====> '2'

ENTER DECIMAL PERCENT OF 24-HOUR RAINFALL FOR DATA PAIR #  3
=====> '.3'

ENTER THE TIME ORDINATE(HOURS) FOR DATA PAIR #  4
=====> '5'

ENTER DECIMAL PERCENT OF 24-HOUR RAINFALL FOR DATA PAIR #  4
=====> '.5'
```

8.4.5.3. <u>User specified 24-hour storm pattern with continuous depth-area adjustment</u>

The program is based upon a 24-hour storm duration and is composed of 480 3-minute unit intervals. Using logarithmic interpolation of nested T-year rainfall depths (e.g., 5-minute, 15-minute, ..., 24-hour durations), each depth can be adjusted for depth-area effects and a continuously adjusted rainfall data set is obtained. This plot of adjusted rainfall depth versus time is used to construct the set of 480 3-minute unit rainfalls with interval #1 being the largest 3-minute rainfall, and #480 being the smallest (see Sec. 2.15.2).

The user can then arrange these 480 unit rainfalls in an arbitrary storm pattern by assigning the pattern unit interval number to each sequential unit rainfall, starting with unit interval #1.

```
--------------------------------------------------------------
ENTER A [1] TO CONTINUE
=====> '1'
```

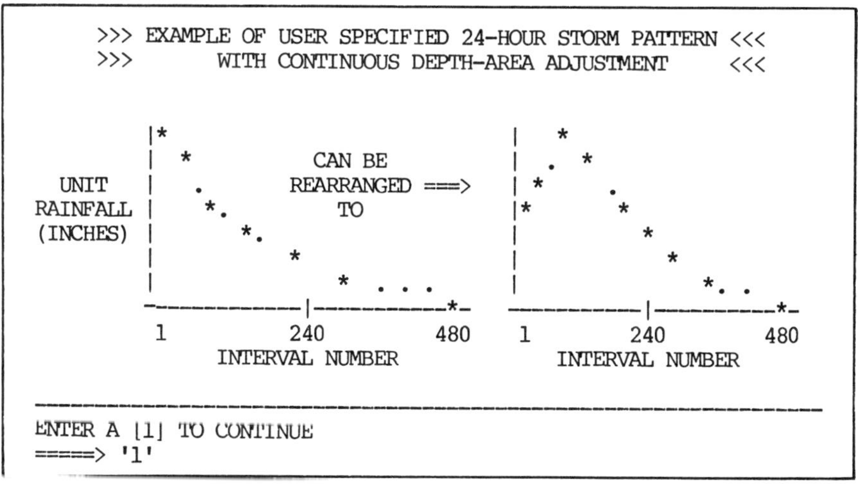

```
>>> EXAMPLE OF USER SPECIFIED 24-HOUR STORM PATTERN <<<
>>>       WITH CONTINUOUS DEPTH-AREA ADJUSTMENT      <<<

         |*                                  |   *
         |  *          CAN BE                | .  *
 UNIT    |   .      REARRANGED ===>          | *     .
RAINFALL |    *.         TO                  |*       *
(INCHES) |      *.                           |          *
         |         *                         |            *
         |            *   . . .              |              *. .
 --------------|-------------*-   ------------|------------*-
  1           240           480   1          240          480
        INTERVAL NUMBER                 INTERVAL NUMBER

-----------------------------------------------------------------
ENTER A [1] TO CONTINUE
=====> '1'
```

An example of a partial input data set is shown as follows:

```
*** USER SPECIFIED STORM PATTERN WITH
    CONTINUOUS DEPTH-AREA ADJUSTMENT ***

ENTER THE NEW INTERVAL NUMBER FOR INTERVAL [    1]
   NOTE: A [0] WILL TERMINATE THE PROCESS
=====> '120'

ENTER THE NEW INTERVAL NUMBER FOR INTERVAL [    2]
   NOTE: A [0] WILL TERMINATE THE PROCESS
=====> '121'

ENTER THE NEW INTERVAL NUMBER FOR INTERVAL [    3]
   NOTE: A [0] WILL TERMINATE THE PROCESS
=====> '130'

ENTER THE NEW INTERVAL NUMBER FOR INTERVAL [    4]
   NOTE: A [0] WILL TERMINATE THE PROCESS
=====> '260'

ENTER THE NEW INTERVAL NUMBER FOR INTERVAL [    5]
   NOTE: A [0] WILL TERMINATE THE PROCESS
=====> '450'
```

8.4.5.4. S-graphs

S-graphs are a plot of the percent of catchment lag versus the percent of the ultimate discharge (0% to 100%). The program sets 0% lag time to 0% ultimate discharge, and 100% lag time to 50% ultimate discharge. The user enters successively increasing values of the % lag as prompted by the % of ultimate discharge, given in 2% increments.

ENTER A [1] TO CONTINUE
====> '1'

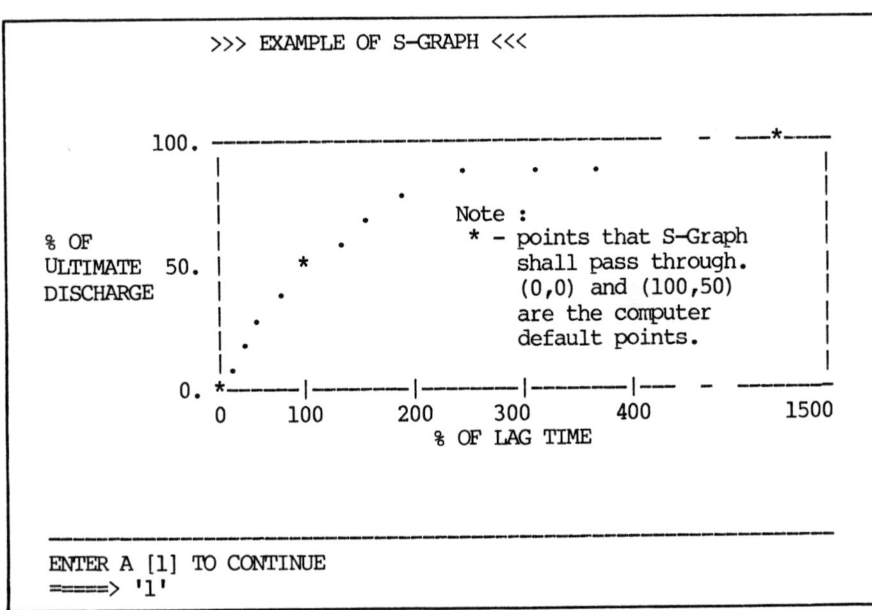

An example of a partial input data set is shown as follows:

```
   *** S-GRAPH ***
NOTE: MAXIMUM ALLOWABLE NUMBER IS "10"

ENTER THE # 1 S-GRAPH AS %LAG VS. %ULTIMATE Q
   NOTE: S-GRAPH PASSES THROUGH
        (0.,0.) AND (100%LAG,50%ULTIMATE Q)

ENTER THE %LAG FOR    2.%ULTIMATE DISCHARGE
=====> '12.5'

ENTER THE %LAG FOR    4.%ULTIMATE DISCHARGE
=====> '20.'

ENTER THE %LAG FOR    6.%ULTIMATE DISCHARGE
=====> '28.'

ENTER THE %LAG FOR    8.%ULTIMATE DISCHARGE
=====> '34.'

ENTER THE %LAG FOR   10.%ULTIMATE DISCHARGE
=====> '40.'
```

8.4.5.5. Data bank summary

If the user tries to add data to a existing data set or tries to delete a non-existing data set, the following error messsage will be shown on the user's CRT:

```
   *** DEPTH-AREA CURVE SETS ***
  (NOTE: MAXIMUM ALLOWABLE NUMBER OF SETS IS '4')

[ 1 ] DEPTH-AREA CURVE SETS EXIST
  THE FOLLOWING DEPTH-AREA CURVE SETS ARE DEFINED :
    1
ENTER THE DEPTH-AREA CURVE SET NUMBER
FOR ADDITION TO MEMORY
=====> '1'
ERROR : INVALID DATA ENTRY....TRY AGAIN

ENTER A [1] TO CONTINUE
=====> '1'
```

After terminating the data bank development process, a summary of information stored in the data bank will be displayed on the user's CRT as follows:

```
---------------------------------------------------------------
     DATA BANK SUMMARY
---------------------------------------------------------------

[ 1 ] DEPTH-AREA CURVE SETS EXIST

[ 1 ] USER SPECIFIED 24-HOUR STORM PATTERNS EXIST

USER SPECIFIED 24-HOUR STORM PATTERN WITH
 CONTINUOUS DEPTH-AREA ADJUSTMENT
(ALREADY DEFINED)

[ 1  ] S-GRAPHS EXIST

...............................................................
   *** CAUTION : MAKE SURE THAT DATA IS AVAILABLE
   ***           BEFORE PROCEEDING WITH THE LINK-NODE MODEL
...............................................................

TYPE "RUNOFF" TO PROCEED TO THE LINK-NODE MODEL DEVELOPMENT
```

8.4.5.6. PROGRAM 8.1

```
C
C  THIS IS A DATA BANK DEVELOPMENT PROGRAM
C  FOR THE RUNOFF COMPUTER PROGRAM
C
C   IBM/PC-XT VERSION DATE : 11/19/1986
C
      COMMON/BLK 1/ DARF(4,110),NDARF(4)
      COMMON/BLK 2/ NEW(480)
      COMMON/BLK 3/ ST(4,50,2),NST(4)
      COMMON/BLK 4/ S(10,52),KNS(10)
C
C------------------------------------------------------------------
C     DEFINITION OF ARRAYS
C
C     DARF   : DEPTH-AREA CURVES
C     ST     : USER SPECIFIED 24-HOUR STORM PATTERN
C     NEW    : NEW INTERVAL NUMBER FOR USER SPECIFIED STORM
C              PATTERN WITH CONTINUOUS DEPTH-AREA ADJUSTMENT
C     S      : S-GRAPH
C------------------------------------------------------------------
C
C  INTRODUCTION OF RUNOFF MODEL
C
      KTX=0
 5    WRITE(0,521)
 521  FORMAT(24(/),12('>'),' RUNOFF MODEL ',12('<'),//,
     1 '     .         .    .         .    .         .',/,
     2 '      \___^___/      \___^___/      \___^___/',//,
     3 '          :         :    :    :    :         :    :',/,
     4 '      :     ^  :      :    :    :         :    :',/,
     5 '         / \    :      :         :      :    :',/,
     6 '        /   \               :         :    :',/,
     7 '       /     \   _/‾\          :    :',/,
     8 '      /       _/       \ _     ',/,
     9 '     /     _/‾ \          ‾ \  ',/,
     2 '                                                      _________',/,
     3 '          - - - - - - - - - - - - - .              /\_________\',/,
     4 '           ~   ~   ~   ~   ~   ~   ~   ~ ~ \       | |_________|',/,
     6 '          - - - - - - - - - - - - - - . ~ \_ _',/,
     7 '                                          \  ~ ~  \__',/,
     8 '                              /\_________\  \   ~   ~ ~ \',/,
     9 '                              | |________|         \ ~ ~ \ ',/,
     C /,72('='),/,'  ENTER A [1] TO SKIP THE INTRODUCTION',
     C ' TEXT',/,5X,'OR A [2] TO REVIEW THE INTRODUCTION TEXT')
      WRITE(0,99)
 99   FORMAT(' =====> ',$)
      READ(0,*)KX
      IF(KX.EQ.2)CALL TEXT
      IF(KX.EQ.1 .OR. KX.EQ.2)GO TO 10
      GO TO 5
10    WRITE(0,511)
 511  FORMAT(24(/),72('*'),/,10('='),'> RUNOFF MODEL <',10('='),/,
     C 72('.'),/,'  SELECT PROGRAM OPERATING MODE:',
```

```
     C /,3X,'[1] = DATA BANK DEVELOPMENT',
     C /,3X,'[2] = EXIT PROGRAM')
      WRITE(0,99)
      READ(0,*)KRUN
      IF(KRUN.EQ.1)GO TO 210
      IF(KRUN.EQ.2)GO TO 220
      GO TO 10
C
C  READ DATA FROM DATA BANK
C
 210  OPEN(UNIT=1,FILE='CONTR.DAT',STATUS='OLD')
      READ(1,515)KKK
 515  FORMAT(///,I3)
      READ(1,516)((DARF(J,I),I=1,110),J=1,4)
 516  FORMAT(11(1X,F8.4))
      READ(1,515)NS
      READ(1,518)((ST(I,J,1),ST(I,J,2),J=1,50),I=1,4)
 518  FORMAT(10(1X,F8.4))
      READ(1,515)KDUM
      READ(1,517)(NEW(I),I=1,480)
 517  FORMAT(10I4)
      READ(1,515)NSS
      READ(1,518)((S(I,J),J=1,52),I=1,10)
      CLOSE(UNIT=1)
C
C  DATA BANK DEVELOPMENT
C
      CALL CONTROL
C
 220  STOP' TYPE "RUNOFF" TO PROCEED TO THE LINK-NODE MODEL DEVELOPMENT'
      END
C
C-----------------------------------------------------------------------
C     THIS PROGRAM DEVELOPS DATA BANK INFORMATION INTERACTIVELY
C-----------------------------------------------------------------------
C
      SUBROUTINE CONTROL
C
      COMMON/BLK 1/ DARF(4,110),NDARF(4)
      COMMON/BLK 2/ NEW(480)
      COMMON/BLK 3/ ST(4,50,2),NST(4)
      COMMON/BLK 4/ S(10,52),KNS(10)
C
      IXX=0
 800  IF(IXX.NE.0)GO TO 880
      WRITE(0,716)
 716  FORMAT(24(/),76('='),/,
     1 10X,'DATA BANK DEVELOPMENT',/,76('='),//,
     2 '          _________________           DATA BANK INCLUDES:',/,
     3 '         |\_________________\         ',/,
     4 '         ||  . . . . . . .  |          *DEPTH-AREA CURVE SETS',/,
     5 '         ||  .            . |          *USER SPECIFIED 24-HOUR',/,
     6 '         ||  .    CRT     . |           STORM PATTERNS',/,
     7 '         ||  .            . |          *USER SPECIFIED STORM',/,
     8 '         \|  . . . . . . .  |           PATTERN WITH CONTINUOUS',/,
```

```
     9 '        ------.---.------       DEPTH-AREA ADJUSTMENT',/,
     9 '             /_____\           *S-GRAPHS',/,
     8 '        ____________________',/,
     7 '        \ = = = = = = = = = = \',/,
     6 '         \ = = = = = = = = = = \',/,
     5 '          \____________________\',//,76('.'),/,
     4 2X,'ENTER A [1] TO CONTINUE')
      WRITE(0,99)
      READ(0,*)KX
      IF(KX.NE.1)GO TO 800
      IXX=1
880   K1=0
      DO 891 I=1,4
      IF(DARF(I,13).EQ.0.)GO TO 891
      K1=K1+1
891   CONTINUE
      K2=0
      DO 892 I=1,4
      IF(ST(I,2,1).EQ.0.)GO TO 892
      K2=K2+1
892   CONTINUE
      K=0
      IF(NEW(1).NE.0)K3=1
      K4=0
      DO 894 I=1,10
      IF(S(I,2).EQ.0.)GO TO 894
      K4=K4+1
894   CONTINUE
      WRITE(0,700)K1,K2
700   FORMAT(24(/),76('-'),/,'  *** PROGRAM DATA BANK DEVELOPMENT',
     1 ' ***',/,76('-'),/,'   DATA BANK INCLUDES : ',/,
     2 5X,'.MAXIMUM OF [4] DEPTH-AREA CURVE SETS',/,
     3 5X,'([',I2,'] ARE DEFINED)',/,
     4 5X,'.MAXIMUM OF [4] USER SPECIFIED 24-HOUR STORM PATTERNS',/,
     5 5X,'([',I2,'] ARE DEFINED)',/,
     6 5X,'.[1] USER SPECIFIED STORM PATTERN WITH CONTINUOUS ',
     7 'DEPTH-AREA ADJUSTMENT')
      IF(K3.EQ.1)WRITE(0,802)
802   FORMAT(5X,'(ALREADY DEFINED)')
      IF(K3.EQ.0)WRITE(0,803)
803   FORMAT(5X,'(CURRENTLY IS UNDEFINED)')
      WRITE(0,804)K4
804   FORMAT(5X,'.MAXIMUM OF [10] S-GRAPHS',/,
     1 5X,'([',I3,'] ARE DEFINED)',/,
     2 76('.'),//,'  SELECT PROGRAM INSTRUCTION NUMBER : ',/,
     3 '    [1] = ADD TO DATA BANK STORAGE',/,'    [2] = DELETE',
     4 ' DATA FROM STORAGE',/,'    [3] = EXIT PROGRAM',
     5 ' AND PROCEED TO LINK-NODE MODEL DEVELOPMENT')
      WRITE(0,99)
      READ(0,*)KDATA
      IF(KDATA.EQ.1 .OR. KDATA.EQ.2)GO TO 810
      IF(KDATA.EQ.3)GO TO 860
801   WRITE(0,999)
999   FORMAT('  ERROR : INVALID DATA ENTRY....TRY AGAIN',/,
     1 '    ENTER A [1] TO CONTINUE')
```

```
      WRITE(0,99)
 99   FORMAT(2X,'=====> ',$)
      READ(0,*)KX
      IF(KX.NE.1)GO TO 801
      GO TO 800
C
 810  IF(KDATA.EQ.1)WRITE(0,717)
 717  FORMAT(24(/),'  SELECT A SET OF DATA TO BE ADDED',
     1 ' TO DATA BANK:')
      IF(KDATA.EQ.2)WRITE(0,718)
 718  FORMAT(24(/),'  SELECT A SET OF DATA TO BE DELETED',
     1 ' FROM DATA BANK:')
      WRITE(0,701)
 701  FORMAT('    [1] = DEPTH-AREA CURVE SETS',/,'    [2] = ',
     1 'USER SPECIFIED 24-HOUR STORM PATTERNS',/,'    [3] = ',
     2 'USER SPECIFIED STORM PATTERN WITH CONTINUOUS DEPTH-',
     3 'AREA ADJUSTMENT',/,'    [4] = S-GRAPHS')
      WRITE(0,99)
      READ(0,*)KMOD
      IF(KMOD.EQ.1)GO TO 820
      IF(KMOD.EQ.2)GO TO 830
      IF(KMOD.EQ.3)GO TO 840
      IF(KMOD.EQ.4)GO TO 850
 811  WRITE(0,999)
      WRITE(0,99)
      READ(0,*)KX
      IF(KX.NE.1)GO TO 811
      GO TO 810
C
C  DEPTH-AREA CURVE SETS
C
 820  IF(KTX1.EQ.0. .AND. KDATA.EQ.1)CALL TEXT1
      KTX1=1
      WRITE(0,98)
      WRITE(0,62)
 98   FORMAT(24(/))
 62   FORMAT(/,2X,'*** DEPTH-AREA CURVE SETS ***')
      WRITE(0,66)
 66   FORMAT('    (NOTE: MAXIMUM ALLOWABLE NUMBER OF SETS IS "4")')
C
      KKK=0
      DO 205 I=1,4
      IF(DARF(I,13).EQ.0.)GO TO 205
      KKK=KKK+1
      NDARF(KKK)=I
 205  CONTINUE
      IF(KKK.EQ.0 .AND. KDATA.EQ.2)GO TO 824
      IF(KKK.EQ.0 .AND. KDATA.EQ.1)GO TO 200
C
      WRITE(0,61)KKK,(NDARF(I),I=1,KKK)
 61   FORMAT(/,2X,' [',I2,' ] DEPTH-AREA CURVE SETS EXIST',/,
     1 5X,'THE FOLLOWING DEPTH-AREA CURVE SETS ARE DEFINED :',/,
     2 7X,I2,:,I2,:,I2,:,I2)
      WRITE(0,713)
 713  FORMAT('  ENTER THE DEPTH-AREA CURVE SET NUMBER')
```

```
      IF(KDATA.EQ.1)WRITE(0,711)
 711  FORMAT('  FOR ADDITION TO MEMORY')
      IF(KDATA.EQ.2)WRITE(0,712)
 712  FORMAT('  FOR DELETION FROM MEMORY')
      WRITE(0,99)
      READ(0,*)KNO
      IF(KNO.LT.0 .OR. KNO.GT.4)GO TO 824
      IF(KDATA.EQ.1)GO TO 821
      DO 822 I=1,KKK
      IF(KNO.EQ.NDARF(I))GO TO 823
 822  CONTINUE
 824  WRITE(0,999)
      WRITE(0,99)
      READ(0,*)KX
      IF(KX.NE.1)GO TO 824
      GO TO 800
 823  DO 825 I=1,110
      DARF(KNO,I)=0.
 825  CONTINUE
      KKK=KKK-1
      GO TO 800
 821  DO 826 I=1,KKK
      IF(KNO.EQ.NDARF(I))GO TO 824
 826  CONTINUE
C
C  INPUT THE DEPTH-AREA CURVE SETS
C
 200  IF(KKK.EQ.0)KNO=1
      IJK=KNO
      DARF(IJK,1)=0.
      DARF(IJK,2)=5.
      DARF(IJK,3)=10.
      DARF(IJK,4)=15.
      DARF(IJK,5)=20.
      DARF(IJK,6)=40.
      DARF(IJK,7)=60.
      DARF(IJK,8)=80.
      DARF(IJK,9)=100.
      DARF(IJK,10)=125.
      DARF(IJK,11)=150
      DO 20 I=1,10
      J=11*I+1
      DARF(IJK,J)=1.
 20   CONTINUE
 72   WRITE(0,51)
 51   FORMAT(24(/),2X,'ENTER 5-MINUTE DEPTH-AREA CURVE DATA:',/,
     1 5X,'ENTER THE 5-MINUTE ADJUSTMENT FACTORS CORRESPONDING TO'
     1 ,/,11X,'5-, 10-, 15-, 20-, 40-, 60-, 80-, 100-,',
     2 /11X,'125-, AND 150-SQUARE MILES OF CATCHMENT AREA')
      READ(0,*)(DARF(IJK,I),I=13,22)
      IF(DARF(IJK,13).GT.1.)GO TO 72
      DO 71 I=12,21
      IF(DARF(IJK,I+1) .GT. DARF(IJK,I))GO TO 72
 71   CONTINUE
 74   WRITE(0,52)
```

```
52   FORMAT(/2X,'ENTER 15-MINUTE DEPTH-AREA CURVE DATA:',/,
    1 5X,'ENTER THE 15-MINUTE ADJUSTMENT FACTORS CORRESPONDING TO'
    1 ,/,11X,'5-, 10-, 15-, 20-, 40-, 60-, 80-, 100-,',
    2 /11X,'125-, AND 150-SQUARE MILES OF CATCHMENT AREA')
     READ(0,*)(DARF(IJK,I),I=24,33)
     IF(DARF(IJK,24).GT.1.)GO TO 74
     DO 73 I=23,32
     IF(DARF(IJK,I+1) .GT. DARF(IJK,I))GO TO 74
     IF(DARF(IJK,I+1) .LT. DARF(IJK,I-10))GO TO 74
73   CONTINUE
76   WRITE(0,53)
53   FORMAT(/2X,'ENTER 30-MINUTE DEPTH-AREA CURVE DATA:',/,
    1 5X,'ENTER THE 30-MINUTE ADJUSTMENT FACTORS CORRESPONDING TO'
    1 ,/,11X,'5-, 10-, 15-, 20-, 40-, 60-, 80-, 100-,',
    2 /11X,'125-, AND 150-SQUARE MILES OF CATCHMENT AREA')
     READ(0,*)(DARF(IJK,I),I=35,44)
     IF(DARF(IJK,35).GT.1.)GO TO 76
     DO 75 I=34,43
     IF(DARF(IJK,I+1) .GT. DARF(IJK,I))GO TO 76
     IF(DARF(IJK,I+1) .LT. DARF(IJK,I-10))GO TO 76
75   CONTINUE
78   WRITE(0,54)
54   FORMAT(/2X,'ENTER 1-HOUR DEPTH-AREA CURVE DATA:',/,
    1 5X,'ENTER THE 1-HOUR ADJUSTMENT FACTORS CORRESPONDING TO'
    1 ,/,11X,'5-, 10-, 15-, 20-, 40-, 60-, 80-, 100-,',
    2 /11X,'125-, AND 150-SQUARE MILES OF CATCHMENT AREA')
     READ(0,*)(DARF(IJK,I),I=46,55)
     IF(DARF(IJK,46).GT.1.)GO TO 78
     DO 77 I=45,54
     IF(DARF(IJK,I+1) .GT. DARF(IJK,I))GO TO 78
     IF(DARF(IJK,I+1) .LT. DARF(IJK,I-10))GO TO 78
77   CONTINUE
82   WRITE(0,55)
55   FORMAT(/2X,'ENTER 2-HOUR DEPTH-AREA CURVE DATA:',/,
    1 5X,'ENTER THE 2-HOUR ADJUSTMENT FACTORS CORRESPONDING TO'
    1 ,/,11X,'5-, 10-, 15-, 20-, 40-, 60-, 80-, 100-,',
    2 /11X,'125-, AND 150-SQUARE MILES OF CATCHMENT AREA')
     READ(0,*)(DARF(IJK,I),I=57,66)
     IF(DARF(IJK,57).GT.1.)GO TO 82
     DO 81 I=56,65
     IF(DARF(IJK,I+1) .GT. DARF(IJK,I))GO TO 82
     IF(DARF(IJK,I+1) .LT. DARF(IJK,I-10))GO TO 82
81   CONTINUE
84   WRITE(0,56)
56   FORMAT(/2X,'ENTER 3-HOUR DEPTH-AREA CURVE DATA:',/,
    1 5X,'ENTER THE 3-HOUR ADJUSTMENT FACTORS CORRESPONDING TO'
    1 ,/,11X,'5-, 10-, 15-, 20-, 40-, 60-, 80-, 100-,',
    2 /11X,'125-, AND 150-SQUARE MILES OF CATCHMENT AREA')
     READ(0,*)(DARF(IJK,I),I=68,77)
     IF(DARF(IJK,68).GT.1.)GO TO 84
     DO 83 I=67,76
     IF(DARF(IJK,I+1) .GT. DARF(IJK,I))GO TO 84
     IF(DARF(IJK,I+1) .LT. DARF(IJK,I-10))GO TO 84
83   CONTINUE
86   WRITE(0,57)
```

```
 57   FORMAT(/2X,'ENTER 6-HOUR DEPTH-AREA CURVE DATA:',/,
     1 5X,'ENTER THE 6-HOUR ADJUSTMENT FACTORS CORRESPONDING TO'
     1 ,/,11X,'5-, 10-, 15-, 20-, 40-, 60-, 80-, 100-,',
     2 /11X,'125-, AND 150-SQUARE MILES OF CATCHMENT AREA')
      READ(0,*)(DARF(IJK,I),I=79,88)
      IF(DARF(IJK,79).GT.1.)GO TO 86
      DO 85 I=78,87
      IF(DARF(IJK,I+1) .GT. DARF(IJK,I))GO TO 86
      IF(DARF(IJK,I+1) .LT. DARF(IJK,I-10))GO TO 86
 85   CONTINUE
 88   WRITE(0,58)
 58   FORMAT(/2X,'ENTER 12-HOUR DEPTH-AREA CURVE DATA:',/,
     1 5X,'ENTER THE 12-HOUR ADJUSTMENT FACTORS CORRESPONDING TO'
     1 ,/,11X,'5-, 10-, 15-, 20-, 40-, 60-, 80-, 100-,',
     2 /11X,'125-, AND 150-SQUARE MILES OF CATCHMENT AREA')
      READ(0,*)(DARF(IJK,I),I=90,99)
      IF(DARF(IJK,90).GT.1.)GO TO 88
      DO 87 I=89,98
      IF(DARF(IJK,I+1) .GT. DARF(IJK,I))GO TO 88
      IF(DARF(IJK,I+1) .LT. DARF(IJK,I-10))GO TO 88
 87   CONTINUE
 92   WRITE(0,59)
 59   FORMAT(/2X,'ENTER 24-HOUR DEPTH-AREA CURVE DATA:',/,
     1 5X,'ENTER THE 24-HOUR ADJUSTMENT FACTORS CORRESPONDING TO'
     1 ,/,11X,'5-, 10-, 15-, 20-, 40-, 60-, 80-, 100-,',
     2 /11X,'125-, AND 150-SQUARE MILES OF CATCHMENT AREA')
      READ(0,*)(DARF(IJK,I),I=101,110)
      IF(DARF(IJK,101).GT.1.)GO TO 92
      DO 91 I=100,109
      IF(DARF(IJK,I+1) .GT. DARF(IJK,I))GO TO 92
      IF(DARF(IJK,I+1) .LT. DARF(IJK,I-10))GO TO 92
 91   CONTINUE
      KKK=KKK+1
      GO TO 800
C
C  USER SPECIFIED STORM PATTERN
C
 830  IF(KTX2.EQ.0. .AND. KDATA.EQ.1)CALL TEXT2
      KTX2=1
      WRITE(0,98)
      WRITE(0,63)
 63   FORMAT(/,2X,'*** USER SPECIFIED 24-HOUR STORM PATTERNS ***')
      WRITE(0,66)
C
      NS=0
      DO 831 I=1,4
      IF(ST(I,2,1).EQ.0.)GO TO 831
      NS=NS+1
      NST(NS)=I
 831  CONTINUE
      IF(NS.EQ.0 .AND. KDATA.EQ.2)GO TO 824
      IF(NS.EQ.0 .AND. KDATA.EQ.1)GO TO 310
      WRITE(0,702)NS,(NST(I),I=1,NS)
 702  FORMAT(/,2X,' [',I2,' ] 24-HOUR STORM PATTERNS EXIST',/,
     1 5X,'THE FOLLOWING 24-HOUR STORM PATTERNS ARE DEFINED :',/,
```

```
   2 7X,I2,:,I2,:,I2,:,I2)
     WRITE(0,714)
714  FORMAT('  ENTER STORM NUMBER')
     IF(KDATA.EQ.1)WRITE(0,711)
     IF(KDATA.EQ.2)WRITE(0,712)
     WRITE(0,99)
     READ(0,*)KNO
     IF(KNO.LT.0 .OR. KNO.GT.4)GO TO 824
     IF(KDATA.EQ.1)GO TO 832
     DO 833 I=1,NS
     IF(KNO.EQ.NST(I))GO TO 834
833  CONTINUE
     GO TO 824
834  DO 835 I=1,50
     ST(KNO,I,1)=0.
     ST(KNO,I,2)=0.
835  CONTINUE
     NS=NS-1
     GO TO 800
832  DO 836 I=1,NS
     IF(KNO.EQ.NST(I))GO TO 824
836  CONTINUE
310  IF(NS.EQ.0)KNO=1
     I=KNO
     J=2
     WRITE(0,32)I
32   FORMAT(/,2X,'ENTER THE #',I2,' 24-HOUR STORM PATTERN AS TIME',
   1 ' VS. PERCENT OF 24-HOUR RAINFALL',/,5X,'NOTE:',
   2 ' 1. 0 < TIME < 24  AND 0 < DECIMAL PERCENT < 1',/,11X,
   3 '2. NO MORE THAN 50 DATA PAIRS PER SET ARE ALLOWED',/,11X,
   4 '3. THE FIRST DATA PAIR IS DEFINED TO BE (0-HOUR,0-%)')
340  WRITE(0,34)J
34   FORMAT(/,'  ENTER THE TIME ORDINATE(HOURS) FOR DATA PAIR #',I3)
     WRITE(0,99)
     READ(0,*)ST(I,J,1)
     IF(ST(I,J,1) .LT. ST(I,J-1,1))GO TO 340
     IF(ST(I,J,1) .GE. 24.)GO TO 355
350  WRITE(0,39)J
39   FORMAT(/,2X,'ENTER DECIMAL PERCENT OF 24-HOUR RAINFALL FOR DATA',
   1 ' PAIR #',I3)
     WRITE(0,99)
     READ(0,*)ST(I,J,2)
     IF(ST(I,J,2) .GT. 1.)GO TO 350
     IF(ST(I,J,2) .LT. ST(I,J-1,2))GO TO 350
     IF(ST(I,J,2) .EQ. 1.)GO TO 357
     J=J+1
     IF(J .GT. 50)GO TO 330
     GO TO 340
355  ST(I,J,2)=1.
     ST(I,J,1)=24.
     GO TO 330
357  ST(I,J+1,1)=24.
     ST(I,J+1,2)=1.
330  CONTINUE
     NS=NS+1
```

```
      GO TO 800
C
C  USER SPECIFIED STORM PATTERN WITH CONTINUOUS DEPTH-AREA ADJUSTMENT
C
 840  IF(KTX3.EQ.0. .AND. KDATA.EQ.1)CALL TEXT3
      KTX3=1
      WRITE(0,98)
      IF(KDATA.EQ.1)WRITE(0,64)
 64   FORMAT(/,2X,'*** USER SPECIFIED STORM PATTERN WITH',
     1 /,6X,'CONTINUOUS DEPTH-AREA ADJUSTMENT ***')
C
      IF(KDATA.EQ.2)GO TO 841
      IF(NEW(1).NE.0 .AND. KDATA.EQ.1)GO TO 842
 841  DO 440 I=1,480
 440  NEW(I)=0
      IF(KDATA.EQ.2)GO TO 800
      DO 420 I=1,480
 44   WRITE(0,42)I
 42   FORMAT(/,2X,'ENTER THE NEW INTERVAL NUMBER FOR INTERVAL [',
     1 I4,']',/,5X,'NOTE: A [0] WILL TERMINATE THE PROCESS')
      WRITE(0,99)
      READ(0,*)NEW(I)
      IF(NEW(I).GT.480)GO TO 44
      IF(NEW(I).EQ.0)GO TO 410
 420  CONTINUE
 410  CONTINUE
      GO TO 800
 842  WRITE(0,43)
 43   FORMAT(/,'  DELETE THE DATA FROM THE MEMORY BEFORE ADDING',/,
     1 '  NEW DATA TO THE MEMORY',/,'  ENTER A [1] TO CONTINUE')
      WRITE(0,99)
      READ(0,*)KX
      IF(KX.NE.1) GO TO 842
      GO TO 800
C
C  S-GRAPHS
C
 850  IF(KTX4.EQ.0. .AND. KDATA.EQ.1)CALL TEXT4
      KTX4=1
      WRITE(0,98)
      WRITE(0,65)
 65   FORMAT(/,2X,'*** S-GRAPH ***')
      WRITE(0,171)
 171  FORMAT('   NOTE: MAXIMUM ALLOWABLE NUMBER IS "10"')
C
      NSS=0
      DO 851 I=1,10
      IF(S(I,2).EQ.0.)GO TO 851
      NSS=NSS+1
      KNS(NSS)=I
 851  CONTINUE
      IF(NSS.EQ.0 .AND. KDATA.EQ.2)GO TO 824
      IF(NSS.EQ.0 .AND. KDATA.EQ.1)GO TO 510
      WRITE(0,703)NSS,(KNS(I),I=1,NSS)
 703  FORMAT(/,2X,' [',I2,' ] S-GRAPHS EXIST',/,
```

```
     1 5X,'THE FOLLOWING S-GRAPHS ARE DEFINED :',/,
     2 7X,I2,:,I2,:,I2,:,I2,:,I2,:,I2,:,I2,:,I2,:,I2,:,I2)
      WRITE(0,715)
 715  FORMAT('  ENTER S-GRAPH NUMBER')
      IF(KDATA.EQ.1)WRITE(0,711)
      IF(KDATA.EQ.2)WRITE(0,712)
      WRITE(0,99)
      READ(0,*)KNO
      IF(KNO.LT.0 .OR. KNO.GT.10)GO TO 824
      IF(KDATA.EQ.1)GO TO 852
      DO 853 I=1,NSS
      IF(KNO.EQ.KNS(I))GO TO 854
 853  CONTINUE
      GO TO 824
 854  DO 855 I=1,52
      S(KNO,I)=0.
 855  CONTINUE
      NSS=NSS-1
      GO TO 800
 852  DO 856 I=1,NSS
      IF(KNO.EQ.KNS(I))GO TO 824
 856  CONTINUE
 510  IF(NSS.EQ.0)KNO=1
      I=KNO
      WRITE(0,172)I
 172  FORMAT(/,2X,'ENTER THE #',I2,' S-GRAPH AS %LAG',
     1 ' VS. %ULTIMATE Q',/,5X,'NOTE: S-GRAPH PASSES THROUGH',/,11X,
     2 ' (0.,0.) AND (100%LAG,50%ULTIMATE Q)')
      XX=2.
      DO 540 J=2,52
      IF(XX.EQ.50.)S(I,J)=100.
      IF(XX.EQ.50.)GO TO 530
 520  WRITE(0,173)XX
 173  FORMAT(/,2X,'ENTER THE %LAG FOR ',F4.0,'%ULTIMATE DISCHARGE')
      WRITE(0,99)
      READ(0,*)S(I,J)
      IF(XX.LT.50. .AND. S(I,J).GE.100.)GO TO 520
      IF(S(I,J).LE.S(I,J-1))GO TO 520
 530  IF(XX.LT.98.)GO TO 535
      XX=XX+1.
      GO TO 540
 535  XX=XX+2.
 540  CONTINUE
      NSS=NSS+1
      GO TO 800
C
C  OUTPUT DATA BANK TO DISK
C
 860  KKK=0
      DO 861 I=1,4
      IF(DARF(I,13).EQ.0.)GO TO 861
      KKK=KKK+1
      NDARF(KKK)=I
 861  CONTINUE
      WRITE(0,704)
```

```
 704  FORMAT(24(/),76('-'),/,'   DATA BANK SUMMARY',/,76('-'))
      WRITE(0,705)KKK
 705  FORMAT(/,'  [',I2,' ] DEPTH-AREA CURVE SETS EXIST')
 871  NS=0
      DO 862 I=1,4
      IF(ST(I,2,1).EQ.0.)GO TO 862
      NS=NS+1
      NST(NS)=I
 862  CONTINUE
      WRITE(0,706)NS
 706  FORMAT(/,'  [',I2,' ] USER SPECIFIED 24-HOUR STORM PATTERNS',
     1 ' EXIST')
 872  KI=0
      IF(NEW(1).NE.0)KI=1
      WRITE(0,707)
 707  FORMAT(/,'  USER SPECIFIED 24-HOUR STORM PATTERN WITH',
     1 /,'   CONTINUOUS DEPTH-AREA ADJUSTMENT')
      IF(KI.EQ.1)WRITE(0,721)
 721  FORMAT('  (ALREADY DEFINED)')
      IF(KI.EQ.0)WRITE(0,722)
 722  FORMAT('  (CURRENTLY IS UNDEFINED)')
 873  NSS=0
      DO 863 I=1,10
      IF(S(I,2).EQ.0.)GO TO 863
      NSS=NSS+1
      KNS(NSS)=I
 863  CONTINUE
      WRITE(0,708)NSS
 708  FORMAT(/,'  [',I2,' ] S-GRAPHS EXIST')
      IF(NSS.NE.0)GO TO 870
      WRITE(0,710)
 710  FORMAT(/,'  ** WARNING...AT LEAST ONE S-GRAPH SHOULD BE ',
     1 'DEFINED IN THE DATA BANK **')
 874  WRITE(0,723)
 723  FORMAT(/,'  ENTER A [1] TO CONTINUE')
      WRITE(0,99)
      READ(0,*)KX
      IF(KX.NE.1)GO TO 874
      GO TO 880
 870  WRITE(0,709)
 709  FORMAT(/,76('.'),/,'   *** CAUTION : MAKE SURE THAT ',
     1 'SUFFICIENT DATA IS AVAILABLE ',/,3X,'***',11X,
     2 'BEFORE PROCEEDING WITH THE LINK-NODE MODEL',/,
     3 76('.'),//)
C
      OPEN(UNIT=1,FILE='CONTR.DAT',STATUS='NEW')
      WRITE(1,62)
      WRITE(1,113)KKK
 113  FORMAT(/,I3)
      WRITE(1,11)((DARF(J,I),I=1,110),J=1,4)
 11   FORMAT(11(1X,F8.4))
C
      WRITE(1,63)
      WRITE(1,113)NS
      WRITE(1,13)((ST(I,J,1),ST(I,J,2),J=1,50),I=1,4)
```

```
 13   FORMAT(10(1X,F8.4))
C
      WRITE(1,64)
      IJ=1
      WRITE(1,114)IJ
 114  FORMAT(I3)
      WRITE(1,12)(NEW(I),I=1,480)
 12   FORMAT(10I4)
C
      WRITE(1,65)
      WRITE(1,113)NSS
      WRITE(1,13)((S(I,J),J=1,52),I=1,10)
      CLOSE(UNIT=1)
C
      IF(KDATA.EQ.4)STOP
      RETURN
      END
C
C-----------------------------------------------------------------------
C     INTRODUCTION OF RUNOFF MODEL
C-----------------------------------------------------------------------
C
      SUBROUTINE TEXT
C
      WRITE(0,10)
 10   FORMAT(24(/),16X,'>>>>>> PROGRAM RUNOFF <<<<<<',//,
     1 '   This is a computer-aided-design program in which the user',
     2 /,' interacts with the development of the link-node model.',
     3 /,' The user must first define a DATA BANK which stores',/,
     4 ' various hydrologic parameters. (This DATA BANK will be',/,
     5 ' stored permanently, unless changed at a later time.)',/,
     6 ' Then the user proceeds to the link-node model development'
     7 ,/,' for a catchment.')
      WRITE(0,25)
 25   FORMAT(/,10X,15('-'),/,10X,'   DATA BANK   ',/,10X,
     1 '  DEVELOPMENT  ',/,10X,15('-'),/,16X,' ',/,16X,' ',/,
     2 16X,'V',/,8X,19('-'),/,8X,'  LINK-NODE MODEL  ',/,
     3 8X,'    DEVELOPMENT    ',/,8X,19('-'))
 21   WRITE(0,30)
      READ(0,*)KX
      IF(KX.NE.1)GO TO 21
      WRITE(0,20)
 20   FORMAT(24(/),'   The link-node model is first conceptually',
     1 ' developed on paper by :',/,
     2 5X,'(1) subdividing the catchment into subareas,',/,
     3 5X,'(2) linking the subareas by channel routing elements,',/,
     4 5X,'(3) identifying retention basin characteristics,',/,
     5 5X,'(4) developing catchment hydrologic parameters, and',/,
     6 5X,'(5) preparing a catchment hydrology map with node ',
     7 'numbers assigned.')
      WRITE(0,40)
 40   FORMAT(/,10X,'NODE #101 ===>   *...',/,
```

```
      1 26X,'.     .',/,
      1 25X,'.      . <------Subarea #1',/,
      1 24X,'.       .',/,
      1 24X,'.       ...        Legend  ',/,
      2 10X,'NODE #102 ===> ....*.    .       * : nodal point',/,
      2 9X,'(Detention Basin).  \    .      .. : watershed boundary',/,
      3 26X,'.    \      .      \: routing link',/,10X,
      4 'Channel Routing ----->\   .',/,
      5 14X,'Element      .        \  .<--Subarea #2',/,
      5 26X,'.       \  .',/,
      6 27X,'........* <====NODE #103',/)
 50   WRITE(0,30)
 30   FORMAT(76('-'),/,2X,'ENTER A [1] TO CONTINUE',/,
      1 '   =====> ',$)
       READ(0,*)KX
       IF(KX.NE.1)GO TO 50
       WRITE(0,60)
 60   FORMAT(24(/),3X,'The link-node model is developed',
      1 ' INTERACTIVELY',/,' because the user "accepts" or',
      2 ' "rejects" all',/,' intermediate computational results',
      2 ' as the model',/,
      3 ' is constructed in the downstream flow direction.',/,
      4 '   After using a certain process (such as channel',/,
      5 ' routing, developing a subarea runoff hydrograph, etc.),'
      6 ,/,' the computed results are displayed for the user to',/,
      7 ' "accept" (entering a [1]) or "reject" (entering a [2]).',
      8 /,'   Rejecting intermediate results returns the program',/,
      9 ' to the previous study point so that the user may make',/,
      9 ' another attempt to solve this part of the problem.',/,
      9 '   After building the link-node model, the data file is',/,
      9 ' now available for other editing.  The user in now in a',/,
      9 ' BATCH MODE.',/,'   Particulars of the program are '
      9 ,'contained in the book.',//,30('-'),'END',43('-'))
 65   WRITE(0,30)
       READ(0,*)KX
       IF(KX.NE.1)GO TO 65
C
       RETURN
       END
C
C-----------------------------------------------------------------
C      TEXT1 - DEPTH-AREA CURVE SET
C-----------------------------------------------------------------
C
       SUBROUTINE TEXT1
C
       WRITE(0,10)
 10   FORMAT(24(/),'   Depth-area curve sets provide a reduction',
      1 ' factor to',/,' area-averaged point rainfall data for water'
      2 ,'shed area',/,' size effects.  Each curve set is composed ',
      3 'of 9 curves,',/,' with a curve for the 5-, 15-, 30-minute,',
      3 ' 1-,',
      4 ' 2-, 3-,',/,' 6-, 12-, and 24-hour storm durations.  Each',
      5 ' curve is',/,' defined for ten catchment area sizes of 5-,',
      6 ' 10-, 15-,',/,' 20-, 40-, 60-, 80-, 100-, 125-, and 150-',
```

```
     7 'square miles.',/,' (Each curve factor is set to 1 for 0-',
     8 'square miles.)',/,'   Data input are checked to guarantee '
     9 'that:',/,5X,'(1) all factors are between 0 and 1;',/,5X,
     9 '(2) all factors',
     9 ' decrease with increasing catchment area size; and'
     9 ,/,5X,'(3) all factors decrease with decreasing '
     9 'precipitation duration.')
 20   WRITE(0,15)
 15   FORMAT(/,76('-'),/,'  ENTER A [1] TO CONTINUE',/,
     1 '  =====> ',$)
      READ(0,*)KX
      IF(KX.NE.1)GO TO 20
      WRITE(0,60)
 60   FORMAT(24(/),9X,'>>> EXAMPLE OF DEPTH-AREA CURVE SETS <<<',//,
     1 15X,'1. #',24('-'),/,18X,'|**',21X,'|',/,18X,'| * *.',18X,
     2 '|',/,18X,'|   *   * .',14X,'|',/,15X,'.8 |   *     *  *  *',
     3 ' . * . *<== 24-HOUR CURVE',/,4X,'REDUCTION',5X,'|      .',
     4 17X,'|',/,5X,'FACTORS',6X,'|',6X,'*',16X,'|  Note:',/,
     5 18X,'|',7X,'.',15X,'|   # - computer default point.',/,
     6 15X,'.6 |',9X,'*.',12X,'|   * - points shall be entered',/,
     7 18X,'|',12X,'*.',9X,'|',7X,'interactively by user.',/,
     8 18X,'|',15X,'*.',6X,'|',/,
     9 18X,'|',19X,'* . *<== 5-MINUTE CURVE',/,
     9 15X,'.4 --------|-------|--------',/,
     9 18X,'0',6X,'50',6X,'100',5X,'150',/,25X,'AREA SIZE',/,
     9 23X,'(SQUARE MILES)'/)
  70  WRITE(0,15)
      READ(0,*)KX
      IF(KX.NE.1)GO TO 70
      RETURN
      END
C
C-----------------------------------------------------------------
C     TEXT2 - 24-HOUR FIXED STORM PATTERN
C-----------------------------------------------------------------
C
      SUBROUTINE TEXT2
C
      WRITE(0,10)
 10   FORMAT(24(/),'   Up to FOUR 24-hour fixed storm patterns can',
     1 ' be defined',/,' in the DATA BANK.  Each pattern is a plot',
     2 ' of storm',/,' time (hours) versus percent of total 24-hour',
     3 ' storm',/,' rainfall.  The program assumes 0% rainfall for'
     4 ,/,' 0-hour storm time as the first data pair.',/,
     5 ' Then the user enters successively increasing storm '
     5 ,/,' [time (hours), % (24-hour total rainfall)]',
     6 ' data pairs.',/,' The pattern must terminate on or before'
     7 ,' 24-hours in time.')
 15   WRITE(0,20)
 20   FORMAT(/,76('-'),/,' ENTER A [1] TO CONTINUE',/,
     1 ' =====> ',$)
      READ(0,*)KX
      IF(KX.NE.1)GO TO 15
      WRITE(0,30)
 30   FORMAT(24(/),7X,'>>> EXAMPLE OF 24-HOUR FIXED STORM PATTERN',
```

```
     1 ' <<<',//,13X,'100. ',19('-'),'.*...*',/,18X,'|',15X,'.',7X,
     2 '|',4X,'Note:',/,18X,'|',13X,'.',9X,'|',5X,'* - points',
     3 ' that 24-hour',/,18X,'|',11X,'.',11X,'|',9X,'storm pattern',
     4 ' shall',/,7X,'% OF',7X,'|',9X,'.',13X,'|',9X,'pass through.'
     5 ,/,5X,'24-HOUR',6X,'|',7X,'.',15X,'|',9X,'(0,0) and (24,100)'
     6 ,/,6X,'TOTAL   50. |',5X,'.',17X,'|',9X,'are the computer',/,
     7 5X,'RAINFALL',5X,' ',4X,'.',18X,' ',9X,'default points.',/,
     8 18X,'|   .',19X,'|',5X,'(In this example, 100%',/,18X,'|',
     9 '  .',20X,'|',6X,'total rainfall was',/,18X,'| .',21X,'|',6X,
     9 'obtained at storm time',/,18X,'|.',22X,'|',6X,
     9 'of 20 hours).',/,15X,'0. *',11('-'),'|',12('-'),
     9 /,18X,'0',10X,'12',10X,'24',/,22X,'STORM TIME (HOURS)'/)
 40   WRITE(0,20)
      READ(0,*)KX
      IF(KX.NE.1)GO TO 40
      RETURN
      END
C
C-----------------------------------------------------------------------
C     TEXT3 - USER SPECIFIED 24-HOUR STORM PATTERN
C             WITH CONTINUOUS DEPTH-AREA ADJUSTMEMT
C-----------------------------------------------------------------------
C
      SUBROUTINE TEXT3
C
      WRITE(0,10)
 10   FORMAT(24(/),'   The program is based upon a 24-hour storm'
     1 ,' duration',/,' and is composed of 480 3-minute unit',
     2 ' intervals.  Using',/,' logarithmic interpolation of',
     3 ' nested T-year rainfall depths',/,' (e.g., 5-minute, 15-',
     4 'minute, ..., 24-hour durations), each',/,' depth can be',
     5 ' adjusted for depth-area effects and a',/,' continuously'
     6 ,' adjusted rainfall data set is obtained.',/,' This plot',
     7 ' of adjusted rainfall depth versus time',/,' is used to',
     8 ' construct the set of 480 3-minute unit',/,' rainfalls,',
     9 ' with interval #1 being the largest 3-minute',/,' rainfall',
     9 ' and #480 being the smallest (see book text, sec. 2.15.2.).',
     9 /,'  The user can then arrange these 480 unit rainfalls',
     9 ' in',/,' an arbitrary storm pattern by assigning the ',
     9 'storm pattern',/,' unit interval number to each',
     9 ' sequential unit rainfall,',/,' starting with unit',
     9 ' interval #1.')
 15   WRITE(0,20)
 20   FORMAT(/,76('-'),/,'  ENTER A [1] TO CONTINUE',/,
     1 ' =====> ',$)
      READ(0,*)KX
      IF(KX.NE.1)GO TO 15
      WRITE(0,30)
 30   FORMAT(24(/),9X,'>>> EXAMPLE OF USER SPECIFIED 24-HOUR STORM'
     1 ,' PATTERN <<<',/,9X,'>>>',7X,'WITH CONTINUOUS DEPTH-AREA',
     2 ' ADJUSTMENT     <<<',///,10X,' *',37X,'|    *',/,10X,'|  *',
     3 10X,'CAN BE',19X,'|   .  *',/,'   UNIT    |    .',7X,'REARRANG',
     4 'ED ===>     UNIT    | *     .',/,' RAINFALL |     *.',9X,'TO'
     5 ,12X,'RAINFALL |*',7X,'*',/,' (INCHES)  ',7X,'*.',20X,
     6 '(INCHES) |',10X,'*',/,10X,'|',11X,'*',26X,'|',12X,'*',/,
```

```
     7 10X,'|',15X,'*  . . .',15X,'|',15X,'*. .',/,10X,13('-'),'|',
     8 11('-'),'*-',12X,11('-'),'|',10('-'),'*-',/,11X,'1',10X,
     9 '240',9X,'480',13X,'1',8X,'240',8X,'480',/,16X,'INTERVAL ',
     9 'NUMBER',22X,'INTERVAL NUMBER',/)
 40   WRITE(0,20)
      READ(0,*)KX
      IF(KX.NE.1)GO TO 40
      RETURN
      END
C
C-----------------------------------------------------------------------
C     TEXT4 - S-GRAPH
C-----------------------------------------------------------------------
C
      SUBROUTINE TEXT4
C
      WRITE(0,10)
 10   FORMAT(24(/),'   S-graphs are a plot of the percent of catch',
     1 'ment lag',/,' versus the percent of the ultimate discharge (0%',
     2 ' to 100%).',/,' The program sets 0% lag time to 0%',
     3 ' ultimate discharge,',/,' and 100% lag time to 50%',
     4 ' ultimate discharge.  The user',/,' enters successively'
     5 ,' increasing values of % lag as',/,' prompted by the %',
     6 ' of ultimate discharge, given in 2%',/,' increments.')
 15   WRITE(0,20)
 20   FORMAT(/,76('-'),/,' ENTER A [1] TO CONTINUE',/,
     1 ' =====> ',$)
      READ(0,*)KX
      IF(KX.NE.1)GO TO 15
      WRITE(0,30)
 30   FORMAT(24(/),28X,'>>> EXAMPLE OF S-GRAPH <<<',//,13X,'100. '
     1 ,38('-'),'   -    ---*----',/,18X,'|',20X,'.     .    .    .',
     2 14X,'|',/,18X,'|',14X,'.',36X,'|',/,18X,'|',11X,'.',39X,'|'
     3 ,/,18X,'|',9X,'.',8X,'Note:',28X,'|',/,6X,'% OF',8X,'|',7X,'.',
     4 11X,'* - points that S-Graph',9X,'|',/,4X,'ULTIMATE  50.'
     5 ,' |      *',16X,'shall pass through.',9X,'|',/,4X,'DISCHA'
     6 ,'RGE     |      .',18X,'(0,0) and (100,50)',10X,'|',/,18X,'|',
     7 '    .',19X,'are the computer',12X,'|',/,18X,'|   .',20X,
     8 'default points.',13X,'|',/,18X,'|  .',49X,'|',/,18X,'|.',
     9 50X,'|',/,15X,'0. *-------|--------|--------|--------|---   ',
     9 ' -    --------',/,18X,'0      100       200       300        ',
     9 '400',14X,'1500',/,36X,'% OF LAG TIME',/)
 40   WRITE(0,20)
      READ(0,*)KX
      IF(KX.NE.1)GO TO 40
      RETURN
      END
C
```

8.5. Link-Node Model Development

A watershed can be discretized into several subareas linked together by channel flow routing submodels. Five working streams are available in the RUNOFF program which serve as a computational scratch file when developing the watershed link-node model. A stream stores a runoff hydrograph that can either be generated from the synthetic unit hydrograph model or received flow from other subarea link-node model. Each subarea link-node model can be constructed according to the specified option (Fig. 8.2) that was chosen by the program user.

8.5.1. Interactive Procedure

After executing the RUNOFF model, the user can select one of the following operating modes as follows:

```
****************************************************************
  ------------> RUNOFF MODEL <============
................................................................
 SELECT PROGRAM OPERATING MODE:
  [1] = CREATE A NEW DATA FILE INTERACTIVELY
  [2] = EXECUTE AN EXISTING DATA FILE
  [3] = EXIT PROGRAM
 =====> '1'

ENTER THE INPUT FILE NAME
=====> 'TEST.DAT'
```

After the input file has been specified, the summary of the data bank storage is displayed again for the program user's convenience.

```
----------------------------------------------------------------
     DATA BANK SUMMARY
----------------------------------------------------------------

 [ 1 ] DEPTH-AREA CURVE SETS EXIST

 [ 1 ] USER SPECIFIED 24-HOUR STORM PATTERNS EXIST

 USER SPECIFIED 24-HOUR STORM PATTERN WITH
  CONTINUOUS DEPTH-AREA ADJUSTMENT
 (ALREADY DEFINED)

 [ 1  ] S-GRAPHS EXIST
................................................................
  *** CAUTION : MAKE SURE THAT DATA IS AVAILABLE
  ***           BEFORE PROCEEDING WITH THE LINK-NODE MODEL
................................................................
ENTER A [1] TO CONTINUE
=====> '1'
```

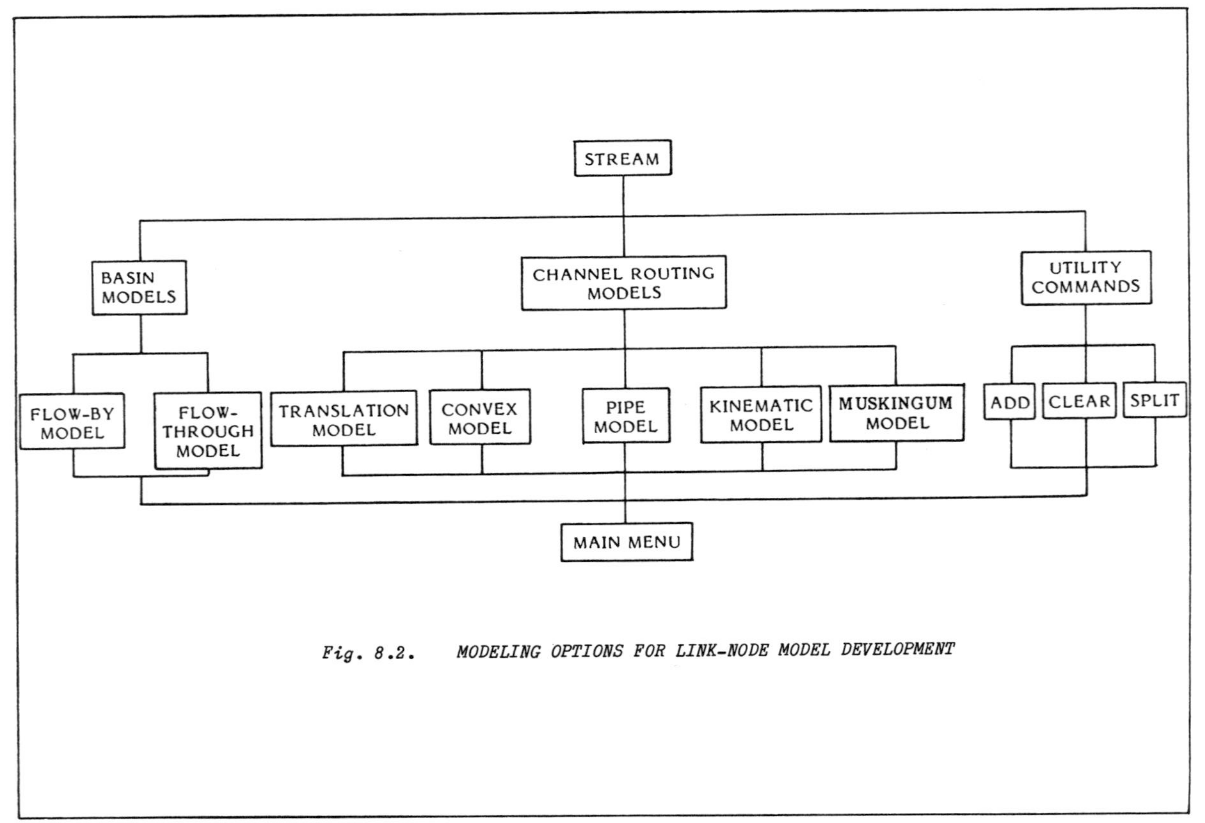

Fig. 8.2. MODELING OPTIONS FOR LINK-NODE MODEL DEVELOPMENT

Next the main menu of the RUNOFF model is displayed :

```
   LINK-NODE MODEL OPTIONS:
*** UNIT HYDROGRAPH MODEL
   [1]  = GENERATE A RUNOFF HYDROGRAPH
*** BASIN ROUTING MODELS
   [2]  = FLOW-BY BASIN ROUTING (HYDROGRAPH SEPARATION)
   [3]  = FLOW-THROUGH BASIN ROUTING
*** CHANNEL ROUTING MODELS
   [4]  = HYDROGRAPH TRANSLATION
   [5]  = CONVEX ROUTING
   [6]  = PIPE ROUTING
   [7]  = MUSKINGUM ROUTING
   [8]  = KINEMATIC WAVE ROUTING
*** UTILITY COMMANDS
   [9]  = ADD A STREAM TO ANOTHER STREAM
   [10] = SPLIT PORTION OF A STREAM AND ADD TO ANOTHER STREAM
   [11] = CLEAR A STREAM
   [12] = VIEW A STREAM HYDROGRAPH
   [13] = PRINT A STREAM HYDROGRAPH
   [14] = TERMINATE LINK-NODE MODEL DEVELOPMENT
=====> 'KODE'
```

By selecting the proper routing model (see Chapter 6) and utility commands (see Section 8.5.2), the program user can construct a link-node model for any watershed. Subarea runoff hydrographs (or a single area unit hydrograph model) are developed using the programs from Chapters 2,3, and 5.

8.5.2. Utility Commands

The utility commands for the RUNOFF model are defined on Table 8.4.

TABLE 8.4
UTILITY COMMANDS FOR PROGRAM

Command	Definition
ADD	A hydrograph of a specified stream is added to another specified stream.
CLEAR	A specified stream is clear by setting its values to zeroes.
SPLIT	A specified stream is split into two portions. One portion remains in the original stream. The other portion is added to another specified stream.
VIEW	A specified stream hydrograph is being shown on the user's CRT.
PRINT	A specified stream hydrograph is being printed to a specified disk file.

Note: VIEW and PRINT use subroutine HGRAPH to generate a line printer plot of a specified stream hydrograph.

8.5.2.1. ADD Command

```
  ENTER STREAM NUMBER WHOSE HYDROGRAPH IS TO BE ADDED TO
  ANOTHER STREAMS HYDROGRAPH
  (ALLOWABLE VALUE IS BETWEEN [1] AND [5])
=====> 'NA'
  ENTER NODE NUMBER
=====> 'INODE'
  ENTER STREAM NUMBER THAT STREAM NUMBER
  'NA' IS TO BE ADDED ON
=====> 'NB'
  ENTER DECIMAL PERCENTAGE OF STREAM NUMBER 'NA' TO BE ADDED
  TO STREAM NUMBER 'NB'
=====> 'PB'
  ENTER [1] = WHOLE HYDROGRAPH
        [2] = PARTIAL HYDROGRAPH
=====> 'KK'
```

If (KK.EQ.2), then

```
  ENTER MODEL TIME (HOURS) FOR BEGINNING OF RESULTS
=====> 'TIME1'
  ENTER MODEL TIME (HOURS) GREATER THAN ['TIME1 '] FOR END OF
  RESULTS
=====> 'TIME2'
```

After the hydrograph has been displayed, an accept or reject statement will appear on the CRT screen as previously discussed.

```
*** FLOW PROCESS AT NODE 'INODE' STREAM NUMBER 'NA' IS ADDED TO
  STREAM NUMBER 'NB'

  ENTER A [1] TO ACCEPT THE PROCESS
=====> 'KOPT'
```

If (KOPT.NE.1), then the previous flow process will be ignored and the main menu will appear on the CRT screen for next flow process.

8.5.2.2. <u>SPLIT Command</u>

```
  ENTER STREAM NUMBER THAT THE FLOW WILL BE SPLIT TO ANOTHER
  STREAM
  (ALLOWABLE VALUE IS BETWEEN [1] AND [5] )
=====> 'NA'
  ENTER NODE NUMBER
=====> 'INODE'
  ENTER STREAM NUMBER THAT RECEIVES THE FLOW FROM STREAM NUMBER
  'NA'
=====> 'NB'
  ENTER DECIMAL PERCENTAGE OF STREAM NUMBER 'NA' TO BE ADDED TO
  STREAM NUMBER 'NB'
=====> 'PB'
```

8.5.2.3. <u>CLEAR Command</u>

```
  ENTER STREAM NUMBER THAT IT'S DATA BANK IS TO BE CLEARED
  (ALLOWABLE NUMBER IS BETWEEN [1] AND [5])
=====> 'NA'
---------------------------------------------
  STREAM NUMBER 'NA' IS SET TO ZERO
---------------------------------------------
  ENTER A [1] TO ACCEPT THE PROCESS
=====> 'KOPT'
```

If (KOPT.NE.1), then the previous flow process will be ignored and the main menu will appear on the CRT screen for the next flow process selection.

8.5.2.4. VIEW and PRINT Commands

```
  ENTER STREAM NUMBER
  (ALLOWABLE NUMBER IS BETWEEN [1] AND [5])
=====> 'NA'
```

After the runoff hydrograph has been displayed or printed, the main menu will appear on the CRT screen for the next flow process selection.

8.5.2.5. EXECUTION and EXIT RUNOFF Program

After the link-node model has been developed, the program execution modes will appear on the CRT as:

```
****************************************************************
        ==============> RUNOFF MODEL <===================
................................................................

SELECT PROGRAM OPERATING MODE:
[1] = CREATE A NEW DATA FILE INTERACTIVELY
[2] = EXECUTE AN EXISTING DATA FILE
[3] = EXIT PROGRAM
======>`KRUN'
```

If (KRUN.EQ.2), then

```
ENTER THE INPUT FILE NAME
=====>'TEST.DAT'
ENTER THE OUTPUT FILE NAME
=====>'TEST.ANS'
```

After the execution of the new data file, the execution options will reappear on the CRT screen. If (KRUN.EQ.3), then the RUNOFF program is terminated.

8.5.2.6. PROGRAM 8.2

```
C
C  THIS PROGRAM GENERATES THE RUNOFF HYDROGRAPH FOR
C  A WATERSHED BY USING THE UNIT HYDROGRAPH ANALYSIS
C  AND LINK-NODE MODEL DEVELOPMENT PROCEDURES
C
C   IBM/PC-XT VERSION DATE : 12/03/1986
C
      COMMON/BLK 1/ PRECIP(480)
      COMMON/BLK 2/ DARF(4,110),NDARF(4)
      COMMON/BLK 3/ XF(9)
      COMMON/BLK 4/ NEW(480)
      COMMON/BLK 5/ SMASS(49)
      COMMON/BLK 6/ R(9)
      COMMON/BLK 7/ ST(4,50,2),NST(4)
      COMMON/BLK 8/ S(10,52),KNS(10)
      COMMON/BLK 9/ PERCENT(150)
      COMMON/BLK 10/ AREA
      COMMON/BLK 11/ H(1000)
      COMMON/BLK 12/ STREAM(5,1000)
      COMMON/BLK 13/ NUMB(5)
      COMMON/BLK 14/ QNOR(21),VNOR(21)
      COMMON/BLK 15/ KTX1,KTX2,KTX3,KTX4,KTX
      COMMON/BLK 16/ A(3000),B(3000)
C
      CHARACTER IFILE*15,OFILE*15
C
C-----------------------------------------------------------------
C     DEFINITION OF ARRAYS
C
C     PRECIP : RAINFALL IN 3-MINUTE INTERVAL
C     DARF   : DEPTH-AREA CURVES
C     XF     : DEPTH-AREA ADJUSTMENT FACTORS
C     SMASS  : DISTRIBUTION CURVES OF S.C.S. 24-HOUR STORM
C     R      : WATERSHED AREA-AVERAGED POINT RAINFALL
C     ST     : USER SPECIFIED 24-HOUR STORM PATTERN
C     NEW    : NEW INTERVAL NUMBER FOR USER SPECIFIED STORM
C              PATTERN WITH CONTINUOUS DEPTH-AREA ADJUSTMENT
C     S      : S-GRAPH
C     PERCENT: PRECENT OF S-GRAPH
C     H      : OUTFLOW HYDROGRAPH
C     STREAM : STREAM HYDROGRAPH
C     NUMB   : NUMBER OF STREAM HYDROGRAPH INTERVALS
C     QNOR   : DISCHARGE TABLE FOR INTERPOLATING NORMAL VELOCITY
C     VNOR   : VELOCITY TABLE FOR INTERPOLATING NORMAL VELOCITY
C     A & B  : TEMPORARY STORAGES
C-----------------------------------------------------------------
```

```
C
      KTX1=0
      KTX2=0
      KTX3=0
      KTX4=0
      KTX=0
 10   WRITE(0,511)
 511  FORMAT(24(/),72('*'),/,10('='),'> RUNOFF MODEL <',10('='),/,
     C 72('.'),/,'  SELECT PROGRAM OPERATING MODE:',
     C /,3X,'[1] = CREATE A NEW DATA FILE INTERACTIVELY',
     C /,3X,'[2] = EXECUTE AN EXISTING DATA FILE',
     C /,3X,'[3] = EXIT PROGRAM')
      WRITE(0,99)
 99   FORMAT('  =====> ',$)
      READ(0,*)KRUN
      IF(KRUN.EQ.1)GO TO 220
      IF(KRUN.EQ.2)GO TO 230
      IF(KRUN.EQ.3)STOP
      GO TO 10
C
C  INPUT/OUTPUT FILES SPECIFICATION
C
 220  WRITE(0,512)
 512  FORMAT(/,'  ENTER THE INPUT FILE NAME')
      WRITE(0,99)
      READ(0,513)IFILE
 513  FORMAT(A15)
      OPEN(UNIT=7,FILE=IFILE,STATUS='NEW')
      KRD=0
      KWT=0
      GO TO 15
C
 230  WRITE(0,512)
      WRITE(0,99)
      READ(0,513)IFILE
      WRITE(0,514)
 514  FORMAT(/,'  ENTER THE OUTPUT FILE NAME')
      WRITE(0,99)
      READ(0,513)OFILE
      OPEN(UNIT=8,FILE=IFILE,STATUS='OLD')
      OPEN(UNIT=2,FILE=OFILE,STATUS='NEW')
      KRD=8
      KWT=2
C
C  READ DATA FROM DATA BANK
C
 15   OPEN(UNIT=1,FILE='CONTR.DAT',STATUS='OLD')
      READ(1,515)KKK
 515  FORMAT(///,I3)
      READ(1,516)((DARF(J,I),I=1,110),J=1,4)
 516  FORMAT(11(1X,F8.4))
      READ(1,515)NS
      READ(1,518)((ST(I,J,1),ST(I,J,2),J=1,50),I=1,4)
 518  FORMAT(10(1X,F8.4))
      READ(1,515)KDUM
```

```
      READ(1,517)(NEW(I),I=1,480)
 517  FORMAT(10I4)
      READ(1,515)NSS
      READ(1,518)((S(I,J),J=1,52),I=1,10)
      CLOSE(UNIT=1)
C
C  DISPLAY DATA BANK INFORMATION TO USER
C
 860  KKK=0
      DO 861 I=1,4
      IF(DARF(I,13).EQ.0.)GO TO 861
      KKK=KKK+1
      NDARF(KKK)=I
 861  CONTINUE
      IF(KRUN.EQ.2)GO TO 871
      WRITE(0,704)
 704  FORMAT(24(/),76('-'),/,'   DATA BANK SUMMARY',/,76('-'))
      WRITE(0,705)KKK
 705  FORMAT(/,'  [',I2,' ] DEPTH-AREA CURVE SETS EXIST')
 871  NS=0
      DO 862 I=1,4
      IF(ST(I,2,1).EQ.0.)GO TO 862
      NS=NS+1
      NST(NS)=I
 862  CONTINUE
      IF(KRUN.EQ.2)GO TO 872
      WRITE(0,706)NS
 706  FORMAT(/,'  [',I2,' ] USER SPECIFIED 24-HOUR STORM PATTERNS',
     1 ' EXIST')
 872  KI=0
      IF(NEW(1).NE.0)KI=1
      IF(KRUN.EQ.2)GO TO 873
      WRITE(0,707)
 707  FORMAT(/,'  USER SPECIFIED 24-HOUR STORM PATTERN WITH',
     1 /,'  CONTINUOUS DEPTH-AREA ADJUSTMENT')
      IF(KI.EQ.1)WRITE(0,717)
 717  FORMAT('  (ALREADY DEFINED)')
      IF(KI.EQ.0)WRITE(0,718)
 718  FORMAT('  (CURRENTLY IS UNDEFINED)')
 873  NSS=0
      DO 863 I=1,10
      IF(S(I,2).EQ.0.)GO TO 863
      NSS=NSS+1
      KNS(NSS)=I
 863  CONTINUE
      IF(KRUN.EQ.2)GO TO 17
      WRITE(0,708)NSS
 708  FORMAT(/,'  [',I2,' ] S-GRAPHS EXIST')
      IF(NSS.NE.0)GO TO 864
      WRITE(0,710)
 710  FORMAT(/,76('.'),/,'   *** WARMING : AT LEAST ONE S-GRAPH',
     1 ' SHOULD BE PROVIDED BY THE USER ***',/,76('.'),/)
      STOP ' USE DATA BANK DEVELOPMENT PROGRAM TO ENTER ONE S-GRAPH'
C
 864  WRITE(0,709)
```

```
 709  FORMAT(/,76('.'),/,'   *** CAUTION : MAKE SURE THAT ',
     1 'SUFFICIENT DATA IS AVAILABLE ',/,3X,'***',11X,
     2 'BEFORE PROCEEDING WITH THE LINK-NODE MODEL',/,76('.'),/,
     3 '  ENTER A [1] TO CONTINUE')
      WRITE(0,99)
      READ(0,*)KX
      IF(KX.NE.1)GO TO 864
C
C  INITIALIZE ARRAY
C
 17   DO 12 I=1,5
      NUMB(I)=0
      DO 12 J=1,1000
      STREAM(I,J)=0.
      IF(I.GT.3)GO TO 12
      K=(I-1)*1000+J
      A(K)=0.
      B(K)=0.
 12   CONTINUE
C
C  LINK-NODE MODEL DEVELOPMENT
C
 20   IF(KRUN.NE.1)GO TO 35
      IF(KTX.NE.0)GO TO 27
      WRITE(0,508)
 508  FORMAT(24(/),76('='),/,10X,'LINK-NODE MODEL DEVELOPMENT',/,
     C 76('='),/)
      WRITE(0,509)
 509  FORMAT(/,10X,'NODE #101 ===>   *...',/,
     1 26X,'.    .',/,
     1 25X,'.     . <------Subarea #1',/,
     1 24X,'.       .',/,
     1 24X,'.      ...        Legend  ',/,
     2 10X,'NODE #102 ===> ....*.    .       * : nodal point',/,
     2 9X,'(Detention Basin).  \     .     .. : watershed boundary',/,
     3 26X,'.    \     .      \: routing link',/,10X,
     4 'Channel Routing ----->\   .',/,
     5 14X,'Element      .       \  .<--Subarea #2',/,
     5 26X,'.       \  .',/,
     6 27X,'........* <====NODE #103',/)
 25   WRITE(0,510)
 510  FORMAT(/,76('.'),/,'  ENTER A [1] TO CONTINUE')
      WRITE(0,99)
      READ(0,*)KX
      IF(KX.NE.1)GO TO 25
      KTX=1
 27   WRITE(0,506)
 506  FORMAT(24(/),5X,'LINK-NODE MODEL OPTIONS:',/
     C ,3X,'*** UNIT HYDROGRAPH MODEL',/,6X,'[1]  = GENERATE A ',
     C 'SUBAREA RUNOFF HYDROGRAPH',/,'   *** BASIN ROUTING MODELS',/,
     C 6X,'[2]  = FLOW-BY BASIN ROUTING (HYDROGRAPH SEPARATION)'
     C ,/,6X,'[3]  = FLOW-THROUGH BASIN ROUTING',/,3X,
     C '*** CHANNEL ROUTING MODELS',/,
     C 6X,'[4]  = HYDROGRAPH TRANSLATION',/,
     C 6X,'[5]  = CONVEX ROUTING',/,6X,'[6]  = PIPE ROUTING',/,6X,
```

```
     C '[7]  = MUSKINGUM ROUTING',/,6X,
     C '[8]  = KINEMATIC WAVE ROUTING',/,'   *** UTILITY COMMANDS',/,
     C 6X,'[9]  = ADD A STREAM TO ANOTHER STREAM',/,6X,'[10] =',
     C ' SPLIT PORTION OF A STREAM AND ADD TO ANOTHER STREAM',/,6X,
     C '[11] = CLEAR A STREAM',/,6X,'[12] = VIEW A STREAM HYDROGRAPH',
     C /,6X,'[13] = PRINT A STREAM HYDROGRAPH ',
     C /,6X,'[14] = TERMINATE LINK-NODE MODEL DEVELOPMENT')
      WRITE(0,99)
 35   READ(KRD,*)KODE
      IF(KODE.LT.1 .OR. KODE.GT.14)GO TO 27
      IF(KODE.EQ.1)GO TO 110
      IF(KODE.EQ.2)CALL FLOWBY(KRUN,KODE,KRD,KWT)
      IF(KODE.EQ.3)CALL FLOWTR(KRUN,KODE,KRD,KWT)
      IF(KODE.EQ.4)CALL MOVE(KRUN,KODE,KRD,KWT)
      IF(KODE.EQ.5)CALL CONVEX(KRUN,KODE,KRD,KWT)
      IF(KODE.EQ.6)CALL PIPE(KRUN,KODE,KRD,KWT)
      IF(KODE.EQ.7)CALL MUSK(KRUN,KODE,KRD,KWT)
      IF(KODE.EQ.8)CALL KROUT(KRUN,KODE,KRD,KWT)
      IF(KODE.EQ.9)CALL ADD(KRUN,KODE,KRD,KWT)
      IF(KODE.EQ.10)CALL SPLIT(KRUN,KODE,KRD,KWT)
      IF(KODE.EQ.11)CALL CLEAR(KRUN,KODE,KRD,KWT)
      IF(KODE.EQ.12)GO TO 40
      IF(KODE.EQ.13 .AND. KRUN.EQ.1)WRITE(7,*)KODE
      IF(KODE.EQ.13)GO TO 40
      IF(KODE.EQ.14)GO TO 30
      GO TO 20
C
C  UNIT HYDROGRAPH
C
 110  CALL RAIN(NS,KKK,KRUN,KODE,KRD,KWT)
      CALL SLOSS(KKST,KRUN,KRD,KWT)
      CALL SGRAPH(NSS,INTERV,UNIT,KRUN,KRD,KWT)
      IF(KRUN.EQ.1)CLOSE(UNIT=10)
C
C  HYDROGRAPH DISPLAY
C
      IF(KRUN.EQ.2)GO TO 235
      WRITE(KWT,501)
 501  FORMAT(/,2X,'*** VIEW THE HYDROGRAPH ***')
      CALL HGRAPH(0,INTERV,KWT,UNIT,1,KRUN,KRD)
 215  WRITE(KWT,502)
 502  FORMAT(/,2X,'ENTER A [1] TO ACCEPT THE HYDROGRAPH',/,
     C 5X,'OR A [2] TO REJECT THE HYDROGRAPH')
      WRITE(KWT,99)
 235  READ(KRD,*)KODE
      IF(KODE.EQ.1)GO TO 225
      IF(KODE.EQ.2)GO TO 27
      GO TO 215
 225  IF(KRUN.EQ.1)CALL TRANSF
      IF(KRUN.EQ.1)WRITE(7,*)KODE
      IF(KRUN.EQ.1)UNIT=3.
      IF(KODE.EQ.1)CALL HGRAPH(0,INTERV,KWT,UNIT,2,KRUN,KRD)
C
 205  IF(KRUN.EQ.1)WRITE(0,503)
 503  FORMAT(/,2X,'ASSIGN A STREAM NUMBER THAT RECEIVES THE ',
```

```
     C 'PREVIOUS HYDROGRAPH',
     C /,2X,'(ALLOWABLE VALUE IS BETWEEN [1] AND [5])')
      IF(KRUN.EQ.1)WRITE(0,99)
      READ(KRD,*)NA
      IF(NA.LT.1 .OR. NA.GT.5)GO TO 205
      IF(KRUN.EQ.1)WRITE(7,*)NA
      IF(KRUN.EQ.2)WRITE(KWT,201)NA
 201  FORMAT(/,5X,'HYDROGRAPH IS ADDED TO STREAM NO. ',I2,//,
     C 76('='),/)
      DO 100 I=1,INTERV
 100  STREAM(NA,I)=STREAM(NA,I)+H(I)
      NUMB(NA)=INTERV
C
      GO TO 20
C
C  VIEW/PRINT HYDROGRAPH
C
 40   IF(KRUN.EQ.1)WRITE(0,507)
 507  FORMAT(/,2X,'ENTER THE STREAM NUMBER',/,2X,'(ALLOWABLE',
     C ' NUMBER IS BETWEEN [1] AND [5])')
      IF(KRUN.EQ.1)WRITE(0,99)
      READ(KRD,*)NA
      IF(NA.LT.1 .OR. NA.GT.5)GO TO 40
      IF(KRUN.EQ.1.AND.KODE.EQ.13)WRITE(7,*)NA
      UNIT=3.
      INTERV=NUMB(NA)
      DO 50 I=1,1000
      H(I)=0.
      H(I)=STREAM(NA,I)
 50   CONTINUE
      IF(KODE.EQ.12)CALL HGRAPH(NA,INTERV,KWT,UNIT,1,KRUN,KRD)
      IF(KODE.EQ.13)CALL HGRAPH(NA,INTERV,KWT,UNIT,2,KRUN,KRD)
      GO TO 20
C
 30   IF(KRUN.EQ.1)WRITE(7,*)KODE
      IF(KRUN.EQ.1)CLOSE(UNIT=7)
      IF(KRUN.EQ.2)CLOSE(UNIT=8)
      CLOSE(UNIT=2)
      GO TO 10
C
      END
C
C-----------------------------------------------------------------------
      SUBROUTINE TRANSF
C-----------------------------------------------------------------------
C     THIS PROGRAM TRANSFERS ACCEPTABLE DATA TO INPUT DATA FILE
C
      CHARACTER IVAR*16
C
      OPEN(UNIT=10,FILE='TEMP.DAT',STATUS='OLD')
C
      DO 100 I=1,1000
      READ(10,70,END=200)IVAR
      WRITE(7,70)IVAR
 70   FORMAT(A16)
```

```
 100  CONTINUE
 200  CLOSE(UNIT=10)
C
      RETURN
      END
C
```

8.5.2.7. PROGRAM 8.3

```
C
C-----------------------------------------------------------------------
C     THIS SUBROUTINE ADDS A STREAM DATA BANK
C     TO ANOTHER STREAM DATA BANK
C-----------------------------------------------------------------------
C
      SUBROUTINE ADD(KRUN,KXY,KRD,KWT)
C
      COMMON/BLK 12/ STREAM(5,1000)
      COMMON/BLK 13/ NUMB(5)
      COMMON/BLK 16/ A(3000),B(3000)
C
      DO 100 I=1,1000
      A(I)=0.
      B(I)=0.
 100  CONTINUE
C
 105  IF(KRUN.NE.1)GO TO 115
      WRITE(KWT,501)
 501  FORMAT(24(/),2X,'ENTER STREAM NUMBER WHOSE HYDROGRAPH IS'
     C ,/,2X,'TO BE ADDED TO ANOTHER STREAMS HYDROGRAPH',/,
     C 2X,'(ALLOWABLE VALUE IS BETWEEN [1] AND [5])')
      WRITE(KWT,99)
 99   FORMAT('  =====> ',$)
 115  READ(KRD,*)NA
      IF(NA.LT.1 .OR. NA.GT.5)GO TO 105
      IF(NUMB(NA).NE.0)GO TO 110
 116  WRITE(KWT,503)NA
 503  FORMAT(/,'  *** STREAM NUMBER ',I2,' IS EMPTY ***')
      WRITE(KWT,505)
 505  FORMAT(/,'  ENTER A [1] TO CONTINUE')
      WRITE(KWT,99)
      READ(KRD,*)KX
      IF(KX.NE.1)GO TO 116
      RETURN
 110  IF(KRUN.EQ.2)GO TO 125
      WRITE(KWT,504)
 504  FORMAT(/,2X,'ENTER NODE NUMBER')
      WRITE(KWT,99)
 125  READ(KRD,*)INODE
      IF(KRUN.NE.1)GO TO 135
      WRITE(KWT,502)NA
 502  FORMAT(/,2X,'ENTER STREAM NUMBER THAT STREAM NUMBER',I2,
     C ' IS TO BE ADDED TO')
      WRITE(KWT,99)
 135  READ(KRD,*)NB
      IF(NB.EQ.NA)GO TO 110
C
      IF(KRUN.NE.1)GO TO 145
      WRITE(KWT,201)INODE,NA,NB
 201  FORMAT(/,76('-'),/,5X,'*** FLOW PROCESS AT NODE',I4,/,
     C 10X,'STREAM NUMBER',I2,' IS ADDED TO',
     C ' STREAM NUMBER',I2,/,76('-'),/)
```

```
 200  WRITE(KWT,202)
 202  FORMAT(/,2X,'ENTER A [1] TO ACCEPT THE PROCESS',/,
     C 5X,'OR A [2] TO REJECT THE PROCESS')
      WRITE(KWT,99)
      READ(KRD,*)KOPT
      IF(KOPT.LT.1 .OR. KOPT.GT.2)GO TO 200
      IF(KOPT.EQ.2)RETURN
C
      WRITE(7,*)KXY
      WRITE(7,*)NA
      WRITE(7,*)INODE
      WRITE(7,*)NB
C
 145  CALL MREAD(NA,A)
      CALL MREAD(NB,B)
      NUMBA=NUMB(NA)
      NUMBB=NUMB(NB)
      NUMBER=NUMBA
      IF(NUMBB.GT.NUMBA)NUMBER=NUMBB
      DO 120 I=1,NUMBER
 120  B(I)=B(I)+A(I)
      NUMB(NB)=NUMBER
C
      CALL MWRITE(NB,B)
C
      IF(KRUN.EQ.2)WRITE(KWT,201)INODE,NA,NB
C
      RETURN
      END
C
C---------------------------------------------------------------------
C     THIS SUBROUTINE SPLITS STREAM A INTO STREAM A AND STREAM B
C---------------------------------------------------------------------
C
      SUBROUTINE SPLIT(KRUN,KXY,KRD,KWT)
C
C  VARIBALES
C     PB=DECIMAL PERCENTAGE OF STREAM A TO BE ADDED TO STREAM B
C     NA,NB=STREAM NUMBERS
C
      COMMON/BLK 12/ STREAM(5,1000)
      COMMON/BLK 13/ NUMB(5)
      COMMON/BLK 16/ A(3000),B(3000)
C
      DO 100 I=1,1000
      A(I)=0.
      B(I)=0.
 100  CONTINUE
C
 105  IF(KRUN.NE.1)GO TO 145
      WRITE(KWT,501)
 501  FORMAT(24(/),2X,'ENTER STREAM NUMBER THAT THE FLOW',/,2X,
     C 'WILL BE SPLIT TO ANOTHER STREAM',/,
     C 2X,'(ALLOWABLE VALUE IS BETWEEN [1] AND [5])')
      WRITE(KWT,99)
```

```
99   FORMAT('  ====> ',$)
145  READ(KRD,*)NA
     IF(NA.LT.1 .OR. NA.GT.5)GO TO 105
     IF(NUMB(NA).NE.0)GO TO 110
146  WRITE(KWT,521)NA
521  FORMAT(/,'  *** STREAM NUMBER ',I2,' IS EMPTY ***')
     WRITE(KWT,517)
517  FORMAT(/,'  ENTER A [1] TO CONTINUE')
     WRITE(KWT,99)
     READ(KRD,*)KX
     IF(KX.NE.1)GO TO 146
     RETURN
110  IF(KRUN.NE.1)GO TO 155
     WRITE(KWT,504)
504  FORMAT(/,2X,'ENTER NODE NUMBER')
     WRITE(KWT,99)
155  READ(KRD,*)INODE
     IF(KRUN.NE.1)GO TO 165
     WRITE(KWT,502)NA
502  FORMAT(/,2X,'ENTER STREAM NUMBER THAT RECEIVES THE FLOW',
    C ' FROM STREAM NUMBER',I2)
     WRITE(KWT,99)
165  READ(KRD,*)NB
     IF(NB.EQ.NA)GO TO 110
     IF(NB.LT.1 .OR. NB.GT.5)GO TO 110
115  IF(KRUN.NE.1)GO TO 175
     WRITE(KWT,503)NA,NB
503  FORMAT(/,2X,'ENTER DECIMAL PERCENTAGE OF STREAM NUMBER',I2,
    C ' TO BE ADDED TO STREAM NUMBER',I2)
     WRITE(KWT,99)
175  READ(KRD,*)PB
     IF(PB.LE.0. .OR. PB.GT.1.)GO TO 115
125  IF(KRUN.NE.1)GO TO 185
     WRITE(KWT,513)
513  FORMAT(/,'  *** SELECT PRINT OPTIONS: ***',/,
    C 2X,'ENTER A [1] = WHOLE HYDROGRAPH',/,5X,
    C 'OR A [2] = PARTIAL HYDROGRAPH')
     WRITE(KWT,99)
185  READ(KRD,*)KK
     IF(KK.EQ.1)GO TO 130
     IF(KK.EQ.2)GO TO 140
     GO TO 125
130  TIME1=0.
     TIME2=100.
     GO TO 150
140  IF(KRUN.NE.1)GO TO 195
     WRITE(KWT,514)
514  FORMAT(/,'  ENTER MODEL TIME(HOURS) FOR BEGINNING OF',
    C ' RESULTS')
     WRITE(KWT,99)
195  READ(KRD,*)TIME1
     TIME12=TIME1+.05
     IF(KRUN.NE.1)GO TO 205
     WRITE(KWT,516)TIME12
516  FORMAT(/,'  ENTER MODEL TIME(HOURS) GREATER THAN [ ',F8.3,
```

```
     C ' ] FOR END OF RESULTS')
      WRITE(KWT,99)
 205  READ(KRD,*)TIME2
      IF(TIME2 .LE. TIME12)GO TO 140
 150  CONTINUE
C
      WRITE(KWT,22)INODE,NA,NB
 22   FORMAT(/,76('-'),/,5X,'*** FLOW PROCESS AT NODE',I4,/,
     C 10X,'MODEL STREAM SPLIT-FLOW WHERE A ',
     C 'CONSTANT PROPORTION',/,10X,'OF STREAM NUMBER',I2,' IS',
     C ' ADDED TO STREAM TO STREAM NUMBER',I2,':',//)
C
C  GRAPHICS
C
      TIME=0.
      PA=1.-PB
      WRITE(KWT,23)NB,NA,PB,NA
 23   FORMAT(/,20X,'    INFLOW              INFLOW',/,
     C 20X,'  (STREAM',I2,')        (STREAM',I2,')',/,
     C 27X,'|',19X,'|',/,27X,'|',19X,'|',/,
     C 27X,'|   (',F4.3,')(STREAM',I2,') |')
      WRITE(KWT,24)NB,PA,NA,PB,NA
 24   FORMAT(27X,'|<------------------* <= SPILTFLOW MODEL',/,
     C 27X,'|',19X,'|',/,27X,'|',19X,'|',/,
     C 27X,'|',19X,'|',/,
     C 27X,'V',19X,'V',/,24X,'STREAM',I2,12X,'(',F4.3,')(STREAM',
     C I2,')',/,20X,'+ (',F4.3,')(STREAM',I2,')',///)
      WRITE(KWT,25)NA,NB,PA,NA,PB,NB
 25   FORMAT(/,10X,'STREAM NUMBER',I2,' IS SPLIT TOWARDS STREAM',
     C I2,/,10X,'WHERE ',F4.2,'(DECIMAL PERCENT) REMAINS IN',
     C ' STREAM',I2,/,10X,'AND ',F4.2,'(DECIMAL PERCENT) IS ADDED',
     C ' TO STREAM',I2)
      WRITE(KWT,26)NB,NA,NB,NA
 26   FORMAT(//,10X,'STREAM SPLITFLOW MODELING RESULTS:',//,
     C 10X,'  MODEL    INFLOW      INFLOW      OUTFLOW     ',
     C 'OUTFLOW',/,10X,'  TIME    STREAM',I2,3(4X,'STREAM',I2),/,
     C 10X,'  (HRS)     (CFS)       (CFS)       (CFS)       (CFS)')
C
      CALL MREAD(NA,A)
      CALL MREAD(NB,B)
C
C  MODEL SPLIFLOW
C
      NUMBER=NUMB(NA)
      DO 200 I=1,NUMBER
      X=PB*A(I)
      AIN=A(I)
      BIN=B(I)
      B(I)=B(I)+X
      A(I)=A(I)-X
      TIME=TIME+.05
      IF(TIME.LT.TIME1 .OR. TIME.GT.TIME2)GO TO 200
      WRITE(KWT,27)TIME,BIN,AIN,B(I),A(I)
 27   FORMAT(10X,F7.3,3X,4(F8.1,4X))
 200  CONTINUE
```

```
C
      IF(KRUN.NE.1)GO TO 225
 235  WRITE(KWT,28)
 28   FORMAT(/,2X,'ENTER A [1] TO ACCEPT THE PROCESS',/,
     C 5X,'OR A [2] TO REJECT THE PROCESS')
      WRITE(KWT,99)
      READ(KRD,*)KOPT
      IF(KOPT.LT.1 .OR. KOPT.GT.2)GO TO 235
      IF(KOPT.EQ.2)RETURN
C
      WRITE(7,*)KXY
      WRITE(7,*)NA
      WRITE(7,*)INODE
      WRITE(7,*)NB
      WRITE(7,*)PB
      WRITE(7,*)KK
      IF(KK.EQ.2)WRITE(7,*)TIME1
      IF(KK.EQ.2)WRITE(7,*)TIME2
C
 225  NUMB1=NUMB(NB)
      IF(NUMBER.GT.NUMB1)NUMB1=NUMBER
      NUMB(NB)=NUMB1
C
      CALL MWRITE(NA,A)
      CALL MWRITE(NB,B)
C
      RETURN
      END
C
C-----------------------------------------------------------------------
C     THIS SUBROUTINE CLEARS A SPECIFIED STREAM DATA BANK
C-----------------------------------------------------------------------
C
      SUBROUTINE CLEAR(KRUN,KXY,KRD,KWT)
C
      COMMON/BLK 12/ STREAM(5,1000)
      COMMON/BLK 13/ NUMB(5)
C
 100  IF(KRUN.NE.1)GO TO 105
      WRITE(KWT,501)
 501  FORMAT(24(/),2X,'ENTER STREAM NUMBER THAT ITS DATA BANK IS',
     C ' TO BE CLEARED'/,2X,'(ALLOWABLE NUMBER IS BETWEEN [1',
     C '] AND [5])')
      WRITE(KWT,99)
 99   FORMAT('  =====> ',$)
 105  READ(KRD,*)NA
      IF(NA.LT.1 .OR. NA.GT.5)GO TO 100
C
      IF(KRUN.NE.1)GO TO 115
      WRITE(KWT,22)NA
 22   FORMAT(/,76('-'),/,10X,'STREAM NUMBER',I2,' IS SET TO ZERO.',
     C /,76('-'))
 200  WRITE(KWT,23)
 23   FORMAT(/,2X,'ENTER A [1] TO ACCEPT THE PROCESS',/,
     C 5X,'OR A [2] TO REJECT THE PROCESS')
```

```
      WRITE(KWT,99)
      READ(KRD,*)KOPT
      IF(KOPT.LT.1 .OR. KOPT.GT.2)GO TO 200
      IF(KOPT.EQ.2)RETURN
C
      WRITE(7,*)KXY
      WRITE(7,*)NA
C
 115  DO 110 I=1,1000
 110  STREAM(NA,I)=0.
C
      IF(KRUN.EQ.2)WRITE(KWT,22)NA
C
      RETURN
      END
C
C-----------------------------------------------------------------------
C     THIS SUBROUTINE READS A STREAM DATA BANK AND PUTS INTO ARRAY Z
C-----------------------------------------------------------------------
C
      SUBROUTINE MREAD(K,Z)
C
      COMMON/BLK 12/ STREAM(5,1000)
      DIMENSION Z(3000)
C
      DO 100 I=1,1000
 100  Z(I)=STREAM(K,I)
      RETURN
      END
C
C-----------------------------------------------------------------------
C     THIS SUBROUTINE WRITES A STREAM DATA BANK INTO MEMORY
C-----------------------------------------------------------------------
C
      SUBROUTINE MWRITE(K,Z)
C
      COMMON/BLK 12/ STREAM(5,1000)
      DIMENSION Z(3000)
C
      DO 100 I=1,1000
 100  STREAM(K,I)=Z(I)
      RETURN
      END
C
```

8.6. INPUT/OUTPUT UNITS

Table 8.5 shows the units that are assigned for each input/output file.

TABLE 8.5
INPUT/OUTPUT UNITS AND THEIR ACCESSIBLE FILES

INPUT UNIT	OUTPUT UNIT	ACCESSIBLE FILE
1		CONTR.DAT
	2	User specified output file
3		SCSMASS.DAT
	7	User Specified Input File (write only)
8		User Specified Input File (read only)
	10	TEMP.DAT (Temporary data for write only)

The CONTR.DAT file consists of depth-area adjustment curves (maximum of 4 sets), user specified 24-hour storm pattern(s) (maximum of 4 sets), an user specified 24-hour storm pattern with continuous depth-area ajustment curve, and S-graph(s) (maximum of 10 sets). This file name is reserved for the RUNOFF program and its contents will be preserved after each execution of the RUNOFF program. The preserved information can be revised when the DBANK program is executed by the program user. The SCSMASS.DAT consists of the SCS type I, type II, type III and type IA, 24-hour storm patterns, and its contents are stored within the RUNOFF program. The TEMP.DAT file stores the temporary information from the synthetic unit hydrograph model.

8.7. Application of Link-Node Model

Figure 8.3 depicts an example watershed schematic for the link-node model. The RUNOFF computer program results follow the figure.

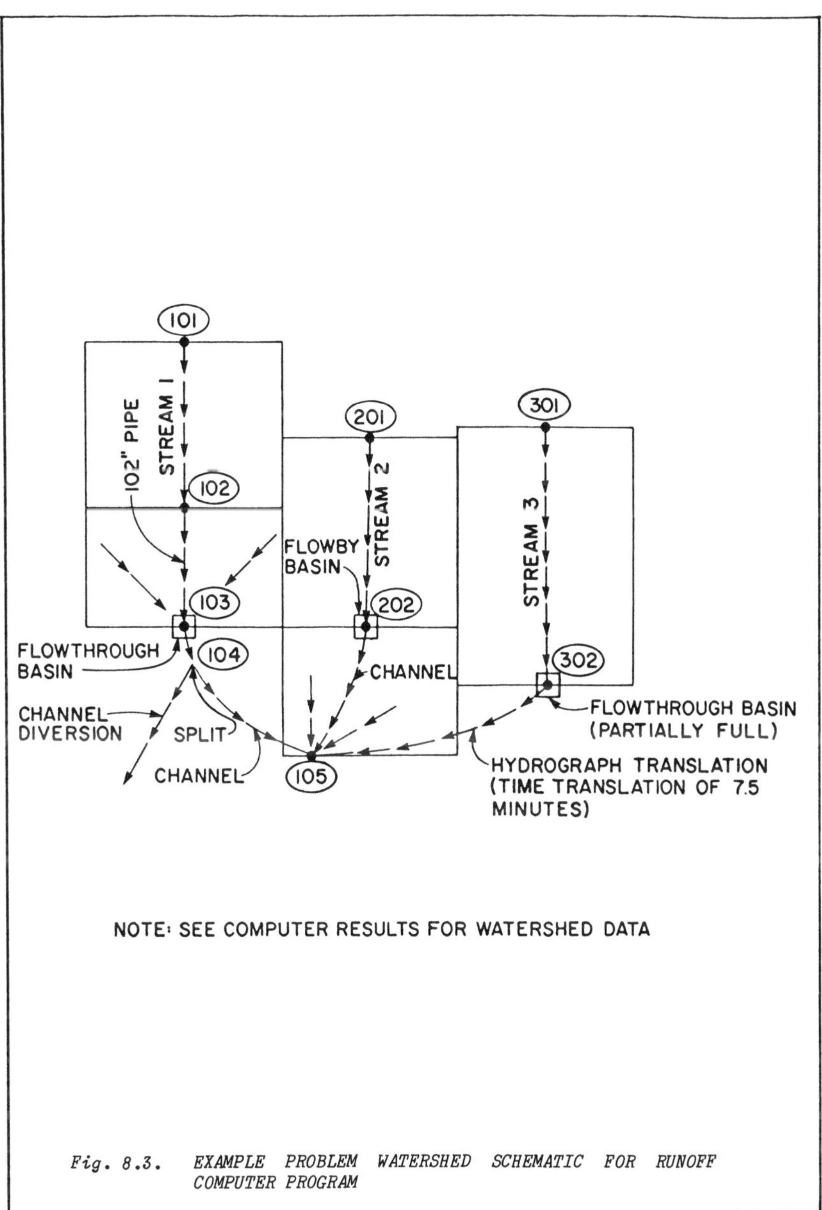

Fig. 8.3. EXAMPLE PROBLEM WATERSHED SCHEMATIC FOR RUNOFF COMPUTER PROGRAM

```
****************************************************************************
----------------------------------------------------------------------------
****************************************************************************

  WATERSHED AREA =    1000.000 ACRES

  24-HOUR NESTED DESIGN STORM PATTERN # 5
  WATERSHED AREA-AVERAGED POINT RAINFALL
      5-MINUTE VALUE =    .52 INCHES
     15-MINUTE VALUE =    .82 INCHES
     30-MINUTE VALUE =  1.09 INCHES
      1-HOUR   VALUE =  1.45 INCHES
      2-HOUR   VALUE =  2.00 INCHES
      3-HOUR   VALUE =  2.43 INCHES
      6-HOUR   VALUE =  3.36 INCHES
     12-HOUR   VALUE =  4.20 INCHES
     24-HOUR   VALUE =  5.63 INCHES

  *** DEPTH-AREA REDUCTION FACTORS :
      5-MINUTE FACTOR =  .950
     15-MINUTE FACTOR =  .950
     30-MINUTE FACTOR =  .950
      1-HOUR   FACTOR =  .950
      2-HOUR   FACTOR =  .975
      3-HOUR   FACTOR =  .994
      6-HOUR   FACTOR =  .996
     12-HOUR   FACTOR =  .997
     24-HOUR   FACTOR =  .998


  SOIL-LOSS MODEL
  PHI INDEX =    .013 INCHES/HOUR
  PERCENT OF PRECIPITATION =    .300
   TOTAL SOIL-LOSS VOLUME (ACRE-FEET) =      124.9270
  WATERCOURSE LENGTH =  10000.000 FEET
  LENGTH FROM CONCENTRATION POINT TO CENTROID =    5000.000 FEET
  ELEVATION VARIATION ALONG WATERCOURSE =     100.000 FEET
  BASIN FACTOR ALONG WATERCOURSE = .020


  WATERCOURSE "LAG" TIME =    .282 HOURS
  UNIT HYDROGRAPH TIME UNIT =    3.00 MINUTES
  UNIT INTERVAL PERCENTAGE OF LAG-TIME =  17.73
----------------------------------------------------------------------------


                         UNIT HYDROGRAPH DETERMINATION
----------------------------------------------------------------------------
     INTERVAL            "S" GRAPH            UNIT HYDROGRAPH
      NUMBER           MEAN VALUES             ORDINATES(CFS)
----------------------------------------------------------------------------
         1                1.500                    302.405
         2                5.776                    861.732
```

3	12.487	1352.872
4	22.811	2080.839
5	35.882	2634.675
6	48.552	2553.681
7	58.051	1914.791
8	64.706	1341.374
9	69.494	965.066
10	73.244	755.852
11	76.210	597.723
12	78.755	513.092
13	80.880	428.368
14	82.846	396.197
15	84.352	303.445
16	85.627	257.017
17	86.925	261.683
18	88.183	253.486
19	89.269	218.997
20	90.280	203.829
21	91.170	179.244
22	92.018	170.937
23	92.685	134.561
24	93.330	129.943
25	93.960	120.965
26	94.486	106.133
27	94.993	102.115
28	95.499	102.070
29	95.984	97.673
30	96.342	72.132
31	96.679	68.048
32	97.017	68.002
33	97.355	68.135
34	97.692	68.002
35	98.012	64.538
36	98.257	49.391
37	98.494	47.615
38	98.730	47.659
39	98.955	45.261
40	99.049	18.921
41	99.142	18.921
42	99.236	18.921
43	99.330	18.921
44	99.424	18.921
45	99.518	18.921
46	99.612	18.921
47	99.706	18.921
48	99.799	18.921
49	99.893	18.921
50	99.987	18.921
51	100.000	2.573

```
-----------------------------------------------------------------------------------
***********************************************************************************

   *** FLOW PROCESS FROM NODE  101 TO NODE  102 WITH
  TOTAL STORM RUNOFF VOLUME(ACRE-FEET) =     342.8886
```

```
*****************************************************************************

                    HYDROGRAPH POINTS GIVEN IN THREE-MINUTE INTERVALS

    TIME      VOLUME        Q   0.      525.0      1050.0    1575.0    2100.0
   (HOURS)     (AF)       (CFS)
-----------------------------------------------------------------------------
   14.00    115.6584     240.80  .   Q     .  V      .         .         .
   14.05    116.6668     244.04  .   Q     .  V      .         .         .
   14.10    117.6928     248.28  .   Q     .  V      .         .         .
   14.15    118.7398     253.38  .   Q     .  V      .         .         .
   14.20    119.8132     259.76  .   Q     .  V      .         .         .
   14.25    120.9170     267.12  .    Q    .   V     .         .         .
   14.30    122.0512     274.48  .    Q    .   V     .         .         .
   14.35    123.2121     280.92  .    Q    .   V     .         .         .
   14.40    124.3964     286.60  .    Q    .   V     .         .         .
   14.45    125.6021     291.79  .    Q    .   V     .         .         .
   14.50    126.8286     296.81  .    Q    .   V     .         .         .
   14.55    128.0751     301.66  .    Q    .   V     .         .         .
   14.60    129.3419     306.56  .    Q    .    V    .         .         .
   14.65    130.6290     311.48  .    Q    .    V    .         .         .
   14.70    131.9371     316.56  .     Q   .    V    .         .         .
   14.75    133.2653     321.41  .     Q   .    V    .         .         .
   14.80    134.6121     325.94  .     Q   .    V    .         .         .
   14.85    135.9768     330.26  .     Q   .    V    .         .         .
   14.90    137.3579     334.21  .     Q   .     V   .         .         .
   14.95    138.7541     337.89  .     Q   .     V   .         .         .
   15.00    140.1671     341.95  .     Q   .     V   .         .         .
   15.05    141.6005     346.87  .     Q   .     V   .         .         .
   15.10    143.0578     352.69  .     Q   .     V   .         .         .
   15.15    144.5421     359.18  .     Q   .     V   .         .         .
   15.20    146.0560     366.36  .     Q   .      V  .         .         .
   15.25    147.6022     374.18  .      Q  .      V  .         .         .
   15.30    149.1835     382.68  .      Q  .      V  .         .         .
   15.35    150.8029     391.89  .      Q  .      V  .         .         .
   15.40    152.4584     400.65  .      Q  .      V  .         .         .
   15.45    154.1440     407.90  .      Q  .      V  .         .         .
   15.50    155.8551     414.10  .      Q  .       V .         .         .
   15.55    157.5836     418.29  .      Q  .       V .         .         .
   15.60    159.3256     421.57  .       Q .       V .         .         .
   15.65    161.0898     426.94  .       Q .       V .         .         .
   15.70    162.8971     437.36  .       Q .        V.         .         .
   15.75    164.7711     453.51  .       Q .        V.         .         .
   15.80    166.7392     476.28  .        Q.        V.         .         .
   15.85    168.8360     507.41  .        Q.        V.         .         .
   15.90    171.1130     551.05  .         Q        V.         .         .
   15.95    173.6552     615.21  .         .Q        V         .         .
   16.00    176.6134     715.88  .         .  Q      V         .         .
   16.05    180.4835     936.57  .         .      Q  .V        .         .
   16.10    185.6287    1245.15  .         .         .V Q      .         .
   16.15    191.9417    1527.75  .         .         . V      Q.         .
   16.20    199.5164    1833.07  .         .         .  V      .   Q     .
   16.25    207.7858    2001.19  .         .         .   V     .       Q .
   16.30    215.6259    1897.31  .         .         .    V    .     Q   .
   16.35    222.1813    1586.41  .         .         .    V    Q         .
```

```
  16.40    227.5317   1294.79   .         .         .   Q V   .         .
  16.45    232.0089   1083.48   .         .         Q      V  .         .
  16.50    235.9160    945.53   .         .       Q .      V  .         .
  16.55    239.4013    843.42   .         .     Q   .      V  .         .
  16.60    242.6132    777.30   .         .   Q     .       V .         .
  16.65    245.5862    719.46   .         .  Q      .       V .         .
  16.70    248.3900    678.51   .         . Q       .       V .         .
  16.75    250.9679    623.87   .         .Q        .        V.         .
  16.80    253.3985    588.21   .         .Q        .        V.         .
  16.85    255.7568    570.70   .         Q         .        V.         .
  16.90    258.0330    550.82   .         Q         .         V         .
  16.95    260.2006    524.57   .        Q.         .         V         .
  17.00    262.2859    504.65   .        Q.         .         V         .
  17.05    264.2784    482.17   .        Q.         .         V         .
  17.10    266.1929    463.32   .       Q .         .         .V        .
  17.15    267.9995    437.20   .       Q .         .         .V        .
  17.20    269.7368    420.42   .       Q .         .         .V        .
  17.25    271.4058    403.91   .      Q  .         .         .V        .
  17.30    272.9897    383.31   .      Q  .         .         .V        .
  17.35    274.5172    369.65   .      Q  .         .         . V       .
  17.40    275.9972    358.14   .     Q   .         .         . V       .
  17.45    277.4254    345.64   .     Q   .         .         . V       .
  17.50    278.7812    328.11   .     Q   .         .         . V       .
  17.55    280.0964    318.29   .     Q   .         .         . V       .
  17.60    281.3801    310.65   .    Q    .         .         . V       .
  17.65    282.6341    303.47   .    Q    .         .         . V       .
  17.70    283.8571    295.96   .    Q    .         .         .  V      .
  17.75    285.0443    287.31   .    Q    .         .         .  V      .
  17.80    286.1835    275.70   .    Q    .         .         .  V      .
  17.85    287.2924    268.35   .    Q    .         .         .  V      .
  17.90    288.3726    261.42   .   Q     .         .         .  V      .
  17.95    289.4196    253.36   .   Q     .         .         .  V      .
  18.00    290.4084    239.31   .   Q     .         .         .  V      .
*****************************************************************************

     HYDROGRAPH IS ADDED TO STREAM NO.  1

=============================================================================



=============================================================================
     *** FLOW PROCESS FROM NODE 102 TO NODE 103
          MODEL PIPEFLOW ROUTING OF STREAM 1 WHERE
          STORAGE EFFECTS ARE NEGLECTED WITHIN THE PIPE, FLOW
          VELOCITIES ARE ESTIMATED BY ASSUMING STEADY FLOW FOR
          EACH UNIT INTERVAL(NORMAL DEPTH), AND FLOWS IN EXCESS
          OF (.02)(DIAMETER) ARE PONDED AT THE UPSTREAM INLET:


          PIPELENGTH(FT) =    5000.00     MANNINGS FACTOR =  .013
          UPSTREAM ELEVATION(FT) =     400.00
          DOWNSTREAM ELEVATION(FT) =     260.00
          PIPE DIAMETER(FT) =   8.50
```

NORMAL DEPTH VELOCITY PIPE ROUTING RESULTS:

TIME (HRS)	INFLOW (CFS)	VELOCITY (FPS)	OUTFLOW (CFS)	UPSTREAM PONDING(AF)
14.000	240.8	21.8	238.5	.000
14.050	244.0	21.9	241.2	.000
14.100	248.3	22.0	244.5	.000
14.150	253.4	22.1	249.0	.000
14.200	259.8	22.3	254.2	.000
14.250	267.1	22.5	260.9	.000
14.300	274.5	22.7	268.6	.000
14.350	280.9	22.9	275.5	.000
14.400	286.6	23.0	281.7	.000
14.450	291.8	23.1	287.3	.000
14.500	296.8	23.2	292.5	.000
14.550	301.7	23.4	297.6	.000
14.600	306.6	23.5	302.4	.000
14.650	311.5	23.6	307.4	.000
14.700	316.6	23.7	312.3	.000
14.750	321.4	23.8	317.5	.000
14.800	325.9	23.9	322.3	.000
14.850	330.3	24.0	326.8	.000
14.900	334.2	24.1	331.1	.000
14.950	337.9	24.2	335.0	.000
15.000	342.0	24.3	338.7	.000
15.050	346.9	24.4	342.9	.000
15.100	352.7	24.5	348.0	.000
15.150	359.2	24.7	354.1	.000
15.200	366.4	24.8	360.7	.000
15.250	374.2	24.9	367.6	.000
15.300	382.7	25.1	375.6	.000
15.350	391.9	25.3	384.4	.000
15.400	400.6	25.4	393.8	.000
15.450	407.9	25.6	402.6	.000
15.500	414.1	25.7	409.5	.000
15.550	418.3	25.8	415.6	.000
15.600	421.6	25.8	419.3	.000
15.650	426.9	25.9	422.4	.000
15.700	437.4	26.1	428.3	.000
15.750	453.5	26.4	440.0	.000
15.800	476.3	26.8	457.8	.000
15.850	507.4	27.3	482.3	.000
15.900	551.0	28.0	520.1	.000
15.950	615.2	28.7	575.7	.000
16.000	715.9	29.9	645.5	.000
16.050	936.6	32.0	788.7	.000
16.100	1245.1	34.1	1043.9	.000
16.150	1527.7	35.4	1344.8	.000
16.200	1833.1	31.6	1416.0	.161
16.250	2001.2	31.6	1794.1	1.017
16.300	1897.3	31.6	1794.1	1.443
16.350	1586.4	31.6	1794.1	.585
16.400	1294.8	35.6	1923.4	.000

16.450	1083.5	33.1	1405.9	.000
16.500	945.5	32.0	1035.8	.000
16.550	843.4	31.1	908.5	.000
16.600	777.3	30.4	821.7	.000
16.650	719.5	29.9	761.1	.000
16.700	678.5	29.4	706.5	.000
16.750	623.9	28.8	662.5	.000
16.800	588.2	28.4	614.6	.000
16.850	570.7	28.2	583.9	.000
16.900	550.8	28.0	566.1	.000
16.950	524.6	27.6	547.1	.000
17.000	504.6	27.3	521.0	.000
17.050	482.2	26.9	498.5	.000
17.100	463.3	26.6	475.6	.000
17.150	437.2	26.1	458.2	.000
17.200	420.4	25.8	430.4	.000
17.250	403.9	25.5	416.2	.000
17.300	383.3	25.1	400.0	.000
17.350	369.7	24.9	378.9	.000
17.400	358.1	24.6	367.0	.000
17.450	345.6	24.4	356.0	.000
17.500	328.1	24.0	342.6	.000
17.550	318.3	23.7	324.3	.000
17.600	310.0	23.6	316.4	.000
17.650	303.5	23.4	309.2	.000
17.700	296.0	23.2	302.2	.000
17.750	287.3	23.0	294.8	.000
17.800	275.7	22.8	286.0	.000
17.850	268.3	22.6	274.2	.000
17.900	261.4	22.4	267.4	.000
17.950	253.4	22.1	260.0	.000

```
*****************************************************************************
-----------------------------------------------------------------------------
*****************************************************************************

  WATERSHED AREA =       500.000 ACRES

  24-HOUR NESTED DESIGN STORM PATTERN # 5
  WATERSHED AREA-AVERAGED POINT RAINFALL
      5-MINUTE VALUE =    .52 INCHES
     15-MINUTE VALUE =    .82 INCHES
     30-MINUTE VALUE =   1.09 INCHES
      1-HOUR   VALUE =   1.45 INCHES
      2-HOUR   VALUE =   2.00 INCHES
      3-HOUR   VALUE =   2.43 INCHES
      6-HOUR   VALUE =   3.36 INCHES
     12-HOUR   VALUE =   4.20 INCHES
     24-HOUR   VALUE =   5.63 INCHES

  *** DEPTH-AREA REDUCTION FACTORS :
      5 MINUTE FACTOR =  .975
     15-MINUTE FACTOR =  .975
     30-MINUTE FACTOR =  .975
```

```
    1-HOUR   FACTOR =  .975
    2-HOUR   FACTOR =  .988
    3-HOUR   FACTOR =  .997
    6-HOUR   FACTOR =  .998
   12-HOUR   FACTOR =  .998
   24-HOUR   FACTOR =  .999


SOIL-LOSS MODEL
PHI INDEX =    .010 INCHES/HOUR
PERCENT OF PRECIPITATION =    .250
 TOTAL SOIL-LOSS VOLUME (ACRE-FEET) =       51.6391
WATERCOURSE LENGTH =   5000.000 FEET
LENGTH FROM CONCENTRATION POINT TO CENTROID =   2500.000 FEET
ELEVATION VARIATION ALONG WATERCOURSE =     140.000 FEET
BASIN FACTOR ALONG WATERCOURSE = .015


WATERCOURSE "LAG" TIME =    .103 HOURS
UNIT HYDROGRAPH TIME UNIT =    3.00 MINUTES
UNIT INTERVAL PERCENTAGE OF LAG-TIME =  48.67
-----------------------------------------------------------------------------
```

UNIT HYDROGRAPH DETERMINATION

INTERVAL NUMBER	"S" GRAPH MEAN VALUES	UNIT HYDROGRAPH ORDINATES(CFS)
1	5.753	579.757
2	31.006	2545.099
3	60.063	2928.403
4	73.409	1344.992
5	80.551	719.809
6	85.111	459.563
7	88.538	345.333
8	91.241	272.392
9	93.226	200.083
10	94.783	156.933
11	96.083	131.028
12	97.043	96.731
13	97.940	90.366
14	98.628	69.386
15	99.065	44.006
16	99.502	44.006
17	99.938	44.006
18	100.000	6.229

```
-----------------------------------------------------------------------------
*****************************************************************************

   *** FLOW PROCESS FROM NODE  103 TO NODE  103 WITH
  TOTAL STORM RUNOFF VOLUME(ACRE-FEET) =      182.5563

*****************************************************************************
```

```
                    HYDROGRAPH POINTS GIVEN IN THREE-MINUTE INTERVALS

   TIME       VOLUME          Q    0.        450.0     900.0    1350.0    1800.0
  (HOURS)      (AF)         (CFS)
------------------------------------------------------------------------------
  14.00     64.7716     136.45  .  Q      .   V     .         .         .
  14.05     65.3449     138.75  .  Q      .   V     .         .         .
  14.10     65.9365     143.16  .  Q      .   V     .         .         .
  14.15     66.5482     148.04  .  Q      .   V     .         .         .
  14.20     67.1736     151.33  .  Q      .   V     .         .         .
  14.25     67.8101     154.05  .  Q      .   V     .         .         .
  14.30     68.4571     156.57  .  Q      .   V     .         .         .
  14.35     69.1144     159.06  .  Q      .    V    .         .         .
  14.40     69.7820     161.56  .  Q      .    V    .         .         .
  14.45     70.4601     164.09  .  Q      .    V    .         .         .
  14.50     71.1488     166.66  .  Q      .    V    .         .         .
  14.55     71.8485     169.33  .  Q      .    V    .         .         .
  14.60     72.5595     172.07  .  Q      .    V    .         .         .
  14.65     73.2825     174.95  .  Q      .     V   .         .         .
  14.70     74.0177     177.93  .  Q      .     V   .         .         .
  14.75     74.7640     180.60  .   Q     .     V   .         .         .
  14.80     75.5155     181.87  .   Q     .     V   .         .         .
  14.85     76.2718     183.03  .   Q     .     V   .         .         .
  14.90     77.0387     185.59  .   Q     .     V   .         .         .
  14.95     77.8191     188.86  .   Q     .      V  .         .         .
  15.00     78.6149     192.59  .   Q     .      V  .         .         .
  15.05     79.4277     196.68  .   Q     .      V  .         .         .
  15.10     80.2588     201.13  .   Q     .      V  .         .         .
  15.15     81.1100     206.00  .   Q     .      V  .         .         .
  15.20     81.9831     211.28  .   Q     .      V  .         .         .
  15.25     82.8799     217.03  .   Q     .       V .         .         .
  15.30     83.8027     223.32  .   Q     .       V .         .         .
  15.35     84.7540     230.20  .    Q    .       V .         .         .
  15.40     85.7276     235.62  .    Q    .       V .         .         .
  15.45     86.6965     234.47  .    Q    .       V .         .         .
  15.50     87.6584     232.78  .    Q    .        V.         .         .
  15.55     88.6423     238.12  .    Q    .        V.         .         .
  15.60     89.6643     247.32  .    Q    .        V.         .         .
  15.65     90.7360     259.34  .    Q    .        V.         .         .
  15.70     91.8713     274.74  .     Q   .         V         .         .
  15.75     93.0920     295.42  .     Q   .         V         .         .
  15.80     94.4315     324.16  .      Q  .         V         .         .
  15.85     95.9289     362.37  .       Q .         .V        .         .
  15.90     97.6484     416.11  .        Q.         .V        .         .
  15.95     99.7026     497.12  .         .Q        .V        .         .
  16.00    102.3448     639.42  .         .   Q     . V       .         .
  16.05    106.4829    1001.43  .         .         . QV      .         .
  16.10    113.3808    1669.27  .         .         .   V     .      Q  .
  16.15    120.4106    1701.22  .         .         .     V   .      Q  .
  16.20    124.0896    1083.91  .         .         .   Q  V  .         .
  16.25    128.0517     765.22  .         .      Q  .       V .         .
  16.30    130.5096     594.83  .         .  Q      .       V .         .
  16.35    132.5602     496.24  .         .Q        .        V.         .
  16.40    134.3321     428.79  .        Q.         .        V.         .
  16.45    135.8995     379.32  .       Q .         .        V.         .
```

```
 16.50    137.3332    346.96  .        Q  .          .         V         .
 16.55    138.6507    318.84  .        Q  .          .         V         .
 16.60    139.8495    290.10  .       Q   .          .         V         .
 16.65    140.9700    271.16  .       Q   .          .         V         .
 16.70    142.0008    249.45  .      Q    .          .        .V         .
 16.75    142.9481    229.26  .      Q    .          .        .V         .
 16.80    143.8541    219.25  .     Q     .          .        .V         .
 16.85    144.7201    209.57  .     Q     .          .        .V         .
 16.90    145.5013    189.07  .     Q     .          .        .V         .
 16.95    146.2428    179.42  .    Q      .          .        . V        .
 17.00    146.9583    173.16  .    Q      .          .        . V        .
 17.05    147.6492    167.19  .    Q      .          .        . V        .
 17.10    148.3092    159.74  .    Q      .          .        . V        .
 17.15    148.9388    152.34  .    Q      .          .        . V        .
 17.20    149.5461    146.98  .    Q      .          .        . V        .
 17.25    150.1349    142.47  .    Q      .          .        . V        .
 17.30    150.7071    138.48  .    Q      .          .        .  V       .
 17.35    151.2644    134.87  . Q         .          .        .  V       .
 17.40    151.8080    131.55  . Q         .          .        .  V       .
 17.45    152.3390    128.49  . Q         .          .        .  V       .
 17.50    152.8582    125.66  . Q         .          .        .  V       .
 17.55    153.3665    123.01  . Q         .          .        .  V       .
 17.60    153.8645    120.52  . Q         .          .        .  V       .
 17.65    154.3528    118.16  . Q         .          .        .  V       .
 17.70    154.8319    115.95  . Q         .          .        .  V       .
 17.75    155.3024    113.87  . Q         .          .        .   V      .
 17.80    155.7648    111.89  . Q         .          .        .   V      .
 17.85    156.2193    110.01  . Q         .          .        .   V      .
 17.90    156.6667    108.25  . Q         .          .        .   V      .
 17.95    157.1071    106.58  . Q         .          .        .   V      .
 18.00    157.5408    104.98  . Q         .          .        .   V      .
****************************************************************************

    HYDROGRAPH IS ADDED TO STREAM NO.  1

============================================================================

============================================================================

    *** FLOW PROCESS FROM NODE 103 TO NODE 104
    STORAGE BASIN HYDROGRAPH ROUTING MODEL
============================================================================
    ENTERED INFORMATION:
----------------------------------------------------------------------------
    USE STREAM # 1 AS THE INFLOW HYDROGRAPH
    CONSTANT HYDROGRAPH TIME UNIT(MINUTES) =     3.000
    ASSUMED INITIAL DEPTH(FEET) IN STORAGE BASIN =      .00
============================================================================
    DEPTH-VS.-STORAGE AND DEPTH-VS.-DISCHARGE INFORMATION:
----------------------------------------------------------------------------
    TOTAL NUMBER OF BASIN DEPTH INFORMATION ENTRIES =    5
----------------------------------------------------------------------------
    *BASIN-DEPTH  STORAGE    OUTFLOW  **BASIN-DEPTH  STORAGE    OUTFLOW  *
    *   (FEET)  (ACRE-FEET)   (CFS)   **   (FEET)  (ACRE-FEET)   (CFS)   *
    *      .000        .000       .000**       2.000    10.000   200.000*
```

```
*      4.000       20.000     500.000**       6.000      30.000   1000.000*
*     10.000       50.000    2000.000**
==============================================================================
------------------------------------------------------------------------------
    INITIAL BASIN DEPTH(FEET) =      .00
    INITIAL BASIN STORAGE(ACRE-FEET) =          .00
    INITIAL BASIN OUTFLOW(CFS) =           .00
    DEAD STORAGE(ACRE-FEET) =     150.00
------------------------------------------------------------------------------
------------------------------------------------------------------------------
    BASIN STORAGE, OUTFLOW AND DEPTH ROUTING VALUES:
     INTERVAL   DEPTH    {S-O*DT/2}    {S+O*DT/2}
      NUMBER   (FEET)   (ACRE-FEET)   (ACRE-FEET)
         1       .00       .00000        .00000
         2      2.00      9.58678      10.41322
         3      4.00     18.96694      21.03306
         4      6.00     27.93388      32.06612
         5     10.00     45.86777      54.13223
    WHERE S=STORAGE(AF);O=OUTFLOW(AF/MIN.);DT=UNIT INTERVAL(MIN.)
------------------------------------------------------------------------------
------------------------------------------------------------------------------


         *UNIT-HYDROGRAPH STORAGE-BASIN ROUTING*
    NOTE: COMPUTED BASIN DEPTH, OUTFLOW, AND STORAGE QUANTITIES
          OCCUR AT THE GIVEN TIME. BASIN INFLOW VALUES REPRESENT THE
          AVERAGE INFLOW DURING THE RECENT HYDROGRAPH UNIT INTERVAL.
    GRAPH NOTATION: "I"=MEAN UNIT INFLOW; "O"=OUTFLOW AT GIVEN TIME
------------------------------------------------------------------------------
     TIME    INFLOW  OUTFLOW     STORAGE
    (HOURS) (CFS)     (CFS)       ACRE-FT  0.       762.     1523.     2285.    3046.
     14.00     374.9     314.7      163.822 .  O     .         .         .        .
        [BASIN DEPTH(FEET) =      2.8]
     14.05     379.9     322.7      164.091 .  O     .         .         .        .
        [BASIN DEPTH(FEET) =      2.8]
     14.10     387.7     330.9      164.362 .  OI    .         .         .        .
        [BASIN DEPTH(FEET) =      2.9]
     14.15     397.0     339.1      164.636 .  OI    .         .         .        .
        [BASIN DEPTH(FEET) =      2.9]
     14.20     405.6     347.4      164.913 .  OI    .         .         .        .
        [BASIN DEPTH(FEET) =      3.0]
     14.25     415.0     355.9      165.196 .  OI    .         .         .        .
        [BASIN DEPTH(FEET) =      3.0]
     14.30     425.1     364.5      165.484 .  OI    .         .         .        .
        [BASIN DEPTH(FEET) =      3.1]
     14.35     434.6     373.2      165.773 .  OI    .         .         .        .
        [BASIN DEPTH(FEET) =      3.2]
     14.40     443.3     381.9      166.062 .   O    .         .         .        .
        [BASIN DEPTH(FEET) =      3.2]
     14.45     451.4     390.4      166.340 .   O    .         .         .        .
        [BASIN DEPTH(FEET) =      3.3]
     14.50     459.2     398.9      166.630 .   O    .         .         .        .
        [BASIN DEPTH(FEET) =      3.3]
     14.55     466.9     407.3      166.909 .   O    .         .         .        .
        [BASIN DEPTH(FEET) =      3.4]
```

```
14.60     474.5     415.6        167.186 .    O    .         .         .         .
    [BASIN DEPTH(FEET) =        3.4]
14.65     482.3     423.8        167.461 .    OI   .         .         .         .
    [BASIN DEPTH(FEET) =        3.5]
14.70     490.3     432.1        167.735 .    OI   .         .         .         .
    [BASIN DEPTH(FEET) =        3.5]
14.75     498.1     440.1        168.004 .    OI   .         .         .         .
    [BASIN DEPTH(FEET) =        3.6]
14.80     504.2     447.9        168.264 .    OI   .         .         .         .
    [BASIN DEPTH(FEET) =        3.7]
14.85     509.9     455.6        168.519 .    OI   .         .         .         .
    [BASIN DEPTH(FEET) =        3.7]
14.90     516.7     463.1        168.771 .    OI   .         .         .         .
    [BASIN DEPTH(FEET) =        3.8]
14.95     523.9     470.6        169.021 .    OI   .         .         .         .
    [BASIN DEPTH(FEET) =        3.8]
15.00     531.3     478.2        169.273 .     O   .         .         .         .
    [BASIN DEPTH(FEET) =        3.9]
15.05     539.5     485.9        169.531 .     O   .         .         .         .
    [BASIN DEPTH(FEET) =        3.9]
15.10     549.1     493.9        169.798 .     O   .         .         .         .
    [BASIN DEPTH(FEET) =        4.0]
15.15     560.1     503.8        170.076 .     O   .         .         .         .
    [BASIN DEPTH(FEET) =        4.0]
15.20     572.0     517.7        170.355 .     OI .         .         .         .
    [BASIN DEPTH(FEET) =        4.1]
15.25     584.7     531.6        170.632 .     OI .         .         .         .
    [BASIN DEPTH(FEET) =        4.1]
15.30     599.0     545.7        170.914 .     OI .         .         .         .
    [BASIN DEPTH(FEET) =        4.2]
15.35     614.6     560.0        171.200 .     OI .         .         .         .
    [BASIN DEPTH(FEET) =        4.2]
15.40     629.4     573.7        171.474 .       O .         .         .         .
    [BASIN DEPTH(FEET) =        4.3]
15.45     637.0     586.0        171.721 .       O .         .         .         .
    [BASIN DEPTH(FEET) =        4.3]
15.50     642.3     597.7        171.953 .       O .         .         .         .
    [BASIN DEPTH(FEET) =        4.4]
15.55     653.7     609.4        172.187 .       O .         .         .         .
    [BASIN DEPTH(FEET) =        4.4]
15.60     666.6     621.5        172.430 .       OI.         .         .         .
    [BASIN DEPTH(FEET) =        4.5]
15.65     681.7     634.8        172.695 .       OI.         .         .         .
    [BASIN DEPTH(FEET) =        4.5]
15.70     703.0     650.6        173.012 .       OI.         .         .         .
    [BASIN DEPTH(FEET) =        4.6]
15.75     735.5     670.8        173.417 .         O.        .         .         .
    [BASIN DEPTH(FEET) =        4.7]
15.80     781.9     697.5        173.950 .         OI        .         .         .
    [BASIN DEPTH(FEET) =        4.8]
15.85     844.7     733.6        174.673 .         OI        .         .         .
    [BASIN DEPTH(FEET) =        4.9]
15.90     936.3     784.4        175.687 .          OI       .         .         .
    [BASIN DEPTH(FEET) =        5.1]
15.95    1072.9     858.3        177.165 .           .O I    .         .         .
```

```
      [BASIN DEPTH(FEET) =      5.4]
   16.00   1284.9    985.5      179.709 .        . O  I .       .      .
      [BASIN DEPTH(FEET) =     5.9]
   16.05   1790.1   1222.6      184.451 .        .   O   . I    .      .
      [BASIN DEPTH(FEET) =     6.9]
   16.10   2713.1   1532.9      190.657 .        .        O     .   I  .
      [BASIN DEPTH(FEET) =     8.1]
   16.15   3046.0   1765.1      195.302 .        .      . O     .      I
      [BASIN DEPTH(FEET) =     9.1]
   16.20   2499.9   1908.3      198.165 .        .      .   O   . I    .
      [BASIN DEPTH(FEET) =     9.6]
NOTE: BASIN LIMITATIONS ARE EXCEEDED.
                         COMPUTED VALUES ARE EXTRAPOLATIONS
   16.25   2559.3   2014.2      200.285 .        .      .    O  . I    .
      [BASIN DEPTH(FEET) =    10.1]
NOTE: BASIN LIMITATIONS ARE EXCEEDED.
                         COMPUTED VALUES ARE EXTRAPOLATIONS
   16.30   2388.9   2075.2      201.503 .        .      .    O  .I     .
      [BASIN DEPTH(FEET) =    10.3]
NOTE: BASIN LIMITATIONS ARE EXCEEDED.
                         COMPUTED VALUES ARE EXTRAPOLATIONS
   16.35   2290.4   2121.3      202.425 .        .      .     O I      .
      [BASIN DEPTH(FEET) =    10.5]
NOTE: BASIN LIMITATIONS ARE EXCEEDED.
                         COMPUTED VALUES ARE EXTRAPOLATIONS
   16.40   2352.2   2111.4      202.228 .        .      .     O I      .
      [BASIN DEPTH(FEET) =    10.4]
NOTE: BASIN LIMITATIONS ARE EXCEEDED.
                         COMPUTED VALUES ARE EXTRAPOLATIONS
   16.45   1785.2   2012.6      200.253 .        .      . I  O  .      .
      [BASIN DEPTH(FEET) =    10.1]
   16.50   1382.8   1880.1      197.603 .        .     I . O    .      .
      [BASIN DEPTH(FEET) =     9.5]
   16.55   1227.4   1747.1      194.941 .        .  I    . O    .      .
      [BASIN DEPTH(FEET) =     9.0]
   16.60   1111.8   1620.7      192.413 .        . I     .O     .      .
      [BASIN DEPTH(FEET) =     8.5]
   16.65   1032.3   1503.3      190.066 .        . I    O.      .      .
      [BASIN DEPTH(FEET) =     8.0]
   16.70    955.9   1394.8      187.896 .        . I   O .      .      .
      [BASIN DEPTH(FEET) =     7.6]
   16.75    891.7   1295.2      185.903 .        .I   O  .      .      .
      [BASIN DEPTH(FEET) =     7.2]
   16.80    833.8   1205.0      184.100 .        I   O   .      .      .
      [BASIN DEPTH(FEET) =     6.8]
   16.85    793.5   1124.3      182.487 .        I  O    .      .      .
      [BASIN DEPTH(FEET) =     6.5]
   16.90    755.1   1052.5      181.051 .       I.  O    .      .      .
      [BASIN DEPTH(FEET) =     6.2]
   16.95    726.6    988.5      179.769 .       I. O     .      .      .
      [BASIN DEPTH(FEET) =     6.0]
   17.00    694.2    930.7      178.614 .       I.O      .      .      .
      [BASIN DEPTH(FEET) =     5.7]
   17.05    665.7    878.2      177.564 .      I .O      .      .      .
      [BASIN DEPTH(FEET) =     5.5]
```

```
  17.10    635.4    830.4      176.608 .      I O         .       .       .
     [BASIN DEPTH(FEET) =     5.3]
  17.15    610.5    786.1      175.723 .      I O         .       .       .
     [BASIN DEPTH(FEET) =     5.1]
  17.20    577.4    745.3      174.906 .      IO.         .       .       .
     [BASIN DEPTH(FEET) =     5.0]
  17.25    558.6    708.4      174.169 .     I O.         .       .       .
     [BASIN DEPTH(FEET) =     4.8]
  17.30    538.5    674.3      173.486 .     I O.         .       .       .
     [BASIN DEPTH(FEET) =     4.7]
  17.35    513.7    642.8      172.856 .     IO .         .       .       .
     [BASIN DEPTH(FEET) =     4.6]
  17.40    498.5    614.5      172.290 .     IO .         .       .       .
     [BASIN DEPTH(FEET) =     4.5]
  17.45    484.5    588.6      171.772 .     IO .         .       .       .
     [BASIN DEPTH(FEET) =     4.4]
  17.50    468.3    564.1      171.283 .   IO   .         .       .       .
     [BASIN DEPTH(FEET) =     4.3]
  17.55    447.3    541.3      170.825 .   IO   .         .       .       .
     [BASIN DEPTH(FEET) =     4.2]
  17.60    436.9    520.8      170.417 .   IO   .         .       .       .
     [BASIN DEPTH(FEET) =     4.1]
  17.65    427.4    502.5      170.049 .   IO   .         .       .       .
     [BASIN DEPTH(FEET) =     4.0]
  17.70    418.2    491.1      169.705 .   IO   .         .       .       .
     [BASIN DEPTH(FEET) =     3.9]
  17.75    408.6    480.9      169.363 .   IO   .         .       .       .
     [BASIN DEPTH(FEET) =     3.9]
  17.80    397.9    470.4      169.013 .   O    .         .       .       .
     [BASIN DEPTH(FEET) =     3.8]
  17.85    384.2    459.8      168.662 .   O    .         .       .       .
     [BASIN DEPTH(FEET) =     3.7]
  17.90    375.7    449.5      168.316 .  IO    .         .       .       .
     [BASIN DEPTH(FEET) =     3.7]
  17.95    366.6    439.2      167.975 .  IO    .         .       .       .
     [BASIN DEPTH(FEET) =     3.6]
  18.00    356.8    428.6      167.619 .  IO    .         .       .       .
     [BASIN DEPTH(FEET) =     3.5]
-----------------------------------------------------------------------------


=============================================================================
    *** FLOW PROCESS AT NODE 104
         MODEL STREAM SPLIT-FLOW WHERE A CONSTANT PROPORTION
         OF STREAM NUMBER 1 IS ADDED TO STREAM NUMBER 4:
```

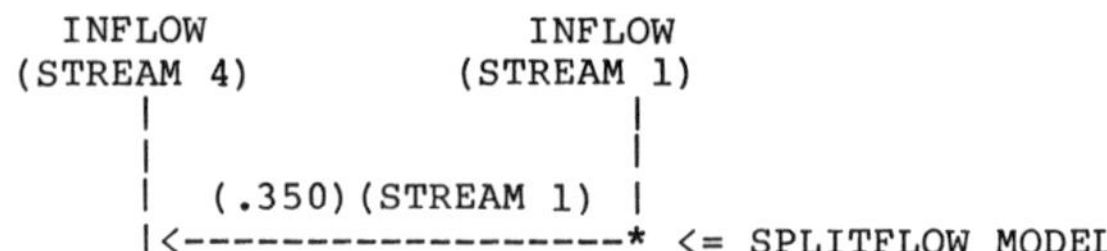

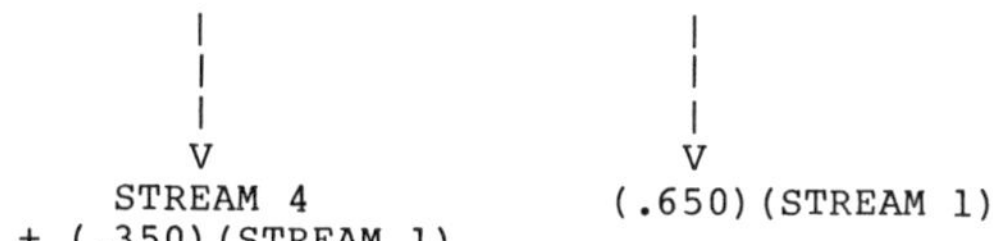

```
STREAM NUMBER 1 IS SPLIT TOWARDS STREAM 4
WHERE  .65(DECIMAL PERCENT) REMAINS IN STREAM 1
AND  .35(DECIMAL PERCENT) IS ADDED TO STREAM 4
```

STREAM SPLITFLOW MODELING RESULTS:

MODEL TIME (HRS)	INFLOW STREAM 4 (CFS)	INFLOW STREAM 1 (CFS)	OUTFLOW STREAM 4 (CFS)	OUTFLOW STREAM 1 (CFS)
14.000	.0	314.7	110.1	204.5
14.050	.0	322.7	113.0	209.8
14.100	.0	330.9	115.8	215.1
14.150	.0	339.1	118.7	220.4
14.200	.0	347.4	121.6	225.8
14.250	.0	355.9	124.6	231.3
14.300	.0	364.5	127.6	236.9
14.350	.0	373.2	130.6	242.6
14.400	.0	381.9	133.6	248.2
14.450	.0	390.4	136.6	253.8
14.500	.0	398.9	139.6	259.3
14.550	.0	407.3	142.5	264.7
14.600	.0	415.6	145.5	270.1
14.650	.0	423.8	148.3	275.5
14.700	.0	432.1	151.2	280.8
14.750	.0	440.1	154.0	286.1
14.800	.0	447.9	156.8	291.2
14.850	.0	455.6	159.4	296.1
14.900	.0	463.1	162.1	301.0
14.950	.0	470.6	164.7	305.9
15.000	.0	478.2	167.4	310.8
15.050	.0	485.9	170.1	315.8
15.100	.0	493.9	172.9	321.1
15.150	.0	503.8	176.3	327.5
15.200	.0	517.7	181.2	336.5
15.250	.0	531.6	186.1	345.5
15.300	.0	545.7	191.0	354.7
15.350	.0	560.0	196.0	364.0
15.400	.0	573.7	200.8	372.9
15.450	.0	586.0	205.1	380.9
15.500	.0	597.7	209.2	388.5
15.550	.0	609.4	213.3	396.1
15.600	.0	621.5	217.5	404.0
15.650	.0	634.8	222.2	412.6
15.700	.0	650.6	227.7	422.9

15.750	.0	670.8	234.8	436.0
15.800	.0	697.5	244.1	453.4
15.850	.0	733.6	256.8	476.9
15.900	.0	784.4	274.5	509.8
15.950	.0	858.3	300.4	557.9
16.000	.0	985.5	344.9	640.5
16.050	.0	1222.6	427.9	794.7
16.100	.0	1532.9	536.5	996.4
16.150	.0	1765.1	617.8	1147.3
16.200	.0	1908.3	667.9	1240.4
16.250	.0	2014.2	705.0	1309.3
16.300	.0	2075.2	726.3	1348.9
16.350	.0	2121.3	742.4	1378.8
16.400	.0	2111.4	739.0	1372.4
16.450	.0	2012.6	704.4	1308.2
16.500	.0	1880.1	658.0	1222.1
16.550	.0	1747.1	611.5	1135.6
16.600	.0	1620.7	567.2	1053.4
16.650	.0	1503.3	526.2	977.2
16.700	.0	1394.8	488.2	906.6
16.750	.0	1295.2	453.3	841.9
16.800	.0	1205.0	421.7	783.2
16.850	.0	1124.3	393.5	730.8
16.900	.0	1052.5	368.4	684.1
16.950	.0	988.5	346.0	642.5
17.000	.0	930.7	325.7	604.9
17.050	.0	878.2	307.4	570.8
17.100	.0	830.4	290.6	539.8
17.150	.0	786.1	275.1	511.0
17.200	.0	745.3	260.9	484.4
17.250	.0	708.4	248.0	460.5
17.300	.0	674.3	236.0	438.3
17.350	.0	642.8	225.0	417.8
17.400	.0	614.5	215.1	399.4
17.450	.0	588.6	206.0	382.6
17.500	.0	564.1	197.4	366.7
17.550	.0	541.3	189.4	351.8
17.600	.0	520.8	182.3	338.5
17.650	.0	502.5	175.9	326.6
17.700	.0	491.1	171.9	319.2
17.750	.0	480.9	168.3	312.6
17.800	.0	470.4	164.6	305.8
17.850	.0	459.8	160.9	298.9
17.900	.0	449.5	157.3	292.2
17.950	.0	439.2	153.7	285.5

```
=============================================================================
    *** FLOW PROCESS FROM NODE 104 TO NODE 105
        MODEL CHANNEL ROUTING BY THE CONVEX METHOD WHERE
        A MODIFIED C-ROUTING COEFFICIENT IS ESTIMATED IN ORDER
        TO ROUT THE STREAM 1 INFLOW HYDROGRAPH BY 3-MINUTE
        INTERVALS(reference: the National Engineering Handbook,
        Hydrology, Chapter 17, page 17-52, August,1972,
```

```
U.S. Department of Commerce).

ASSUMED REGULAR CHANNEL INFORMATION:
  BASEWIDTH(FT) =   10.00     CHANNEL Z =   .00
  UPSTREAM ELEVATION =     260.00
  DOWNSTREAM ELEVATION =     200.00
  CHANNEL LENGTH(FT) =    5000.00     MANNINGS FACTOR =  .015


CHANNEL ROUTING COEFFICIENT ESTIMATED:
  MAXIMUM INFLOW(CFS) =  1378.82
  AVERAGE FLOWRATE IN EXCESS OF 50% MAXIMUM INFLOW =  1091.01
  CHANNEL NORMAL VELOCITY FOR Q =  1091.01 CFS =  20.46 FPS
  ESTIMATED CHANNEL ROUTING COEFFICIENT =  .923

MODIFIED CHANNEL ROUTING COEFFICIENT FOR 3-MINUTE
UNIT INTERVALS IS CSTAR =  .892


CONVEX METHOD CHANNEL ROUTING RESULTS:

 MODEL         INFLOW      OUTFLOW
  TIME       (STREAM 1)  (STREAM 1)
 (HRS)         (CFS)        (CFS)
 14.000         204.5        197.0
 14.050         209.8        202.5
 14.100         215.1        207.8
 14.150         220.4        213.1
 14.200         225.8        218.4
 14.250         231.3        223.8
 14.300         236.9        229.3
 14.350         242.6        234.8
 14.400         248.2        240.5
 14.450         253.8        246.1
 14.500         259.3        251.7
 14.550         264.7        257.2
 14.600         270.1        262.7
 14.650         275.5        268.1
 14.700         280.8        273.5
 14.750         286.1        278.8
 14.800         291.2        284.1
 14.850         296.1        289.2
 14.900         301.0        294.2
 14.950         305.9        299.2
 15.000         310.8        304.1
 15.050         315.8        309.0
 15.100         321.1        314.0
 15.150         327.5        319.1
 15.200         336.5        325.1
 15.250         345.5        333.2
 15.300         354.7        342.2
 15.350         364.0        351.3
 15.400         372.9        360.5
 15.450         380.9        369.5
 15.500         388.5        377.9
```

15.550	396.1	385.6
15.600	404.0	393.2
15.650	412.6	401.0
15.700	422.9	409.4
15.750	436.0	419.1
15.800	453.4	431.2
15.850	476.9	447.1
15.900	509.8	468.3
15.950	557.9	497.9
16.000	640.5	540.5
16.050	794.7	611.0
16.100	996.4	739.9
16.150	1147.3	922.9
16.200	1240.4	1088.8
16.250	1309.3	1202.9
16.300	1348.9	1282.1
16.350	1378.8	1332.7
16.400	1372.4	1367.0
16.450	1308.2	1373.3
16.500	1222.1	1329.8
16.550	1135.6	1253.2
16.600	1053.4	1167.9
16.650	977.2	1084.4
16.700	906.6	1006.0
16.750	841.9	933.4
16.800	783.2	866.4
16.850	730.8	805.5
16.900	684.1	750.8
16.950	642.5	701.9
17.000	604.9	658.4
17.050	570.8	619.2
17.100	539.8	583.8
17.150	511.0	551.6
17.200	484.4	521.9
17.250	460.5	494.5
17.300	438.3	469.6
17.350	417.8	446.7
17.400	399.4	425.6
17.450	382.6	406.4
17.500	366.7	389.0
17.550	351.8	372.7
17.600	338.5	357.4
17.650	326.6	343.6
17.700	319.2	331.1
17.750	312.6	322.2
17.800	305.8	315.1
17.850	298.9	308.3
17.900	292.2	301.5
17.950	285.5	294.7

```
*************************************************************************
-------------------------------------------------------------------------
*************************************************************************
```

```
WATERSHED AREA =      800.000 ACRES

24-HOUR NESTED DESIGN STORM PATTERN # 5
WATERSHED AREA-AVERAGED POINT RAINFALL
    5-MINUTE VALUE =    .52 INCHES
   15-MINUTE VALUE =    .82 INCHES
   30-MINUTE VALUE =   1.09 INCHES
    1-HOUR   VALUE =   1.45 INCHES
    2-HOUR   VALUE =   2.00 INCHES
    3-HOUR   VALUE =   2.43 INCHES
    6-HOUR   VALUE =   3.36 INCHES
   12-HOUR   VALUE =   4.20 INCHES
   24-HOUR   VALUE =   5.63 INCHES

*** DEPTH-AREA REDUCTION FACTORS :
    5-MINUTE FACTOR =   .960
   15-MINUTE FACTOR =   .960
   30-MINUTE FACTOR =   .960
    1-HOUR   FACTOR =   .960
    2-HOUR   FACTOR =   .980
    3-HOUR   FACTOR =   .995
    6-HOUR   FACTOR =   .997
   12-HOUR   FACTOR =   .998
   24-HOUR   FACTOR =   .998


SOIL-LOSS MODEL
PHI INDEX =    .015 INCHES/HOUR
PERCENT OF PRECIPITATION =    .400
 TOTAL SOIL-LOSS VOLUME (ACRE-FEET) =      131.7458
WATERCOURSE LENGTH =  12000.000 FEET
LENGTH FROM CONCENTRATION POINT TO CENTROID =   6000.000 FEET
ELEVATION VARIATION ALONG WATERCOURSE =     200.000 FEET
BASIN FACTOR ALONG WATERCOURSE = .025


WATERCOURSE "LAG" TIME =    .368 HOURS
UNIT HYDROGRAPH TIME UNIT =    3.00 MINUTES
UNIT INTERVAL PERCENTAGE OF LAG-TIME =  13.60
```

UNIT HYDROGRAPH DETERMINATION

INTERVAL NUMBER	"S" GRAPH MEAN VALUES	UNIT HYDROGRAPH ORDINATES(CFS)
1	1.093	176.271
2	4.077	481.093
3	8.014	634.872
4	14.134	986.846
5	22.294	1315.850
6	31.972	1560.616
7	42.451	1689.600
8	51.435	1448.823

9	58.251	1098.952
10	63.544	853.544
11	67.489	636.174
12	70.873	545.661
13	73.571	435.085
14	75.808	360.688
15	77.942	344.142
16	79.547	258.744
17	81.261	276.360
18	82.735	237.620
19	83.941	194.596
20	84.949	162.544
21	85.924	157.210
22	86.926	161.502
23	87.916	159.733
24	88.770	137.671
25	89.595	132.970
26	90.346	121.120
27	91.026	109.682
28	91.706	109.614
29	92.281	92.760
30	92.776	79.774
31	93.271	79.774
32	93.765	79.750
33	94.204	70.815
34	94.593	62.689
35	94.982	62.643
36	95.370	62.689
37	95.759	62.689
38	96.098	54.586
39	96.358	41.901
40	96.617	41.855
41	96.876	41.762
42	97.135	41.762
43	97.394	41.808
44	97.653	41.762
45	97.911	41.623
46	98.120	33.614
47	98.302	29.307
48	98.483	29.261
49	98.664	29.216
50	98.846	29.260
51	98.998	24.540
52	99.050	8.379
53	99.102	8.379
54	99.154	8.379
55	99.206	8.379
56	99.258	8.379
57	99.310	8.379
58	99.362	8.379
59	99.414	8.379
60	99.466	8.379
61	99.518	8.379
62	99.570	8.379
63	99.622	8.379

```
        64                 99.674                   8.379
        65                 99.725                   8.379
        66                 99.777                   8.379
        67                 99.829                   8.379
        68                 99.881                   8.379
        69                 99.933                   8.379
        70                 99.985                   8.379
        71                100.000                   2.373
------------------------------------------------------------------------------
******************************************************************************

     *** FLOW PROCESS FROM NODE  201 TO NODE  202 WITH
    TOTAL STORM RUNOFF VOLUME(ACRE-FEET) =     242.7110

******************************************************************************

              HYDROGRAPH POINTS GIVEN IN THREE-MINUTE INTERVALS

   TIME     VOLUME        Q    0.      350.0      700.0    1050.0    1400.0
  (HOURS)    (AF)       (CFS)
------------------------------------------------------------------------------
  14.00     77.7184    160.41  .   Q      . V       .          .         .
  14.05     78.3898    162.46  .   Q      . V       .          .         .
  14.10     79.0713    164.93  .   Q      .  V      .          .         .
  14.15     79.7640    167.63  .   Q      .  V      .          .         .
  14.20     80.4698    170.81  .   Q      .  V      .          .         .
  14.25     81.1906    174.42  .   Q      .  V      .          .         .
  14.30     81.9277    178.38  .    Q     .  V      .          .         .
  14.35     82.6820    182.55  .    Q     .  V      .          .         .
  14.40     83.4528    186.52  .    Q     .  V      .          .         .
  14.45     84.2386    190.17  .    Q     .  V      .          .         .
  14.50     85.0388    193.64  .    Q     .   V     .          .         .
  14.55     85.8525    196.93  .    Q     .   V     .          .         .
  14.60     86.6797    200.18  .    Q     .   V     .          .         .
  14.65     87.5203    203.42  .    Q     .   V     .          .         .
  14.70     88.3743    206.69  .    Q     .   V     .          .         .
  14.75     89.2418    209.93  .    Q     .   V     .          .         .
  14.80     90.1220    213.01  .     Q    .   V     .          .         .
  14.85     91.0152    216.14  .     Q    .   V     .          .         .
  14.90     91.9207    219.14  .     Q    .    V    .          .         .
  14.95     92.8382    222.03  .     Q    .    V    .          .         .
  15.00     93.7675    224.91  .     Q    .    V    .          .         .
  15.05     94.7092    227.89  .     Q    .    V    .          .         .
  15.10     95.6649    231.29  .     Q    .    V    .          .         .
  15.15     96.6368    235.18  .     Q    .    V    .          .         .
  15.20     97.6265    239.52  .     Q    .     V   .          .         .
  15.25     98.6360    244.31  .     Q    .     V   .          .         .
  15.30     99.6671    249.52  .      Q   .     V   .          .         .
  15.35    100.7215    255.18  .      Q   .     V   .          .         .
  15.40    101.7991    260.76  .      Q   .     V   .          .         .
  15.45    102.8974    265.80  .      Q   .     V   .          .         .
  15.50    104.0171    270.96  .      Q   .      V  .          .         .
  15.55    105.1558    275.57  .      Q   .      V  .          .         .
  15.60    106.3125    279.91  .      Q   .      V  .          .         .
  15.65    107.4883    284.55  .       Q  .      V  .          .         .
```

```
  15.70    108.6884    290.43   .       Q .      V  .         .         .
  15.75    109.9256    299.41   .       Q .       V .         .         .
  15.80    111.2190    312.99   .       Q .       V .         .         .
  15.85    112.5917    332.20   .        Q.       V .         .         .
  15.90    114.0772    359.49   .         Q       V .         .         .
  15.95    115.7254    398.85   .         .Q       V.         .         .
  16.00    117.6243    459.53   .         .  Q     V.         .         .
  16.05    120.0574    588.81   .         .     Q  V.         .         .
  16.10    123.2031    761.28   .         .         VQ        .         .
  16.15    126.8958    893.63   .         .         V    Q    .         .
  16.20    131.3387   1075.17   .         .         .V        Q         .
  16.25    136.4105   1227.38   .         .         . V       .    Q    .
  16.30    141.8503   1316.45   .         .         .  V      .      Q  .
  16.35    147.3149   1322.42   .         .         .   V     .      Q  .
  16.40    152.2128   1185.29   .         .         .    V    .  Q      .
  16.45    156.3654   1004.93   .         .         .    V  Q .         .
  16.50    159.9184    859.83   .         .         .   Q V   .         .
  16.55    162.9642    737.09   .         .         .Q    V   .         .
  16.60    165.7042    663.09   .         .       Q .      V  .         .
  16.65    168.1607    594.47   .         .     Q   .      V  .         .
  16.70    170.4105    544.45   .         .    Q    .       V .         .
  16.75    172.5401    515.37   .         .   Q     .       V .         .
  16.80    174.4857    470.84   .         .  Q      .       V .         .
  16.85    176.3765    457.57   .         .  Q      .        V.         .
  16.90    178.1475    428.58   .         . Q       .        V.         .
  16.95    179.8017    400.32   .         .Q        .        V.         .
  17.00    181.3663    378.64   .         Q         .        V.         .
  17.05    182.8842    367.32   .         Q         .         V         .
  17.10    184.3689    359.31   .         Q         .         V         .
  17.15    185.8098    348.69   .        Q.         .         V         .
  17.20    187.1819    332.05   .        Q.         .         V         .
  17.25    188.5048    320.14   .        Q.         .         .V        .
  17.30    189.7706    306.31   .       Q .         .         .V        .
  17.35    190.9809    292.90   .       Q .         .         .V        .
  17.40    192.1505    283.06   .       Q .         .         .V        .
  17.45    193.2630    269.22   .      Q  .         .         .V        .
  17.50    194.3284    257.82   .      Q  .         .         . V       .
  17.55    195.3662    251.14   .      Q  .         .         . V       .
  17.60    196.3771    244.64   .     Q   .         .         . V       .
  17.65    197.3518    235.87   .     Q   .         .         . V       .
  17.70    198.2939    228.00   .     Q   .         .         . V       .
  17.75    199.2153    222.99   .     Q   .         .         . V       .
  17.80    200.1169    218.17   .     Q   .         .         . V       .
  17.85    200.9966    212.90   .     Q   .         .         .  V      .
  17.90    201.8446    205.21   .    Q    .         .         .  V      .
  17.95    202.6584    196.94   .    Q    .         .         .  V      .
  18.00    203.4560    193.02   .    Q    .         .         .  V      .
*****************************************************************************

     HYDROGRAPH IS ADDED TO STREAM NO.  2

=============================================================================
```

```
===========================================================================
    *** FLOW PROCESS AT NODE 202
         MODEL STREAM NUMBER 2 FLOWING PAST A FLOW-BY STRUCTURE:
         FLOW EXCESS IS ASSUMED TO BE PERMANENTLY STORED.
```

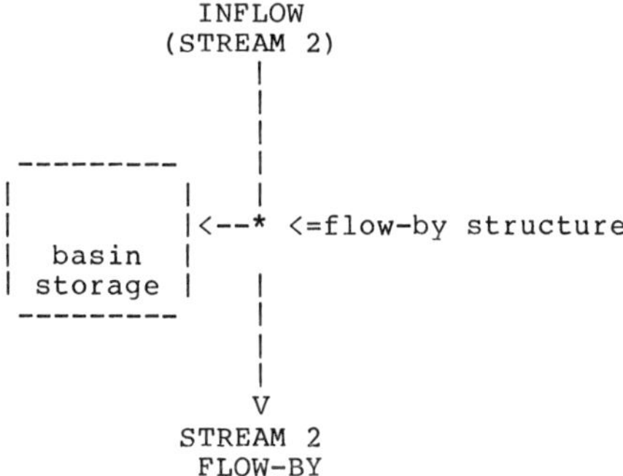

INFLOW-VS.-OUTFLOW INFORMATION :

INFLOW (CFS)	OUTFLOW(CFS)
.000	.000
500.000	400.000
850.000	800.000
900.000	800.000
3000.000	800.000

FLOW-BY BASIN MODELING RESULTS:

MODEL TIME (HRS)	STREAM 2 INFLOW (CFS)	STREAM 2 FLOW-BY (CFS)	BASIN VOLUME (AF)
14.000	160.4	128.3	15.544
14.050	162.5	130.0	15.678
14.100	164.9	131.9	15.814
14.150	167.6	134.1	15.953
14.200	170.8	136.6	16.094
14.250	174.4	139.5	16.238
14.300	178.4	142.7	16.386
14.350	182.5	146.0	16.536
14.400	186.5	149.2	16.691
14.450	190.2	152.1	16.848
14.500	193.6	154.9	17.000
14.550	196.9	157.5	17.171
14.600	200.2	160.1	17.336
14.650	203.4	162.7	17.504
14.700	206.7	165.3	17.675

14.750	209.9	167.9	17.848
14.800	213.0	170.4	18.024
14.850	216.1	172.9	18.203
14.900	219.1	175.3	18.384
14.950	222.0	177.6	18.568
15.000	224.9	179.9	18.754
15.050	227.9	182.3	18.942
15.100	231.3	185.0	19.133
15.150	235.2	188.1	19.327
15.200	239.5	191.6	19.525
15.250	244.3	195.4	19.727
15.300	249.5	199.6	19.933
15.350	255.2	204.1	20.144
15.400	260.8	208.6	20.360
15.450	265.8	212.6	20.579
15.500	271.0	216.8	20.803
15.550	275.6	220.5	21.031
15.600	279.9	223.9	21.262
15.650	284.5	227.6	21.498
15.700	290.4	232.3	21.738
15.750	299.4	239.5	21.985
15.800	313.0	250.4	22.244
15.850	332.2	265.8	22.518
15.900	359.5	287.6	22.815
15.950	398.9	319.1	23.145
16.000	459.5	367.6	23.525
16.050	588.8	501.5	23.886
16.100	761.3	698.6	24.145
16.150	893.6	800.0	24.532
16.200	1075.2	800.0	25.669
16.250	1227.4	800.0	27.435
16.300	1316.4	800.0	29.569
16.350	1322.4	800.0	31.727
16.400	1185.3	800.0	33.320
16.450	1004.9	800.0	34.166
16.500	859.8	800.0	34.414
16.550	737.1	671.0	34.687
16.600	663.1	586.4	35.004
16.650	594.5	508.0	35.361
16.700	544.4	450.8	35.748
16.750	515.4	417.6	36.152
16.800	470.8	376.7	36.542
16.850	457.6	366.1	36.920
16.900	428.6	342.9	37.274
16.950	400.3	320.3	37.605
17.000	378.6	302.9	37.918
17.050	367.3	293.9	38.221
17.100	359.3	287.4	38.518
17.150	348.7	278.9	38.806
17.200	332.1	265.6	39.081
17.250	320.1	256.1	39.345
17.300	306.3	245.0	39.598
17.350	292.9	234.3	39.841
17.400	283.1	226.4	40.074
17.450	269.2	215.4	40.297

```
        17.500     257.8      206.3        40.510
        17.550     251.1      200.9        40.718
        17.600     244.6      195.7        40.920
        17.650     235.9      188.7        41.115
        17.700     228.0      182.4        41.303
        17.750     223.0      178.4        41.487
        17.800     218.2      174.5        41.668
        17.850     212.9      170.3        41.844
        17.900     205.2      164.2        42.013
        17.950     196.9      157.5        42.176


=============================================================================
    *** MUSKINGUM ROUTING MODEL ***

  FLOW PROCESS FROM NODE 202 TO NODE 105
  STREAM NUMBER [ 2 ] IS ROUTED BY THE MUSKINGUM METHOD
   ROUTING PARAMETERS:
       X =  .200
       K =   500.0 SEC.
       USER-SELECTED TIMESTEP =   270.0 SEC.

   THE ROUTING EQUATION IS: O2 = C0*I2 + C1*I1 + C2*O1
       WITH C0 =  .065 ;C1 =  .439 ;C2 =  .495


        TIME      INFLOW RATE    OUTFLOW RATE
       (HOURS)       (CFS)           (CFS)

       14.00         128.33         124.79
       14.05         129.97         125.90
       14.10         131.94         127.01
       14.15         134.10         129.41
       14.20         136.64         130.92
       14.25         139.54         132.44
       14.30         142.71         135.88
       14.35         146.04         138.02
       14.40         149.22         140.17
       14.45         152.14         144.52
       14.50         154.91         146.80
       14.55         157.54         149.07
       14.60         160.15         153.19
       14.65         162.73         155.27
       14.70         165.35         157.36
       14.75         167.95         161.23
       14.80         170.41         163.24
       14.85         172.91         165.25
       14.90         175.31         168.96
       14.95         177.62         170.86
       15.00         179.92         172.75
       15.05         182.31         176.25
       15.10         185.03         178.14
       15.15         188.14         180.02
       15.20         191.61         183.98
```

15.25	195.45	186.40
15.30	199.61	188.82
15.35	204.14	194.05
15.40	208.61	197.16
15.45	212.64	200.27
15.50	216.77	206.27
15.55	220.46	209.39
15.60	223.93	212.50
15.65	227.64	218.11
15.70	232.34	221.17
15.75	239.53	224.23
15.80	250.39	231.94
15.85	265.76	238.70
15.90	287.59	245.46
15.95	319.08	266.65
16.00	367.62	287.94
16.05	501.50	309.23
16.10	698.61	401.77
16.15	800.00	487.41
16.20	800.00	573.05
16.25	800.00	687.59
16.30	800.00	715.95
16.35	800.00	744.32
16.40	800.00	772.42
16.45	800.00	779.38
16.50	800.00	786.34
16.55	670.96	782.95
16.60	586.38	744.02
16.65	507.97	705.09
16.70	450.80	612.63
16.75	417.57	567.38
16.80	376.67	522.13
16.85	366.06	453.51
16.90	342.87	428.49
16.95	320.26	403.48
17.00	302.91	363.46
17.05	293.86	347.08
17.10	287.45	330.70
17.15	278.95	308.97
17.20	265.64	299.77
17.25	256.11	290.58
17.30	245.05	273.62
17.35	234.32	265.10
17.40	226.45	256.59
17.45	215.38	241.60
17.50	206.25	233.90
17.55	200.92	226.20
17.60	195.71	213.73
17.65	188.70	208.30
17.70	182.40	202.88
17.75	178.39	193.12
17.80	174.54	188.90
17.85	170.32	184.68
17.90	164.17	177.51
17.95	157.55	173.37

```
****************************************************************************
----------------------------------------------------------------------------
****************************************************************************

  WATERSHED AREA =     110.000 ACRES

  24-HOUR NESTED DESIGN STORM PATTERN # 5
  WATERSHED AREA-AVERAGED POINT RAINFALL
      5-MINUTE VALUE =    .52 INCHES
     15-MINUTE VALUE =    .82 INCHES
     30-MINUTE VALUE =   1.09 INCHES
      1-HOUR   VALUE =   1.45 INCHES
      2-HOUR   VALUE =   2.00 INCHES
      3-HOUR   VALUE =   2.43 INCHES
      6-HOUR   VALUE =   3.36 INCHES
     12-HOUR   VALUE =   4.20 INCHES
     24-HOUR   VALUE =   5.63 INCHES

  *** DEPTH-AREA REDUCTION FACTORS :
      5-MINUTE FACTOR =   .994
     15-MINUTE FACTOR =   .994
     30-MINUTE FACTOR =   .994
      1-HOUR   FACTOR =   .994
      2-HOUR   FACTOR =   .997
      3-HOUR   FACTOR =   .999
      6 HOUR   FACTOR =  1.000
     12-HOUR   FACTOR =  1.000
     24-HOUR   FACTOR =  1.000


  SOIL-LOSS MODEL
  PHI INDEX =    .013 INCHES/HOUR
  PERCENT OF PRECIPITATION =    .300
   TOTAL SOIL-LOSS VOLUME (ACRE-FEET) =       13.6383
  WATERCOURSE LENGTH =   3000.000 FEET
  LENGTH FROM CONCENTRATION POINT TO CENTROID =   1500.000 FEET
  ELEVATION VARIATION ALONG WATERCOURSE =      15.000 FEET
  BASIN FACTOR ALONG WATERCOURSE = .020


  WATERCOURSE "LAG" TIME =    .129 HOURS
  UNIT HYDROGRAPH TIME UNIT =    3.00 MINUTES
  UNIT INTERVAL PERCENTAGE OF LAG-TIME =  38.80
----------------------------------------------------------------------------


                         UNIT HYDROGRAPH DETERMINATION
----------------------------------------------------------------------------
     INTERVAL             "S" GRAPH           UNIT HYDROGRAPH
      NUMBER             MEAN VALUES          ORDINATES(CFS)
----------------------------------------------------------------------------
         1                  4.104                 90.991
         2                 20.779                369.716
```

```
        3                 47.757                 598.161
        4                 65.267                 388.229
        5                 74.244                 199.039
        6                 79.901                 125.424
        7                 83.921                  89.125
        8                 86.814                  64.153
        9                 89.369                  56.651
       10                 91.418                  45.421
       11                 92.996                  34.981
       12                 94.313                  29.198
       13                 95.428                  24.728
       14                 96.358                  20.614
       15                 97.097                  16.385
       16                 97.827                  16.195
       17                 98.402                  12.753
       18                 98.896                  10.952
       19                 99.104                   4.614
       20                 99.312                   4.614
       21                 99.521                   4.614
       22                 99.729                   4.614
       23                 99.937                   4.614
       24                100.000                   1.401
----------------------------------------------------------------------------
****************************************************************************

    *** FLOW PROCESS FROM NODE  105 TO NODE  105 WITH
   TOTAL STORM RUNOFF VOLUME(ACRE-FEET) =       37.9361

****************************************************************************

               HYDROGRAPH POINTS GIVEN IN THREE-MINUTE INTERVALS

   TIME     VOLUME        Q     0.      100.0     200.0     300.0     400.0
  (HOURS)    (AF)       (CFS)
----------------------------------------------------------------------------
  14.00     13.2092     27.77  . Q       .  V       .         .         .
  14.05     13.3255     28.15  . Q       .   V      .         .         .
  14.10     13.4440     28.68  . Q       .   V      .         .         .
  14.15     13.5653     29.35  . Q       .   V      .         .         .
  14.20     13.6890     29.92  . Q       .   V      .         .         .
  14.25     13.8146     30.42  .  Q      .   V      .         .         .
  14.30     13.9423     30.88  .  Q      .   V      .         .         .
  14.35     14.0718     31.36  .  Q      .   V      .         .         .
  14.40     14.2034     31.83  .  Q      .   V      .         .         .
  14.45     14.3369     32.33  .  Q      .    V     .         .         .
  14.50     14.4726     32.83  .  Q      .    V     .         .         .
  14.55     14.6104     33.36  .  Q      .    V     .         .         .
  14.60     14.7505     33.90  .  Q      .    V     .         .         .
  14.65     14.8929     34.47  .  Q      .    V     .         .         .
  14.70     15.0378     35.06  .  Q      .    V     .         .         .
  14.75     15.1850     35.63  .  Q      .     V    .         .         .
  14.80     15.3340     36.05  .  Q      .     V    .         .         .
  14.85     15.4844     36.38  .  Q      .     V    .         .         .
  14.90     15.6367     36.87  .  Q      .     V    .         .         .
  14.95     15.7918     37.52  .  Q      .     V    .         .         .
```

```
15.00     15.9498      38.25   .  Q      .     V   .         .         .
15.05     16.1113      39.08   .  Q      .     V   .         .         .
15.10     16.2765      39.96   .  Q      .      V  .         .         .
15.15     16.4456      40.94   .   Q     .      V  .         .         .
15.20     16.6191      41.97   .   Q     .      V  .         .         .
15.25     16.7972      43.12   .   Q     .      V  .         .         .
15.30     16.9805      44.35   .   Q     .      V  .         .         .
15.35     17.1694      45.72   .   Q     .       V .         .         .
15.40     17.3634      46.95   .   Q     .       V .         .         .
15.45     17.5600      47.56   .   Q     .       V .         .         .
15.50     17.7570      47.67   .   Q     .       V .         .         .
15.55     17.9579      48.63   .   Q     .       V .         .         .
15.60     18.1660      50.35   .    Q    .        V.         .         .
15.65     18.3837      52.68   .    Q    .        V.         .         .
15.70     18.6132      55.56   .    Q    .        V.         .         .
15.75     18.8589      59.45   .    Q    .        V.         .         .
15.80     19.1266      64.77   .     Q   .         V         .         .
15.85     19.4252      72.28   .      Q  .         V         .         .
15.90     19.7662      82.52   .       Q .         V         .         .
15.95     20.1718      98.15   .        Q.         .V        .         .
16.00     20.6841     123.97   .         . Q       .V        .         .
16.05     21.4610     188.01   .         .       Q . V       .         .
16.10     22.6958     298.83   .         .         .  V     Q.         .
16.15     24.1957     362.96   .         .         .    V    .     Q   .
16.20     25.3264     273.64   .         .         .     VQ   .         .
16.25     26.1025     187.80   .         .       Q .      V  .         .
16.30     26.6946     143.29   .         .   Q     .       V .         .
16.35     27.1769     116.73   .         .Q        .       V .         .
16.40     27.5842      98.56   .        Q.         .        V.         .
16.45     27.9518      88.96   .       Q .         .        V.         .
16.50     28.2840      80.37   .       Q .         .        V.         .
16.55     28.5838      72.56   .      Q  .         .         V         .
16.60     28.8588      66.55   .     Q   .         .         V         .
16.65     29.1124      61.39   .     Q   .         .         V         .
16.70     29.3471      56.80   .    Q    .         .         V         .
16.75     29.5647      52.64   .    Q    .         .         .V        .
16.80     29.7719      50.14   .    Q    .         .         .V        .
16.85     29.9658      46.93   .   Q     .         .         .V        .
16.90     30.1483      44.18   .   Q     .         .         .V        .
16.95     30.3151      40.35   .   Q     .         .         .V        .
17.00     30.4753      38.76   .  Q      .         .         . V       .
17.05     30.6296      37.35   .  Q      .         .         . V       .
17.10     30.7777      35.84   .  Q      .         .         . V       .
17.15     30.9189      34.18   .  Q      .         .         . V       .
17.20     31.0501      31.74   .  Q      .         .         . V       .
17.25     31.1746      30.14   .  Q      .         .         . V       .
17.30     31.2952      29.18   . Q       .         .         . V       .
17.35     31.4124      28.35   . Q       .         .         .  V      .
17.40     31.5265      27.63   . Q       .         .         .  V      .
17.45     31.6379      26.96   . Q       .         .         .  V      .
17.50     31.7468      26.35   . Q       .         .         .  V      .
17.55     31.8533      25.77   . Q       .         .         .  V      .
17.60     31.9575      25.23   . Q       .         .         .  V      .
17.65     32.0597      24.72   . Q       .         .         .  V      .
17.70     32.1598      24.24   . Q       .         .         .  V      .
```

```
  17.75     32.2581     23.79  . Q        .         .         .   V     .
  17.80     32.3546     23.36  . Q        .         .         .   V     .
  17.85     32.4495     22.95  . Q        .         .         .   V     .
  17.90     32.5427     22.57  . Q        .         .         .   V     .
  17.95     32.6345     22.20  . Q        .         .         .   V     .
  18.00     32.7248     21.86  . Q        .         .         .   V     .
****************************************************************************

     HYDROGRAPH IS ADDED TO STREAM NO.  2

============================================================================


============================================================================
     *** FLOW PROCESS AT NODE 105
          STREAM NUMBER 2 IS ADDED TO STREAM NUMBER 1
----------------------------------------------------------------------------



****************************************************************************
----------------------------------------------------------------------------
****************************************************************************

  WATERSHED AREA =     1350.000 ACRES

  24-HOUR NESTED DESIGN STORM PATTERN # 5
  WATERSHED AREA-AVERAGED POINT RAINFALL
      5-MINUTE VALUE =   .52 INCHES
     15-MINUTE VALUE =   .82 INCHES
     30-MINUTE VALUE =  1.09 INCHES
      1-HOUR   VALUE =  1.45 INCHES
      2-HOUR   VALUE =  2.00 INCHES
      3-HOUR   VALUE =  2.43 INCHES
      6-HOUR   VALUE =  3.36 INCHES
     12-HOUR   VALUE =  4.20 INCHES
     24-HOUR   VALUE =  5.63 INCHES

  *** DEPTH-AREA REDUCTION FACTORS :
      5-MINUTE FACTOR =  .933
     15-MINUTE FACTOR =  .933
     30-MINUTE FACTOR =  .933
      1-HOUR   FACTOR =  .933
      2-HOUR   FACTOR =  .966
      3-HOUR   FACTOR =  .992
      6-HOUR   FACTOR =  .995
     12-HOUR   FACTOR =  .996
     24-HOUR   FACTOR =  .997


  SOIL-LOSS MODEL
  PHI INDEX =    .025 INCHES/HOUR
  PERCENT OF PRECIPITATION =    .650
   TOTAL SOIL-LOSS VOLUME (ACRE-FEET) =      363.9334
  WATERCOURSE LENGTH =  13500.000 FEET
```

```
LENGTH FROM CONCENTRATION POINT TO CENTROID =   7000.000 FEET
ELEVATION VARIATION ALONG WATERCOURSE =    250.000 FEET
BASIN FACTOR ALONG WATERCOURSE = .030

WATERCOURSE "LAG" TIME =   .479 HOURS
UNIT HYDROGRAPH TIME UNIT =   3.00 MINUTES
UNIT INTERVAL PERCENTAGE OF LAG-TIME =  10.43
```

UNIT HYDROGRAPH DETERMINATION

INTERVAL NUMBER	"S" GRAPH MEAN VALUES	UNIT HYDROGRAPH ORDINATES(CFS)
1	.834	227.067
2	2.860	551.251
3	5.563	735.319
4	8.883	903.643
5	13.829	1345.832
6	19.915	1656.038
7	26.448	1777.663
8	34.762	2262.143
9	42.633	2141.953
10	49.587	1892.295
11	55.632	1644.669
12	59.982	1183.802
13	63.982	1088.562
14	66.953	808.216
15	69.703	748.450
16	71.936	607.477
17	73.959	550.600
18	75.626	453.644
19	77.325	462.164
20	78.711	377.304
21	79.888	320.126
22	81.235	366.643
23	82.426	324.099
24	83.375	258.169
25	84.252	238.501
26	84.999	203.422
27	85.744	202.812
28	86.507	207.599
29	87.280	210.250
30	88.030	204.084
31	88.675	175.578
32	89.307	172.009
33	89.931	169.566
34	90.472	147.247
35	90.993	141.899
36	91.515	141.951
37	92.013	135.631
38	92.406	106.793
39	92.785	103.326

40	93.164	103.177
41	93.544	103.226
42	93.919	102.259
43	94.238	86.566
44	94.535	81.063
45	94.834	81.113
46	95.131	81.065
47	95.430	81.113
48	95.728	81.115
49	96.009	76.427
50	96.216	56.453
51	96.414	53.906
52	96.613	54.211
53	96.812	54.110
54	97.011	54.008
55	97.210	54.110
56	97.408	54.112
57	97.607	54.008
58	97.806	54.110
59	97.996	51.766
60	98.142	39.843
61	98.281	37.705
62	98.420	37.906
63	98.560	37.908
64	98.698	37.805
65	98.837	37.807
66	98.971	36.277
67	99.028	15.591
68	99.085	15.591
69	99.143	15.591
70	99.200	15.591
71	99.257	15.591
72	99.315	15.591
73	99.372	15.591
74	99.429	15.591
75	99.486	15.591
76	99.544	15.591
77	99.601	15.591
78	99.658	15.591
79	99.716	15.591
80	99.773	15.591
81	99.830	15.591
82	99.887	15.591
83	99.945	15.591
84	100.000	15.028

****

*** FLOW PROCESS FROM NODE 301 TO NODE 302 WITH
TOTAL STORM RUNOFF VOLUME(ACRE-FEET) = 267.1649

****

HYDROGRAPH POINTS GIVEN IN THREE-MINUTE INTERVALS

```
   TIME      VOLUME          Q    0.       425.0     850.0    1275.0    1700.0
 (HOURS)      (AF)         (CFS)
-------------------------------------------------------------------------------
  14.00     74.6279     152.41  .  Q      .V        .         .         .
  14.05     75.2659     154.39  .  Q      .V        .         .         .
  14.10     75.9135     156.74  .  Q      .V        .         .         .
  14.15     76.5718     159.31  .  Q      .V        .         .         .
  14.20     77.2417     162.10  .  Q      .V        .         .         .
  14.25     77.9251     165.40  .  Q      .V        .         .         .
  14.30     78.6237     169.06  .  Q      .V        .         .         .
  14.35     79.3383     172.93  .   Q     .V        .         .         .
  14.40     80.0711     177.35  .   Q     .V        .         .         .
  14.45     80.8220     181.72  .   Q     . V       .         .         .
  14.50     81.5901     185.87  .   Q     . V       .         .         .
  14.55     82.3747     189.86  .   Q     . V       .         .         .
  14.60     83.1741     193.46  .   Q     . V       .         .         .
  14.65     83.9884     197.06  .   Q     . V       .         .         .
  14.70     84.8168     200.47  .   Q     . V       .         .         .
  14.75     85.6590     203.82  .   Q     . V       .         .         .
  14.80     86.5143     206.97  .   Q     . V       .         .         .
  14.85     87.3825     210.11  .   Q     .  V      .         .         .
  14.90     88.2635     213.20  .    Q    .  V      .         .         .
  14.95     89.1570     216.22  .    Q    .  V      .         .         .
  15.00     90.0625     219.14  .    Q    .  V      .         .         .
  15.05     90.9804     222.12  .    Q    .  V      .         .         .
  15.10     91.9105     225.09  .    Q    .  V      .         .         .
  15.15     92.8539     228.30  .    Q    .  V      .         .         .
  15.20     93.8117     231.78  .    Q    .   V     .         .         .
  15.25     94.7854     235.63  .    Q    .   V     .         .         .
  15.30     95.7768     239.93  .    Q    .   V     .         .         .
  15.35     96.7875     244.59  .    Q    .   V     .         .         .
  15.40     97.8171     249.16  .    Q    .   V     .         .         .
  15.45     98.8642     253.40  .    Q    .   V     .         .         .
  15.50     99.9290     257.69  .     Q   .   V     .         .         .
  15.55    101.0121     262.11  .     Q   .    V    .         .         .
  15.60    102.1117     266.10  .     Q   .    V    .         .         .
  15.65    103.2279     270.11  .     Q   .    V    .         .         .
  15.70    104.3644     275.04  .     Q   .    V    .         .         .
  15.75    105.5264     281.20  .     Q   .    V    .         .         .
  15.80    106.7285     290.91  .     Q   .    V    .         .         .
  15.85    107.9945     306.37  .      Q  .     V   .         .         .
  15.90    109.3582     330.01  .      Q  .     V   .         .         .
  15.95    110.8825     368.88  .       Q .     V   .         .         .
  16.00    112.6645     431.25  .         Q     V   .         .         .
  16.05    115.0521     577.80  .         .  Q   V  .         .         .
  16.10    118.1816     757.34  .         .      Q  .         .         .
  16.15    121.8993     899.67  .         .       V .Q        .         .
  16.20    126.2156    1044.56            .       V .   Q     .         .
  16.25    131.4086    1256.69  .         .        V.        Q.         .
  16.30    137.2483    1413.21  .         .         V         .  Q      .
  16.35    143.4299    1495.97  .         .         .V        .    Q    .
  16.40    150.1812    1633.81  .         .         . V       .       Q .
  16.45    156.6148    1556.93  .         .         .  V      .     Q   .
  16.50    162.4422    1410.24  .         .         .   V     .  Q      .
  16.55    167.6092    1250.42  .         .         .    V   Q.         .
```

```
 16.60     171.8998    1038.32  .         .         .  QV      .          .
 16.65     175.7865     940.57  .         .         . Q    V   .          .
 16.70     179.1148     805.46  .         .        Q .     V   .          .
 16.75     182.1893     744.04  .         .       Q  .      V  .          .
 16.80     184.9522     668.62  .         .     Q    .      V  .          .
 16.85     187.5256     622.76  .         .   Q      .        V.          .
 16.90     189.8896     572.08  .         . Q        .        V.          .
 16.95     192.1703     551.95  .         . Q        .        V.          .
 17.00     194.2740     509.08  .         .Q         .         V.         .
 17.05     196.2506     478.33  .         .Q         .         V.         .
 17.10     198.2200     476.60  .         .Q         .         V.         .
 17.15     200.0718     448.14  .         Q          .         V.         .
 17.20     201.7821     413.91  .        Q.          .          V         .
 17.25     203.4118     394.38  .        Q.          .          V         .
 17.30     204.9551     373.49  .       Q .          .          V         .
 17.35     206.4615     364.53  .       Q .          .          V         .
 17.40     207.9385     357.44  .       Q .          .          .V        .
 17.45     209.3821     349.36  .       Q .          .          .V        .
 17.50     210.7784     337.90  .      Q  .          .          .V        .
 17.55     212.1033     320.64  .      Q  .          .          .V        .
 17.60     213.3891     311.16  .      Q  .          .          .V        .
 17.65     214.6369     301.99  .      Q  .          .          . V       .
 17.70     215.8272     288.04  .     Q   .          .          . V       .
 17.75     216.9828     279.66  .     Q   .          .          . V       .
 17.80     218.1102     272.82  .     Q   .          .          . V       .
 17.85     219.1998     263.70  .     Q   .          .          . V       .
 17.90     220.2291     249.09  .    Q    .          .          . V       .
 17.95     221.2321     242.72  .    Q    .          .          .  V      .
 18.00     222.2164     238.21  .    Q    .          .          .  V      .
****************************************************************************

     HYDROGRAPH IS ADDED TO STREAM NO.  3

============================================================================

============================================================================

     *** FLOW PROCESS FROM NODE 302 TO NODE 302
     STORAGE BASIN HYDROGRAPH ROUTING MODEL
============================================================================
     ENTERED INFORMATION:
----------------------------------------------------------------------------
     USE STREAM # 3 AS THE INFLOW HYDROGRAPH
     CONSTANT HYDROGRAPH TIME UNIT(MINUTES) =    3.000
     ASSUMED INITIAL DEPTH(FEET) IN STORAGE BASIN =      .00
============================================================================
     DEPTH-VS.-STORAGE AND DEPTH-VS.-DISCHARGE INFORMATION:
----------------------------------------------------------------------------
     TOTAL NUMBER OF BASIN DEPTH INFORMATION ENTRIES =   5
----------------------------------------------------------------------------
     *BASIN-DEPTH  STORAGE     OUTFLOW  **BASIN-DEPTH  STORAGE     OUTFLOW  *
     *   (FEET)  (ACRE-FEET)    (CFS)   **   (FEET)  (ACRE-FEET)    (CFS)   *
     *      .000       .000        .000**     2.000       5.000    100.000*
     *     4.000     10.000     200.000**     6.000      15.000    800.000*
     *     8.000     25.000    1700.000**
```

```
===========================================================================
---------------------------------------------------------------------------
     INITIAL BASIN DEPTH(FEET) =     .00
     INITIAL BASIN STORAGE(ACRE-FEET) =        .00
     INITIAL BASIN OUTFLOW(CFS) =        .00
     DEAD STORAGE(ACRE-FEET) =        .00
---------------------------------------------------------------------------
---------------------------------------------------------------------------
     BASIN STORAGE, OUTFLOW AND DEPTH ROUTING VALUES:
      INTERVAL   DEPTH    {S-O*DT/2}    {S+O*DT/2}
       NUMBER   (FEET)    (ACRE-FEET)   (ACRE-FEET)
          1       .00        .00000        .00000
          2      2.00       4.79339       5.20661
          3      4.00       9.58678      10.41322
          4      6.00      13.34711      16.65289
          5      8.00      21.48760      28.51240
     WHERE S=STORAGE(AF);O=OUTFLOW(AF/MIN.);DT=UNIT INTERVAL(MIN.)
---------------------------------------------------------------------------
---------------------------------------------------------------------------


         *UNIT-HYDROGRAPH STORAGE-BASIN ROUTING*
     NOTE: COMPUTED BASIN DEPTH, OUTFLOW, AND STORAGE QUANTITIES
           OCCUR AT THE GIVEN TIME. BASIN INFLOW VALUES REPRESENT THE
           AVERAGE INFLOW DURING THE RECENT HYDROGRAPH UNIT INTERVAL.
     GRAPH NOTATION: "I"=MEAN UNIT INFLOW; "O"=OUTFLOW AT GIVEN TIME
---------------------------------------------------------------------------
      TIME   INFLOW  OUTFLOW     STORAGE
     (HOURS) (CFS)    (CFS)       ACRE-FT  0.      408.     817.    1225.    1634.
       14.00    152.4    134.8        6.739 . O      .        .        .        .
          [BASIN DEPTH(FEET) =     2.7]
       14.05    154.4    136.4        6.822 . OI     .        .        .        .
          [BASIN DEPTH(FEET) =     2.7]
       14.10    156.7    138.2        6.908 . OI     .        .        .        .
          [BASIN DEPTH(FEET) =     2.8]
       14.15    159.3    139.9        6.997 . OI     .        .        .        .
          [BASIN DEPTH(FEET) =     2.8]
       14.20    162.1    141.8        7.092 . OI     .        .        .        .
          [BASIN DEPTH(FEET) =     2.8]
       14.25    165.4    143.8        7.192 . OI     .        .        .        .
          [BASIN DEPTH(FEET) =     2.9]
       14.30    169.1    146.0        7.300 . OI     .        .        .        .
          [BASIN DEPTH(FEET) =     2.9]
       14.35    172.9    148.3        7.416 . OI     .        .        .        .
          [BASIN DEPTH(FEET) =     3.0]
       14.40    177.3    150.8        7.540 . OI     .        .        .        .
          [BASIN DEPTH(FEET) =     3.0]
       14.45    181.7    153.4        7.671 .  O     .        .        .        .
          [BASIN DEPTH(FEET) =     3.1]
       14.50    185.9    156.1        7.807 .  O     .        .        .        .
          [BASIN DEPTH(FEET) =     3.1]
       14.55    189.9    159.0        7.948 .  O     .        .        .        .
          [BASIN DEPTH(FEET) =     3.2]
       14.60    193.5    161.8        8.092 .  O     .        .        .        .
          [BASIN DEPTH(FEET) =     3.2]
```

```
14.65     197.1     164.8          8.239 .  O     .       .       .       .
   [BASIN DEPTH(FEET) =       3.3]
14.70     200.5     167.7          8.387 .  O     .       .       .       .
   [BASIN DEPTH(FEET) =       3.4]
14.75     203.8     170.7          8.536 .  O     .       .       .       .
   [BASIN DEPTH(FEET) =       3.4]
14.80     207.0     173.7          8.687 .  OI    .       .       .       .
   [BASIN DEPTH(FEET) =       3.5]
14.85     210.1     176.7          8.837 .  OI    .       .       .       .
   [BASIN DEPTH(FEET) =       3.5]
14.90     213.2     179.8          8.988 .  OI    .       .       .       .
   [BASIN DEPTH(FEET) =       3.6]
14.95     216.2     182.8          9.138 .  OI    .       .       .       .
   [BASIN DEPTH(FEET) =       3.7]
15.00     219.1     185.8          9.288 .  OI    .       .       .       .
   [BASIN DEPTH(FEET) =       3.7]
15.05     222.1     188.8          9.439 .  OI    .       .       .       .
   [BASIN DEPTH(FEET) =       3.8]
15.10     225.1     191.8          9.589 .  OI    .       .       .       .
   [BASIN DEPTH(FEET) =       3.8]
15.15     228.3     194.8          9.741 .  OI    .       .       .       .
   [BASIN DEPTH(FEET) =       3.9]
15.20     231.8     197.9          9.895 .  OI    .       .       .       .
   [BASIN DEPTH(FEET) =       4.0]
15.25     235.6     205.4         10.045 .   O    .       .       .       .
   [BASIN DEPTH(FEET) =       4.0]
15.30     239.9     220.0         10.167 .   O    .       .       .       .
   [BASIN DEPTH(FEET) =       4.1]
15.35     244.6     230.7         10.256 .   O    .       .       .       .
   [BASIN DEPTH(FEET) =       4.1]
15.40     249.2     238.9         10.324 .   O    .       .       .       .
   [BASIN DEPTH(FEET) =       4.1]
15.45     253.4     245.5         10.379 .   O    .       .       .       .
   [BASIN DEPTH(FEET) =       4.2]
15.50     257.7     251.2         10.427 .   OI   .       .       .       .
   [BASIN DEPTH(FEET) =       4.2]
15.55     262.1     256.3         10.470 .    O   .       .       .       .
   [BASIN DEPTH(FEET) =       4.2]
15.60     266.1     261.0         10.508 .    O   .       .       .       .
   [BASIN DEPTH(FEET) =       4.2]
15.65     270.1     265.6         10.547 .    O   .       .       .       .
   [BASIN DEPTH(FEET) =       4.2]
15.70     275.0     270.6         10.588 .    O   .       .       .       .
   [BASIN DEPTH(FEET) =       4.2]
15.75     281.2     276.7         10.639 .    O   .       .       .       .
   [BASIN DEPTH(FEET) =       4.3]
15.80     290.9     285.4         10.712 .    O   .       .       .       .
   [BASIN DEPTH(FEET) =       4.3]
15.85     306.4     298.5         10.820 .    OI  .       .       .       .
   [BASIN DEPTH(FEET) =       4.3]
15.90     330.0     318.7         10.989 .     O  .       .       .       .
   [BASIN DEPTH(FEET) =       4.4]
15.95     368.9     351.0         11.259 .     OI.       .       .       .
   [BASIN DEPTH(FEET) =       4.5]
16.00     431.2     412.0         11.767 .        O       .       .       .
```

```
    [BASIN DEPTH(FEET) =     4.7]
16.05     577.8      513.6      12.613 .        . OI      .       .      .
    [BASIN DEPTH(FEET) =     5.0]
16.10     757.3      638.7      13.656 .        .   O I .       .      .
    [BASIN DEPTH(FEET) =     5.5]
16.15     899.7      771.2      14.760 .        .      O.I      .      .
    [BASIN DEPTH(FEET) =     5.9]
16.20    1044.6      896.2      16.069 .        .       .O  I   .      .
    [BASIN DEPTH(FEET) =     6.2]
16.25    1256.7     1033.8      17.598 .        .       .   O   I      .
    [BASIN DEPTH(FEET) =     6.5]
16.30    1413.2     1165.8      19.064 .        .       .     O .  I   .
    [BASIN DEPTH(FEET) =     6.8]
16.35    1496.0     1290.9      20.455 .        .       .       .O   I .
    [BASIN DEPTH(FEET) =     7.1]
16.40    1633.8     1386.4      21.516 .        .       .       . O    I
    [BASIN DEPTH(FEET) =     7.3]
16.45    1556.9     1416.9      21.854 .        .       .       . O  I .
    [BASIN DEPTH(FEET) =     7.4]
16.50    1410.2     1389.7      21.553 .        .       .       . O    .
    [BASIN DEPTH(FEET) =     7.3]
16.55    1250.4     1312.8      20.698 .        .       .       IO     .
    [BASIN DEPTH(FEET) =     7.1]
16.60    1038.3     1211.4      19.571 .        .       .   I  O.      .
    [BASIN DEPTH(FEET) =     6.9]
16.65     940.6     1105.3      18.392 .        .       . I  O  .      .
    [BASIN DEPTH(FEET) =     6.7]
16.70     805.5     1001.6      17.240 .        .      I.  O    .      .
    [BASIN DEPTH(FEET) =     6.4]
16.75     744.0      909.0      16.211 .        .     I .O     .      .
    [BASIN DEPTH(FEET) =     6.2]
16.80     668.6      826.4      15.294 .        .    I  O     .      .
    [BASIN DEPTH(FEET) =     6.1]
16.85     622.8      742.5      14.521 .        .   I O .     .      .
    [BASIN DEPTH(FEET) =     5.8]
16.90     572.1      670.8      13.923 .        .  I O  .     .      .
    [BASIN DEPTH(FEET) =     5.6]
16.95     552.0      615.1      13.459 .        . I O   .     .      .
    [BASIN DEPTH(FEET) =     5.4]
17.00     509.1      566.8      13.057 .        .I O    .     .      .
    [BASIN DEPTH(FEET) =     5.2]
17.05     478.3      531.3      12.761 .        .IO     .     .      .
    [BASIN DEPTH(FEET) =     5.1]
17.10     476.6      503.9      12.533 .        .O      .     .      .
    [BASIN DEPTH(FEET) =     5.0]
17.15     448.1      475.0      12.291 .        IO      .     .      .
    [BASIN DEPTH(FEET) =     4.9]
17.20     413.9      446.8      12.057 .        O       .     .      .
    [BASIN DEPTH(FEET) =     4.8]
17.25     394.4      421.8      11.849 .       IO       .     .      .
    [BASIN DEPTH(FEET) =     4.7]
17.30     373.5      400.0      11.674 .       O.       .     .      .
    [BASIN DEPTH(FEET) =     4.7]
17.35     364.5      385.0      11.542 .       O.       .     .      .
    [BASIN DEPTH(FEET) =     4.6]
```

```
17.40     357.4     372.4          11.437 .       O.         .        .        .
    [BASIN DEPTH(FEET) =       4.6]
17.45     349.4     361.0          11.342 .      IO.         .        .        .
    [BASIN DEPTH(FEET) =       4.5]
17.50     337.9     348.4          11.237 .      O .         .        .        .
    [BASIN DEPTH(FEET) =       4.5]
17.55     320.6     335.5          11.129 .      O .         .        .        .
    [BASIN DEPTH(FEET) =       4.5]
17.60     311.2     324.0          11.033 .      O .         .        .        .
    [BASIN DEPTH(FEET) =       4.4]
17.65     302.0     312.5          10.937 .     IO .         .        .        .
    [BASIN DEPTH(FEET) =       4.4]
17.70     288.0     301.1          10.843 .    O   .         .        .        .
    [BASIN DEPTH(FEET) =       4.3]
17.75     279.7     291.2          10.760 .    O   .         .        .        .
    [BASIN DEPTH(FEET) =       4.3]
17.80     272.8     282.1          10.684 .    O   .         .        .        .
    [BASIN DEPTH(FEET) =       4.3]
17.85     263.7     271.9          10.599 .    O   .         .        .        .
    [BASIN DEPTH(FEET) =       4.2]
17.90     249.1     261.6          10.513 .   IO   .         .        .        .
    [BASIN DEPTH(FEET) =       4.2]
17.95     242.7     253.2          10.443 .   O    .         .        .        .
    [BASIN DEPTH(FEET) =       4.2]
18.00     238.2     246.3          10.386 .   O    .         .        .        .
    [BASIN DEPTH(FEET) =       4.2]
----------------------------------------------------------------------------
```

```
============================================================================
   *** FLOW PROCESS FROM NODE 302 TO NODE 105
       TRANSLATE STREAM NUMBER 3 FORWARD IN TIME BY    .125 HOURS:


        MODEL      STREAM 3        STREAM 3
         TIME       (CFS)        TRANSLATED    .125 HOURS
        14.000        134.8       130.8
        14.050        136.4       132.4
        14.100        138.2       134.0
        14.150        139.9       135.6
        14.200        141.8       137.3
        14.250        143.8       139.0
        14.300        146.0       140.9
        14.350        148.3       142.8
        14.400        150.8       144.9
        14.450        153.4       147.2
        14.500        156.1       149.6
        14.550        159.0       152.1
        14.600        161.8       154.8
        14.650        164.8       157.6
        14.700        167.7       160.4
        14.750        170.7       163.3
        14.800        173.7       166.3
        14.850        176.7       169.2
```

14.900	179.8	172.2
14.950	182.8	175.2
15.000	185.8	178.2
15.050	188.8	181.3
15.100	191.8	184.3
15.150	194.8	187.3
15.200	197.9	190.3
15.250	205.4	193.3
15.300	220.0	196.4
15.350	230.7	201.6
15.400	238.9	212.7
15.450	245.5	225.4
15.500	251.2	234.8
15.550	256.3	242.2
15.600	261.0	248.4
15.650	265.6	253.8
15.700	270.6	258.7
15.750	276.7	263.3
15.800	285.4	268.1
15.850	298.5	273.7
15.900	318.7	281.1
15.950	351.0	291.9
16.000	412.0	308.6
16.050	513.6	334.9
16.100	638.7	381.5
16.150	771.2	462.8
16.200	896.2	576.1
16.250	1033.8	704.9
16.300	1165.8	833.7
16.350	1290.9	965.0
16.400	1386.4	1099.8
16.450	1416.9	1228.3
16.500	1389.7	1338.7
16.550	1312.8	1401.6
16.600	1211.4	1403.3
16.650	1105.3	1351.3
16.700	1001.6	1262.1
16.750	909.0	1158.3
16.800	826.4	1053.5
16.850	742.5	955.3
16.900	670.8	867.7
16.950	615.1	784.5
17.000	566.8	706.6
17.050	531.3	642.9
17.100	503.9	590.9
17.150	475.0	549.1
17.200	446.8	517.6
17.250	421.8	489.4
17.300	400.0	460.9
17.350	385.0	434.3
17.400	372.4	411.3
17.450	361.0	392.9
17.500	348.4	378.7
17.550	335.5	366.7
17.600	324.0	354.7

```
          17.650         312.5      341.9
          17.700         301.1      329.7
          17.750         291.2      318.2
          17.800         282.1      306.8
          17.850         271.9      296.2
          17.900         261.6      286.7
          17.950         253.2      277.0

===========================================================================
    *** FLOW PROCESS AT NODE 105
         STREAM NUMBER 3 IS ADDED TO STREAM NUMBER 1
---------------------------------------------------------------------------

---------------------------------------------------------------------------
***************************************************************************

  *** STREAM NUMBER  1

    *** FLOW PROCESS FROM NODE  105 TO NODE  105 WITH
   TOTAL STORM RUNOFF VOLUME(ACRE-FEET) =     739.6653

***************************************************************************

                HYDROGRAPH POINTS GIVEN IN THREE-MINUTE INTERVALS

   TIME      VOLUME        Q     0.       900.0     1800.0    2700.0    3600.0
  (HOURS)     (AF)       (CFS)
---------------------------------------------------------------------------
  14.00     150.7943     480.35  .    Q  V .        .         .         .
  14.05     152.8147     488.94  .    Q  V .        .         .         .
  14.10     154.8704     497.48  .    Q  V .        .         .         .
  14.15     156.9673     507.45  .    Q  V .        .         .         .
  14.20     159.1017     516.54  .    Q  V .        .         .         .
  14.25     161.2739     525.68  .    Q  V .        .         .         .
  14.30     163.4926     536.90  .    Q  V .        .         .         .
  14.35     165.7531     547.05  .     Q V .        .         .         .
  14.40     168.0563     557.38  .     Q  V.        .         .         .
  14.45     170.4121     570.10  .     Q  V.        .         .         .
  14.50     172.8123     580.86  .     Q  V.        .         .         .
  14.55     175.2576     591.74  .     Q  V.        .         .         .
  14.60     177.7557     604.55  .     Q  V.        .         .         .
  14.65     180.2987     615.40  .     Q  V.        .         .         .
  14.70     182.8867     626.30  .     Q  V.        .         .         .
  14.75     185.5272     639.00  .      Q  V        .         .         .
  14.80     188.2117     649.66  .      Q  V        .         .         .
  14.85     190.9395     660.11  .      Q  V        .         .         .
  14.90     193.7176     672.30  .      Q  V        .         .         .
  14.95     196.5390     682.79  .      Q  V        .         .         .
  15.00     199.4041     693.34  .      Q  V        .         .         .
  15.05     202.3197     705.58  .      Q  V        .         .         .
  15.10     205.2797     716.33  .      Q  .V       .         .         .
  15.15     208.2853     727.34  .       Q .V       .         .         .
  15.20     211.3486     741.33  .       Q .V       .         .         .
  15.25     214.4728     756.06  .       Q .V       .         .         .
  15.30     217.6617     771.71  .       Q .V       .         .         .
```

```
15.35    220.9373     792.68  .       Q .V        .         .         .
15.40    224.3145     817.30  .        Q. V       .         .         .
15.45    227.7969     842.72  .        Q. V       .         .         .
15.50    231.3779     866.60  .        Q. V       .         .         .
15.55    235.0383     885.82  .        Q. V       .         .         .
15.60    238.7756     904.44  .         Q V       .         .         .
15.65    242.6004     925.59  .         Q  V      .         .         .
15.70    246.5045     944.79  .         Q  V      .         .         .
15.75    250.4965     966.08  .         Q  V      .         .         .
15.80    254.6124     996.03  .         .Q V      .         .         .
15.85    258.8755    1031.69  .         .Q V      .         .         .
15.90    263.3275    1077.38  .         .Q  V     .         .         .
15.95    268.0987    1154.62  .         . Q V     .         .         .
16.00    273.3094    1260.99  .         .   Q     .         .         .
16.05    279.2727    1443.11  .         .    VQ   .         .         .
16.10    286.8017    1822.02  .         .    V    Q         .         .
16.15    296.0418    2236.10  .         .     V   .   Q     .         .
16.20    306.4206    2511.67  .         .     V   .      Q  .         .
16.25    317.9216    2783.23  .         .      V  .         Q         .
16.30    330.2154    2975.10  .         .      V  .         .  Q      .
16.35    343.2681    3158.75  .         .       V .         .    Q    .
16.40    357.0607    3337.81  .         .        V.         .      Q  .
16.45    371.3994    3469.97  .         .         V         .       Q .
16.50    386.0075    3535.17  .         .         V         .        Q.
16.55    400.5133    3510.39  .         .         .V        .        Q.
16.60    414.4876    3381.79  .         .         . V       .      Q  .
16.65    427.7197    3202.17  .         .         .  V      .    Q    .
16.70    439.8583    2937.55  .         .         .  V      . Q       .
16.75    451.0638    2711.71  .         .         .   V     Q         .
16.80    461.3619    2492.16  .         .         .   V  Q  .         .
16.85    470.7061    2261.29  .         .         .    Q    .         .
16.90    479.3474    2091.19  .         .         .  Q V    .         .
16.95    487.3235    1930.23  .         .         .Q    V   .         .
17.00    494.6261    1767.22  .         .        Q.     V   .         .
17.05    501.4301    1646.58  .         .       Q .      V  .         .
17.10    507.7991    1541.29  .         .      Q  .      V  .         .
17.15    513.7652    1443.79  .         .     Q   .      V  .         .
17.20    519.4306    1371.03  .         .    Q    .       V .         .
17.25    524.8218    1304.67  .         .   Q     .       V .         .
17.30    529.9180    1233.28  .         .  Q      .       V .         .
17.35    534.7712    1174.50  .         .  Q      .       V .         .
17.40    539.4041    1121.14  .         . Q       .        V.         .
17.45    543.8169    1067.90  .         .Q        .        V.         .
17.50    548.0647    1027.96  .         .Q        .        V.         .
17.55    552.1614     991.39  .         .Q        .        V.         .
17.60    556.0916     951.10  .         Q         .         V         .
17.65    559.8872     918.55  .         Q         .         V         .
17.70    563.5566     888.00  .        Q.                   V         .
17.75    567.0001     857.34  .        Q.        .         V         .
17.80    570.5464     834.18  .        Q.        .         V         .
17.85    573.9023     812.12  .        Q.        .        .V         .
17.90    577.1594     788.21  .       Q .        .        .V         .
17.95    580.3299     767.27  .       Q .        .        .V         .
18.00    583.4119     745.83  .       Q .        .        .V         .
****************************************************************************
```

bibliography

1. American Society of Civil Engineers and Water Pollution Control Federation, "Design and Construction of Sanitary and Storm Sewers," Manual of Practice, No. 9, New York, 1970.

2. Aron, G. and White, E., "Fitting A Gamma Distribution Over a Synthetic Unit Hydrograph," Water Resources Bulletin, Vol. 18(1) 95-98, Feb., 1982.

3. Beard, L., "Impact of Hydrologic Uncertainities on Flood Insurance," Journal of The Hydraulics Division, ASCE, Vol. 104, HY11, 1978.

4. Beard, L., "Hypothetical Floods," HEC Water Resources Publication, Volume 5, March 1975.

5. Beard, L. and Chang, S., "Urbanization Impact of Streamflow," Journal of The Hydraulics Division, ASCE, Vol. 105, HY6, 647-659, 1979.

6. Beven, K., "On the Generalized Kinematic Routing Method," Water Resources Research, Vol. 15, No. 5, October, 1979.

7. California, State of, Division of Highways, "California Culvert Practice," Sacramento, California, 1942.

8. Cermak, R. and Feldman, A., "Urban Hydrologic Modeling Using HEC-1/Kinematic Wave," Presented at the 19th Annual AWRA Conference, October 9-13, 1983, San Antonio, Texas.

9. Dawdy, D. and O'Donnell, T., "Mathematical Models of Catchment Behavior," Journal of The Hydraulics Division, ASCE, Vol. 91, HY4, 1965.

10. Dawdy, D. and Bergmann, J., "Effect of Rainfall Variability On Streamflow Simulation," Water Resources Research, Vol. 5 No. 5, October, 1969.

11. Dickinson, W., Holland, M. E. and Smith, G. L., "An Experimental Rainfall-Runoff Facility," No. 25, Hyidrology Papers, Colorado State University, Fort Collins, Colorado, September, 1967.

12. Diskin, M. H. and Simon, E., "A Procedure for the Selection of Objective Functions for Hydrologic Simulation Models," Journal of Hydrology, 34, 129-149, 1977.

13. Eagleson, P. S., Dynamic Hydrology, McGraw-Hill, 1970.

14. Espey, W. H. and Altman, D. G., "Nomographs for Ten-Minute Unit Hydrographs for Small Urban Watersheds," Rep. EPA-500/9-78-035 Environmental Protection Agency, Washington, DC, 1978.

15. Freeze, R. A., "The Mechanism of Natural Groundwater Recharge and Discharge: 1. One-Dimensional, Vertical, Unsteady, Unsaturated Flow Above a Recharging or Discharging Groundwater Flow System," Water Resources Research, Vol. 5, 1969.

16. Garen, D. C. and Burges, S. J., "Approximate Error Bounds for Simulated Hydrographs," Journal of The Hydraulics Division, ASCE, Vol. 107, HY11, 1981.

17. Gburek, W. J., "Discussion of Hydrologic Consequences of Rainfall Augmentation," Journal of The Hydraulic Division, ASCE, Vol. 97, HY12, 2114-2115, 1971.

18. Green, W. H. and Ampt, G. A., "Studies on Soil Physics, I, the Flow of Air and Water Through Soils," J. Agr.Sci.,(9), 1911.

19. Gupta, V. and Sorooshian S., "Uniqueness and Observability of Conceptual Rainfall-Runoff Model Parameter: The Percolation Process Examined," Water Resources Research, Vol. 19, 269-276, 1983.

20. Guymon, G. L., Harr, M.E., Berg, R. L. and Hromadka II, T.V., "A Probabilistic-Deterministic Analysis of One-Dimensional Ice Segregation in a Freezing Soil Column," Cold Regions Science and Technology, (5), 1981.

21. Hall, M. J., Urban Hydrology, Elsevier, London, 1984.

22. Hershfield, D. M., "Extreme Relationships," Journal of The Hydraulics Division, ASCE, Vol. 88, HY6, 1962.

23. Hjelmfelt, A. and Burwell, R., "Spatial Variability of Runoff," Journal of Irrigation and Drainage Engineering, Vol. 110, No. 1, March, 1984.

24. Holtan, H. N., "A Concept for Infiltration Estimates in Watershed Engineering," U.S. Dept. of Agr., ARS 41-51, 1961.

25. Hornberger, G. M., Beven, K. J., Cosby, B. J. and Sappington, D.E., "Shenandoah Watershed Study: Calibration of A Topography-Based, Variable Contributing Area Hydrological Model to a Small Forested Catchment," Water Resources Research, Vol. 21, 1841-1850, 1985.

26. Horton, R. E., "Rainfall Interception," Mon. Weather Rev., 47, 602-623, 1919.

27. Horton, R. E., "The Role of Infiltration in the Hydrologic Cycle," Trans. A.G.U., (A), 1933.

28. Hromadka II, T. V., Nestlinger, A. J., and Schwarze, J. W., "A Modified S.C.S. Runoff Hydrograph Method," 10th Urban Drainage Symposium, Univ. of Kentucky, Lexington, Kentucky, 1983a.

29. Hromadka II, T. V., Clements, J. C. and Guymon, G. L., "Guidelines for Interactive Software in Water Resources Engineering," Water Resources Bulletin, Feb. 1983b.

30. Hromadka II, T. V. and McCuen, R. H., "San Bernardino County Hydrology Manual," County of San Bernardino, CA, 1986a.

31. Hromadka II, T. V. and McCuen, R. H., "Orange County Hydrology Manual," County of Orange, CA, 1986b.

32. Hromadka II, T. V. and Yen, C. C., "A Diffusion Hydrodynamic Model: (DHM)," Advances in Water Resources, Vol. 9, No. 3, 118-170, 1986.

33. Huff, F. A., "Time Distribution of Rainfall in Heavy Storms," Water Resources Research, (4), 1967.

34. Johnston, P. and Pilgrim, D., "Parameter Optimization for Watershed Models," Water Resources Research, Vol. 12, 477-486, 1976.

35. Keifer, C. J. and Chu, H. H., "Synthetic Storm Pattern for Drainage Design," Journal of The Hydraulics Division, Vol. 83, HY4, 1957.

36. Kelway, P., "The Rainfall Recorder Problem," Journal of Hydrology, 26, 55-77, 1975.

37. Kirpich, Z. P., "Time of Concentration of Small Agricultural Watersheds," Civ. Eng., 10(6), 1940.

38. Klemes, V. and Bulu, A., "Limited Confidence in Confidence Limits Derived by Operational Stochastic Hydrologic Models," Journal of Hydrology, 42, 9-22, 1979.

39. Loague, K. and Freeze, R., "A Comparison of Rainfall-Runoff Modeling Techniques on Small Upland Catchments," Water Resources Research, Vol. 21, 229-248, 1985.

40. McCuen, R. H. and Bondelid, T. R., "Estimating Unit Hydrograph Peak Rate Factors," Journal of Irrigation and Drainage Division, Vol. 109, IR2, 1983.

41. McCuen, R. H., Statistical Methods for Engineers, Prentice-Hall, Inc., Englewood Cliffs, NJ, 439 pp, 1985.

42. McCuen, R. H. and Snyder, W. M., Hydrologic Modeling, Prentice-Hall, Inc., Englewood Cliffs, NJ, 568 pp, 1986.

43. Mein, R. G. and Brown, B. M., "Sensitivity of Optimized Parameters in Watershed Models," Water Resources Research, Vol. 14, 1978.

44. Miller, J. F., Frederic, R. H. and Tracey, R. J., "Precipitation Frequency Atlas of the Conterminous Western United States," NOAA Atlas 2, Nat. Weather Serv., Silver Springs, MD, 1973.

45. Mockus, V. J., "National Engineering Handbook, Section 4, Hydrology," U.S.Dept. of Agri., S.C.S., Washington, DC, 1972.

46. Myers, V. A. and Zehr, R. M., "A Methodology for Point-to-Area Rainfall Frequency Ratios," NOAA Tech. Rep. NWS 24, U.S. Dept. of Commerce, Washington, DC 1980.

47. Nash, J. and Sutcliffe, J., "River Flow Forecasting Through Conceptual Models Part 1 - A Discussion of Principles," Journal of Hydrology, Vol. 10, 282-290, 1970.

48. Neff, E., "How Much Rain Does a Rain Gage Gage?," Journal of Hydrology, 35, 213-220, 1977.

49. Philip, J. R., "The Theory of Infiltration: 1. The Infiltration Equation and its Solution," Soil Sci., (83) 1957a.

50. Philip, J. R., "The Theory of Infiltration: 2. The Profile at Infinity," Soil Sci. (83), 1957b.

51. Philip, J. R., "The Theory of Infiltration: 3. Moisture Profiles and Relation to Experiment," Soil Sci. (84), 1957c.

52. Philip, J. R., "The Theory of Infiltration: 4. Sorptivity and Algebraic Infiltration Equations," Soil Sci. (84), 1957d.

53. Philip, J. R., "The Theory of Infiltration: 5. The Influence of the Initial Moisture Content," Soil Sci. (84), 1957e.

54. Philip, J. R., "The Theory of Infiltration: 6. Effect of Water Depth over Soil," Soil Sci. (85), 1958.

55. Pilgrim, D., "Travel Times and Nonlinearity of Flood Runoff From Tracer Measurements on a Small Watershed," Water Resources Research, Vol. 12, 487-496, 1976.

56. Pilgrim, D. H. and Cordery, I., "Rainfall Temporal Patterns for Design Floods," Journal of The Hydraulics Division, ASCE, Vol. 101, HY1, 1975.

58. Rallison, R. E., "Origin and Evolution of the S.C.S. Runoff Equation," ASCE Watershed Management Symposium, Boise, Idaho, 1980.

59. Riverside County Hydrology Manual, County of Riverside, CA, 1978.

60. Rubin, J. and Steinhardt, R., "Soil Water Relations During Rain Infiltration: I. Theory," Soil Sci. Soc. Amer. Proc., (27), 1963.

61. Rubin, J., Steinhardt, R. and Reiniger, P., "Soil Water Relations During Rain Infiltration: II. Moisture Content Profiles During Rains of Low Intensities," Soil Sci. Soc. Amer. Proc., (28), 1964.

62. San Diego County Hyrology Manual, County of San Diego, CA, 1975.

63. Schilling, W. and Fuchs, L., "Errors in Stormwater Modeling-A Quantitative Assessment," Journal of The Hydraulics Division, ASCE, Vol. 112, HY2, 111-123, 1986.

64. Scully, D. and Bender, D., "Separation of Rainfall Excess from Total Rainfall," Water Resource Research, Vol. 5, 877-883, 1969.

65. Sherman, L. K., "Streamflow from Rainfall by the Unit-Graph Method," Eng. News Rec., (108), 1932.

66. Singh, V. P., "Unit Hydrographs - A Comparative Study," Water Resources Bulletin 12, 1976.

67. Singh, V. P., "Sensitivity of Some Runoff Models to Errors in Rainfall Excess," Journal of Hydrology, 33, 301-318 pgs., 1977.

68. Snyder, F. F., Snythetic Unit Hydrographs," EOS Trans. A.G.U., 19(1), 1938.

69. Sorooshian, S. and Gupta, V., "Automatic Calibration of Conceptual Rainfall-Runoff Models: The Question of Parameter Observability and Uniqueness," Water Resources Research, Vol. 19, 260-268, 1983.

70. Terstriep, M. L. and Stall, J. B., "The Illinois Urban Drainage Area Simulator," ILLUDAS, Bul. 58, State Wat. Surv., Urbana, Illinois, 1974.

71. Troutman, B., "An Analysis of Input of Perception-Runoff Models Using Regression with Errors in the Independent Variables," Water Resources Research, Vol. 18, 947-964, 1982.

72. Unver, O. and Mays, L. W., "Optimal Determination of Loss Rate Functions and Unit Hydrographs," Water Resources Research, Vol. 20, No. 2, 203-214, 1984.

73. U.S. Army Corps of Engineers, Los Angeles District, "Improved Procedure for Determining Drainage Area Lag Values," 1962.

74. U.S. Army Corps of Engineers, Los Angeles District, "Interim Report on Survey for Flood Control Tahquitz Creek," CA, 1963.

75. U.S. Army Corps of Engineers, The Hydrologic Engineering Center, "Hydrologic Engineering Methods for Water Resources Development: Volume 5 - Hypothetical Floods," March, 1975.

76. U.S. Army Corps of Engineers, The Hydrologic Engineering Center, "Testing of Several Runoff Models on an Urban Watershed," Technical Paper No. 59, 1978.

77. U.S. Army Corps of Engineers, The Hydrologic Engineering Center, "Continuous Hydrologic Simulation of the West Branch Dupage River Above West Chicago: An Application of Hydrocomp's HSP," Research Note No. 6, 1979.

78. U.S. Army Corps of Engineers, The Hydrologic Engineering Center, "Adoption of Flood Flow Frequency Estimates at Ungaged Locations," Training Document No. 11, February, 1980.

79. U.S. Army Corps of Engineers, The Hydrologic Engineering Center, "Hydrologic Analysis of Ungaged Watersheds Using HEC-1," Training Document No. 15, April, 1982.

80. U.S. Department of Transportation, Federal Aviation Agency, "Adv. Circular on Airport Drainage," Rep. A/C 050-5320-5B, Washington, DC, 1975.

81. U.S. Department of Transportation, Federal Highway Administration, "Hydrology," Hydraulic Engineering Circular No. 19, October, 1984.

82. U.S. Bureau of Reclamation, "Design of Small Dams," p.544, U.S. Government Printing Office, Washington, DC, 1973.

83. U.S. Department of Agriculture, Soil Conservation Service, "A Method for Estimating Volume and Rate of Runoff in Small Watersheds," Tech. Paper 149, Washington, DC, 1973.

84. U.S. Department of Agriculture, Soil Conservation Service, "Urban Hydrology for Small Watersheds," Tech. Release No. 55, Second Ed., Washington, DC, 1986.

85. U.S. Department of Commerce, National Oceanic and Atmospheric Administration, National Weather Service, "NOAA Atlas 2, Precipitation-Frequency Atlas of the Western United States, Volume XI-California," 1973.

86. Viessman, W., Jr., Knapp, J. W., Lewis, G. L. and Harbaugh, T. E., Introduction to Hydrology, 2nd ed., IEP Publishers, New York, 1977.

87. Watt, W. and Kidd, C., "Quurm-A Realistic Runoff Model," Journal of Hydrology, 27, 225-235, 1975.

88. Williams, D. W., Cameron, R. J. and Evans, G. P., "TRRL and Unit Hydrograph Simulations Compared with Measurements in an Urban Catchment," Journal of Hydrology, 48, 63-70, 1980.

89. Wood, E. F., "An Analysis of the Effects of Parameter Uncertainty in Deterministic Hydrologic Models," Water Resources Research, Vol. 12, 1976.

author index

Altman, D. G., 247
American Society of Civil Engineers
and Water Pollution Control Federation, 102
Ampt, G. A., 102
Aron, G., 231, 387

Bear, L., 21, 22, 23, 38, 39
Bender, D., 104, 105
Berg, R. L., 15, 17
Germann, J., 35
Beven, B. J., 22
Beven, K., 37
Bondelid, T. R., 244
Brown, B. M., 15, 34
Bulu, A., 33
Burges, S. J., 23
Burwell, R., 37

California, State of, Division of Highways, 143
Cameron, R. J., 104, 105
Cermak, R., 36
Chang, S., 21, 22, 23, 39
Chu, H. H., 63
Clements, J. C., 175, 403
Cordery, I., 63
Cosby, B.J., 22

Dawdy, D., 23, 35
Dickinson, W., 37
Diskin, M. H., 35

Eagleson, P. S., 2
Espey, W. H., 247
Evans, G. P., 104, 105

Feldman, A., 36
Freeze, R. A., 21, 23, 28, 100
Frederic, R. H., 48, 59
Fuchs, L., 22, 23, 104

Garen, D. C., 23
Gburek, W. J., 33
Green, W. H., 102
Gupta, V., 33, 34
Guymon, G. L., 15, 17, 175, 403

Hall, M. J., 28
Harbaugh, T. E., 102
Harr, M. E., 15, 17
Hershfield, D. H., 63, 66
Hjelmfelt, A., 37
Holland, M. E., 37
Holtan, H. N., 102
Hornberger, G. M., 22
Horton, R. E., 100, 103
Hromadka II, T. V., 1,2, 15, 17, 36, 66, 107, 175, 228
Huff, F. A., 63

Johnston, P., 34

Keifer, C. J., 63
Kelway, P., 35
Kidd, C., 33
Kirpich, Z. P., 143

Klemes, V., 33
Knapp, J. W., 102

Lewis, G. L., 102
Loague, K., 21, 23, 28

Mays, L. W., 107
McCuen, R. H., 1, 2, 66, 107, 244, 379, 393, 395
Mein, R. G., 15, 34
Miller, J. F., 48, 59
Mockus, V. J., 295, 298
Myers, V. A., 59

Nash, J., 24, 36
Neff, E., 35
Nestlinger, A. J., 1, 2, 228

O'Donnell, T., 23

Philip, J. R., 102
Pilgrim, D., 34, 37, 63

Rallison, R. E., 113
Reiniger, P., 100
Rubin, J., 100

Sappington, R. E., 22
Schilling, W., 22, 23, 104
Schwarze, J. W., 1, 2, 228
Scully, D., 104, 105
Sherman, L. K., 227
Simon, E., 35
Singh, K. P. 37
Smith, G. L., 37
Snyder, F. F., 227, 393
Snyder, W. M., 277, 393
Sorooshian, S., 33, 34
Stall, J. B., 63
Steinhardt, R., 100
Sutcliffe, J., 24, 36

Terstriep, M. L., 63
Tracey, R. J.,48, 59
Troutman, B., 35

Unver, O., 107
U.S. Army Corps of Engineers

Los Angeles District, 61
The Hydrologic Engineers Center, 1, 3, 21, 24, 28, 38, 39, 40,41, 45, 61, 63
U.S. Department of Transportation
Federal Aviation Agency, 143
Federal Highway Administration, 21
U.S. Bureau of Reclamation, 113
U.S. Department of Agriculture, 21, 64, 243
U.S. Department of Commerce, 48, 55, 57

Viessman, W., Jr., 102

Watt, W., 33
White, E., 233, 387
Williams, D.W., 104, 105
Wood, E.F., 12

Yen, C. C., 36

Zehr, R. M., 59

subject index

Antecedent moisture condition (AMC), 17
Area-averaged point rainfall, 57
Arithmetic mean, 57

Balanced flood, 43
Baseflow, 18
Basin
    factors, 236
    routing model, 18
Bias, 371-373
Black box models, 33, 36
Box-and-Whisker plots, 373

Calibrating unit hydrograph and loss functions, 379
Calibration
    unit hydrographs, 384
    S-hydrographs, 388
    flood frequency discharges, 391
    time of concentration models, 393
    channel routing parameters, 397
    convex method, 397
    modified Att-kin method, 400
    Muskigum method, 400
Channel routing models, 18
Computer-aided design interaction, 174
Constant
    proportion loss rate, 104
    rational model, 81
Continuous simulation model, 24
Correlation coefficient, 376
Cumulative probability, 54

Depth
area adjustment, 17, 39, 59, 73
duration-frequency, 48, 57
Depression storage capacity, 18, 102
Design storm pattern, 15, 21, 38, 55, 73
Detention storage, 288
Direct runoff, 19
Discretization errors, 18

Effective rainfall, 99
Exceedance proability, 54
Exponential loss rate function, 107, 126

Flood control system, 20
Flood flow frequency analysis, 4
Floy-by channel model, 295
Flow-through basin, 291

Groundcover condition, 17

Histogram, 373
Horton loss rate, 107, 123
Hydraulic modeling errors, 19
Hydrograph translation, 306
Hydrologic cycle, 2
Hydrologic processes, 19
Hypothetical floods, 40

Impervious area, 19
Infiltration capacity, 17, 19, 20, 100
Initial abstraction, 19, 107, 121
Intensity, 54
Intensity-duration-frequency curves, 55
Isohyetal method, 57,. 59

Junction Analysis, 159

Kinematic wave
models, 23,36
routing approximation, 306

Linear model, 36
Linear interpolation method, 81
Link-node model, 15, 157, 402, 435
Loss models, 123

Master plan of drainage, 162-170
Mechanisms for coding, 47
Modified
    Att-kin method, 301
    composite storm pattern, 66
    Convex channel routing method, 295
Model calibration, 371

Nodal point method, 158
Non-linear model, 36

Phi-index method, 103, 123
Physically-based models, 33, 36
Pipeflow routing model, 305
Point rainfall, 48
Precipitation, 46

Rainfall intensity, 15
Rainfall-runoff models, 28, 33
Rational method, 2, 148
Return period, 54
Runoff coefficient, 149
Runoff hydrograph
    model uncertainty, 18
    routing techniques, 288, 306

SCS
    loss separation, 109, 121, 123
    storm pattern, 73
Subarea summation method, 159
Submodel combination, 18
Statistical analysis, 20
Stochastic hydrologic methods, 33
Synthetic
    critical storm patterns, 59, 71
    unit hydrograph methods, 2, 227-255

Thiessen polygon method, 57, 59
Time of concentration, 142
Travel time, 146

Unit hydrograph method, 22, 36, 39, 227-255

Watershed
    lag time, 236
    runoff, 19